Biogas: Technology and Principles

Biogas: Technology and Principles

Edited by **Brad Hill**

NY RESEARCH
PRESS
New York

Published by NY Research Press,
23 West, 55th Street, Suite 816,
New York, NY 10019, USA
www.nyresearchpress.com

Biogas: Technology and Principles
Edited by Brad Hill

International Standard Book Number: 978-1-63238-061-6 (Hardback)

Printed in China.

Contents

Preface

Introduction to biogas, including its technology as well as principles, has been provided in this well-structured and descriptive book. It encompasses numerous research reports demonstrating sequential production of biogas and also discusses the principles and mechanisms involved in its operation. The efficacy of technologies adopted for the production of biogas, specifically anaerobic digestion of waste and gas production systems, its structure, mechanism and other aspects of the process, applications and purification of biogas, architecture of pipes and plants, evaluation of environmental facets, economic issues and a few other topics which hold importance in any aspect of biogas generation have been examined in the book.

The information shared in this book is based on empirical researches made by veterans in this field of study. The elaborative information provided in this book will help the readers further their scope of knowledge leading to advancements in this field.

Finally, I would like to thank my fellow researchers who gave constructive feedback and my family members who supported me at every step of my research.

Editor

1

Anaerobic Biogas Generation for Rural Area Energy Provision in Africa

B. Amigun[1,*], W. Parawira[2], J. K. Musango[3], A. O. Aboyade[4] and A. S. Badmos[1]
[1]*Renewable Energy Group, National Biotechnology Development Agency (NABDA)- an Agency Under the Federal Ministry of Science and Technology, Abuja,*
[2]*Department of Applied Biology, Kigali Institute of Science and Technology (KIST), Kigali,*
[3]*Gauteng City-Region Observatory, a Partnership Between University of Johannesburg, University of Witwatersrand and Gauteng Provincial Government, Wits,*
[4]*Process Engineering Department, Stellenbosch University, Stellenbosch,*
[1]*Nigeria*
[2]*Rwanda*
[3,4]*South Africa*

1. Introduction

1.1 Energy overview in Africa

Energy plays a central role in national development process as a domestic necessity and major factor of production, whose cost directly affects price of other goods and services (Amigun and von Blottnitz, 2008). It affects all aspects of development, such as social, economic, political and environmental, including access to health, water, agricultural productivity, industrial productivity, education and other vital services that improve the quality of life. Currently, many African countries experience frequent blackouts and the cost of electricity blackouts is not known. The continent's energy consumption and demand is expected to continue to grow as development progresses at rates faster than those of developed countries. The desire for improved quality of life and rises in population together with energy demands from the transport, industrial and domestic sectors will continue to drive this growth. Ensuring the provision of adequate, affordable, efficient and reliable high-quality energy services with minimum adverse effect on the environment in sustainable way is not only pivotal for development, but crucial for African countries most of which are struggling to meet present energy demands (Amigun et al., 2008). African countries need sustainable energy supplies to be in a position to improve their overall net productivity and become major players in global technological and economic progress. Unreliable energy supply may account for the low levels of private investment the African continent attracts and the poor economic productivity of its limited industries. Improvement

* Corresponding Author

in the quality and magnitude of energy services in developing countries is required for them to meet developmental objectives including the Millennium Development Goals (MDGs). Africa is not only the poorest continent in the world but it was the only major developing region with negative growth in income per capita during 1980-2000 (World Bank, 2003).

Although reliable regional energy statistics are not readily available, existing estimates of energy use in Eastern and Southern Africa indicate a significant and persistent dependence on traditional biomass energy technologies and limited use of modern, sustainable energy technologies (Karekezi, 1994a). Biomass in the form of mainly wood-fuel and charcoal is the dominant energy source used in sub-Saharan Africa

Because of the shortage in commercial modern energy and current economic situation in most African countries, the fuel substitution away from biomass is less likely because of declining disposable incomes for both urban and rural population. There is fuel-switch back to traditional fuels as modern fuels become scarce in some areas but the wood fuels are also becoming scarce in some countries. Biomass is cheap but when used in an unplanned (unsustainable) manner leads to consumption beyond regenerative limits with serious environmental consequences. On average, about 40% of total commercial energy is consumed in six countries in the Northern sub-region and a similar share in Southern Africa with over 80% by South Africa. The other 45 or more countries share the remaining 20%. Similarly, the major oil and gas producers are limited to about ten countries in the North and West regions while about 95% coal (anthracite in nature) is produced in South Africa. This uneven distribution of the fossil energy resources (crude oil and natural gas) on the African continent is reflected in the energy production and consumption patterns (Table 1). As a result, 70% of countries on the continent depend on imported energy resources, which support the need to harness the available abundant renewable energy resources (Amigun, 2008).

Major energy exporter[a]	Net energy exporter	Importers[b]
Nigeria	Angola	Benin
Algeria	Cameroon	Eritrea
Libya	Congo	Ethiopia
South Africa	Democratic Republic of Congo	Ghana
Egypt	Cote d' Ivoire	Kenya
Gabon	Gabon	Morocco
Congo	Sudan	Mozambique
		Namibia
		Senegal
		Tanzania
		Togo
		Zambia
		Zimbabwe

[a]Major energy exports are in excess of 0.5 quads
[b]Most of the African countries energy imports are very small (less than 0.3 quads)

Table 1. The energy distribution in Africa indicating countries which export and import energy (Amigun et al., 2008)

Africa is a net energy exporter, but the majority of its population lacks access to modern fuels, and many countries rely on imported energy. More than 500 million people living in sub-Saharan Africa do not have electricity in their homes and rely on solid forms of biomass (firewood, agricultural residues, animal wastes, etc) to meet basic energy needs for cooking, heating and lighting. The disadvantages of these traditional fuels are many: they are inefficient energy carriers and their heat is difficult to control, they produce dangerous emissions and their current rate of extraction is not sustainable. The unsustainable use of fuel wood biomass can accelerate deforestation and lead to soil erosion, desertification and increased risk of flooding and biodiversity loss. The low levels of modern (commercial) energy consumption prevalent in Africa besides the heavy usage of traditional (non-commercial) biofuels- primarily biomass is also due to largely underdeveloped energy resources, poorly developed commercial energy infrastructure, widespread and severe poverty which makes it impossible for people to pay for conventional energy resources and the landlocked status of some African countries that make the cost of importing commercial energy more expensive (World Bank, 2003; Amigun et al., 2008). The existing aging and neglected facilities for thermal and hydro energy production need rehabilitation and expansion for the efficient delivery of useful energy services. Upgrading the abundant biomass in Africa to higher-quality energy carriers could help change the energy situation in the continent. The problems arising from non-sustainable use of fossil fuels and traditional biomass fuels have led to increased awareness and widespread research on the accessibility of new and renewable energy resources, such as biogas. The development of renewable energy technologies and in particular biogas technology can help reduce the dependence on non-renewable resources and minimise the social impacts and environmental degradation problems associated with fossil fuel (Amigun and von Blottnitz, 2008).

1.2 The role of renewable energy: Biogas technology (anaerobic digestion)

As mentioned above, the economic prosperity and quality of life of a country are closely linked to the level of its per capita energy consumption and the strategy adopted to use energy as a fundamental tool to achieve the same (Amigun et al. 2008; Singh & Sooch 2004). This is illustrated in Figure 1.

Renewable energy could provide the much desired sustainable rural revitalization in most developing countries. It is an ideal alternative because it could be a less expensive option for low income communities. An ideal renewable energy source is one which is locally available, affordable and can be easily used and managed by local communities. Anaerobic digestion is one of a number of technologies that offers the technical possibility of decentralized approaches to the provision of modern energy services using resources such as; cow dung, human waste and agricultural residues to produce energy. Anaerobic digestion of the large quantities of municipal, industrial and agricultural solid waste in Africa can provide biogas that can be used for heat and electricity production and the digester residue can be recycled to agriculture as a secondary fertilizer. Anaerobic digestion systems are relatively simple, economical, and can operate from small to large scales in urban and rural locations (Amigun & von Blottnitz, 2009). In this regard, many African governments have realised that renewable energies could play a very important role in supplementing other existing energy sources.

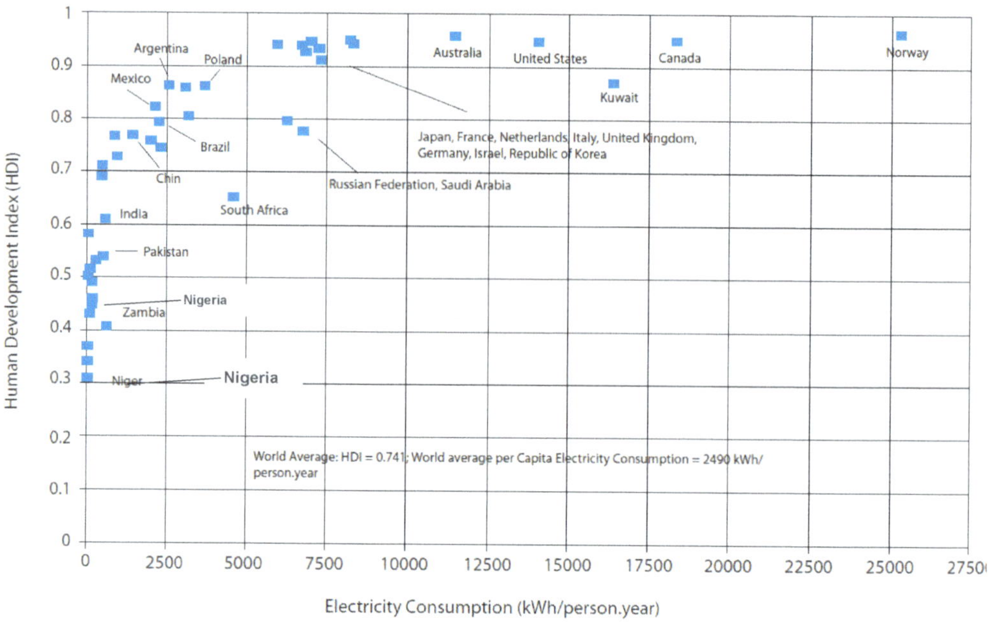

Fig. 1. Human development index (HDI) and per capita electricity consumption, 2003 - 2004, (Source: UNDP, 2006)

Anaerobic digestion describes the natural breakdown of organic matter in the absence of oxygen into a methane rich gas (biogas) via the complex and synergistic interactions of various micro-organisms types including hydrolytic, fermentative, acidogenic, and methanogenic bacteria (Lusk et al. 1996, Parawira, 2004b). The first group of microorganism secretes enzymes, which hydrolyses polymeric materials such as proteins and polysaccharides to monomers such as glucose and amino acids. The fermentative bacteria convert these monomers to organic acids, primarily propionic and acetic acid. The acidogenic bacteria convert these acids to hydrogen, carbon dioxide, and acetate, which the methanogens utilize via two major pathways to produce methane and carbon dioxide (Lusk et al. 1996; Verma 2002). The potential for organic matter decomposition to generate a flammable gas has been recognized for more than 400 years. In 1808, it was determined that methane was present in the gases produced during the anaerobic digestion of cattle manure. In 1868, Bechamp, a student of Pasteur attempted to isolate the microorganism responsible for the anaerobic bioconversion of ethanol to methane.

The first practical application of anaerobic digestion for energy production took place in England in 1896 when biogas from sewage sludge digestion was used to fuel street lamps. As is the case for many other renewable technologies, interests in anaerobic digestion suffered with the rise of the dependence of petroleum. However some developing countries, mainly in Asia, embraced the technology for the small scale provision of energy and sanitation services (Monnet 2003). Since that time, anaerobic digestion has received considerable interest to harness its waste disposal and energy producing capabilities, with municipal sewage disposal attracting the widest application (Lusk et al. 1996).

The anaerobic digestion process will occur at most temperatures below 70°C, but in the commercial operation of digesters two main temperature ranges are typically employed; the mesophilic range (30-44°C) and thermophilic range (45-60°C). In addition to sewage sludge, organic farm wastes, municipal solid waste, green botanical waste and organic industrial waste have also been used as feedstock in various small to large scale digesters across the world. Current commercial anaerobic digestion processes generally involve the following steps; pre-treatment (including size reduction and the separation of non-biodegradable substances), digestion, biogas cleaning and conditioning (to remove CO_2, water vapour and other undesirables), and subsequently biogas utilization (via internal combustion engines, or the more efficient combined heat and power plant (CHP)). The solid residue from the digestion process (called digestate) can be used as compost.

Various types of small to medium scale biogas digesters have been developed including the floating drum, fixed dome, and plastic bag design (Amigun & Blottnitz 2007). The amount of biogas produced from a specific digester depends on factors such as the amount of material fed, the type of material, the carbon/nitrogen ratio, and digestion time and temperature (Omer & Fadalla 2003; Schwart et al. 2005; Chynoweth et al. 2001). Depending on the context, any type may be used. However, most of the small to medium scale biogas plants built so far are of the fixed dome type (Amigun & von Blottnitz 2009). The technology is gradually gaining popularity in developing countries, especially in Africa where the lack of clean and sustainable energy source represents damage to the environment and its people (Amigun & von Blottnitz 2009). In addition, Sub-Saharan Africa with its warm climates is well-suited for the biogas digester technology (Aboyade 2004).

In the subsequent sections of this chapter, the current state of status of biogas technology in sub-Saharan Africa will be presented, along with a discussion of opportunities and challenges faced. The socio-economic benefits of biogas digesters is also been investigated through the use of case studies of commercial and demonstration plants on the continent. The economics of biogas technology in terms of investment and maintenance in the rural African context is discussed.

2. Biogas technology overview and status in Africa

Biogas technology is viewed as one of the renewable technologies in Africa that can help eases its energy and environmental problems. To date, some digesters have been installed in several sub-Saharan countries, utilising a variety of waste such as from slaughterhouses, municipal wastes, industrial waste, animal dung and human excreta. Small-scale biogas plants are located all over the continent but very few of them are operational. In most African countries, for example, Burundi, Ivory Coast, and Tanzania, biogas is produced through anaerobic digestion of human and animal excreta using the Chinese fixed-dome digester and the Indian floating-cover biogas digester, which are not reliable and have poor performance in most cases (Omer and Fadalla, 2003). These plants were built for schools, health clinics and mission hospitals and small-scale farmers, in most cases by non-governmental organisations. In Africa the interest in biogas technology has been further stimulated by the promotional efforts of various international organisations and foreign aid agencies through their publications, meetings and visits. Most of the plants have only operated for a short period due to poor technical quality. Table 3 gives a list of the African countries with biogas production units as at 2007. There is thus a need to introduce more efficient reactors to improve both the biogas yields and the reputation of the technology. The development of large-scale anaerobic digestion technology in Africa is still embryonic, but with a lot of potentials.

Country	Geographical characteristic		Region	No of small/ medium digester (≤ 100m³)	No of Large scale digester (>100m³)	Level of technology development
	Landlocked	Coastal				
Botswana	*		Southern Africa	Several	Few	Low
Burkina Faso	*		West Africa	Few	-	Low
Burundi	*		Central Africa	Several	Several	High
Cameroon		*	Central Africa	Few	-	Low
Congo-Brazzaville		*	Central Africa	Several	Few	Low
Côte d'Ivoire		*	West Africa	Several	Few	Low
Egypt		*	North Africa	Several	Few	High
Eritrea		*	East Africa	Few	-	Low
Ethiopia	*		East Africa	Few	-	Low
Ghana		*	West Africa	Several	Few	High
Guinea		*	West Africa	Few	-	Low
Kenya		*	East Africa	Several	Several	High
Lesotho	*		Southern Africa	Few	-	Medium
Malawi	*		Southern Africa	Few	-	Low
Mali	*		West Africa	Several	Few	High
Morocco		*	North Africa	Several	-	Medium
Namibia		*	Southern Ãfrica	Few	-	Low
Nigeria		*	West Africa	Few	Few	Low
Rwanda	*		Central Africa	Several	Few	High
Sierra Leone		*	West Africa	Few	-	Low
South Africa		*	Sothern Africa	Several	Several	High
Sudan		*	East Africa	Few	-	Low
Swaziland	*		Southern Africa	Several	-	Medium
Tanzania		*	East Africa	Several	Several	High
Tunisia		*	North Africa	Few	-	Low
Uganda	*		East Africa	Few	-	Low
Zimbabwe	*		Southern Africa	Several	Few	Medium

Sources: Karekezi, (2002), Allafrica.com, (2000), Akinbami *et al*, (2001), Spore, (2004), Amigun and von Blottnitz, (2007).

Table 2. Countries with documented biogas producing units in Africa as at 2007

Some of the first biogas digesters were set up in Africa in the 1950s in South Africa and Kenya. In other countries such as in Tanzania, biogas digesters were first introduced in 1975 and in others even more recently (South Sudan in 2001). To date, biogas digesters have been installed in several sub-Saharan countries including Burundi, Botswana, Burkina Faso, Cote d'Ivoire, Ethiopia, Ghana, Guinea, Lesotho, Namibia, Nigeria, Rwanda, Zimbabwe, South Africa and Uganda (Winrock International, 2007). Biogas digesters have utilized a variety of inputs such as waste from slaughterhouses, waste in urban landfill sites, industrial waste (such as bagasse from sugar factories), water hyacinth plants, animal dung and human excreta. Biogas digesters have been installed in various places including commercial farms (such as in chicken and dairy farms in Burundi), a public latrine block (in Kibera, Kenya), prisons in Rwanda, and health clinics and mission hospitals (in Tanzania) (Winrock International, 2007). However, by far the most widely attempted model is the household biogas digester - largely using domestic animal excreta (Table 2). This is due to the fact that this technology is closely linked to poverty alleviation and rural development. The biogas produced from these household-level systems has been used mostly for cooking, with some use for lighting.

Global experience shows that biogas technology is a simple and readily usable technology that does not require overtly sophisticated capacity to construct and manage. It has also been recognized as a simple, adaptable and locally acceptable technology for Africa (Gunnerson and Stuckey, 1986; Taleghani and Kia, 2005). There are some cases of successful biogas intervention in Africa, which demonstrate the effectiveness of the technology and its relevance for the region. The lessons learned from biogas experiences in Africa suggest that having a realistic and modest initial introductory phase for Biogas intervention; taking into account the convenience factors in terms of plant operation and functionality; identifying the optimum plant size and subsidy level; and; having provision for design adaptation are key factors for successful biogas implementation in Africa (Biogas for better life, 2007). Biogas technology has multiple beneficial effects.

2.1 Challenges to biogas commercialisation in Africa and possible measures to overcome them

The implementation of the biogas technology on large scale may be prevented or slowed down by a number of constraints. They may be grouped as follows: political, social-cultural, financial, informational, institutional, technical and training (Omer and Fadalla, 2003, Ni and Nyns, 1996). Some of the difficulties encountered in the development of anaerobic treatment for biogas production in developing countries are in Table 3.

There is lack of coherent biogas technology strategy in many sub-Saharan African countries despite the increase in the price of conventional fuel on a daily basis, and their rising demand mainly to technical and non-technical factors. The main contentious problems of biogas commercialisation in sub-Saharan African countries relate to economics and political will and many site-specific issues. Some of these issues are informed by local dynamics of perceptions; influenced by personal, social and institutional factors and beliefs, as well as internal conflicts, due to perceived environmental, social and ecological risks, that were aggravated by miscommunication and the lack of understanding.

- Inexperienced contractors and consultants, resulting in poor-quality plants, and poor choice of materials.
- Lack of reliable information on the potential benefits of the technology by financial institutions.
- Complete absence of academic, bureaucratic, legislation and commercial infrastructure in the region/country.
- Lack of knowledge on the system in practice, sometimes even in research institutes and universities.
- Community acceptance issues and poor ownership responsibility by users.
- Complete absence of pilot studies, and no full-scale experience.
- No properly educated operators, lack of credibility, lack of technical knowledge on maintenance and repair.
- Uninformed or poorly informed authorities and policy makers.
- Failure by government to support biogas technology through focussed energy policy.
- Research at universities is frequently considered to be too academic in nature, even when it is quite applied.

Table 3. Some of the difficulties thwarting development of biogas technology in Africa (Mwakaje, 2007; Murphy, 2001, Lettinga, 2001; Lettinga, 1995, Switzenbum, 1995; Tafdrup, 1995; Iza *et al.*, 1991)

2.1.1 Economic factors which affect biogas production and commercialisation

The economy of a biogas plant consists of large investments costs, some operation and maintenance costs, mostly free raw materials, e.g., animal dung, water, aquatic weeds, terrestrial plants, sewage sludge, industrial wastes, agricultural wastes and income from sale of biogas or electricity and heat (Amigun and von Blottnitz, 2007). The economics of biogas production and consumption is dependent on a number of factors specific to the local situation, as shown in Table 4. The economics of biogas production and use, therefore, depends upon the specific country and project situation

a. Cost of biomass material, which varies among countries depending on land availability, agricultural productivity, labour costs, etc
b. Biogas production costs, which depends on the plant location, size and technology, which vary among countries
c. The cost of corresponding fossil fuel (gasoline, diesel) in individual countries
d. The strategic benefit of substituting imported petroleum with domestic resources

Table 4. Economic factors which affect biogas production and commercialisation

The main limitations to the adoption of large-scale biogas technology are both institutional and economic. Establishing a self-sustaining institutional system that can collect and process urban waste and effectively market the generated biogas fuel is a complex activity that calls for sophisticated organisational capability and initiative (Karekezi, 1994b). The energy transition in Africa is an incremental process and not a leapfrog process, dependent upon household, national and regional accumulations of technological capabilities. Biogas technology absorption, therefore, cannot occur without the proper social, cultural, political and economic institutions to support adoption, dissemination and appropriate contextual innovation (Murphy, 2001). The Taka Gas Project in Tanzania (Mbuligwe and Kassenga, 2004) is a very good example of how large-scale biogas technology projects have failed to take off in Africa. The main objective of the Taka Gas Project was to obtain biogas through anaerobic digestion of municipal solid waste from Dar es Salaam city and serve as a model for other urban areas in Africa to emulate. The project was well prepared with analysis of solid waste as feedstock for the project, strategies for operationalising the project, environmental impacts and economic feasibility and other technical and non-technical and socio-economic issues studied for the project but it has never took off the ground due to bureaucracy.

The investment cost of even the smallest of the biogas units is prohibitive for most rural households of sub-Saharan Africa. Evidence from the experiences in Eastern and Southern African countries is still limited, but the general consensus is that the larger combined septic tank/biogas units that are run by institutions such as hospitals and schools have proved to be more viable than the small-scale household bio-digesters. There is need for subsidy-led programmes which will be demand-driven and market-oriented to increase the adoption of biogas plants. Subsidies are justified to make up for the difference between ability to pay and the higher societal benefits (maintenance of forest cover, prevention of land degradation, and reduction in emissions of greenhouse gases) and private benefits (reduction in expenditure for firewood and kerosene, savings in time for cooking and firewood collection and health) accruing to users. Besides the expense, many consumers are hesitant to adopt the biogas technology reflecting the lack of public awareness of the relevant issues. To date, this combination of factors has largely stifled the use of biogas technology in Africa.

2.1.2 Political factors affecting biogas production and commercialisation

The political barriers that exist are mainly in the area of sovereignty rights and the will to initiate national biogas technology programmes. Another problem is the high number of armed conflicts and political instability in the continent which together with the region's debt burden have reduced the region's credibility. Hence, providing capital even for modest investments will prove difficult. African governments need to commit themselves to renewable energy programmes. Government constant commitments to the development and promotion of renewable energy sources have been instrumental in promoting an ambitious alcohol fuel in Brazil, biogas programmes in Europe, China and India. It could be helpful to learn from the experiences gained in the developed world but adapted to the needs and situation in developing countries. However, in some African countries, the hostile social climate and political instability prevent opportunities of international collaboration and support.

2.1.3 Technical factors affecting biogas production and commercialisation

There are three major types of digesters that have been in use in developing countries: Chinese fixed dome digester, the Indian floating drum digester and the more recent tube digesters. These reactors are small in size (5-10 m^3) and mostly used at household level to deliver the energy demand for household cooking and lighting. The advantages of these reactors are that they are inexpensive compared to sophisticated systems, can be built with locally available material, are easy to handle and do not have moving parts which are prone to failure. The working principle of these reactors is the same although there are substantial differences between them. The substrate enters through the inlet pipe into the digester tank where the substrate has an average retention time of 10-30 days. The biogas is collected above the slurry and leaves the tank through a gas pipe into the top cover. In the fixed dome digester, the top is made of concrete or bricks as the rest of the digester below ground. The floating cover type has steel cover floating on the slurry, which is above ground, whereas the rest of the digester is also below the ground. The digested slurry leaves the digester through an outlet pipe and is collected in outlet pit. However, these digesters have several limitations. Each of the digester type does not have facilities for mixing the slurry or for maintaining a certain temperature in the digester and controlling it. There are also no facilities to remove sand, stones and other non-digestible materials, which will over the years, accumulate and decrease the volume of the digester and hence will reduce its efficiency. The accumulation of inert and non-degradable material makes it necessary to stop the process from time to time and remove the materials, thereby increasing labour and maintenance cost of the technology.

There is also lack of adequate coordinating framework as one of the most important weakness of energy institutions in Africa. Lack of coordination among institutions and conflicting interests are obstacles to good penetration of biogas technology into the African market. Rationalising functions and building institutions around them will improve the situation (Davidson, 1992). Constant persuasion and active campaigns can help reduce institutional inertia and resistance to adoption of biogas technology. Most renewable energy technologies require long development periods and dedicated stakeholders are important for building up experiences and competencies. New technologies often need to be nurtured for over decades, before sufficient socio-technical momentum emerges. Alignment between the technical, economic, regulatory and social context can provide the basis for building up momentum, until the biogas technology is able to survive on its own. Many African countries have a National programme having a three-pronged focus: sanitation, rural energy, and organic fertilizer usage, aimed at promoting domestic and agricultural based plants and this will help in promoting and implementing biogas plants. There are also now many biogas service providers in many African countries that specialize in the construction of biogas plants. The major focus of the biogas service providers is on sanitation. The service providers have used the hygiene-promoting aspect of biogas plants to market the technology.

There is need for continuous improvement of the biogas technology because its implementation is intrinsically the exploitation of the technical advantages. In some instances biogas plants have not worked effectively because of lack of support, lack of repairs and poor design. Lack of knowledge about biogas technology is often cited as a reason for non-adoption of biogas in some countries in Africa. Where people have installed biogas reactors, problems arising from the bad quality of the installed units and the poor operations and maintenance capacity of users have led to poor performance and even abandonment of biogas digesters. In some instance, the demonstration effect has been one of failure and has served to deter rather than enhance biogas adoption. A survey in Kenya of about 21 existing plants in 1986 found only 8 out of 21 functional and 13 out of 21 not functional or never finished (Day et al., 1990). According to the authors, the major problems associated with existing biogas plants in Kenya include inadequate design and construction, poor maintenance, and poor social acceptance. The effect of individual economic status is also important to consider in the assessments of biogas technology. Ni and Nyns (1996) reported that most surveys have revealed that biogas is more accepted by upper and middle-income farmers. The obvious effect of the income of individuals is the ability of investment to install a digester system and above all to maintain it operational. The regular operation of a biogas plant is more difficult to achieve than its initial installation. The routine operation and maintenance of the digester system need much physical work that is usually laborious and messy, making the biogas benefits less attractive.

2.1.4 Other factors affecting biogas production and commercialisation

The site-specific issues that have limited the scope of biogas technology in sub-Saharan Africa include the availability of water and organic materials for effective biodigester operation. Limited water availability poses a constraint for biogas operation in some countries because biogas plants typically require water and substrates such as manure to be mixed in an equal ratio. Small-scale farmers frequently lack sufficient domestic animals to obtain enough manure for the biodigester to produce sufficient gas for lighting and cooking. Even where households keep sufficient numbers of animals, semi nomadic or the free grazing system of many communities in sub-Saharan Africa makes it difficult to collect dung to feed digesters (Abbey, 2005). In countries where houses are clustered together as in Nigeria, a community plant might be more feasible (Akinbami et al., 2001).

In assessing the economic viability of biogas projects one should distinguish four major areas of applications: individual household units, community plants, large-scale commercial plants and industrial plants. In each of these cases, the financial feasibility of the facility depends largely on whether outputs in the form of gas and slurry can substitute for costly feeds which were previously purchased, the efficiencies with which the fuel is used or possible equipment which could lead to higher efficiencies. If 'externalities' such as employment, import substitution, energy security, environmental protection, and so on are considered then the economics change usually in favour of the biogas technology (Hall et al., 1992).

All too often, projects intended to introduce new energy technologies are conceived without proper understanding of the needs, problems, capabilities and priorities of the targeted users. Most of the Chinese and Indian biogas plants introduced in Africa are not functional due to many reasons. One of the major reasons of the failure is the separation of national interests and individual family/community interests (Ni and Nyns, 1996). There is need to learn from the past experiences and adapt the biogas technology from Europe and Asia for local African circumstances. There is also the need for bottom-up approach that takes the user interest into account. The Botswana biogas water pumping programme of the mid-1980s is a good example of how a misunderstanding of the target communities' needs and problems lead to project failure. The Botswana government's effort was to introduce biogas as the main pumping fuel in some areas. Water supply is a priority in Botswana due to its arid climate. The problems that arose were not technical but rather socio-economic. The villages targeted to 'benefit' from the biogas-pumped water felt disadvantaged in that they had to pay for the water they collected with cattle dung while other villages paid nothing by using the usual government or donor-supplied diesel engines. The benefits of biogas were important to the government as a means of reducing dependence on imported diesel. The perception from the point of view of the intended project beneficiaries was different. Today the biogas plants are disused. The principal reason is that real acceptance of the biogas technology depends on individual interests that do not totally respond to those at the national level. This suggests the necessity of understanding fully the individual interests of a project.

Renewable energy projects conceived without carefully consulting the intended recipients and beneficiaries face serious acceptance problems and fail prematurely due to abandonment. Numerous large-scale demonstration projects such as a sophisticated integrated biogas engine generator system at Kushinga Phikelela near Marondera in Zimbabwe collapsed when weaned from donor support. The reasons of failure had mainly to do with supply of spare parts which had to be procured with scarce foreign currency and lack of local capacity and funds to maintain demonstration installation. The host institute did not need the biogas technology since it has grid electricity and hence neglected it.

Some potential users are reluctant to try the biogas digesters out of concern about sanitation. Use of human wastes from for biogas production and the subsequent digested sludge, for example in schools, as a source of fertiliser faces cultural and health resistance. Even though the anaerobic digestion process naturally reduces the pathogen load, handling biogas feedstock particularly human excreta and using biogas slurry as fertiliser does pose some risk of infection (Brown, 2006). A major difficulty is utilising manure sources properly. There is usually lack of enough supply of manure for efficient and sustainable biogas production. Liquid manure is preferred for most biogas plants, but households may not be accustomed to storing and handling it. People also find it difficult to collect, store and deliver fresh manure to the digester. Liquid manure must be stored in pits or other installations that require investment of time and labour. Therefore promotion of liquid manure digesters requires additional education and training to ensure sustainability. The problems also include that animals must be penned for effective collection of animal dung, farmers must own a sufficient number of livestock to generate continuous flows of biogas, and the initial costs for the required infrastructure may be deterrent (Karekezi, 1994b). The effort of maintenance and control on biogas plants often does not meet the level of literacy skills of rural population.

It is also important to realise that lack of information on improved technologies such as biogas technology at all levels, government, energy institutions, and consumers, poses a very serious problem for technology penetration. Poor infrastructures prevent access to even the vast information available in the public domain about biogas technology and its application. Generating interest among the various stakeholders and setting up information systems using relatively cheap devices now available can assist greatly. Setting up or strengthening existing information systems is very important for the use of renewable energy technologies such as biogas. These systems should be capable of coordinating energy and energy-related information activities with appropriate means for collection, filtering, storage, retrieval and dissemination. In order to promote the implementation and proper use of anaerobic digestion technology, it is important to initiate long-term anaerobic digestion and other renewable energy training and capacity-building programmes, and to perform scientific work in this field (through appropriate research). It is important to establish contacts between research and university groups and experienced contractors, and to initiate collaboration with polluting industries, i.e., to interest them in the system, either for use as an environmental protection method, or for energy production. In addition, experts should provide reliable and pertinent information about the biogas technology and its potential to local authorities, politicians, and the public in general. It demands a lot of efforts in achieving an efficient transfer of knowledge from research centres and universities to state sanitation companies, consulting engineers firms and government environmental control agencies. There is also need and to obtain grants from the government or international organisations, and industry for pilot-plant and/or demonstration-scale projects (Foresti, 2001; Karekezi, 1994a).

To overcome some of the socio-cultural barriers, intensive educational and campaign programmes may have to be mounted to raise the awareness consciousness of the benefits of this technology. A case in point is that of a full-scale digester installed to treat opaque beer brewery wastewater in Harare which is just being used to treat the wastewater but the biogas from plant is currently vented to the atmosphere (Parawira et al., 2005). Further benefits of the plant could be realised by tapping the energy generated by the anaerobic process in the form of methane.

2.2 Possible measures to improve biogas production and commercialisation

The economics of large-scale biogas plants, probably to serve communities, could also be investigated since they may have a much higher benefit-cost ratio compared to family sized plants. In order to launch commercial biogas systems in Africa, it is therefore necessary to introduce incentives in the form of policies, legislation, taxes and financial subsidies and weaken the barriers. This has been the practice in India, China and even in European countries. Presumably community biogas plants which permit higher efficiency rather than household plants should be set up than family units in rural communities. The chances of success of a village biogas plant would be higher for villages with clustered dwellings rather than with dwellings scattered over large distances.

A list summarising the priority issues which must be tackled by most African nation for the development of biogas technology is given in Table 5.

- Evaluation or re-evaluation of the energy demand and supply patterns and their sectoral distributions at national level in order to estimate the contribution of biogas technology and other renewable energy make to the nation.
- Assessment of the potential of new and renewable energy sources such as biogas so as to tailor their use to the actual needs, and to substitute them for conventional sources wherever appropriate.
- Support of investigations, application, development, training and demonstration for the development of biogas technology.
- Establishment of a technical and scientific information network connected with international sources to diffuse the latest technological advances and applications, and enhancement of the research in accordance with national needs.
- Encouragement of joint research and development activities of mutual benefit.
- Organisation of a number of demonstration and pilot projects to illustrate the potential of new and renewable sources and to disseminate technological information.

• Provision of several economic incentives to accelerate and increase the biogas applications in the country Preparation of large scale projects to obtain support from non-profit international organisations and agencies such as United Nations, World Bank, Organization of Arab Petroleum Exporting Countries (OAPEC), Islamic Foundation for Science, Technology and Development (IFSTAD, Jordan), etc.
• Increasing public awareness of biogas technology by distributing simple explanatory pamphlets or using other media.
• Encouragement of potential private investors by offering a governmental partnership in the production of biogas technology.

Table 5. Priority issues which must be tackled by African countries for the development of biogas technology.

2.2.1 Biogas technology research in sub-Saharan countries

In developing countries, biogas energy research should be planned and conducted as the main factor leading to its contribution to the solution of energy problems. Keeping this in mind, the results of the research should be applicable on a nation-wide scale and constitute a part of the country's development plan. In many of the developing countries, there is remain some basic research areas mostly on the quantity and potential biogas yield of fermentable organic wastes available, the size and type of biogas digesters which can be economically viable for the potential consumers of the biogas technology.

Biogas technology research in selected sub-Saharan African countries has recently been reviewed by Mshandete and Parawira (2009). The review provided an insight and update of the state of biogas technology research in some selected sub-Saharan African countries in peer reviewed literature. An attempt was made to pinpoint future research in critically reviewing the biogas technology research. The methane-producing potential of various agriculturally sourced feedstocks has been researched, as has the advantages of co-digestion to improve carbon-to-nitrogen ratios and the use of pretreatment to improve the hydrolysis rates. Some optimisation techniques associated with anaerobic digestion including basic design considerations of single or two-stage systems, pretreatment, co-digestion, environmental conditions within the reactor such as temperature, pH, buffering capacity have been attempted in some of the researches in Nigeria, Tanzania, and Zimbabwe. However, there appears to be little research in biogas technology in many sub-Saharan African countries in internationally peer reviewed literature. However, biogas technology research will only have an impact if relevant and appropriate areas of research are identified and prioritised.

3. Gender implication of rural energy technology

Generally, rural women are greatly involved in managing household energy systems. Rural women are also the ones who are directly affected by the rural energy crisis. As mentioned in previous sections, traditional firewood cooking causes faster depletion of biomass resources and increases the time that women require in collecting firewood. These activities consume a great deal of the time and labour of women and increase the drudgery of women. In addition, the use of traditional energy technologies has a negative impact on women's health due to the smoke from firewood and their heavy workload. There is therefore the need for an intervention, that help to reduce women's labour and time, which could be used for other productive purposes, and to improve the health conditions of women. In this regards, an intervention with anaerobic digestion is needed. Such an intervention should be based on gender concerns both at macro and micro levels in terms of recognizing women's roles and responsibilities and their priorities regarding rural energy. The focus should be on reducing expenditure of human energy rather than only saving fuel. Hence, it is very important here to consider the practical gender needs, which fulfil the regular energy needs at household level while saving the time and labour of women, and the strategic gender needs, which provides the opportunities for women to be involved in social and economic activities for their self-enhancement and empowerment.

3.1 Biogas in rural communities and its benefit

Households in Africa, particularly in the rural areas are increasingly facing energy supply problems. According to United Nations (2010) there are approximately 60% of the total African population living in the rural areas. Biomass in form of wood, cow dung, and crop residues biomass constitutes 30% of the energy used in Africa and over 80% used in many sub-Saharan countries such as Burundi (91%), Rwanda and Central Africa Republic (90%), Mozambique (89 %), Burkina Faso (87%), Benin (86%), Madagascar and Niger (85%) (cited in United Nations Economic and Social Council, 2007). The availableness of these traditional fuels (wood, dried dung and agricultural waste) is declining (Deutsche Gesellschaft für Internationale Zusammenarbeit (GIZ) GmbH and Integrated Science and Technology ISAT, undated), while the commercialised fuels (e.g. charcoal) are very expensive and their availability unreliable. Domestic biogas provides an opportunity to overcome these challenges in the rural areas. This is because biogas production makes use of domestic resources such as agricultural crop wastes and animal wastes such as pigs, cattle, and poultry as well as human excreta. Biogas production using the existing domestic resources therefore, has a potential to provide a number of benefits to the rural communities in Africa. Biogas plants that are well functioning can provide a wide range of direct benefits to the users particularly in the rural areas. Many of these benefits are directly linked to the Millennium Development Goals of reducing income poverty, promoting gender quality, promoting health and environmental sustainability.

3.1.1 Renewable energy generation

The bulk of the rural population in Africa have no access to electricity. According to World Economic Outlook (2010), only 14% of sub-Saharan African has access to electricity. It is thus estimated that 582 million rural people in sub-Saharan Africa did not have access to electricity in 2009 (World Economic Outlook, 2010). North Africa is an exception because 98.4% of rural population is electrified and only 2 million did not have access to electricity in 2009 (World Economic Outlook, 2010). Biogas is a potential off-grid, clean energy fuel solution for rural areas of Africa (Amigun and von Blottnitz, 2010), that can provide energy services such as cooking, heating and lighting.

3.1.2 Environmental benefits

Fuel wood consumption is often portrayed as a cause of environmental degradation, and may lead to energy insecurity for rural African households, especially where the resource is commercialized (Hiemstra-van der Horst and Hovorka, 2009). The high dependence on woodfuel in the sub-Saharan Africa has resulted in an alarming rate of tree felling and deforestation (cited in United Nations Economic and Social Council, 2007). According to the United Nations Environmental Programme (2011), nearly half of the forest loss in Africa is due to removal of wood fuel. The estimated deforestation rate in Africa is twice the world rate (AfriNews, 2008). More than 15 million hectares of tropical forests are depleted or burnt every year in order to provide for small-scale agriculture or cattle ranching or for use as fuel wood for heating and cooking (United Nations Convention to Combat Desertification, 2004). Some alarming and worrying deforestation facts in Africa include (AfriNews, 2008): loss of over 90% of West Africa's original forest - currently, only a small proportion remains; between 1980 and 1995, an area of 1.1 million ha was cleared every year; only one tree is replanted for every 28 trees cut down. In Uganda where 90% of the population lives in rural areas and directly depends on land for cultivation and grazing, forestland has shrunk dramatically. In Nigeria, it is feared that the country will be left without forest due to the present level of deforestation activities.

Forests are required in order to build a resilient natural ecosystem as they moderate climate, act as water reservoirs and are habitat to wildlife. The loss of ground cover due to deforestation thus results in secondary problems such as exposing the soil to erosion during heavy rainfall, flooding, increased evaporation, drought, and increase in the greenhouse gas emissions. Familiar country specific example is the recent frequent droughts and floods experienced in East African countries, particularly Kenya, Somali, Uganda and Ethiopia, that have been associated with deforestation (IRIN, 2006; Mekonnen, 2006). Similarly, the declining rainfall in the West African countries is also attributed to deforestation. The use of alternative energy such as biogas has a potential to reduce the demand for wood and charcoal use, hence reducing greenhouse gas emissions improving water quality, conserving of resources - particularly trees and forests - and producing wider macroeconomic benefits to the nation (Amigun and Blottnitz, 2010) due to reduced deforestation. In addition, the slurry and waste from the biogas plants provides a high quality fertiliser that can be used to improve the soil fertility and increase productivity in agriculture dependent rural communities in Africa.

3.1.3 Improving quality of life in rural areas

The use of biogas has a potential improve the quality of life in the rural areas through reduced drudgery in women and children, reduced indoor smoke, improved sanitation and better lighting (Amigun and Blottnitz, 2011). Wood fuel gathering is a hard and time consuming duty for women. For instance, it is estimated that women can spend 2-6 hours in collecting wood fuel (DFID, 2002) depending on the country and region. For instance, one study in Limpopo, South Africa found that the rural women spend 5-6 hours (Masekoameng et al., 2005), while another study in a different region of South Africa report that the women spend over two hours. This takes away time that could be better utilized in other productive activities such as income generation or education particularly for girls who have to be absent from school to undertake such task. Biogas plants thus can help in reducing the workload of women and girls in collecting firewood.

Burning traditional fuel releases smoke which contains toxic pollutants such as carbon monoxide, hydrocarbons and particulate matter (Smith et al., 2005). Some of the prevalent health problems caused by the smoke inherent to traditional ways of cooking and heating, particularly open fires include: sneezing, nausea, headache, dizziness, eye irritation and respiratory illnesses (Onguntoke et al., 2010). Biogas improves health of the rural people by providing a cleaner cooking fuel thus avoiding these health problems. Women and children have the greatest risk of these health problems and children under 5 years are at high risk of contracting acute respiratory illnesses such as, pneumonia. Often, the rural population are also faced with lack of sanitation, resulting in water borne diseases affecting mainly women and children. Operating a biogas plant implies that manure is directly fed to the plant keeping the kitchen smoke free and farmyard cleaner.

3.2 Lesson from some biogas initiatives (case studies) in Africa

As indicated in Table 3, there are some digesters have been installed in a number of sub-Saharan Africa. These have mainly been pilot or demonstration projects aimed at testing the technical viability of small-scale biogas technology at a limited scale (Hivos, 2009a). These pilot projects have mostly been funded by non-governmental organizations and built for health clinics, schools, and small-scale farmers. While the small-scale biogas plants are located throughout Africa, only a few of them are operational (Parawira, 2009). There is also limited documentation on whether the existing biogas digesters have been successful in achieving the benefits highlighted in section 3.1. Some country specific examples is Tanzania, Ivory Cost and Burundi, which have produced biogas from animal and human waste using the Chinese fixed-dome digester and the Indian floating-cover digester (Omer and Fadalla, 2003). These have not been reliable and in many cases, poor performance has been reported (Omer and Fadalla, 2003). Thus, the plants have only operated for a short period due to poor technical quality (Mshandate and Parawira, 2009).

Currently, a number of different organizations are establishing biogas initiatives in Africa, particularly in rural areas, in order to supply cleaner burning energy solutions. These initiatives are at different stages of development such as: prefeasibility, feasibility, design and implementation to a limited extent. For instance, Burkard (2009) reports on five biogas case studies in Kenya which were to utilize agricultural leaves, residues from floriculture, and residues from vegetable production and canning. In 2010, it was reported that the Dutch government was to spend 200 million Kenyan Shilling to set up 8000 biogas digesters throughout the country. The initiative was targeting farmers practising zero grazing (Daily Nation, 2010). Similar projects are being implemented in Ethiopia, Uganda, Senegal, Burkina Faso, and Tanzania. There are also some other initiatives such as biogas for better life, which is at various stages of biogas development in Ethiopia, Kenya, Uganda, Sudan, Zambia, Malawi, South Africa, Lesotho, Swaziland, Mali, Senegal, and Ghana[1]. The Netherland Development Organization (SNV) has been supporting the development of National Biogas programmes in East Africa (Ethiopia, Kenya, Uganda, Tanzania, Rwanda) and West Africa (Senegal and Burkina Faso)[2]. While there are few documented successful small-scale biogas plants in the rural areas of Africa, this section will present some selected country specific biogas projects.

---

[1] http://www.biogasafrica.org/

[2] http://www.snvworld.org/en/ourwork/Pages/Renewable%20Energy.aspx

3.2.1 Rwanda

Rwanda has a population of 10.2 million people of which 81% of this population reside in the rural areas in 2010 (United Nations, 2007). One of the famous biogas programmes is the Kigali Institute of Science and Technology (KIST) large-scale biogas plants developed and installed in prisons. The aim of these plants was to treat toilet wastes and generate biogas for cooking. The first plant prison which was operational in 2001, and by 2011, KIST has managed to build and operationalize biogas plants in 10 prisons. Each prison is supplied with a linked series of underground biogas digesters, in which the waste decomposes to produce biogas. After this treatment, the bio-effluent is safe to be used as fertiliser for production of crops and fuel wood. The project was funded by Red Cross and the plant consists of five interlocking chambers. KIST's project saves 50% of wood for cooking and it won Ashden Award in 2006. The projects construction is managed by KIST, who also provides training to both civilians and prisoners.

Another biogas programme is the National Biogas Programme which is promoted by the Rwanda Ministry of Infrastructure, through the support by the Netherlands development organization. The programme aims at reducing firewood use by the households. The Ministry of Infrastructure estimates that 441 units have been installed to date, and approximately 15 000 households will be using biogas by end of 2011 for cooking and lighting[3]. The Ministry of Infrastructure of Rwanda is also collaborating with other ministries (e.g. Ministry of Education) in order to develop biogas plants in schools, clinics and community institutions.

3.2.2 Ghana

Ghana has a population of 24.8 million of which 48.5% live in the rural areas (United Nations, 2007). Netherlands Development Organization (2007) estimates that Ghana has a potential to realise 280 000 domestic biogas plants, that is capable of producing 6000 m^3 of liquid fertiliser, which would increase yield by 25%. However, low perception of biogas has modern energy has made Ghana not to realise the full potential of biogas utilisation (Bensah and Brew-Hammond, 2008). Bensah and Brew-Hammond (2010) highlights the status of the biogas development in Ghana, in which only about 200 units have been installed, thus lagging behind in comparison with other African countries such as Rwanda, Kenya and Tanzania. Some initiatives such as Biogas Technology West Africa Ltd[4] , funded by UNIDO has implemented a number of biogas digesters in Ghana for schools, hospitals and colleges. These are mainly underground masonry dome systems in the range of 60 m^3 to 160 m^3 volume. One example of these projects is Keta secondary school plant for 1200 users, and has a capacity of 80 m^3. The plant is built in sandy, water logged area and it makes use of human waste. The gas is used for cooking. The future development of biogas in Ghana will however not be left to private investors and initiatives if the benefit to the rural communities is to be realised. Bensah and Brew-Hammond (2010) argues that, for successful future development of biogas in Ghana, there is a need for establishing a government body that solely focuses on promoting biogas.

---

[3] http://mininfra.gov.rw/index.php?option=com_content&task=view&id=115&Itemid=143
[4] http://www.biogasonline.com/projects.asp

3.2.3 Mozambique

Mozambique has a population of 22.6 million people, in which 61.6% reside in rural area in 2010 (United Nations, 2007). Similar to Ghana, the Mozambique government does not have an agency solely supporting the development of biogas. Some initiative such as Biogas Technology West Africa Ltd[5] is however, also undertaking a biogas power project in the country. The project is an electric power system powered by a biogas-fired internal combustion engine generator at Mpunsa Village, Chicualacuala District, Gaza Province in Mozambique.

3.2.4 South Africa

Philippi biogas project is funded by the working for energy programme of South Africa, and it is situated in horticultural area zoned as agricultural land[6]. Two digesters have been constructed on site and each of them is 10 m^3. The total plant capacity could 12000 - 15000 litres of biogas per batch load. This is equivalent to 25kw per batch or more than 100 hours of cooking time. This project is still in its early stages of implementation.

3.2.5 Tanzania

Tanzania has a population of 42.5 million people of which 75% live in the rural area (United Nations, 2007). This is one of the countries that has progressed well in terms of biogas development and has several case studies. The first one is in the region of Tanga, which is known for sisal production as a cash crop. The sisal is sold to a number of sisal processing companies to produce fibre. Using the available production methods, only 4 % of the sisal biomass is recovered as fibre and the rest is waste, which is either burnt, producing carbon dioxide or left to decompose, producing Methane (The Bioenergy Site, 2009). Utilising sisal waste for bioenergy can thus be environmentally beneficial since 80% of the plant mass is suitable for biogas production, and can also increase profit to the sisal growing farmers (The Bioenergy Site, 2009). With this opportunity in place, UNIDO, through its initiative on "Rural Energy for productive use" established a biogas pilot demonstration project, with the support from Common Fund for Commodities (CFC)[7]. The plant situated at the Katani Sisal estate in Hale, and utilises the sisal waste generated from the sisal processing plant. The biogas power plant has installed capacity of 300 kW, and was inaugurated by the Tanzanian President in 2008 (UNIDO, 2008). The electricity generated from this plant is used for lighting and running small-scale industries. The company, Katani Limited, also provides energy services to local schools and hospitals in the area (PISCES and FAO, 2009). The company currently plans to expand the capacity to 7000 kW that will be connected to the grid (The East African, 2011).

A Tanzanian Domestic Biogas programme was also initiated in 2007, following a feasibility study by the GTZ. The programme set an ambitious goal of developing 3500 to 4000 units per annum. However, it was estimated that the current construction rate is only 200 to 400 per year (Sika, 2010).

---

[5] http://www.biogasonline.com/projects.asp

[6] http://www.smart2energy.co.za/index.php/pilot-projects/western-cape

[7] http://www.unido.org/index.php?id=6464

3.2.6 Kenya

Kenya has population of 40.6 million people, of which 77.8% reside in the rural areas (United Nations, 2007). Kenya similarly has a programme for promoting domestic biogas development, in which the Kenya National Federation of Agricultural Producer is the implementing agency[8]. The programme targets to install 8000 domestic biogas plants of between $6m^3$ - $12\ m^3$ capacity by 2013, and prioritizes the high agricultural potential regions. A number of demonstration plants have currently been constructed and launched.

3.2.7 Ethiopia

Ethiopia has a population of 89.6 million people, of which 82.4% live in the rural areas (United Nations, 2007). Through the Ethiopia Rural Energy Development and Promotion Centre (EREDPC) the National Biogas Program (NBP) was also launched. The aim of the programme is to establish 14000 biogas plants between 2008 and 2012, in four regions of Ethiopia (EREDPC, 2008). The NBP utilises cattle manure as the feedstock for biogas production (EREDPC, 2008). In 2009, some households had already started experiencing the benefits of the project such as: use of clean cooking fuel; income savings made in terms of time and money to search for fuel and purchase other traditional fuels (wood, charcoal and kerosene) respectively; and income generation from the sale of biogas to the neighbouring towns (Hivos, 2009b).

4. Biogas economics

The economy of a biogas plant is characterised by initial high investments costs, some operation and maintenance costs, mostly free raw materials (animal dung, aquatic weeds, terrestrial plants, sewage sludge, industrial wastes, poultry litter etc.) and income from the sale of biogas or electricity and heat (Amigun and von Blottnitz, 2007). Sometimes, other values can be added, e.g. for improved value of sludge as a fertilizer. The installation cost of a typical biogas plant is site specific (it depends on the topography of the area, labour cost at the site location, community participation, learning curve, use of the biogas product). Also, the economic performance of a biogas system will be very site specific and will depend on current markets for the input and outputs, the nature of agricultural practices and the system of organisation adopted by the community involved (Taleghani and Kia, 2005).

Good understanding of the relation between capital costs and plant size can provide useful information in assessing economic viability of biogas plants, and providing means whereby decisions are taken on developmental of a new project. In a developing economy, local market opportunities frequently restrict the size of a process plants. Scale effects influence costs per unit of capacity (specific cost). The scale economies concept is therefore of key concern because it can help in determining the optimal size of a biogas digester (Amigun and von Blottnitz, 2010).

Higher capital cost is experienced in African biogas industry. This is due to the fact that the current market for biogas in Africa is slow. Contractors therefore tend to lump all of their

[8] http://www.kenfapbiogas.org/index.php?option=com_content&view=category&layout=blog&id=36&Itemid=57

costs into the unit they are constructing because they may not get another order for months (Biogas for better life, 2007). Biogas technology in Africa appears to be implemented by technologically driven oligopolies - an economic situation where there are so few suppliers of a particular product that one supplier's action can have a significant impact on price and its competitors (Butare, 2005; Cawood, 2006, Mojaki Biogas Technology, 2008). The price which the typical firm charges depends on the number of firms in the industry. The less the number of suppliers, the less the competition, and hence the higher the charge. This concept is represented in the equation 1. The higher capital cost experienced in African biogas industry is aggravated by the fact that the current market for biogas in Africa is slow. Contractors therefore tend to lump all of their costs into the unit they are constructing because they may not get another order for months (Biogas for better life, 2007).

$$Q=\left(S/n+S\times b\times \overline{P}\right)-S\times b\times P \quad (1)$$

where:

Q = firm sales; S = total sales of the industry; n = number of firms in the industry;
b = constant term representing the responsiveness of a firm's sales to its price;
P = price charged by the firm itself; $\overline{P}$ = average price charged by its competitors.

Substantial cost reduction could be obtained through design optimisations and efficiencies created through economies of scale, as well as smart implementation and planning. In planning, the concept of clustering installations, where a number of orders for digesters within a defined geographic area would accumulate until a threshold is reached could provide substantial reduction of costs.

There is evidence that higher location factors are partly due to the need of importing specialized equipment (World Bank, 2007). In heavily industrialized countries, the equipment is often fabricated in the same area where the plant is constructed; in developing countries, depending on level of technology needed, equipment is generally imported along with specialised personnel to install it, at premium prices leading to increased investment costs. The investment costs are believed to be affected by the geographical location of the country viz: coastal and landlocked locations. However, a recent report by Amigun and von Blottnitz (2010) on the influence of geographical location (coastal and landlocked biogas plants) on biogas economics revealed that the cost of biogas technology is largely independent of geographical location of the plant, which is probably explained by the use of local construction materials in most small-medium scale biogas plants in Africa. The lower the import content of the total plant costs (for example, amount of steel), the less the external diseconomies which may arise in consequence of sliding exchange rates and transportation construction of materials.

5. Conclusion

Biogas technology represents one of a number of village-scale technologies that offer the technical possibility of more decentralised approaches to development. In addition, this technology offers a very attractive route to utilise certain categories of biomass such as agricultural organic waste or manure in rural areas for partially meeting energy needs (e.g. heating, electricity). This technology can therefore serve as a means to overcome energy poverty which poses a constant barrier to social and economic development in developing countries such as Africa.

Biogas initiatives in Africa is characterised by small to medium scale plants. Biogas technology is however, still beyond the reach of rural poor due to its high initial investment costs. There also exist a number of constraints affecting the implementation of the biogas technology on large scale such as: political, social-cultural, financial, informational, institutional, technical and training constraints. Priority issues which must be tackled by African countries for the development of biogas technology include: evaluation or re-evaluation of the energy demand and supply patterns and their sectoral distributions at national level in order to estimate the contribution of biogas technology and other renewable energy make to the nation; assessment of the potential of new and renewable energy sources such as biogas so as to tailor their use to the actual needs, and to substitute them for conventional sources wherever appropriate; support of investigations, application, development, training and demonstration for the development of biogas technology.

Sub-Saharan Africa with its warm climates is well-suited for the biogas digester technology. However, it is very important to consider the practical gender needs, which fulfil the regular energy needs at household level while saving the time and labour of women, and the strategic gender needs, which provides the opportunities for women to be involved in social and economic activities for their self-enhancement and empowerment at the planning phase of biogas development.

6. References

Abbey, A.T. 2005. Biogas in Uganda: A new experience. *Leisa,* 21:13-15

AfriNews 2008. Deforestation reaches worrying level - UN. http://www.africanews.com/site/list_messages/18831 [accessed 2 October 2011].

Ahring B, Angelidaki I, Johansen K. 1992.Anaerobic treatment of manure together with industrial waste. *Water Science Technology,*25:311-318.

Akinbami J.-F.K., Ilori M.O., Oyebisi T.O., Akinwuni I.O., Adeoti O. 2001. Biogas energy use in Nigeria: current status, future prospects and policy implications. *Renewable and Sustainable Energy Reviews,* 5:97-112.

Akinbami J.-F.K., Salami, A.T., Siyanbola, W.O. 2003. An integrated strategy for sustainable forest-energy-environment interactions in Nigeria. *Journal of environmental Management,* 69:115-128.

Amigun B., Sigamoney R., von Blottnitz H. 2008. Commercialisation of biofuel industry in Africa: A review. *Renewable and Sustainable Energy Reviews,* 12:690-711.

Amigun B., von Blottnitz H. 2007. Investigation of scale economies for African biogas installations. *Energy Conversion and Management,* 48:3090-3094.

Amigun, b. von Blottnitz, H. 2010. Capacity-cost and location-cost analyses for biogas plants in Africa. *Resources, Conservation and Recycling,* 55, 63-73.

Angelidaki, I., Boe, K. and Ellegaard L. 2004 Effect of operating conditions and reactor configuration on efficiency of full-scale biogas plants. Proceedings of the 10th World Congress in Anaerobic Digestion, Montreal Canada, 29th August-2nd September.

Angelidaki, I., and Ellegaard L. 2003. Co digestion of manure and organic wastes in centralised biogas plants. *Applied Biochemistry and Biotechnology,* 109:95-105.

Ayalon, O., Avnimelech, Y., and Shechter, M. 2001 Solid waste treatment as a high-priority and low-cost alternative for greenhouse gas mitigation. *Environmental Management*, 27, 697-704.

Barber, W.P. and Stuckey, D.C. 1999 The use of anaerobic baffled reactor (ABR) for wastewater treatment: A Review. *Water Research*, 33, 1559 -1578.

Bensah E. C and Brew-Hammond, A .2008. Biogas effluent and food production in Ghana. 4th National Conference of Ghana Society of Agricultural Engineering. http://energycenter.knust.edu.gh/downloads/6/64.pdf [accessed 10 October 2011].

Bensah E. C and Brew-Hammond, A. 2010. Biogas technology dissemination in Ghana: history, current status, future prospects, and policy significance. *International Journal of Energy and Environment*, 1: 277-294.

Björnsson, L., Mattiasson, B. and Henrysson, T. 1997 Effects of support material on the pattern of volatile fatty acid accumulation at overload in anaerobic digestion of semi-solid waste. *Applied Microbiology and Biotechnology*, 47, 640-644.

Borja, R., Martin, A., Luque, M., and Alonso, V. 1994 Kinetics of methane production from wine distillery wastewater in an immobilized cell bioreactor using sepiolite as support medium. *Resources Conservation and Recycling*, 10, 317-327.

Bouallagui, H., Cheikh, R.B., Marouuani, L. and Hamdi M. 2003. Mesophilic biogas production from fruit and vegetable waste in tubular digester. *Bioresource Technology*, 86, 85-89.

Brown, V. 2006. Biogas: a bright idea for Africa. *Environmental Health Perspectives*, 114:300-303.

Brummeler, E.T. 2000 Full scale experience with the BIOCEL process. *Water Science and Technology*, 41, 299-304.

Burkard T.2009. Project cases of biogas-plants in Kenya. http://www.akut-umwelt.de/bin/news/AKUT-Project-cases-in-Kenya.pdf [accessed 29 September 2011].

Callaghan, F.J, Luecke K, Wase, D.A.J, Thayanithy K., Forster, C.F. 1997. Co-digestion of cattle slurry and waste milk under shock loading conditions. *Journal Chemical Technology and Biotechnology*,68:405- 410.

Callaghan, F,J, Wase D,A,J, Thayanithy K, Forster C,F. 2002. Continuous co-digestion of cattle slurry with fruit and vegetable wastes and chicken manure. *Biomass Bioenergy*;27:71-77.

Chynoweth, D.P., Owens, J.M. and Legrand R. 2001 Renewable methane from anaerobic digestion of biomass. *Renewable Energy*, 22, 1-8.

Daily Nation 2010. Kenya: Dutch to inject Sh200 million into biogas project, 27 May. http://allafrica.com/stories/201005271360.html [accessed 28 September 2011].

DFID 2002. Energy for the poor. http://www.ecn.nl/fileadmin/ecn/units/bs/JEPP/energyforthepoor.pdf [accessed 29 September 2011].

Davidson, O. R. 1992. Energy issues in Sub-Saharan Africa: Future Directions. *Annual Reviews Energy Environment* 17:359-403.

Day, D.L., Chen, T.H., Anderson, J.C., Steinberg, M.P. 1990. Biogas plants for small farms in Kenya. *Biomass* 21:83-99.

de Baere, L. 2000 Anaerobic digestion of solid waste: state of the art. *Water Science and Technology*, 41, 283-290.

Edelmann, W., Schleiss, K. and Joss, A. 2000 Ecological, energetic and economic comparison of anaerobic digestion with different competing technologies to treat biogenic wastes. *Water Science and Technology*, 41, 263-273.

EREDPC 2008. National Biogas Programme Ethiopia. Programme Implementation Document. http://www.snvworld.org/en/Documents/NBP_implementation_document_Ethiopia_2008.pdf [accessed 10 October 2011].

Francese A.P., Aboagye-Mathiesen G., Olesen T., Co'rdoba P.R. and Sineriz F. 2000 Feeding approaches for biogas production from animal wastes and industrial effluents. *World Journal of Biotechnology*, 16:147-150.

Foresti, E. 2001. Perspectives on anaerobic treatment in developing countries. *Water Science and Technology*, 44, 1-6.

Frankin, R.J. 2001. Full-scale experiences with anaerobic treatment of industrial wastewater. *Water Science and Technology*, 44, 141-148.

Gijzen, H.J. 2001 Anaerobes, aerobes and phototrophs, a winning team for wastewater management. *Water Science and Technology*, 44, 123-132.

Gijzen, H.J. 2002 Anaerobic digestion for sustainable development: a natural approach. *Water Science and Technology*, 45, 321-328.

Grommen, R. and W. Verstraete 2002 Environmental biotechnology: the ongoing quest. *Journal of Biotechnology*, 98(1), 113-123.

GTZ and ISAT (undated). Biogas digest volume 1. http://www.gtz.de/de/dokumente/en-biogas-volume1.pdf[accessed28 September2011].

Gujer, W. and Zehnder, A.J.B. 1983 Conversion processes in anaerobic digestion. *Water Science and Technology*, 15, 127-167.

Gunaseelan, V.N. 1997 Anaerobic digestion of biomass for methane production: A review. *Biomass and Bioenergy*, 13, 83-114.

Hall D.O., Rosillo-Calle F., de Groot P. 1992. Lessons from case studies in developing countries. *Energy Policy*, Janaury:62-73.

Hiemstra-van der Horst, G. Hovorka, A. J. 2009. Fuelwood: The "other" renewable energy source for Africa? *Biomass and Bioenergy*, 33, 1605-1616.

Hivos 2009a. Africa Biogas Partnership Initiative. http://www.hivos.nl/eng/community/partner/10009895 [accessed 28 September 2011].

Hivos 2009b. Ethiopian households now enjoy the benefit of domestic biogas. http://www.hivos.nl/dut/Users/Members/jmsika/Jean-Marc-Sika/Ethiopian-Households-now-enjoy-the-benefit-of-domestic-biogas [accessed 10 October 2011].

IRIN 2006. Deforestation exacerbates droughts, floods. 10 November. http://www.irinnews.org/report.aspx?reportid=61528 [accessed 8 October 2011].

Iza, J., Colleeran, E., Paris, J.M. and Wu, W.M. 1991 International workshop on anaerobic treatment technology for municipal and industrial wastewaters: summary paper. *Water Science and Technology*, 24, 1-7.

Kaparaju P, Luostarinen S, Kalmari J, Rintala J. 2002. Co-digestion of energy crops and industrial confectionery by-products with cow manure: batch-scale and farm-scale evaluation. *Water Science and Technology*,45:275-280.

Karekezi S. 1994a. Renewable energy technologies as an option for a low-carbon energy future for developing countries: case examples from Eastern and Southern Africa. International Conference on National Action to Mitigate Global Climate Change. 7-9 June, Copenhagen, Denmark. Organised by UNEP Collaborating Centre on Energy and Environment.

Karekezi, S. 1994b. Disseminating renewable energy technologies in sub-Saharan Africa. *Annual Reviews*, 19:387-421.

Kashyap, D.R., Dadhich, K.S. and Sharma, S.K. 2003 Biomethanation under psychrophilic conditions: a review. *Bioresource Technology*, 87, 147-153.

Kivaisi A.K. and Rubindamayugi, M.S.T. 1996. The potential of agro-industrial residues for production of biogas and electricity in Tanzania. *WREC*, 917-921.

Laichena J.K., Wafula J.C. 1997. Biogas technology for rural households in Kenya. Organisation of the Petroleum Exporting Countries (OPEC) *Energy Review*, 21: 223-244.

Lantz, M., Svensson, M., Björnsson, L., and Börjesson, P. 2007. The prospects for an expansion of biogas systems in Sweden - Incentives, barriers and potentials. Energy Policy, 35:1830-1843.

Lema, J.M. and Omil, F. 2001 Anaerobic treatment: a key technology for sustainable management of wastes in Europe. *Water Science and Technology*, 44, 133-140.

Lissens, G., Vandevivere, P., de Baere, L., Biey, E.M. and Verstraete, W. 2001 Solid waste digesters: process performance and practice for municipal solid waste digestion. *Water Science and Technology*, 44, 91-102.

Lettinga, G. 1995 Anaerobic digestion and wastewater treatment systems. *Antonie van Leeuwenhoek*, 67, 3-28.

Lettinga, G., Hulshoff Pol, L.W., Zeeman, G., Field, J., van Lier, J.B., van Buuren, J.C.L. 1997. Anaerobic treatment in sustainable environmental production concepts. In: *8th International Conference on Anaerobic Digestion. Sendai*, pp. 32-39.

Lettinga, G. 2001 Digestion and degradation, air for life. *Water Science and Technology*, 44(8), 157-176.

Lettinga, G. 2004 With anaerobic treatment approach towards a more sustainable and robust environmental protection. In: *10th International Conference on Anaerobic Digestion. Montreal, Canada, 29th August-3rd September*, Vol 1 pp. 2-12.

Masekoameng KE, Simaleng TE and Saidi T.2005. Journal of Energy in Southern Africa 3: 4-9.

Mata-Alvarez, J., Macé, S. and Llabre`s, P. 2000 Anaerobic digestion of organic solid wastes. An overview of research achievements and perspectives. Review paper. *Bioresource Technology*, 74, 3-16.

Mbohwa, C and Fukuda, S. 2003 Electricity from baggasse in Zimbabwe. *Biomass and Bioenergy*, 25, 197-207.

Mbuligwe, S.E. 2002. Institutional solid waste management practices in developing countries: a case study of three academic institutions in Tanzania. Resources, Conservation and Recycling 35:131-146.

Mbuligwe S.E., and Kassenga, G.R. 2004. Feasibility and strategies for anaerobic digestion of solid waste for energy production in Dar es Salaam city, Tanzania. *Resources, Conservation and Recycling*, 42:183-203.

Mekonnen, M 2006. Drought, famine and conflict: the case from the horn of Africa. http://www.beyondintractability.org/case_studies/drought_famine_conflict.jsp?nid=6796 [accessed 9 October, 2011].

Misi S.N, Forster CF. 2001. Batch co-digestion of two component mixtures of agro-wastes. *Process Safety and Environment Protection*,79(B6):365-371.

Mshandete, A., Björnsson, L., Kivaisi, A.K., Rubindamayugi, M.S.T., and Mattiasson B. 2006. Effect of particle size on biogas yield from sisal fibre waste. *Renewable Energy*, 31:2385-2392.

Mshandete A. and Parawira W. 2009. Biogas technology research in selected sub-Saharan African countries - a review. *African Journal of Biotechnology*, 8:116-125.

Mwakaje A.G. 2007. Dairy farming and biogas use in Rungwe district, South-west Tanzania: A study of opportunities and constraints. *Renewable and Sustainable Energy Reviews*, Article in press.

Murphy, J.T. 2001. Making the energy transition in rural East Africa: Leapfrogging an alternative? *Technological Forecasting and Social Change*, 68:73-193.

Ni, J. Q., and Nyns, E.J. 1996. New concept for the evaluation of rural biogas management in developing countries. *Energy Convers. Mgmt*, 37:1525:1534.

Noykova, N., Muller, T.G., Gyllenberg, M., and Timmer, J. 2002 Quantitative analysis of anaerobic wastewater treatment processes: Identifiability and Parameter Estimation. *Biotechnology and Bioengineering*, 78, 89-103.

Oguntoke O, Opeolu B O and Babatunde N.2010. Indoor air pollution and health risks among rural dwellers in Odeda area, South-Western Nigeria. Ethiopian Journal of Environmental Studies and Management 3: 39-46.

Omer, A. M., and Fadalla, Y. 2003. Biogas energy technology in Sudan. *Renewable Energy*, 28:499-507.

Parawira W. 2004a. Anaerobic batch digestion of solid potato waste alone and in combination with sugar beet leaves. Renewable Energy, 29:1811-1823.

Parawira W. 2004b. Anaerobic treatment of agricultural residues and wastewater: Application of high-rate reactors. Doctoral dissertation, Department of Biotechnology, Lund University, Sweden ISBN: 91-89627-28-8.

Parawira, W., Kudita, I., Nyandoroh M.G., Zvauya, R. 2005. A study of industrial anaerobic treatment of opaque beer brewery wastewater in a tropical climate using a full-scale UASB reactor seeded with activated sludge. *Process Biochemistry*, 40:593-599.

Parawira, W., Murto, M., Read, J.S. and Mattiasson, B. 2007. Study of two-stage anaerobic digestion of solid potato waste using reactors under mesophilic and thermophilic conditions. Environmental Technology, 28: 1205-1216.

Parawira, W., Read, J. S., Mattiasson, B., and Björnsson, L. 2008. Energy production from agricultural residues: high methane yields in pilot scale two-stage anaerobic digestion. *Biomass and Bioenergy*, 32:44-50.

Pavlostathis, S.G. and Giraldo-Gomez, E. 1991 Kinetics of anaerobic treatment. *Water Science Technology*, 24, 35 - 59.

PISCES and FAO 2009. Small-scale bioenergy initiatives. http://www.impattoambientale.it/documenti/Rapporto%20FAO%202009.pdf [accessed 10 October 2011].

Rajeshwari, K.V., Balakrishnan, M., Kansal, A., Kusum Lata, and Kishore, V.V.N. 2000 State-of-the-art of anaerobic digestion technology for industrial wastewater treatment. *Renewable and Sustainable Energy Reviews*, 4, 135-156.

Rebac, S., van Lier, J.B., Janssen, M.G.J., Dekkers, F., Swinkels, K.T.M., and Lettinga, G.1997 High rate anaerobic treatment of malting waste water in a pilot-scale EGSB system under psychrophilic conditions. *Journal of Chemical Technology and Biotechnology*, 68, 135-146.

Rushbrook PE. Co-disposal of industrial wastes with municipal solid wastes. *Resource Conservation Recycling* 1990;4:33-49.

Sika, JM (2010). The Tanzania Domestic Biogas Programme gets up to speed! http://www.hivos.nl/english/News/News/The-Tanzania-Domestic-Biogas-Programme-gets-up-to-speed [accessed 10 October 2011].

Singh K.J., Sooth S.S. 2004. Comparative study of economics of different models of family size biogas plants for state of Punjab, India. *Energy Convers Manage*, 45:1329-41.

Smith, K.R., Rogers, J. & Cowlin, S.C. 2005. Household fuels and ill-health in developing countries: what improvements can be brought by LP gas (LPG)? Paris, France, World LP Gas Association & Intermediate Technology Development Group.

Switzenbaum, M.S. 1995. Obstacles in the implementation of anaerobic treatment technology. *Bioresource Technology*, 53, 255-262.

Tafdrup, S. 1995. Viable energy production and waste recycling from anaerobic digestion of manure and other biomass materials. *Biomass and Bioenergy*, 9, 303-314.

Tafdrup S. 1994. Centralised biogas plants combine agricultural and environmental benefits with energy production. *Water Science and Technology*,30:133-141.

Taleghani G, Kia AS. 2005. Technical–economical analysis of the Saveh biogas power plant. Renewable Energy. 30:441–6.

Tarlton JE (Undated). Woodfuel in Ciskei: a headload study. http://www.odi.org.uk/resources/libraries/forestry-grey-literature/headload.pdf [accessed 9 September 2011].

The Bioenergy Site ,2009. Tanzania sisal biogas. http://www.thebioenergysite.com/articles/308/tanzania-sisal-biogas [accessed 10 October 2011].

The East African 2011. Tanzania to double sisal-fired biogas capacity, 3 July. http://allafrica.com/stories/201107041813.html [accessed 25 September 2011].

UNIDO 2008. Inauguration of the first sisal biogas electricity pilot plan in the world. http://www.unido.org/index.php%3Fid%3D4835%26ucg_no64%3D1/DATA1/DATA1/Vacancy/fileadmin/user_media/PMO/IDB38/doc/index.php?id=7881&tx_ttnews[tt_news]=244&cHash=f33e93f0c4a4efc60ed7f7ee5f149b47 [accessed 10 October 2011].

United Nations 2007. World urbanization prospects: the 2007 revision population database. http://esa.un.org/unup/p2k0data.asp [accessed 11 October 2011].

United Nations 2010. Number of city dwellers grows amidst continuing disparities in level and pace of urbanization worldwide. http://esa.un.org/unpd/wup/Documents/WUP2009_Press-Release_Final_Rev1.pdf [accessed 29 September 2011].

United Nations Convention to Combat Desertification 2004. A carrying pillar in the global combat against land degradation and food insecurity. Background paper for the San Rossore meeting 'Climate change: a new global vision' Pisa, Italy,15 - 16 July 2004.

United Nations Economic and Social Council 2007. Africa review report on drought and desertification. http://www.un.org/esa/sustdev/csd/csd16/rim/eca_bg3.pdf [accessed 9 September 2011

van Lier, J.B., Van der Zee, F.P., Tan, N.C.G., Rebac, S. and Kleerebezem, R. 2001 Advances in high rate anaerobic treatment: staging of reactor systems. *Water Science and Technology*, 44, 15-25.

Viéitez, E.R. and Ghosh, S. 1999 Biogasification of solid wastes by two-phase anaerobic fermentation. *Biomass and Bioenergy*, 16, 299-309.

Weiland, P. 2000 Anaerobic waste digestion in Germany - Status and recent developments. *Biodegradation*, 11, 415-421.

World Bank, World development indicators, 2003, World Bank, Washington DC, 2003.

World Bank, 1980. Alcohol production from biomass in developing countries, Washington DC, USA.

World Economic Outlook 2010. Access to electricity. http://www.iea.org/weo/electricity.asp [accessed 30 September, 2011]

Zeeman, G. and Saunders, W. 2001. Potential of anaerobic digestion of complex waste (water)s. *Water Science and Technology*, 44, 115-122.

2

Potentials of Selected Tropical Crops and Manure as Sources of Biofuels

Babajide A. Adelekan
Federal College of Agriculture Ibadan,
Institute of Agricultural Research & Training, Ibadan,
Nigeria

1. Introduction

The chapter presents comprehensive and up-to-date knowledge on the themes of biogas, bioethanol, biodiesel as obtained from cassava, cocoyam, jatropha, grasses and manure. The author's research findings as well as those reported by other researchers are used for the discussion. Recommendations as regards how to benefit much more from these biofuels derived from selected tropical crops are presented. It is anticipated that these recommendations will be of immense help to academics and industry specialists working in such areas.

2. Contemporary focus on renewable energy

In contemporary times, a great deal of interest has been generated worldwide regarding the use of biofuels namely biogas, bioethanol and biodiesel for energy supply. The most ambitious goal thus far in respect of the development and exploitation of renewable energy sources appear to be that articulated by the European Renewable Energy Council. According to European Renewable Energy Council EREC (2010) in March 2007, the Heads of States and Governments of the 27 EU Member States adopted a binding target of 20% renewable energy in final energy consumption by 2020 and 100% by 2050. Combined with the commitment to improve energy efficiency by 20% until 2020 and to reduce greenhouse gas emissions by 20% (or respectively 30% in case of a new international climate agreement) against the 1990 level, Europe's political leaders paved the way for a more sustainable energy future for the European Union and for the next generations. In order to reach the binding overall target of at least 20% renewable energy by 2020, the development of all existing renewable energy sources as well as a balanced deployment in the heating and cooling, electricity and transport sectors is needed. According to estimates of the European renewable energy industry around 40% of electricity demand will be generated with renewable energy sources by 2020 (EREC, 2010). Furthermore, the new Renewable Energy Directive (RED) will undoubtedly stimulate the renewable energy heating and cooling market, and according to EREC's projections, up to 25% of heating and cooling consumption can come from renewable energy by 2020. Similar kind of awareness is evident in other regions of the world and cogent efforts are being made to increase the renewable energy share of the energy profile and reduce overdependence on fossil fuels.

For about 3 decades, Brazil has been in the forefront of using renewable energy in the form of bioethanol derived mainly from sugarcane to power fuel-flex vehicles or as oxygenate to gasoline and has made a remarkable success of it. Likewise, the USA has also to some extent used bioethanol to power vehicles. Bioethanol is the biofuel most widely used for transportation worldwide. The global annual production of fuel ethanol is around 40 to 50 billion litres, of which 90 percent is produced by the USA and Brazil from maize and sugarcane respectively (World Bank, 2008). Global ethanol production has seen steady growth since the search for alternatives to petroleum was prompted by the oil crisis of 1973/1974. The USA is now the largest consumer of bioethanol, followed by Brazil. Together they consume 30 billion litres, or three quarters of global production (Licht, 2005). The Economist (2005) reported that as at that time Germany was raising its output of biodiesel by 50% per year; USA was boosting its ethanol production by 30% per year; France aimed to triple its output of biodiesel and ethanol by 2007; China had just built the largest ethanol plant in the world; and also that Brazil was producing around 4 billion litres of ethanol per year, and hoped to export 8 billion litres per year by 2010. China's Ministry of Science and Technology plans that the country would attain 12 million tonnes of biodiesel production by the year 2020 (GTZ, 2006).

According to OECD (2008), the global ethanol and biodiesel production in 2007 is given in Table 1. Certainly, successes recorded as regards exploitation and use of other biomass for energy supply, will further enhance global energy security. Some of the themes involved in this are discussed in this chapter.

Country	Ethanol	Biodiesel
USA	26,500	1,688
Canada	1,000	97
European Union	2,253	6,109
Brazil	19,000	227
China	1,840	114
India	400	45
Indonesia	0	409
Malaysia	0	330
Others	1,017	1,186
World	52,009	10,204

Source: OECD (2008)

Table 1. Global Ethanol and Biodiesel Production for 2007 (in million litres)

3. History of anaerobic biodigestion

Sparse evidence suggests that biogas was known to the Assyrians and Persians centuries before Jesus Christ was born. Further evidence is traceable to count Alessandro Volta who in 1776 concluded that there was a direct link between the amount of decaying organic matter and the amount of flammable gas produced. Sir Humphrey Davy determined in 1808 that methane was present in the gasses produced during the anaerobic decomposition of cattle manure. Helmont recorded the emanation of an inflammable gas from decaying organic matter in the 17th century (Brakel, 1980). It was not until towards the end of the 19th century that methanogenesis was found to be connected to microbial activity. In 1868, Bechamp named the organism responsible for methane production from ethanol. This organism could more accurately be described as a mixed population. Bechamp was able to show that, depending on the substrate different fermentation products were formed. Zehnder et al (1982) stated that it was in 1876 when Herter reported that acetate in sewage was converted to equal amounts of methane and carbon dioxide. Meynell (1976) noted that the first anaerobic digestion plant was built in Bombay, India in 1859. The first notable use of biogas in England occurred in 1859 when gas derived from a sewage treatment facility was used to fuel street lamps in Exeter (McCabe and Eckenfelder, 1957). Then in 1904, Travis put into operation a new two-stage process in which the suspended material was separated from the wastewater and allowed to pass into a separate 'hydrolyzing' chamber (Carcelon and Clark, 2002). Buswell and Hatfield (1936) and some other researchers in the 1930s identified anaerobic bacteria and the conditions that promote the production of methane. Their works also explained such issues as the fate of nitrogen in the anaerobic digestion process, stoichiometry of the reactions, as well as the production of energy from farm and industrial wastes through the anaerobic digestion process. Regarding anaerobic technology, farm-based facilities are the most common. In contemporary times low-technology biogas digesters have been most extensively used in China and India. Bui-Xuan (2004) pointed out that low cost biogas technology has been well received by small holder farms in many developing countries for producing a clean fuel to replace firewood, within the recent ten years. Stating that more than twenty thousand digesters have been installed in Vietnam, mainly paid for by the farmers; however biodigesters are still not fully integrated into the farming system as there is only limited use of by-products (effluent) as fertilizer for vegetables, fruit trees, fish pond and water plants. The paper further stated that the use of effluent from digester can be studied as a resource for small scale farmers. Interest in the technology is increasing in several other parts of the world.

4. Overview of biogas production

Biomass is basically used as fuel, fertilizer, and feed. One fact which is evident in the literature is that the use of biomass, particularly livestock manure as fertilizer and feed has not grown with the continuously increasing rate of production of the manure itself. For instance, Wadman et al., (1987) pointed out that in the Netherlands, the total production of manure from housed cattle (during the winter period only) pigs, poultry, and fattening calves increased from 10 tonnes/ha in 1950 to 26 tonnes/ha in 1982. Neeteson and Wadman (1990) observed that within that same period however in the same country, the need to use animal manures as fertilizers decreased due to the widespread adoption of cheap inorganic fertilizers. These inorganic fertilizers have a number of advantages over manure namely; their composition is known, they are easier to store, transport, and apply and have a more predictive effect on crop growth than manures. Therefore, livestock manure was increasingly regarded as a waste product rather than a fertilizer.

The situation reported for the United Kingdom is another example. Using agricultural census data, Smith and Chambers (1993) estimated that around 190 million tonnes of livestock excreta per year are produced on U.K farms. Some 80 million tonnes of this is collected in buildings and yards where they are stored and hopefully applied to land later. However, land application of all the collected manure has not always been possible over the years. Chalmers (2001) in a review of fertilizer, lime and organic manure use on farms in Great Britain noted that the proportion of UK land receiving organic manures remained at 16% for tillage cropping but increased slightly for grassland, from a mean of 40% in 1983-1987 to 44% in 1993-1997. Just as in the case of the Netherlands referred to previously, livestock manure produced on UK farms constituted a burden since land application of all of it was increasingly impossible. Against the background that Netherlands used as an example has 5, 95, and 14 millions of cattle, poultry, birds, and pigs respectively, implications for other countries which have higher livestock populations are quite significant.

The option of using livestock manure as fuel merits closer investigations for its evident biogas-generation potential. Heltberg et al., (1985) pointed out that biomass could potentially contribute about 3.2 billion GJ to the United states energy resources, which is roughly the amount of energy expected to be supplied from nuclear and hydroelectric power plants in the USA as at that time. Within the USA itself, some projects are already operational. Thomas (1990) reported the case of a commercial project in California which was generating about 17.5 MW of electricity from cattle manure. Biogas typically refers to methane produced by fermentation of manure or other biomass under anaerobic conditions. Mata-Alvarezi (2002) focused on the state of research on the subject in Europe and noted that the process is popular in the rural areas, particularly in the Netherlands and Denmark because it provides a convenient way of turning waste into electricity. The use of biogas is encouraged because methane burns with a clean flame and produces little pollution or no pollution.

The use of manure to produce biogas for energy supply also has attractive prospects in developing countries. According to Akinbami et al., (2001), Nigeria produced about 227,500 tons of fresh animal waste daily. The paper noted that since 1kg of fresh animal waste produces about 0.03 m^3 of biogas, then Nigeria can produce about 6.9 million m^3 of biogas everyday. In addition to all this, 20kg per capita of municipal solid waste (MSW) has been estimated to be generated in the country annually. Going by the census figures 140 million inhabitants, the total generated MSW would be at least 2.8 million tonnes every year. With increasing urbanization and industrialization, the annual MSW generated will continue to increase. Biogas production can therefore be a profitable means of reducing or even eliminating the menace and nuisance of urban waste in many cities by recycling them; while at the same time contributing towards providing adequate solution to the seemingly intractable problem of energy security. In the case of Nigeria, a few small scale biogas plants have been constructed by the Sokoto Energy Research center (SERC) and the Federal Institute of Industrial Research (FIIRO) Oshodi, Lagos. As of now contributions of these small - scale biogas plants to aggregate energy supply are yet to become significant (Energy Commission of Nigeria, 1998). Similar potential as this exists in many countries across the developing world.

Processes for the conversion of biomass to biogas may be classified into two categories namely thermal processes (as in biomass gasification), and biological processes (as in anaerobic digestion). As observed by Chynoweth and Isaacson (1987), the major advantage of thermal processes is their ability to effect total conversion of organic matter at rapid rates.

The major disadvantage however is that they produce a mixture of gaseous products that must be upgraded to methane and are only economic at larger scales. Biological processes on the other hand, have the major advantages of producing biogas composed primarily of methane and carbon dioxide with traces of hydrogen sulfide, and are also low - temperature processes which are economical at a variety of scales. Biomass gasification is a process in which solid fuels are broken down by the use of heat to produce a combustible gas (Foley and Barnard, 1985). Fuels that can be gasified include wood, charcoal, coal, and a variety of other organic materials. In the sense used in this chapter, gasification should be distinguished from biogas production which uses wet organic feed stock and works by means of microbial action. Biological processes of biogas production may be aerobic (Evans and Svoboda, 1985) or anaerobic (Voermans, 1985). However, because of the high cost of aerobic processes particularly as regards the provision of energy to sustain the processes, anaerobic processes are preferred. As noted by Voermans (1985) biogas is the main purpose of anaerobic digestion and it comprises ± 55–70% CH_4; 30–45% CO_2, water vapor, and 0.0–0.5% H_2S Anaerobic digestion is brought about in anaerobic digesters.

5. Biofuels for the production of energy

Biomass represents a continuously renewable potential source of biogas and other biofuels and thus is certainly an option to inevitable fossil fuel depletion. Biogas can be economically converted to methane at facilities ranging from smallholder utility equipments to large scale plants and therefore can be tailored to supply rural and urban gas needs as well as meet regional and nationwide energy demands. According to Shoemaker and Visser (2000), the composition of biogas produced by anaerobic digestion as compared to natural gas is given in Table 2. It is readily seen from the table that overall, biogas is of a better quality than natural gas and possesses much less potential for polluting the environment. Biogas therefore constitutes a good alternative to natural gas.

Component	Natural gas (%)	Biogas (%)
CH_4	85	50-80
CO_2	0.89	20-45
C_2H_6	2.85	-
C_3H_8	0.37	-
C_4H_{10}	0.14	-
N_2	14.32	-
O_2	<0.5	-
H_2S	<0.5	0-1.5
NH_3	-	0-0.45

Table 2. Compositions of Natural Gas and Biogas by Volume

However, the present potential of biofuels to enhance energy security is limited. Globally, the huge volume of biofuels required to substitute for fossil fuels is beyond the present overall capacity of global agriculture. For example in the year 2006/2007, the United States used 20 percent of its maize harvest for ethanol production, which replaced only three percent of its petrol consumption (World Bank, 2008). The possibility of more significant displacement of fossil fuels should be possible with second and third generation biofuels.

Theoretically, biomass includes every material of plant or animal origin. However, the focus of research and use of biomass in practical terms is on those materials from which biogas, ethanol and biodiesel may be derived at economic scales. Earlier researchers reported successes which have been advanced by more recent works. Hill (1984) conducted experiments to investigate methane productivity of some animal waste types at low temperatures and very low volatile solids concentrations. Results indicated that there are large differences between the waste types and that poultry waste produced the highest biogas yield for animal live weight (LW) while dairy waste was the least productive on a LW and total solids (TS) basis. This result corroborates those of Huang and Shin (1981), Huang et al., (1982), and Shih (1984). These studies evaluated the potential of methane generation from chicken manure and also assessed the performance of poultry waste digesters. Of further interest is the finding of the last paper, which showed that a high rate of gas produced at 4.5 v/v/day (methane 3.0 v/v/day) can be reached at 5^0C, 4-day retention time (RT) and 6% volatile solids (VS) concentration. Shih (1984) further pointed out that if this potential can be obtained on a poultry farm, the process of anaerobic digestion for waste treatment and energy production would be economically attractive. The potentials of other kinds of livestock waste for biogas production have also been investigated for example dairy manure (Lindley and Haughen, 1985), beef cattle manure (Hamiton et al., 1985) and pig manure (Fedler and Day, 1985). A common result however, is that these particular livestock waste types did not produce biogas as much as poultry manure in the experiments conducted. In experiments conducted on a digester (Ghederim et al., 1985) gas yields related to the organic matter fed to the digester were 0.5 to $0.6m^3$/kg for pig farm sludge and 0.2 to $0.3m^3$/kg in the case of beef cattle waste. Methane content varied between 60 and 70%.

The possibility of manure-straw mixtures producing more gas than manure alone continues to engage the interests of researchers. Jantrania and White (1985) found that high-solids anaerobic fermentation of poultry manure mixed with corn stover at 30% to 35% initial total solids produced biogas quantitatively comparable to slurry type anaerobic fermentation. However, the retention time of the process was much longer than required in the conventional process. Hills and Roberts (1979) had earlier reported a substantial increase of methane produced from rice-straw manure and barley-straw manure mixtures compared to manure alone. In a comparative study of pig manure and pig manure-corn stover, Fujita et al (1980) concluded that the mixtures produced more methane than manure alone. In a pit-scale study of wheat straw-manure mixture, Hashimoto and Robinson (1985) found a methane production of $0.25m^3$ CH_4/kg of volatile solids (VS).

In more contemporary papers, several researchers have recently reported improvements in biofuel production from various agricultural materials including biogas from mixtures of cassava peels and livestock wastes (Adelekan and Bamgboye, 2009a), biogas from pretreated water hyacinth (Ofuefule et al., 2009), methanol from cow dung (Ajayi, 2009) fuel from indigenous biomass wastes (Saptoadi et al., 2009), ethanol from non-edible plant parts (Inderlwildi and King, 2009), as well as biogas from various livestock wastes (Adelekan and Bamgboye, 2009b). Adelekan (2012) showed that cassava, an often neglected but sturdy crop is a potent energy crop for the production of methane and ethanol, and presented production estimates for these biofuels based on cassava yield from the tropical countries. It has been discovered that, under aerobic conditions, living plants also produce methane which is significantly larger in volume than that produced by dead plants. Although this does not increase global warming because of the carbon cycle (Keppler et al., 2006), it is not readily recoverable for economic purposes. However, the methane which is recoverable for the direct production of energy is from dead plants and other dead biomass under anaerobic conditions.

Prasad et al., (2007) observed that with world reserves of petroleum fast depleting, ethanol has in recent years emerged as the most important alternative resource for liquid fuel and has generated a great deal of research interest in ethanol fermentation. The paper noted that research on improving ethanol production has been accelerating for both ecological and economic reasons, primarily for its use as an alternative to petroleum-based fuels. Based on their genetic diversity, climatic adaptation, biomass and sugar production, field crops have the best potential as large scale fuel sources. Lignocellulosic biomass is the most abundant organic raw material in the world. As observed further, the production of ethanol from renewable lignocellulosic resources will improve energy availability, reduce dependence on petroleum based fuels, decrease air pollution, and diminish atmospheric CO_2 accumulation. Using the by-products of crop processing for ethanol production will also reduce waste disposal problems and lower the risks of polluting the environment.

Adelekan (2011) in laboratory experiments compared the ethanol productivity of selected varieties of cassava, sorghum and maize crops widely grown in West Africa by correlating volumes and masses of ethanol produced to the masses of samples used. The rate of ethanol production were found to be 145 l/tonne, 135 l/tonne and 346 l/tonne for cassava (variety TMS 30555), sorghum and maize respectively. In terms of ethanol productivity, the order observed in the study was maize > cassava > sorghum. The dried mash produced from the process was analysed for its nutritive quality and that from cassava was found to contain 61.8 calories of food energy per 100g; that from maize and sorghum; 59.5 and 58.1 calories respectively, making them good materials for livestock feed composition. Overall, the ethanol produced from these tropical crop varieties is of a good quality. The key advantage is that the ethanol is being produced from renewable sources which are also sustainable. The production and use of ethanol from cassava, sorghum and maize crop is recommended particularly in West African countries which often suffer crucial problems in respect of sourcing and delivery of fossil fuels and also in other tropical countries where these crop varieties are grown. In such places, ethanol can be blended with gasoline. The key production process used is fermentation and this being a natural process is very efficient, safe and not destructive to the environment.

6. Conditions for anaerobic biodigestion

Chynoweth and Isaacson (1987) observed that in any anaerobic digestion process that is not inhibited or kinetically limited, two major factors affecting methane yields are feedstock composition and inoculum characteristics. The composition of the biodegradable organic compounds can influence methane yield in that reduced compounds such as fats and proteins produce a higher percentage of methane than oxidized compounds such as sugars. Ultimate methane yields are however, influenced principally by the biodegradability of the organic components. The same paper noted further that each anaerobic environment may differ in the types of bacteria involved in the methanogenesis, depending on differing factors such as substrate, retention time, temperature, pH, and fluctuations in environmental parameter. Although certain general properties are common from one environment to another, each environment may have its own unique population of bacteria, and associated microbial activities. Key operating factors which have a direct influence on the level and efficiency of biogas include volatile solids loading rate, digester temperature hydraulic retention time, pH and carbon: nitrogen ratio (Vetter et al., 1990).

6.1 Digester temperature

Marchaim (1992) noted that there is a close relationship between the biogas fermentation process and the temperature of the reactor. The higher the temperature, the more biogas is produced but when the temperature is too high, this can cause metabolic process to decline. Hobson et al., (1981) found biogas production to be greatest when the digester temperature was in the range of 32 to 40^{0}C. Hill (1982) also stated that digestion temperatures for optimum design all occur in the mesophilic range of 32^{0}C to 40^0 C. This work suggested that temperature beyond 40^{0}C has little effect on digester performance since the higher volumetric methane productivity is offset by the smaller digestion volume. As observed by the paper these lower temperatures also represent major savings in energy requirements when compared to thermophilic digestion (i.e. 60^{0}C). During the process of anaerobic biodigesiton in order to reach optimum operating temperatures (30-37^{0}C or 85-100^{0}F), some measures must be taken to insulate the digester, especially in high altitudes or cold climates (VITA, 1980). Straw or shredded tree bark can be used around the outside of the digester to provide insulation. According to Carcelon and Clark (2002), anaerobic bacteria communities can endure temperatures ranging from below freezing to above 57.2^{0}C (135^0 F), but they thrive best at temperatures of about 36.7 ^{0}C (98^0 F) (mesophilic) and 54.4^{0}C (130^0 F) thermophilic. Bacteria activity, and thus biogas production falls off significantly between about 39.4^{0}C and 51.7^{0}C (103 ^{0}F and 125 ^{0}F) and gradually from 35 ^{0}C to 0^0 (95 ^{0}F to 32 ^{0}F). To optimize the digestion process, the digester must be kept at a consistent temperature as rapid changes will upset bacterial activity.

The potential of thermophilic digester operating temperatures ($\geq$ 55^{0}C) for anaerobic biogestion of livestock waste has been investigated by several researchers (Converse et. al., 1977; Hashimoto, et. al., 1979; Hashimoto, 1983; Hashimoto, 1984; Hill, 1985; Hill and Bolte, 1985; Hill et. al., 1986) with the technical feasibility being decided in favour of the process. Hill (1990) identified the advantages of thermophilic digestion over conventional mesophilic digestion as reduced hydraulic retention time (HTR), increased loading rate, and smaller physical reactors for identical waste amounts. The major disadvantage identified is the increased use of energy required to heat the feedstock and maintain digester operating temperature. Chen and Hashimoto (1981) however suggested that the development of heat exchangers to recover energy in the effluent somewhat alleviated this advantage.

In cold climates, or during cold weather, optimal temperatures become very expensive to maintain, thus reducing the economic feasibility of the process of anaerobic biodigestion (Cullimore et al., 1985). In view of this, investigations have been conducted into the feasibility of anaerobic biodigesiton at lower temperatures. Stevens and Schulte (1979) thoroughly reviewed the literature regarding low-temperature digestion and found that methanogenesis occurs at temperatures as low as 4^{0}C, and that an increase in temperature from 4^{0}C to 25^{0}C dramatically increased the rate of methanogenesis. Cullimore (1982) reported results which indicated that as digester temperature was reduced from optimal levels, biogas production decreased linearly to extinction at between 0 and 8^0C. Ke-Xin and Nian-Guo (1980) successfully ran several rural digesters at ambient winter temperatures of 12 to 13^0C, and obtained gas yields which were 23 to 40 percent that of the optimal temperature production. Pos et al., (1985) suggested that if the anaerobic digestion process was found to function efficiently at lower temperatures, the use of large digestion units at longer retention times and without heating might be considered. It might then be possible to run full scale digesters at less than optimal temperature in order to increase their economic feasibility.

Safley Jr and Westerman (1990) reported satisfactory digester performance for both winter and summer conditions. However, biogas production was found to fluctuate seasonally with reduced biogas production being noted during the winter. Mean methane yield was found to be 0.34 m^3 CH_4 kg of volatile solids (VS) added. Mean biogas concentration was 69.5% CH_4 and 26.8% CO_2. The loading rate during the 17-month period of study was 0.12 kg VS/m^3-day. Typically, anaerobic digesters are designed to operate in either in the mesophilic (20^0C - 45^0C) or thermophilic (45^0C - 60^0C) temperature ranges. However, as pointed out by Safely Jr and Westerman (1990) the production of methane (called methanogenesis) has been observed at temperatures approaching 0^OC. The anaerobic decomposition of organic matter at low temperature (< 20^0C) is referred to as psychrophilic anaerobic digestion.

6.2 Suitable pH

According to San Thy et al., (2003) biogas fermentation requires an environment with neutral pH and when the value is below 6 or above 8 the process will be inhibited or even cease to produce gas because of toxic effect on the methanogen population. The optimum for biogas production is when the pH value of the input in the digester is between 6 and 7. Increasing the amount of feedstock or a change in the fermentation material is likely to acidify the fermentation system because of the accumulation of volatile fatty acids (VFA). In this way pH can be used to indicate if the system is being overloaded. In the initial period of fermentation, as large amounts of organic acids are produced by the acid-forming bacteria, the pH in the digester may fall below 5 causing inhibition of the growth of the methanogenic bacteria and hence reduced gas generation (Da Silva, 1979). Acetate and fatty acids produced during digestion tend to lower the pH of the digester liquid (Marchaim, 1991). Hansen et al., (1998) stated that acetate-utilizing methanogens are responsible for 70% of the methane produced in biogas reactors.

Buren (1983) pointed out that the micro-organisms involved in anaerobic biodigestion require a neutral or mildly alkaline environment, as a too acidic or too alkaline environment will be detrimental. The work stated that a pH between 7 and 8.5 is best for biodigestion and normal gas production. The pH value for a digester depends on the ratio of acidity and alkalinity and the carbon dioxide content in the digester, the determining factor being the density of the acids. Buren (1983) noted further that for the normal process of digestion, the concentration of volatile acid measured by acetic acid should be below 2000 ppm, as too high a concentration will greatly inhibit the action of the methanogenic micro-organisms. Results of a study by Jantrania and White (1985) further confirm the foregoing. The study compared the performance of a number of digesters processing poultry wastes and found that the pH of the residue from digesters that failed were between 6.1 and 6.7, while the pH from the successful digester was 7.5. The digesters which stopped producing any appreciable amount of gas after 54 days had higher hydrogen sulfide content (over 200 ppm) than the successful digester.

6.3 Volatile solids

The solids concentration of the influent into the biodigester affects the rate of fermentation (Marchaim, 1992). In a reported experiment conducted in China, which is mostly located in the temperate latitudes, the optimum concentration of solids was considered to be 6% in summer but between 10 and 20% in winter and spring. When temperatures are low and materials take longer to decompose; it is better to have a higher total solids concentration, although this might cause a problem with impeded flows through the digesters (San Thy et al., 2003). The loading rate is defined as the amount of volatile solids (fermentable solids) per unit of active biodigester volume per day. Typical values of loading rates are between 0.2 and 2 kg VS/m^3/day. This assumes that total solids (TS) are 17% of the fresh weight of the manure and that the volatile solids content (VS) is 77% (Fulford, 1988). The methane content of the gas can indicate overloading but it is more difficult to measure unless the right equipment is available. If the digester is being overloaded, the gas production will rise up initially and then fall after a while when inhibition occurs. The CH_4 content of the gas will fall while the CO_2 content will rise, because CO_2 is not used by the hydrogen consuming bacteria or because the methanogenic bacteria are inhibited.

The feedstock concentration of volatile solids (VS), the detention time, and the operating temperature are the major design factors, which determine the maximum total daily methane production (Hill, 1982). In a study by Vetter et al., (1990), daily biogas production was determined to be directly proportional to the volatile solids loading rate, given that other factors such as digester temperature and pH stayed relatively the same. Hill (1982) found that maximum VS reduction based on developed kinetic data was 75, 56, 30 and 62 percent for pig, beef, dairy, and poultry waste respectively. No significant increase in VS destruction will occur at temperature greater than approximately 45^{0}C.

6.4 Concentrations of methanogenic microorganisms

Biogas production is not possible without a sufficient quantity of anaerobic bacteria. In fresh manure, the concentration of these is low. Taking some effluent (10 to 30% of daily input) and putting it back into the digester is a way of inoculating the fresh manure with active microbial flora. This inoculation of fresh manure can increase gas production up to 30% and it is very important in a plug flow digester as there is almost no mixing between old and fresh slurry. The main nutrients required by microorganisms involved in anaerobic biodigestion are carbon, nitrogen, and inorganic salts. According to Buren (1983), a specific ratio of carbon to nitrogen must be maintained between 20:1 and 25:1, but this ratio will vary for different raw materials and sometime even for the same ones. The main source of nitrogen is human and animal excrement, while the polymers in crop stalks are the main source of carbon. Buren (1983) noted further that in order to maintain a proper ratio of carbon to nitrogen, there must be proper mixing of excrements with polymer sources. Since there are few common materials with a suitable ratio of carbon to nitrogen, production will generally not go well with only one source of material.

6.5 Retention (or detention) time

The amount of gas produced depends on the slurry in the digester volume (Fulford, 1988). The digester volume is also related to the retention time measured in days and the loading rate, in terms of manure solids per unit liquid volume (San Thy et al., 2003). According to experiences in China, 97% of the total yield of gas from fermenting cattle manure will be produced in a period of 50 days at 35^{0}C. The hydraulic retention time (HTR) in anaerobic digesters is determined by calculating the number of days required for displacement of the fluid volume of the culture. At a given organic loading rate, the HTR is lower when using high water - content feeds than when using those containing less water (Fannin and Biljetina, 1987). The detention time is dependent on all the factors discussed above. Generally a retention time of between 30 and 45 days and in some cases 60 days is enough for substantial gas production (Clanton et al., 1985; Carcelon and Clark, 2002). A study by Hill (1982) found that detention times for digesters designed to produce maximum daily methane volume varied from 7.9 days for dairy waste to 14.8 days for poultry, and similar wide variations in loading rates existed between the two wastes.

6.6 Air-tightness

None of the biological activities of anaerobic microorganisms, including their development, breeding and metabolism, requires oxygen. In fact, they are very sensitive to the presence of oxygen. The breakdown of organic materials in the presence of oxygen will produce carbon dioxide; in airless conditions, it produces methane (Buren, 1983; Voermans, 1985). Ferguson and Mah (1987) pointed out that methane-producing bacteria carry out the terminal step in the formation of biogas from the anaerobic decomposition of biomass. Methane is the final product of mineralizing the organic material in digesters and most anaerobic freshwater habitats. Most of the chemical energy in the starting materials (substrates) actually ends up in the methane released by these anaerobic bacteria. Ferguson and Mah (1987) noted further that in direct contrast, aerobic bacterial metabolism releases most of the chemical energy in the starting substrates by oxidizing them to carbon dioxide and water. Buren (1983) noted that if the digester is not sealed to ensure the absence of air. The action of the microorganisms and the production of biogas will be inhibited and some will escape. It is therefore crucial that the biogas digester be airtight and watertight.

6.7 Moisture content

There must be suitable moisture content of the feedstock as the microorganisms' excretive and other metabolic processes require water. The moisture content should normally be around 90% of the mass of the total contents (Buren 1983). Both too much and too little water are harmful, with too much water the rate of production per unit volume in the digester will fall, preventing optimum use of the digester. If the moisture content is too low, acetic acids will accumulate inhibiting the digestion process and hence production. Furthermore, a rather thick scum will form on the surface of the substrate. This scum may prevent effective mixing of the charge in the digester.

6.8 Carbon: Nitrogen ratio

The carbon:nitrogen (C/N) ratio expresses the relationship between the quantity of carbon and nitrogen present in organic materials. Materials with different C/N ratios differ widely in their yield of biogas. The ideal C/N ratio for anaerobic biodigesiton is between 20:1 and 30:1 (Marchaim, 1992). If C/N ratio is higher than that range, biogas production will be low. This is because the nitrogen will be consumed rapidly by methanogenic bacteria for meeting their protein requirements and will no longer react on the left over carbon remaining in the material. In such case of high C/N ratio, the gas production can be improved by adding nitrogen in farm cattle urine or by fitting latrine to the plant (Fulford, 1988). Materials with high C/N ratio typically are residues of agricultural plants. Conversely if C/N ratio is very low, that is outside the ideal range stated above, nitrogen will be liberated and it will accumulate in the form of ammonia. Ammonia will raise the pH value of the slurry in the digester. A pH value which is higher than 8.5, will be toxic to the methanogenic bacteria in the slurry. The cumulative effect of this is also reduced biogas production. Materials having low C/N ratio could be mixed with those having high C/N ratios so as to bring the average C/N ratio of the mixture to a desirable level. Human excreta, duck dung, chicken dung, and goat dung are some of the materials which typically have low C/N ratios.

According to Karki and Dixit (1984), typical C/N ratios of common organic materials are as shown in Table 3.

#	Organic Materials	C/N ratios
1	Duck dung	8
2	Human excreta	8
3	Chicken dung	10
4	Goat dung	12
5	Pig dung	18
6	Sheep dung	19
7	Cow dung	24
8	Buffalo dung	24
9	Water hyacinth	25
10	Elephant dung	43
11	Maize straw	60
12	Rice straw	70
13	Wheat straw	90
14	Saw dust	200

Source: Karki and Dixit (1984)

Table 3. C/N Ratios of some Organic Materials

7. Cassava (*Manihot* species) as a biofuel

In contemporary times, cassava is being recognized as an important source of biofuel. Research efforts aimed at investigating the potential of this sturdy crop for the production of biogas and bioethanol are currently in progress.

7.1 Global production of cassava

As observed by Adelekan (2012), Cassava (*Manihot esculenta* Cranz) is a very important crop grown for food and industrial purposes in several parts of the tropics. Nigeria, with a 2006 production of 49 million tonnes of cassava is the largest producer of the crop in the world

(National Planning Commission, 2009). Other countries which grow significant quantities of the crop include Brazil, Congo Democratic Republic, Thailand, Indonesia, Ghana and China. A handful of other countries also grow the crop but at much lower production quantities. According to IFAD/FAO (2000) report, cassava is the fourth most important staple crop in the world after rice, wheat and maize. The present annual global production of cassava is estimated at 160 million tonnes. This huge production also results into the discharge of significant cassava-derived solid wastes and liquid wastes into the environment especially during processing. Cassava peels constitute 10–20% by mass of each tuber. Cassava tuber contains 25–30% dry matter by mass, the major portion of which is made up of carbohydrates in the form of starch and sugars. The tuber also contains 70–75% moisture. The ongoing encouragement of cassava cultivation by Governments in Nigeria, Thailand, China and other countries is gradually raising the profile of the crop as a significant cash crop. With increased crop production is also an associated increased production of peels and other cassava-derived wastes. This constitutes an enhanced risk of pollution of the environment. There is therefore a pungent need to find an alternative productive use of the peels. One area of possibility is to investigate the potential of cassava peels for the production of biogas. Finding such an important use for the peel would make it less burdensome on the environment as a pollutant and contribute towards enhancing energy security in the cassava-producing regions.

7.2 Biogas production from cassava waste

Adelekan and Bamgboye (2009a) investigated biogas productivity of cassava peels, mixed with poultry, piggery and cattle waste types in ratios 1:1, 2:1, 3:1 and 4:1 by mass, using 12 Nos. 220l batch type anaerobic digesters in a 3 x 4 factorial experiment using a retention period of 30 days and within the mesophilic temperature range. Biogas yield was significantly ($P \leq 0.05$) influenced by the different mixing ratios of livestock waste with cassava peels. The cumulative average biogas yield from digested cassava peels was 0.6 l/kg-TS. The average cumulative biogas yield increased to 13.7, 12.3, 10.4 and 9.0 l/kg-TS respectively for 1:1, 2:1, 3:1 and 4:1 mixing ratios when cassava peel was mixed with poultry waste. On mixing with piggery waste, the average cumulative biogas yield increased to 35.0, 26.5, 17.1 and 9.3 l/kg-TS respectively for 1:1, 2:1, 3:1 and 4:1 mixing ratios. In the case of mixing with cattle waste, the average cumulative biogas yield increased to 21.3, 19.5, 15.8 and 11.2 l/kg-TS respectively for 1:1, 2:1, 3:1 and 4:1 mixing ratios. Results show that for all livestock waste types, mixing with peels in the ratio 1:1 by mass produced the highest biogas volumes, and highest in piggery waste. Cassava peels have high value of organic carbon and low value of total nitrogen, and this result in a particularly high C/N ratio. According to Karki et al. (1994) high C/N ratio is indicative of the fact that the material is not good for biogas production and will not appreciably yield biogas. However, the work points out that such a material could be mixed with another with a much lower C/N ratio to stabilize the ratio to an optimal value between 22 and 30. Biogas yield was significantly ($P \leq 0.05$) influenced by cassava peels used. The cumulative average biogas yield from digested cassava peels was 0.6 l/kg- TS. This value is low compared with values obtained by Bamgboye (1994) from other lignocellulosic materials such as chopped substrate (1.85 - 3.95 l/kg-TS) and ground water hyacinth substrate (4.01 - 5.55 l/kg-TS). Since cassava peel is a material with a high C/N ratio, it will not yield much biogas. As the paper showed however biogas production from cassava peels was enhanced by mixing with manure.

Bolarinwa and Ugoji (2010) studied biogas production by anaerobic microbial digestion of starchy wastes of *Dioscorea rotundata* (yam) and *Manihot esculenta* (cassava) aided by abattoir liquid effluent using a laboratory digester. The volume of the gas produced at 12hr intervals by feedstock varied for the 72hr of study. The cassava substrate mixture produced the highest daily average volume of gas (397ml), mixture of cassava and effluent 310.4ml; mixture of cassava, yam and effluent 259ml; mixture of cassava and yam produced 243.6ml; yam 238ml; mixture of yam and effluent 169.4ml while abattoir effluent produced the lowest volume of gas (144.4ml). The average pH of digester varied between 5.6 and 6.7 while the temperature varied between 32.3^{0}C and 33.3^{0}C. The microbial load of digester samples was determined at 12hr-intervals. Two groups of bacteria were isolated. Acid-formers isolated included *Staphylococcus aureus, Pseudomonas aeruginosa, Bacillus subtilis, Escherichia coli, Serratia liquefaciens, Micrococcus pyogenes* and *Streptococcus pyogenes* while the methane-formers were *Methanobacterium* sp. and *Methanococcus* sp. This study concluded that spoilt yam and cassava, which are otherwise of no apparent use, could provide a cheap source of renewable energy for domestic use.

7.3 Bioethanol production from cassava

Cassava is the best energy crop used to produce ethanol. This is because the ethanol yield of cassava per unit land area is the highest among all known energy crops. The comparison of ethanol yield produced from different energy crops shows that cassava has the highest ethanol yield of 6,000 kg/ha/yr and highest conversion rate of 150 L/tonne of all the energy crops. Though sugar cane and carrot have higher crop yield of 70 and 45 tonnes/ha/yr respectively compared to 20 tonnes/ha/yr for cassava, the huge quantities of water which they require during their growth periods is a strong limitation when compared to cassava which can actually grow under much drier conditions. Kuiper et al., (2007) noted that a tonne of fresh cassava tubers yields about 150 litres of ethanol.

Adelekan (2010) investigated ethanol productivity of cassava crop in a laboratory experiment by correlating volumes and masses of ethanol produced to the masses of samples used. Cassava tubers (variety TMS 30555) were peeled, cut and washed. 5, 15, 25 and 35 kg samples of the tubers were weighed in three replicates, soaked in water for a period of a day, after which each sample was dried, crushed and the mash mixed with 500, 650, 800, and 950 ml of N-hexane (C_6H_{14}) respectively. This crushed mash was then allowed to ferment for a period of 8 days and afterwards pressed on a 0.6 mm aperture size and sieved to yield the alcohol contained in it. The alcohol was heated at 79°C for 10 h at intervals of 2 h followed by an h cooling. Ethanol yield was at average volumes of 0.31, 0.96, 1.61 and 2.21 litres, respectively, for the selected masses of cassava samples. This study found that a total of 6.77 million tonnes or 1338.77 million gallons of ethanol are available from total cassava production from tropical countries. The production and use of ethanol from cassava crop in the cassava-growing tropical countries of the world certain holds much promise for energy security and is therefore recommended.

Some benefits of using ethanol are that it is not poisonous and neither causes pollution nor any environmental hazard. It does not contribute to the greenhouse effect. It has a higher octane value than gasoline and is therefore an octane booster and anti-knock agent. It reduces a country's dependence on petroleum and it is an excellent raw material for synthetic chemicals. The main crops presently being used for ethanol production are maize, sugarcane and cassava, and among these, cassava has a competitive advantage because of its lower cost of raw material and a simpler ethanol processing technology. Nguyen and Gheewala (2008) conducted a well-to-wheel analysis for cassava-based ethanol in Thailand. The aim of the analysis was to assess the potentials of cassava-based ethanol in the form of gasohol E10 for promoting energy security and reducing environmental impacts in comparison with conventional gasoline. The results showed that cassava-based ethanol in the form of E10, along its whole life cycle, reduced certain environmental loads compared to conventional gasoline. The percentage reductions relative to conventional gasoline are 6.1% for fossil energy use, 6.0% for global warming potential, 6.8% for acidification, and 12.2% for nutrient enrichment. The paper concluded that using biomass in place of fossil fuels for process energy in the manufacture of ethanol leads to improved overall life cycle energy and environmental performance of ethanol blends relative to conventional gasoline.

8. Cocoyam (*Colocasia* and *Xanthosoma* species) as a biofuel

Over the recent past, cocoyam has received inadequate attraction from researchers. Relatively few works reported on considered principally as a food crop. However, as will be seen in this subsection, some papers are beginning to point out the potentials of this crop as a source of biofuel.

8.1 Global production of cocoyam

The world has focused entirely on a comparatively small number of crops to meet the various needs for food and industrial fiber; the total number of economic crops of significance to global trade hovering just above one hundred. The consequence is that thousands of plant species with a considerably larger number of varieties fall into the category of underutilised or neglected crops. These crops are marginalized by agricultural, nutritional and industrial research (Global Forum for Underutilized Species, 2009). One of such neglected crops is cocoyam which over the years has received minimal attention from researchers and other stakeholders of interest. Cocoyam (*Colocasia* and *Xanthosoma* species), a member of the Aracea family of plants, is one of the oldest crops known. It is grown largely in the tropics, for its edible corms and leaves and as an ornamental plant. On a global scale, it ranks 14th as a vegetable crop going by annual production figures of 10 million tonnes (FAO, 2005). Its production estimates vary. However, one study points out that Africa accounts for at least 60% of world production and most of the remaining 40% is from Asia and Pacific regions (Mitra et al., 2007). Another study opines that coastal West Africa accounts for 90% of the global output of the crop with Nigeria accounting for 50% of this (Opata and Nweze, 2009). Cocoyam thrives in infertile or difficult terrains that are not well suited for large scale commercial agriculture for growing most conventional staple crops. As observed by Williams and Haq (2002), since the poor are frequently the main occupants of such areas, cultivation of neglected crops such as cocoyam constitute practical alternatives for them to augment their meagre incomes. The crop's supposed association with the poor may be a reason while conventional agricultural research has not bothered much to take a closer look at it.

8.2 Biogas production from cocoyam

Adelekan (2011) produced methane from cocoyam corms and related the volumes and masses obtained to the masses of corms used; derived guiding numerical relationships for the processes and extrapolated these values using production quantities of the crop reported globally and finally submitted workable estimates as regards biogas which is derivable from aggregate global production of the crop. The scientific innovation and relevance of the work reported lies in the fact that the fermentation and anaerobic digestion methods used are applicable across countries and regions irrespective of available degree of industrialization and climate. A new vista is opened in the use of this neglected crop as a cheap renewable source of energy in view of the rapid depletion, environmental pollution and high costs of fossil fuels. Results show that the 10 million tonnes annual global production of cocoyam is potentially able to produce 39.5 million cubic metres of methane which on burning would produce 179.3×10^7 MJ of enerrgy. The mash obtained as byproduct of the processes is capable of supplying 59 calories of food energy per 100g which is an excellent feedstock for livestock. The use of cocoyam (*Colocasia* and *Xanthosoma* species) as a renewable source of energy for the production of biogas poses no threat to the environment or food supply and is therefore recommended. Furthermore, doing so helps to enhance energy security.

Adeyosoye et al., (2010) studied biogas yield of peels of sweet potato (SPP) and wild cocoyam (WCP). Buffered and sieved goat's rumen liquor was added to 200 mg of dried and milled SPP and WCP in 100 ml syringes supplied with CO_2 under anaerobic condition and incubated for 24 hr. Total biogas produced was measured at 3 hr intervals till the 24th hr when the fermentation was terminated. The inoculum was also incubated separately. The proximate composition of SPP and WCP were similar except for the higher EE content (12%) of SPP. The SPP and WCP used contained 26.81 and 26.97% DM, 3.06 and 3.83% CP, and 78.94 and 79.17% carbohydrate respectively. Both samples had the same crude fibre (7.00%) content. Total biogas produced from SPP, WCP and the inoculum varied from 13.0, 11.0 and 5.0 ml respectively at the 3rd hr through 66.5, 61.5 and 18.0 ml at the 18th hr to 77.5, 72.0 and 30.0 ml at the 24th hr respectively. The differences in biogas production across the treatments were significant ($p < 0.05$). There were no significant differences ($p > 0.05$) in the volumes of methane produced from SPP (42.5 ml) and WCP (39.5 ml) which were significantly ($p < 0.05$) higher than 20.0 ml produced by the inoculum. The study pointed out that peels of sweet potato and cocoyam wastes can produce significant quantities of biogas for domestic applications. The foregoing studies confirm that ultimate methane yields from biomass are influenced principally by the biodegradability of the organic components. The more putrescible the biomass, the higher is the gas yield from the system (Wis, 2009).

8.3 Bioethanol production from cocoyam

Climate change, crop failures, unpredictable commodity prices, wars, political unrest and other forms of dislocations in the established pattern of global affairs, variously show that overreliance on just a few crops is risky to the world. However, bringing those crop species with underexploited potentials out of the shadows into the mainstream would help to spread this risk and enhance the utility of marginal lands on which many of them are cultivated. Most of the comparatively few number of studies reported in respect of cocoyam have focused largely on enhancing its value as a food crop, principally to supply carbohydrates and starch; a role which it already shares with so many competing crops. However, the paper by Adelekan (2011) looked at cocoyam as an energy crop for the supply of ethanol and biogas; a role which if fully developed can raise the profile of this crop in global energy economics. Points in favour of this research are the fact that it is in line with ongoing global research efforts at discovering more energy crops and developing other sources of renewable energy. Some progress has been reported in the use of cassava (another neglected tropical crop) for the production of ethanol as a sustainable source of biofuel in tropical countries Adelekan (2010). Cocoyam also has similar potential for this, most particularly in the tropical and subtropical countries. According to Adelekan (2011) which investigated the global potential of cocoyam as an energy crop, the yield of bioethanol from cocoyam is 139 L/tonne. This compares very favourably with 145 L/tone obtained for cassava (Adelekan, 2010), 100L/tonne for carrot and 70L/tone for sugar cane. Given a global annual production quantity of cocoyam to be 10million tonnes, 331 million gallons of ethanol is potentially available from this.

The question always arises, with a growing demand for ethanol produced from cocoyam, is there a threat to food security in respect of the crop? The answer to this question is twofold. Firstly, the yield of cocoyam, presently about 30 tonnes per hectare (Ekwe et al., 2009) can be tremendously improved through scientific research directed at producing higher yielding varieties. With success in this area, there may not be a need to cultivate more land to increase production of the crop. The present global cultivated total hectares of the crop can still sustain higher improvements in yield. The second part of the answer hzas to do with the need to husband the crop more efficiently to plug avenues for waste. In many parts of the developing world, between the farm and the consumers, 25 to 50% losses still occur to harvested crops because of poor preservation techniques, inadequate storage facilities, deficient transportation infrastructure, weak market structures and other factors. Therefore there is a pungent need to continue to research options which will enhance preservation and lengthen the storage life of cocoyam. Improvements in the area of preservation of the crop will also increase its supply, making its use as an energy crop less potentially deleterious on its use as a food crop and thereby enhancing food security.

Lee (1997) stated that the biological process of bioethanol production utilizing lignocellulosic biomass as substrate requires: 1) delignification to liberate cellulose and hemicelluloses from their complex with lignin, 2) depolymerization of the carbohydrate polymers (cellulose and hemicelluloses) to produce free sugars, and 3) fermentation of mixed hexose and pentose sugars to produce ethanol. In Europe the consumption of bioethanol is largest in Germany, Sweden, France and Spain. Europe produced 90% of its consumption in 2006. Germany produced about 70% of its consumption, Spain 60% and Sweden 50% in the same year. In 2006, in Sweden, there were 792, 85% ethanol (i.e E85) filling stations and in France 131 E85 service stations with 550 more under construction (European Biomass Association 2007).

9. Barbados nut (*Jatropha curcas*) as a biofuel

9.1 Global production of Jatropha

Jatropha is a shrub, belonging to the Euphorbiaceae family, thriving in various environments and across a wide range of ecosystems. It is a plant that can survive several months with minimal water and can actually live up to 40 years or more. It is not edible to human beings or animals. The jatropha industry is in its very early stages, covering a global area estimated at some 900,000 ha. More than 85 percent of jatropha plantings are in Asia, chiefly Myanmar, India, China and Indonesia. Africa accounts for around 12 percent or approximately 120,000 ha, mostly in Madagascar and Zambia, but also in Tanzania and Mozambique. The West African nations of Mali, Ghana and Senegal have also established lofty production targets for Jatropha notably; to cultivate 320,000 ha of *Jatropha curcas* in Senegal by 2012 and 1 million ha in Ghana in the medium term (OECD, 2008). Latin America has approximately 20,000 ha of jatropha, mostly in Brazil. The area planted with jatropha was projected to grow to 4.72 million ha by 2010 and 12.8 million ha by 2015. By then, Indonesia is expected to be the largest producer in Asia with 5.2 million ha, Ghana and Madagascar together will have the largest area in Africa with 1.1 million ha, and Brazil is projected to be the largest producer in Latin America with 1.3 million ha. Total biogas generation potential from *Jatropha curcas* cakes in India has been estimated as 2,550 million m^3 from 10.2 lakh metric ton of *J. curcas* oil seed cakes.

Jatropha curcas contains about 30% oil leaving behind presscake (75% including about 5% losses of oil in extraction process in the mechanical expeller) with residual oil. The oil is used for preparing bio-diesel (Achten et al., 2008) and in soap preparation. The press cake is rich in organic matter (Abreu, 2009). It can be used as manure, as feedstock for biogas production, animal feed and so forth. (Agarwal, 2007). Also, *Jatropha* oil cake is used for enriching the soil (Reyadh, 1997). Envis (2004) observed that *Jatropha* oil cake is an organic fertilizer that is superior to cattle manure and it is in great demand by farmers.

9.2 Biogas production from Jatropha

Ali et al., (2010), studied the use of *Jatropha curcas* defatted waste as an alternative feed in biogas plant for its bio-methanisation. The paper observed that as it remains as defatted cake after the extraction of non-edible oil from *Jatropha* seeds, it cannot be used directly for any purpose due to presence of toxic substance called 'curcin'. This toxin renders it unsafe for the animal feed and other purposes. It contains 5.73% nitrogen, 1.5% phosphorus and about 1% potassium. On the basis of its chemical composition, its application as substrate to the biogas plant can be a sustainable alternative as compared to the other applications of *Jatropha* press cake. The study was conducted on a floating drum type biogas plant. The study observed that the biogas plant, initially charged with pure cattle dung, when gradually replaced with *Jatropha* oil cake (0 - 100%), increased the biogas production up to approximately 25% in reasonable time duration. A significant increase in the percentage of nitrogen, phosphorus and potassium during the biofermentation process invokes the use of the effluent slurry as organic manure. Simultaneous reduction in the amount of the oil (5.67 to 3.95%) sustains the possibility of degradation of oil during methanisation. The plant has showed higher biogas yields at low temperatures also. Therefore, *Jatropha* defatted waste can successfully be used as an addition as well as substrate in already running cattle dung based biogas plant to get higher yield of biogas in comparison to cattle dung feed.

A laboratory experiment was conducted to find out the biogas production potential of dried, powdered Jatropha cake mixed with buffalo dung at 6% total solids (Prateek, 2009). The experiment was run on daily feeding basis in 5-litre capacity glass digesters for 180 days, while biogas production was recorded at 24 hr interval. Quality of biogas and nutritive value of effluent slurry was also determined. Results show significantly higher (139.20%) biogas production in test (Jatropha cake + Buffalo dung) over control (Buffalo dung only) digesters with methane content of 71.74%. Nutritive value of effluent slurry of test digester was significantly higher in terms of available nitrogen and potassium; calcium; magnesium and carbonate contents than that of control digesters. This co-digestion resulted in 92.94% decrease in chemical oxygen demand.

Dhanya et al., (2009) researched the biogas production potential of Jatropha (*Jatropha curcas*, L) Fruit Coat (JFC) alone and in combination with cattle dung (CD) in various proportions at 15 per cent total solids by batch phase anaerobic digestion for a period of ten weeks HRT (Hydraulic Retention Time) under a temperature of $35^{0}C+1^{0}C$. The maximum biogas production was noticed in cattle dung and Jatropha Fruit Coat in 2:1 ratio with 403.84 L/kg dry matter followed by 3:1,1:2, 1:1 and 1:3 having 329.66, 219.77, 217.79, 203.64 L/kg dry matter respectively as compared to 178.49 L/kg dry matter in CD alone. The JFC alone was found to produce 91% of total biogas of that obtained from cattle dung. The per cent methane content of the biogas in all the treatments was found on par with cattle dung.

9.3 Biodiesel production from Jatropha

Ways and means have been sought for many years to be able to produce oil-substitute fuel. Biodiesel extracted from fresh or used vegetable oil whether edible or not, is one such renewable alternative under consideration. Merits of biodiesel are that it can be directly used in engines with little or no modifications; contains little or no sulphur; no aromatics; has a higher cetane number and contains about 10% built-in oxygen and these properties help it burn fully with the result of having less carbon monoxide production, less unburnt carbon and less particulate matter residues. The production of biodiesel would be cheap as it could preferably be extracted from non edible oil sources. *Jatropha curcas* (Linaeus), a non-edible oil-bearing and drought-hardy shrub with ecological advantages, belonging to the *Euphorbiaceae* family, has been found to be the most appropriate renewable alternative source of biodiesel. Presently, the procedure for biodiesel production from jatropha seeds starts with harvesting whole ripe fruits. These fruits are then opened to remove the typically 3 or 4 seeds contained in each fruit. (A matured plant produces about 20kg of seeds in a year). These seeds are then sundried and afterwards stones, sticks, mouldy or damaged seeds and other foreign materials are handpicked from the batch of dried seeds. Next, this cleaned batch of seeds is crushed in an oil extraction machine to free the oil. This extracted oil is then filtered to remove impediments and the oil is poured in air-tight containers for storage. The extracted and filtered vegetable oil can be used directly as a fuel in suitable diesel engines without undergoing the trans-esterification process (Achten et al., 2008). However, to make it more useful in many engines, this Jatropha oil has to undergo a trans-esterification process of the triglyceride molecules in fats and oils with light weight alcohols like ethanol and methanol in a reactor in order to convert it to biodiesel. After being put into the reactor, the Jatropha oil settles; it is washed and purified by evaporation, and the liquid produced is biodiesel. Under optimal conditions, *Jatropha curcas* produces a higher oil yield per hectare compared to peanuts (*Arachis hypogea*), sunflower (*Helianthus annus*), soyabean (*Glycine max*), maize (*Zea mays*) and cotton (*Gossypium* species) (Kaushik et al., 2007). Biodiesel is a promising alternative because it is a renewable liquid fuel source that can be used alone and alternatively blended with petroleum-based diesel.

Jatropha's potential as a new energy source comes at time when interest in biofuel production is at an all-time high. As observed by Parwira (2010), biofuel production could potentially position developing nations to become net exporters of fuel which could greatly advance their objectives of economic independence. The paper noted further that many international corporations in Scandinavia, China, and Europe are purchasing tracts of land in developing countries (especially African countries) in an attempt to capitalize on this growth industry. New uses are being found for biofuel continually and this creates an impetus to strengthen efforts to produce them. In fact, several wireless communication companies have constructed cellular network base stations that are powered by *Jatropha*-based biofuel (Katembo and Gray, 2007). Presently, corn ethanol has a yield of 3100–4000 L/ha. This is still much higher than *Jatropha curcas* which is approximately 460-680 L/Ha of oil (Dar 2007). However, the production of Jatropha biodiesel is still very attractive largely due to its excellent fuel properties.

Kywe and Oo (2009) obtained a biodiesel yield of 30 gallons/day from a pilot plant which produced oil from Jatropha. The biodiesel demonstrated excellent fuel properties and it was found to be of very good quality. Tomomatsu and Swallow (2007) studied the economics and potential value of *Jatropha curcas* biodiesel production in Kenya and noted that in recent years, the production of *Jatropha curcas* has been widely promoted by private enterprises, non-governmental organizations and development agencies as one of the most viable candidates for biodiesel feedstock in Africa. While multiple benefits of jatropha production such as a petroleum product substitute, greenhouse gas mitigation and rural development are emphasized, the viability of production at farm level is questioned. The study revealed that the profitability of jatropha production for smallholder farmers is expected to be minimal unless farm-level production is accompanied by significant investments and policies targeted at enhancing production of the crop. However another economic study which took place in Mali showed that when all uses of Jatropha were taken into consideration, a rate of return of 135% could be achieved (Dinh et al., 2009).

Veljkovic et al., (2006) noted that biodiesel, which is made from renewable sources, consists of the simple alkyl esters of fatty acids. As a future prospective fuel, biodiesel has to compete economically with petroleum diesel fuels. The use of the less expensive feedstock containing fatty acids such as inedible oils, animal fats, waste food oil and byproducts of the refining vegetables oils reduces the costs of producing biodiesel. Therefore the availability and sustainability of supplies of less expensive feedstock will be a crucial determinant in competitively delivering biodiesel to commercial fuel filling stations. Such less expensive feedstock can come from inedible vegetable oils, mostly produced by seed-bearing trees and shrubs such as Jatropha curcas, a plant which has no competing food uses and which grows widely in tropical and subtropical climates across the world (Openshaw, 2000). Berchmans and Hirata (2008) developed a technique to produce biodiesel from crude Jatropha curcas seed oil having high free fatty acids (15% FFA). The high FFA level of the oil was reduced to less than 1% by a two-step pretreatment process. The first step was carried out with 0.60 w/w methanol-to-oil ratio in the presence of 1% w/w H_2SO_4 as an acid catalyst in 1-hr reaction at 50^0C. After the reaction, the mixture was allowed to settle for 2 hr and the methanol–water mixture which separated at the top layer was removed. The second step involved trans esterification using 0.24 w/w methanol to oil and 1.4% w/w NaOH to oil as alkaline catalyst to produce biodiesel at 65^0C. The final yield for methyl esters of fatty acids was achieved for 90% in 2 hr.

Hawash et al., (2011) investigated the trans-esterification of *Jatropha curcas* oil (JCO) to biodiesel using CaO as a solid base catalyst by determining the effects of molar ratio of methanol to oil, water content, reaction time and mass ratio of catalyst to oil in laboratory experiments. Experimental results revealed that a 12:1 molar ratio of methanol to oil, addition of 1.5% (w/v) CaO catalyst, 70°C reaction temperature, 2% water content in the oil produced more than 95% biodiesel yield after 3 hours reaction time. Calcium oxide activated with ammonium carbonate was an efficient super base catalyst for a high yield transesterification reaction and the base strength of CaO was more than 26.5 after dipping in ammonium carbonate solution followed by calcinations. Transesterification of Jatropha oil using supercritical methanol was also studied under the range of temperature from 120°C to 250°C, and range of pressure from 5 - 37 bars using superbase catalyst CaO and acid catalyst. The reaction products were analyzed for their content of glycerol by high performance liquid chromatography (HPLC) and this revealed that the process of supercritical transesterification achieved a yield of more than 95% after 1 hour.

The typical fuel properties of *Jatropha curcas* L, oil are as shown in Table 4 below. These properties show that jatropha biodiesel is a good quality biofuel.

S/N	Property	Numerical quantity	Reference
1	Calorific value ($MJkg^{-1}$)	39.77	Kumar and Sharma (2008)
2	Cetane number	51	Dinh et al., (2009)
3	Cloud point (°C)	2	Achten et al., (2008)
4	Flash point (°C)	235	Achten et al., (2008)
5	Kinematic viscosity at 40°C (mm^2sec^{-1})	41.51	Kywe and Oo (2009)
6	Relationship C/H (%wt)	13.11	Abreu (2009)
7	Relative density	0.87	Kywe and Oo (2009)
8	Sulphur content (%wt)	0.04	Abreu (2009)
9	Carbon residue (%)	0.02	Dinh et al., (2009)

Table 4. Fuel Properties of *Jatropha curcas* oil

10. Common grasses as biofuels

10.1 Global availability of grasses and other wild plants

The grass family (gramineae or poaceae) is perhaps the most successful taxonomic group in the plant kingdom. Members of this group number about 9000 species distributed in about 635 genera and they grow in all ecosystems and agroclimatic zones. From economic and ecological standpoints, they are the most important species in the plant kingdom. The pea family (leguminosae or fabaceae) is the largest family of flowering plants and also contains a large number of species found flourishing in many ecosystems and agroclimatic zones. Both families of plants contain domesticated crops and wild plants which are being researched for their potentials as reliable sources of biofuels. These plants certainly have a significant role to play in an anticipated global scenario which is 100% dependent on bioenergy in the near future.

10.2 Biogas production from grasses and wild plants

A study by Sidibe and Hashimoto (1990) documented the fact that grass straw can be fermented to methane and the yield can be relatively high. This laboratory experiment showed that the ultimate methane yield of rye grass straw (341 ± 5ml/g VS) and fescue grass straw (356±ml/g VS) are not significantly different but both grass straws had significantly higher yield ($p < 0.01$) than dairy cattle manure (288 ± 3 ml/g VS). The paper noted that nitrogen does not appear to be a limiting nutrient in the fermentation of grass straw to methane; the length of time between inocula feeding does not affect the ultimate methane yield of the straw, and longer acclimation may increase the ultimate methane yield of grass straw. Among plants themselves, differences exist regarding their potentials as feedstock for biogas production. For instance, De–Renzo (1997) reviewed anaerobic digestion of plant materials and concluded that aquatic plants such as algae and moss can be much better digested than terrestrial plants because of their toughness. Ordinarily, more digestion results in more biogas production. Akinbami et al (2001) noted that in the tropics, the identified feedstock substrates for an economically feasible biogas programme include water lettuce, water hyacinth, dung, cassava peelings, cassava leaves, urban refuse, solids (including industrial waste), agricultural residue and sewage.

Uzodinma and Ofoefule (2009) investigated the production of biogas from equal blending of field grass (F-G) with some animal wastes which include cow dung (G-C), poultry dung (G-P), swine dung (G-S) and rabbit dung (G-R). The wastes were fed into prototype metallic biodigesters of 50 L working volume on a batch basis for 30 days. They were operated at ambient temperature range of 26 to 32.8°C and prevailing atmospheric pressure conditions. Digester performance indicated that mean flammable biogas yield from the grass alone system was 2.46±2.28 L/total mass of slurry while the grass blended with rabbit dung, cow dung, swine dung and poultry dung gave average yield of 7.73±2.86, 7.53±3.84, 5.66±3.77 and 5.07±3.45 L/total mass of slurry of gas, respectively. The flash point of each of the systems took place at different times. The field grass alone became flammable after 21 days. The grass-swine (G-S) blend started producing flammable biogas on the 10th day, grass-cow (GC) and grass-poultry (G-P) blends after seven (7) days whereas grass-rabbit (G-R) blend sparked on the 6th day of the digestion period. The gross results showed fastest onset of gas flammability from the G-R followed by the G-C blends, while the highest average volume of gas production from G-R blend was 3 times higher than that of F-G alone. Overall, the results indicated that the biogas yield and onset of gas flammability of field grass can be significantly enhanced when combined with rabbit and cow dung.

Ofuefule et al., (2009) reported a comparative study of the effect of different pre-treatment methods on the biogas yield from Water Hyacinth (WH). The WH charged into metallic prototype digesters of 121 L capacity were pre-treated as: dried and chopped alone (WH-A), dried and treated with KOH (WH-T), dried and combined with cow dung (WH-C), while the fresh water Hyacinth (WH-F) served as control. They were all subjected to anaerobic digestion to produce biogas for a 32 day retention period within a mesophilic temperature range of 25 to 36°C. The results of the study showed highest cumulative biogas yield from the WH-C with yield of 356.3 L/Total mass of slurry (TMS) while the WH-T had the shortest onset of gas flammability of 6 days. The mean biogas yield of the fresh Water Hyacinth (WH-F) was 8.48 ± 3.77 L/TMS. When the water Hyacinth was dried and chopped alone (WH-A), dried and treated with KOH (50% w/v) (WH-T) and dried and combined with cow dung (WH-C), the mean biogas yield increased to 9.75 ± 3.40 L/TMS, 9.51 ± 5.01 L/TMS and 11.88 ± L/TMS respectively. Flammable biogas was produced by the WH-F from the 10th day of the digestion period whereas the WH-A, WH-T and WH-C commenced flammable gas production from the 9th, 6th and 11th day respectively. Gas analysis from WH-F shows

Methane (65.0%), CO_2 (34.94%). WH-A contained Methane (60.0%), CO2 (39.94%). WH-T contained Methane (71.0%), CO_2 (28.94%), while WH-C had Methane (64.0%) CO_2 (35.94%). The other gases were found in the same levels and in trace amounts in all the systems. The overall results showed that treating water Hyacinth with KOH did not have a significant improvement on the biogas yield. It also indicated that water Hyacinth is a very good biogas producer and the yield can be improved by drying and combining it with cow dung.

10.3 Ethanol production from common grasses

As pointed out by Barber et al., (2010) perennial grasses benefit the environment in numerous ways. They help to reduce climate change, increase energy efficiency and will constitute a sustainable energy resource for the world. Switchgrass, the most widely used perennial grass for biofuels, is also in such a manner, beneficial to both farmers as well as energy consumers in general. Perennial grasses are crucial to the ecosystem to create a sustainable energy resource for the world and also to limit the use of fossil fuels. These grasses are important because they can produce ethanol, an energy source that emits much less carbon dioxide than other fossil fuels. Reducing carbon dioxide emissions is important because carbon dioxide emissions in the atmosphere constitute one of the leading causes of climate change. Barry (2008) pointed out that 1 bale of switchgrass can yield up to about 50 gallons of ethanol. As reported by Rinehart (2006), researchers are using switchgrass as a biofuel so that they can successfully reduce carbon dioxide emissions. Switchgrass has a high energy in and out ratio because of lignin, the byproduct of the cellulose conversion that stores internal energy for its energy transformation process. Ethanol reduces carbon dioxide emissions by approximately ninety percent when compared to gasoline and consequently, carbon dioxide in the ozone layer of our atmosphere will slowly begin to deplete itself as biofuels created from switchgrass, other grasses and other ethanol sources are utilized. As a rule, all species of the grass family (poaceae) contain starch and should be able to yield ethanol.

11. Manure as a biofuel

Manure has for a long time been recognised as a renewable source of energy. Earlier practices involved direct combustion of manure to produce heat energy; while latter practices involved gasification of manure, followed by combustion. The more recent practices involved anaerobic biodigestion of manure to produce biogas which is scrubbed to purer methane and then combusted.

11.1 On-farm availability of manure

Huge quantities of manure are produced on farms. In fact a cattle farming operation which has a herd size of 10,000 animals can on a daily basis generate wastes equal to that produced by a city of half a million residents. Considering cattle for instance reported values of daily manure production range from 10kg (VITA, 1980) to 60kg (Safley Jr, et al., 1985) per animal. Legg (1990) reports that the 8.5, 28, 6.9 and 104 million cattle, sheep, pigs and poultry reared in England and Wales produced 80, 11, 11, and 30 million tonnes of manure respectively for the year immediately preceding. Smith and Chambers (1998) noted that manure arising from dairy beef farming comprise the majority (73 million tonnes) of the 90 million tonnes annual animal production of livestock manure in the UK. Yet another estimate, Smith et al., (2001) reported that 4.4 million tonnes of poultry manure are produced annually in the UK; comprising about 2.2 million broiler litter, 0.3 million tonnes of turkey litter, 1.5 million tonnes of layers manure (i.e. from egg producing hens) and 0.4 million tonnes from other sources (mainly breeding hens, cocks, and ducks). Total manure production from pigs in England and Wales is estimated to be about 10.03 million tonnes per year with about 45% as slurry and 55% as farmyard manure (Smith et.al., 2001).

The yearly production of livestock wastes in the Netherlands is estimated to be about 10 million tonnes dry matter (De Boer, 1984). The paper noted that 75 to 80% of it is ruminant waste while the rest is attributed to manure from pig and poultry. Specifically in the case of Nigeria, reported values of animal waste production range from 144 million tonnes/year (Energy Commission of Nigeria, 1998) to 285.1 million tonnes/year (Adelekan, 2002). This huge production of manure from farms can constitute a threat to the environment since it may not be readily returned to or in fact absorbed by land for fertilization. The challenge is how to find effective uses for the livestock wastes out of which the production of biofuels is an attractive option.

11.2 Biogas production from manure

Manure continues to be a promising resource for biogas production. Chynoweth et al. (1993) suggested potential biogas production from cattle waste, buffalo waste, piggery waste, chicken waste and human excreta as 0.360, 0.540, 0.180, 0.011 and 0.028 $m^3\ kg^{-1}$. The right mixing ratio of slurry can further increase the quantity of gas which can be produced from any particular feedstock. Adelekan and Bamgboye (2009b) investigated the effect of mixing ratio of slurry on biogas productivity of wastes from poultry birds, pigs and cattle. The investigation was carried out using 9 Nos. 220-litre batch type anaerobic digesters designed to remove CO_2, H_2S and other soluble gasses from the system. Freshly voided poultry, piggery, and cattle wastes were collected from livestock farms at the Institute of Agricultural Research and Training (IAR&T), Moor Plantation, Ibadan, Nigeria. After being totally freed of foreign matter, the samples were well stirred and digested in a 3x3 factorial experiment using a retention period of 30 days and within the mesophilic temperature range. The waste: water mixing ratios of slurry used were 1:1, 2:1 and 3:1 by mass. Three replicates were used for each ratio. Biogas yield was significantly ($p < 0.05$) influenced by the various factors of animal waste ($F=86.40$, $P< 0.05$), different water mixing rates ($F=212.76$, $P< 0.05$) and the interactions of both factors ($F=45.91$, $P< 0.05$). Therefore, biogas yield was influenced by variations in the mixing ratios as well as the waste types used. The 1:1 mixing ratio of slurry resulted in biogas productions of 20.8, 28.1, and 15.6 l/kgTS for poultry, piggery and cattle wastes respectively. The 2:1 ratio resulted in 40.3, 61.2 and 35.0l/kgTS while the 3:1 ratio produced 131.9, 117.0 and 29.8l/kgTS of biogas respectively. Therefore an increasing trend was observed in biogas production as mixing ratio changed from 1:1 to 3:1. For cattle waste however, production decreased from ratio 2:1 to ratio 3:1. The N, P, K values were highest for poultry waste (3.6, 2.1, and 1.4% respectively) and least for cattle waste (2.2, 0.6, 0.5% respectively). Organic carbon was highest for cattle waste (53.9%) and least for poultry waste (38.9%). Reduction in C/N ratio for each experiment ranged from 1.1 to 1.9%. This study found that for poultry and piggery wastes, slurries mixed in ratios 3:1 waste:water produced more biogas than those of 2:1 and 1:1 ratios. For cattle waste, the 2:1 mixing ratio produced the most biogas. The paper therefore recommended a livestock wastes: water mixing ratio of 3:1 for poultry and piggery slurries, and 2:1 for cattle slurry for maximum biogas production from methane-generating systems, given 30% TS content.

11.3 Potential of digested manure as a fertilizer

After the anaerobic digestion of manure to produce biogas, a nutrient-rich substrate which is still very beneficial to plants remains. This observation is supported by the findings of Thomsen (2000). These studies agree that only small differences of between 0.5 and 2.0% are usually measurable in the aggregate nutrient concentrations when digested manure is compared to the undigested form. Adelekan et al., (2010) did a comparative study of the effects of undigested and anaerobically digested poultry manure and conventional inorganic fertilizer on the growth characteristics and yield of maize at Ibadan, Nigeria. The pot experiment consisted of sixty (60) nursery bags, set out in the greenhouse. The treatments, thoroughly mixed with soil, were: control (untreated soil), inorganic fertilizer, (NPK 20:10:10) applied at the 120 kgN/ha; air-dried undigested and anaerobically digested manure applied at 12.5 g/pot, or 25.0 g/pot or 37.5 g/pot, and or 50.0 g/pot. Plant height, stem girth, leaf area, number of leaves at 2, 4, 6 and 8 weeks after planting (WAP) and stover mass and grain yield were measured. Analysis of variance (ANOVA) at $P \leq 0.05$ was used to further determine the relationships among the factors investigated. Generally, results in respect of plants treated with digested manure, were quite comparable with those treated with undigested manure and inorganic fertilizer, right from 2WAP to 6WAP. Stover yield was increased to as much as 1.58, 1.65 and 2.07 times by inorganic fertilizer, digested and undigested manure, respectively while grain yields were increased by only 200% with inorganic fertilizer, but by up to 812 and 933% by digested and undigested manure, respectively. The paper concluded that digested poultry manure enhanced the growth characteristics of the treated plants for the maize variety used. As observed, the order of grain yield was undigested manure > digested manure > inorganic fertilizer. These results agree with those reported by Agbede et al., (2008) for sorghum (*Sorghum vulgare*), Akanni (2005) for tomato (*Lycopersicon esculentum*) and Adenawoola and Adejoro (2005) for jute (*Corchorus olitorus* L).

Organic manures play a direct role in plant growth as a source of all necessary macro and micronutrients in available forms during mineralization. Thereby, they improve both the physical and physiological properties of soil (El Shakweer et al., 1998; Akanni, 2005), thus enhancing soil water holding capacity and aeration (Kingery et al., 1993; Abou el Magd et al., 2005; Agbede et al., 2008). Organic manures decompose to give organic matter which plays an important role in the chemical behavior of several metals in soil through the fulvic and humic acid contents which have the ability to retain metals in complex and chelate forms (Abou el Magd et al., 2006). They release nutrients rather slowly and steadily over a longer period and also improve soil fertility status by activating soil microbial biomass (Ayuso et al., 1996; Belay et al., 2001). They thus, ensure a longer residual effect (Sherma and Mittra, 1991), support better root development and this leads to higher crop yields (Abou el Magd et al., 2005). Improvement of environmental conditions and public health as well as the need to reduce cost of fertilizing crops are also important reasons for advocating increased use of organic manures (Seifritz, 1982). While the practice of anaerobic digestion of biomass for biogas production is increasing, the use of the digested manure for crop production should concurrently be encouraged, judging by its potential to enhance the growth and yield of crops.

12. Conclusions and recommendations on the use of tropical crops as biofuels

1. Biofuels such as biogas, bioethanol and biodiesel are reliable and renewable options for enhancing global energy security. In view of the on-going depletion of fossil duels, keener global research interest should be directed at developing known and new sources of these fuels.
2. Concerted efforts should be made by stakeholders worldwide to encourage the use of biofuels given their multifarious advantages of protecting the environment and mitigating climate change.
3. Research funds should be further directed at developing the potentials of known energy-yielding plants such as cassava, *Jatropha curcas,* common grasses and others to contribute towards ensuring global energy security.
4. Neglected tropical crops such as cocoyam and others identified should be researched for their energy yielding abilities and bring them to the main stream of interest of conventional agricultural research.
5. The potential of manure as a significant environmental pollutant can be lessened if more concerted efforts are made to produce biogas from it and this will also concurrently result in the production of nutrient-rich digested biomass which can be returned to land to enhance crop production.

13. References

Abou el Magd, MM; Hoda Mohammed, A. & Fawz ZF. (2005). Relationship Growth and Yield of Broccoli With increasing N, P or K Ratio in a Mixture of NPK fertilizers. Ann. Agric. Sci. Moshto 43(2): 791- 805.

Abou el Magd, MM; El Bassiony, M. & Fawzy ZF. (2006). Effect of Organic Manures with or without Chemical Fertilizers on Growth, Yield and Quality of some Varieties of Broccoli Plants. J. Appl. Sci. Res. 2(10): 791-798.

Abreu, F. (2009). Agrienergy, Brazil. A powerpoint presentation on Alternative By-products from *Jatropha*. Available online at http://www.ifad.org/events/*Jatropha*/ harvest/F_Abreu.ppt

Achten, J; Verchot, L; Franken, YJ; Mathijs, E; Singh, VP; Aerts, R. & Muys, B. (2008). Jatropha bio-diesel production and use, Biomass Bioenergy 32(12): 1063-1084.

Adelekan, BA. (2012). Cassava as a Potent Energy Crop for the Production of Ethanol and Methane in Tropical Countries. International Journal of Thermal & Environmental Engineering, International Association for Sharing Knowledge and Sustainability (IASKS), Canada, Vol 4, No.1,pp 25-32. Available online at www.iasks.org/ journals/ijtee

Adelekan, BA. (2011). Comparison of Ethanol Productivity of Selected Varieties of Tropical Crops. Paper presented at CIOSTA/CIGR V Conference: Efficient and Safe Production Processes in Sustainable Agriculture and Forestry, 29 June to 1 July,

University of Natural Resources and Applied Life Sciences, Vienna, Austria. Available online at www.nas.boku. ac.at/ciosta2011.html

Adelekan, BA. (2010). Investigation of Ethanol Productivity of Cassava Crop as a Sustainable Source of Biofuel in tropical countries. African Journal of Biotechnology, Vol. 9, No 35, pp 5643-5650, 30 August. Published online at www.academicjournals.org/AJB

Adelekan, BA. (2002). Assessing Nigeria's Agricultural Biomass Potential as a Supplementary Energy Source through Adoption of Biogas Technology. Nig. J. Renewable Energy. Sokoto Energy Research Centre. Vol. 10, No. 1 & 2. Pp 145-150.

Adelekan, BA. & Bamgboye, AI. (2009a). Comparison of Biogas Productivity of Cassava Peels Mixed with the Major Livestock Waste Types in Selected Ratios. Afr. J. Agric. Res. (AJAR), Vol. 4, No. 7, pp 571- 577. Available online at www.academicjournals. org/AJAR

Adelekan, BA. & Bamgboye, AI. (2009b). Effect of Mixing Ratio of Slurry on Biogas Productivity of Major Animal Waste Types. J. Appl. Biosci. Vol. 22, pp 1333-1343. Available online at www.biosciences.elewa.org/JABS

Adelekan, BA; Oluwatoyinbo, FI. & Bamgboye AI. (2010). Comparative Effects of Undigested and Anaerobically Digested Poultry Manure on the Growth and Yield of Maize (*Zea Mays*, L.) African Journal of Environmental Science and Technology, Vol. 4, No. 2, pp 100-107, February. Available Online at www.academicjournals.org/AJEST

Adenawoola, AR & Adejoro, SA. (2005). Residual Effects of Poultry Manure and NPK Fertilizer Residues on Soil Nutrient and Performance of Jute (*Corchorus olitorus*, L.). Niger. J. Soil Sci. 15: 133-135.

Adeyosoye, OI; Adesokan, IA; Afolabi, KD. & Ekeocha, AH. (2010). Estimation of Proximate Composition and Biogas Production from *in vitro* Gas Fermentation of Sweet potato (*Ipomea batatas*) and Wild cocoyam (*Colocasia esculenta*) Peels. African Journal of Environmental Science and Technology Vol. 4, No. 6, pp. 388-391, June. Available online at Available online at www.academicjournals.org/AJEST

Agarwal, AK; Saravanan, IB. & Kumar, EN. (2007). Biodiesel as an Alternative Urban Transport Fuel in India, Biofuels in India: A New Revolution, First edition (Eds.), The ICFAI University Press, Hyderabad, India, pp 151-157.

Agbede, TM; Ojeniyi, SO. and Adeyemo, AJ. (2008). Effect of Poultry Manure on Soil Physical and Chemical Properties, Growth and Grain Yield of Sorghum in South West Nigeria. Am. Eurasian J. Sustain. Agric., 2(1): 72-77.

Ajayi, O. (2009). The Production of Methanol from Cow Dung. Paper presented at the Alternative Energy Conference: Options or Necessity which held in Kuwait, Nov. 2 to 5. Organised by Kuwait Society of Engineers and World Federation of Engineering Organizations, www.ec2009kuwait.org

Akanni, DI. (2005). Response of Nutrient Composition and Yield Components of Tomato (*Lycopersicon esculentum*, M.) to Livestock Manure. Ph.D thesis, Department of Crop, Soil and Pest Manage., Federal University of Technology, Akure, Nigeria p.120.

Akinbami, JFK; Ilori, MO; Oyebisi, TO; Akinwumi, IO. & Adeoti, O. (2001). Biogas Energy Use in Nigeria. Renewable and Sustainable Energy Review. Vol. 5, pp. 97-112.

Ali, N; Kurchania, AK. & Babel, S. (2010). Bio-methanisation of *Jatropha curcas* defatted waste. Journal of Engineering and Technology Research Vol. 2, No. 3, pp. 038-043,

March, 2010 Available online at http://www.academicjournals.org/JETR Ayuso, M; Paschal, JA; Garcia, C, & Hernandez, T. (1996). Evaluation of Urban Wastes for Agric. Use. Soil Sci. Plant Nutr. Vol. 142, No. 1, pp 105-111.

Bamgboye, AI. (1994). Pretreatment Measures to Accelerate and Maximize Water Hyacinth (*Eichhornia Crassipes*) Biodigestibility for Biogas Production. An Unpublished Ph.D Thesis, Department of Agricultural and Environmental Engineering, University of Ibadan, Ibadan, Nigeria.

Barber, R; Castle, B; Ganesh, S; Herbolsheimer, C & Tran, V. (2010). The Future is Green: The Development of Perennial Grasses as a Source of Bioenergy Governor's School for Agriculture, Virginia Tech, Blacksburg, VA 24061, USA.

Barry, P. (2008). Switchgrass May Yield Biofuel Bounty. *Science News*. Vol. 173, No. 3, pp 1-46.

Belay, A; Classens, AS; Wehner, FC. & De Beer, JM. (2001). Influence of Residual Manure on Selected Nutrient Elements and Microbial Composition of Soil under Long Term Crop Rotation. S. Afr. J. Plant Soil 18: 1-16.

Berchmans, HJ. & Hirata, S (2008). Biodiesel Production from Crude *Jatropha curcas* L. Seed oil with a High Content of Free Fatty Acids. Bioresource Technology, Vol. 99, pp 1716-1721. Available online at www.sciencedirect.com

Bolarinwa, OA. & Ugoji, EO. (2010). Production of Biogas from Starchy Wastes. Journal of Sci. Res. Dev. Vol. 12, pp 34–45.

Brakel, VJ. (1980). The Ignis Fatuus of Biogas Small Scale Anaerobic Digesters. In Biogas Plants; A Critical Review of the Pre-1970 Literature, Delft University Press.

Bui-Xuan, AN. (2004). Experience in Biogas Technology Development in Vietnam. www.mekern.org. Accessed May 21, 2011.

Buswell, AM. & Hatfield, WD. (1936). Anaerobic Fermentation Bulletin 32, Urbana, IL. Illinois State Department of Registration and Education.

Buren, V. (1983). A Chinese Biogas Manual. Popularising Technology in the Country side Intermediate Technology Publications Ltd.

Carcelon, J. & Clark, J. (2002). Methane Biogas from Anaerobic Digesters. The U.S Environmental Production Agency, the U.S Department of Agriculture and the U.S Department of Energy pp 64–68.

Chalmers, A G. (2001). A Review of Fertilizer, Lime, and Organic Manure Use on Farm Crops in Great Britain from 1983 to 1987. Soil Use and Management. British Society of Soil Science Vol. 17, No. 4, Dec. pp 254–262.

Chen, YR. & Hashimoto, AG. (1981). Energy Requirements for Anaerobic Fermentation of Livestock Wastes. In Livestock Wastes: A Renewable Resource. RJ. Smith, ed. ASAE Publication No 2–81, American Society of Agricultural Engineers, St Joseph, MI, U.S.A.

Chynoweth, DP. & Isaacson, R. (1987). Anaerobic Biodigestion of Biomass. Chynoweth, DP and Isaacson, R. ed. Elsevier Applied Science, London and New York.

Converse, JC; Graves, RE. & Evans, GW. (1977). Anaerobic Degradation of Dairy Manure under Mesophilic and Thermophylic Temperatures, Transactions of ASAE, Vol. 20, No. 2, pp. 336–340.

Chynoweth, DP; Turik, CE; Owens, JM; Jerger, DE. & Peck, MW. (1993). Biochemical Methane Potential of Biomass and Waste Feed Stocks. Biomass and Bioenergy Vol. 5, pp. 95-111.

Cullimore, DR. (1982). Potential Development of Ambient (Insulated) Temperature Biogas Generation Using Animal Waste Products (Hog Manure). Proceed. Energex '82, Solar Energy Society of Canada pp. 1133–1175.

Cullimore, D.R; Maule. A. & Mansuy, N. (1985). Ambient Temperature Methanogenesis from Ppig Manure Waste Lagoons. Thermal Methanogenesis Incubator Studies. Agricultural Wastes Vol. 12, pp. 147–157.

Dar, W. (2007) Research Needed to Cut Risks to Biofuel Farmers. Science and Development Network.

Da Silva, AF. (1979). Impacts of Methanogenic Organisms in the Biodegrability of Farm Wastes. Journal of Applied Engineering in Agriculture, Vol. 6, No. 1, pp. 78-84. American Society of Agricultural Engineers, St Joseph, Michigan, USA.

De Boer, F. (1984). By-products and Wastes in Animal Husbandry. Proceed. Of an Internl. Symp, Animals as Waste Converters, Wageningen, Proceed. Pp. 10–11.

De-Renzo, DJ. (1997). Energy from Bioconversion of Waste Materials. Noyes Data Corporation, Park Ride, New Jersey, U.S.A.

Dhanya, MS; Gupta, N; Joshi, HC. & Lata (2009). Biogas Potentiality of Agro-wastes Jatropha Fruit Coat. International Journal of Civil and Environmental Engineering, Vol. 1, No. 3.

Dinh, L; Guo, Y; & Mannan, MS. (2009). Sustainability Evaluation of Biodiesel Production using Multicriteria Decision-Making. Environmental Progress & Sustainable Energy, Vol. 28, No.1, American Institute of Chemical Engineers.

Ekwe, K; Nwosu, K; Ekwe, C. & Nwachukwu, I. (2009). Examining the Underexploited Values of Cocoyam (*Colocasia* and *Xanthosoma* species) for Enhanced Household Food Security, Nutrition and Economy in Nigeria. In: Jaenicke, H; Ganry, J; Zeledon-hoeschle, I. & Kahane R. (Eds.) Proceedings of the International Syposium on Underutilized Plants for Food Security, Income and Sustainable Development, *Acta Horticulturae* Vol. 806, pp. 71-78.

El Shakweer MHA, El Sayad EA, Ewees MSA (1998). Soil and Plant Analysis as a Guide for Interpretation of the Improvement Efficiency of Organic Conditioners added to different Soils in Egypt. Commun. Soil Sci. Plant Anal. Vol. 29, pp. 2067-2088.

Energy Commission of Nigeria, ECN. (1998). Potentials for Renewable Energy Application Nigeria, Energy Commission of Nigeria, Lagos Pp 21. ISBN 978 316 - 502 - 3.

Envis Forestry Bulletin. (2004). Envis Center on Forestry, National Forest Library and Information Centre, Forest Research Institute. Dehradun. Available online at: http://www.frienvis.nic in/Jatropha.htm*Jatropha* system (2006) Home site of bagani.de. Available online at http://www.*Jatropha*.de/faq.htm

Evans, M R. & Svoboda, IF. (1985). The Recovery of Energy from Aeration. Agricultural Waste Utilization and Management. Proceed. 5th Internl. Symp. On Agricultural Wastes. ASAE, pp 22–28.

European Biomass Association, EUBIA. (2007). Creating Markets for Renewable Energy Technologies. EU RES Technology Marketing Campaign. Brochure 5, Bioethanol production and use, European Biomass Industry Association, EUBIA 4/2007, p 12. Available online at www.eubia.org/fileadmin/template/main/res/pdf/Projects/Brochyre5_Bioethanol_low_res.pdf

European Renewable Energy Council, EREC (2010). Renewable Energy in Europe. Renewable Energy Technology Roadmap, 20% by 2020. EREC. Available online at www.erec.org

Food and Agricultural Organization, FAO. (2005). Yearbook of Agricultural Statistics. Food and Agricultural Organization of the United Nations, Rome.

Fannin, KF. & Bilijetina, R. (1987). Reactor Designs. In Anaerobic Digestion of Biomass ed. Chynoweth, D. P and Isaacson. R. Elseviewer Appplied Science, London and New York. pp 142.

Fedler, CB; Hrubant, GR; Berger, LL. & Day, DL. (1985). High-rate Ensiling of Beef Manure and Cracked Corn for Cattle Feed Proceed. 5th Internl symp. On Agric.Wastes, ASAE. pp 175-181.

Ferguson, T. & Mah, R. (1987). Methanogenic Bacteria. In Anarobic Digestion of Biomass ed. Chynoweth, DP. & Isaacson, R. Elsevier Applied Science, London and New York. pp 49.

Foley, G. & Barnard, G. (1985). Biomass Gasification in Developing Countries. International Institute for Environment and Development Energy Information Programme. Technical Report No 1. pp 13.

Fujita, AH; Crawford, TK. & Ames, BS. (1980). Comparison of Methane Productivity of Pig Manure and Corn stover. J. Agricultural Wastes. Internl. Socie. of Biological Sciences, Vol. 4, No. 2, pp 20-25.

Fulford, D. (1988). Running of Biogas Program Handbook, Intermediate Technology Publications, London.

German Technical Cooperation, GTZ. (2006). Liquid Biofuels for Transportation Chinese Potential andImplications for Sustainable Agriculture and Energy in the 21st Century. Beijing: GTZ.

Ghederim, V; Negulescu, AL. & Gueron, IJ. (1985). Present State and Trends in Animal Farm Wastes Management in Romania. Proceed. 5th Internl Symp. On Agricultural Wastes. ASAE, pp 26-38.

Global Forum for Underutilised Species (2009). Cocoyam. Available online at www.underutilisedspecies.org/about_GFU.asp

Hamilton, DW; Sharp, PR. & Smith, RJ. (1985). The Operational Characteristics of a Manure Digester for 60 Beef Cattle. Agricultural Waste Utilization and Management. Proceed. 5th Internl. Symp. on Agric Wastes. American Society of Agricultural Engineers, St. Joseph, MI, USA, pp 500-508.

Hansen, MN; Kai, P. & Henrik, BM. (1998). Effects of Methanogens in the Hydrolysis of Biomass. American Society of Agricultural Engineers, TRANS. of the ASAE, Vol. 27, No. 2, pp 121-128.

Hashimoto, AG. (1983a). Thermophilic and Mesophilic Anaerobic Fermentation of Swine Manure. Agricultural Wastes, Vol. 6, pp. 175-191.

Hashimoto, AG. (1983b). Manure: Effect of Temperature and Influence Substrate Concentration on Kinetic parameter (k) Agricultural wastes Vol. 9, No. 4, pp 299-308.

Hashimoto. A G. & Robinson, SA. (1985). Resources and Conservation. Vol. 12, pp 29 - 45.

Hashimoto, AG; Chen, YR; Varel, VH. & Prior RL. (1979). Anaerobic Fermentation of Animal Manure ASAE Paper No 79-4066. ASAE., St Joseph; MI, USA.

Heltberg, PH; Woods, TS; Ashby, AB & Ait, RC. (1985). Baseline. Projections of the United States Energy Supply and Demand to 2010. Gas Research Institute. Chicago, USA.

Hill, DT. (1982). Optimum Operation Design Criteria for Anaerobic Digestion of Animal Manure. TRANS. of the ASAE. Vol. 25, pp 1029-1032.

Hill, DT. (1984). Methane Productivity of the Major Animal Waste Types. TRANS. of the ASAE. Vol. 27, pp 530-534.

Hill, DT. (1985). Practical and Theoretical Aspects of Engineering Modeling of Anaerobic Digestion for Livestock Waste Utilization Systems. TRANS of the ASAE, Vol. 28, No. 2, March-April, pp 599–605,

Hill, DT. (1990). Steady-State Thermophilic Design Equations for Methane Production from Livestock Wastes. Proceed. 6th Internl symp. On Agric Wastes ASAE, pp 88-95.

Hill, DT. & Bolte, JP. (1985). Comparison of Surface Scraped and Flushed Screened Swine Wastes as Methane Substrates for Thermophilic Digestion. TRANS of ASAE, Vol. 28, No. 3, pp 370-374.

Hill, DT; Bolt, JP. & Prince. TJ. (1986). Thermophilic Anaerobic Digestion of Screened Flushed Swine Waste TRANS of ASAE, Vol. 29, No. 1, pp 229 - 234.

Hills, DJ. & Roberts. (1979). Comparison of Biogas Productivity of Manure, Rice Straw and Barley Straw. J. Agricultural Wastes. Internl. Socie. of Biological Sciences, Vol 3, No. 1, pp 267.

Hobson, PN; Bousfield, S. & Summers, R. (1981). Methane Production from Agricultural and Domestic Wastes. Applied Science Publishers Ltd, London.

Huang, JJH. & Shih, JCH (1981). The Potential of Biological Methane Generation from Chicken Manure. Biotech. Bioeng. Vol. 23, pp 2307-2314.

Huang, JJH; Shin, JCH. & Steinsherger, SC. (1982). Poultry Wastes Digester: from the Laboratory to the Farm

In: Vogt, F. (ed) Energy Conservation and Use of Renewable Energies in Bio-industries, 2 Pergamon Press. Oxford. pp 376-383.

IFAD/FAO. (2000). The World Cassava Forum. International Fund for Agricultural Development, Food and Agricultural Organization of the United Nations, Rome.

Inderlwildi OR, King DA (2009). Quo Vadis Biofuels. J. Energy Environ. Sci. 2: p. 343. Doi: 10.1039/b822951c, Available online at www.dx.doi.org/10.1039/b822951c

Jantrania, A. R. and White R. K. 1985. High-Solids Anaerobic Fermentation of Poultry Wastes. Proceed. 5th Internl Symp. On Agricultural Wastes ASAE, St. Joseph, MI, USA, pp 73-80.

Karki, AB. & Dixit, K. (1984). Biogas Fieldbook, Sahayogi Press, Kathmandu, Nepal, pp 42.

Karki, BA; Gautam, KM. & Karki, A. (1994). Biogas Installation from Elephant Dung at Machan Wildlife Resort, Chitwan, Nepal. Biogas Newsletter, Issue No. 45.

Katembo, BI. & Gray PS. (2007). Africa, Seeds Biofuel. J. Multidisciplinary Res. Vol. 1, No. 1-6, p. 22

Kaushik, N; Kumara, K; Kumara, S; Kaushikb, N. & Roy, S. (2007). Genetic Variability and Divergence Studies in Seed Traits and Oil Content of *Jatropha* (*Jatropha* curcas L.) Accessions. Biomass and Bioenergy.

Keppler F, Hamilton JT, Bra M, Rockmann T (2006) Methane Emissions from Terrestrial Plants Under Aerobic conditions. J. Nature, No. 439: pp187-191. (http://dx.doi.org/10.1038/nature04420)

Ke-xin, L. & Nian-guo, L. (1980). Fermentation Technology for Rural Digesters in China. Proceed. Bioenergy '80, Bioenergy Council, New York, USA, pp 440.

Kingery, WL; Wood, CW. & Delaney, DP. (1993). Impact of Longterm Application of Broiler Litter on Environmentally Related Soil Properties. J. Environ. Qual. Vol. 23, pp 139-147.

Kumar, A; & Sharma, S. (2008). An Evaluation of Multipurpose Oil Seed Crop for Industrial Uses (*Jatropha curcas* L.): A Review. *Industrial Crops and Products*, Vol. 28, pp. 1-10.

Kywe, TT. & Oo, MM. (2009). Production of Biodiesel from Jatropha Oil (*Jatropha curcas*) in pilot plant. Proceedings of World Academy of Science, Engineering and Technology, Vol. 38, pp. 481-487.

Lee, J. (1997). Biological Conversion of Lignocellulosic Biomass to Ethanol, Journal of Biotechnology, Vol. 56, Issue 1, 23 July, pp. 1-24.

Legg. B.J. (1990). Farm and Food Waste Utilization without Pollution. Proceed 6th Internl Symp. On Lives. Wastes American Society of Agricultural Engineers St Joseph, MI, USA. pp xi–xxi.

Licht, F.O., (2009) World Fuel Ethanol Production. Renewable Fuels Association. Accessed on July, 20, 2011. Available online at www.ethanolrfa.org/industry/statistics/#E

Lindley, JA. & Haughen, VJ. (1985). Biogas Production after Solid–Liquid Separation of Daily Manure. Proceed 5th Internl Symp. On Agric Wastes ASAE, pp 483–490.

Marchaim, U. (1992). Biogas Processes for Sustainable Development. FAO Agricultural Services Bulletin 95. Food and Agricultural Organization of the United Nations.

McCabe, J. & Eckenfelder, W. eds. (1975). Biological Treatment of Sewage and Industrial Wastes, New York, Reinhold Publishing.

Meynell, P.J (1976). Methane: Planning a Digester. Schocken Books, New York, USA. pp 3.

Mitra, S; Sinha, B; Pal, H. & Tarafdar, J. (2007). Comparative Studies on Morphological Characters, Yield,

Nutritional Status and Isozymes Activity Of Some Elite Genotypes of *Taro* (*Colocasia esculenta* L. Schott var. *antiquorum*) Grown In West Bengal. In: Chandha ML; Kuo G. & Gowda CLL. (Eds.) Proceedings of the First International Conference on Indigenous Vegetables and Legumes. Acta Horticulturae, 752 ISHS, pp. 219-230.

National Planning Commission, NPC. (2008). Nigeria's crop production for 1999-2006. The Fifth National Development Plan (2008-2011), Federal Republic of Nigeria, Abuja, p 38.

Neeteson, J J. & Wadman, WP. (1990). Ammonia Volatilization and Nitrate Leaching after Animal Manure Application to Agricultural Crops. Proceed. Of the 6th Internl Symp. On Agricultural and Food Processing Wastes. ASAE, St. Joseph, MI, USA, pp 148–157.

Nguyen, TLT. & Gheewala, SH. (2008). Life Cycle Assessment of Fuel Ethanol from Cassava in Thailand. International Journal of Life Cycle Assessment. Vol 13, No. 2, pp 147–154.

Ofoefule AU, Uzodinma EO, Onukwuli OD (2009) Comparative Study of the Effect of Different Pretreatment Methods on Biogas Yield from Water Hyacinth (*Eichhornia crassipes*) International Journal of the Physical Sciences, Vol. 4, No. 8, September, pp. 535-539. Available online at www.academicjournals.org/ijps

Opata, PI. & Nweze, NJ. (2009). Indigenous Technologies in Cocoyam Processing: Implications for Food Security in Nigeria. African Crop Science Conference Procedings. African Crop Science Society. Vol. 9, pp. 325-328.

Openshaw, K. (2000). A Review of *Jatropha curcas*: An Oil Plant of Unfulfilled Promise. Biomass and Bioenergy, Vol. 19, pp 1–15.

Organization for Economic Cooperation and Development, OECD. (2008). Green Fuels for Development? Improving Policy Coherence in West Africa. Sahel and West Africa Club /OECD, Paris, France. Available online at www.westafricaclub.org

Parawira, W. (2010). Biodiesel Production from *Jatropha curcas*: A Review. Scientific Research and Essays, Vol. 5, No. 14, July 18, pp. 1796-1808. Available online at www.academicjournals.org/SRE

Pos, J; Eszes, R. & Pavlic, V. (1985). Evaluation of Two Full-size Mixed Mesophilic Digesters Processing Swine Manure Proceed. 5th Internl Symp on Agric. Wastes ASAE, pp 81-90.

Prasad, S; Singh, A. & Joshi, HC. (2007). Ethanol as an Alternative Fuel from Agricultural, Industrial and Urban Residues. Resources, Conservation and Recycling, Vol. 50, Issue 1, March, pp 1-39.

Prateek S; Gopal R; Mayur, S. & Shilpkar, D. (2009). Biomethanation Potential of *Jatropha* (*Jatropha curcas* Cake along with Buffalo Dung. African Journal of Agricultural Research Vol. 4, No. 10, October, pp. 991-995. Available online at www.academicjournals.org/AJAR

Reyadh M (1997). The cultivation of *Jatropha* In Egypt under Secretary of State for Afforestation Ministry of Agriculture and Land Reclamation. Medicinal, Culinary and Aromatic Plants in Near East, FAO Corporate document Repository, Proceedings of International Expert Meeting at Cairo, Egypt.

Rinehart, L. (2006). Switchgrass as a Bioenergy Crop. Available online at http://attra.ncat.org/attra-pub/switchgrass.html.

Safley Jr, LM. & Westerman, PW. (1990). Psychrophilic Anaerobic Digestion of Dairy Cattle Manure. Proceed. 6th Internl. Symp. On Agric and Food Processing Wastes, ASAE, St. Joseph, MI, USA, pp 250-258.

Safley Jr, LM; Westerman, PW. & Barker, JC. (1985). Fresh Dairy Manure Characteristics and Barnlot Nutrient Losses. 5th Internl . Symp. On Agric Wastes, ASAE, St. Joseph, MI, USA, pp 191-199.

San Thy, JL; Preston, TR. & Ly, J. (2003). Effect of Retention Time on Gas Production and Fertilizer Value of Biodigester Effluent, MEKARN Publishers, Bombay, India. pp 67-70.

Saptoadi, H; Untoro, BS; Nugroho, AP. & Sudarwanto, (2009). Utilization of Indigenous Biomass Wastes as Low Cost Alternative Fuels Paper presented at the conference on Alternative Energy Applications; Options or Necessity which held at Kuwait, Nov. 2-5, Kuwait. Organised by Kuwait Society of Engineers and World Federation of Engineering Organizations, 2009, www.ec2009kuwait.org

Seifritz, W. (1982). Alternative and Renewable Sources of Energy in Optimizing Yields: The Role of Fertilizers. In: Proceedings of 12th IPI Congress, pp.153-163.

Sherma, AR. & Mittra, BN. (1991). Effect of Different Rates of Application of Organic and N Fertilizers in a Rice Based Cropping System. J. Agric Sci. (Cambridge) Vol. 117, pp 313-318.

Shih, JCH. (1984). Biogas and Solid By-Product from a Poultry Waste Digester. Proceed. Internl. Symp. On Animals as Waste Converters. Pudoc. Wagenigen, Netherlands. Shoemaker, AH. & Visser, A. (2000). Treatment of Biogas. Anaerobic Digestion: Making Energy and Solving Modern Waste Problem. Edited by Henrik O. AD-NETT-A Network on Anaerobic Digestion of Agro-industrial Waste, Pp 8-21.

Sidibe, AN. & Hashimoto, AG. (1990). Conversion of Grass Straw to Methane. Proceed. 6th Internl. Symp. On Agric and Process Wastes. ASAE, St. Joseph, MI, USA, pp 131-138.

Smith, KA; Brewer, AJ; Crabb, J. & Dauver, A. (2001). A Survey of the Production and Use of Animal Manures in England and Wales II Poultry Manure Soil Use and Management. Vol. 17, No 3, March, pp 48-56.

Smith, KA. & Chambers, BJ. (1998). Nutrient Losses to Water following Land Application of Farm Manure. In Environmentally Friendly Management of Farm Waste. (ed.) T.

Matsunak. Proceedings ofInternational Workshop, Sapporo Japan. 15-29 November 1997, pp 79 - 83.
Smith, KA. & Chambers, BJ. (1993). Utilising the Nitrogen Content of Organic Manure on Farms–Problems and Practical Solutions. Soil Use and Management, Vol. 9, pp 105 – 112.
Stevens, MA. & Schulte, DD. (1979). Low Temperature Digestion of Swine Manure J. Environmental Engineering Division. American society of Civil Engineers, ASCE, Vol. 105 (EE1), pp 33-45.
The Economist (2005). Stirrings in the Corn Field. The Economist Magazine, London, UK. May 14th issue,pp 59-61.
Thomas, S. (1990). Evaluation of Plant Biomass Research for Liquid Fuels Brighton Science Policy Research Unit, University of Sussex, United Kingdom.
Thomsen, IK. (2000). C and N Transformations in 15N Cross-Labeled Solid Ruminant Manure During Anaerobic Storage., Bioresource Technol. Vol. 73, No. 3, pp 267-274.
Tomomatsu, Y and Swallow, B (2007) *Jatropha curcas* and Biodiesel Production in Kenya. Economics and Potential Value Chain Development for Smallholder Farmers. Working paper #54. WorldAgroforestry Centre,Nairobi, Kenya.
Uzodinma, EO. & Ofoefule, AU. (2009). Biogas Production from Blends of Field Grass (*Panicum maximum*) with some animal wastes. International Journal of the Physical Sciences, Vol. 4, No. 2, February, pp.091-095. Available online at www.academicjournals.org/IJPS
Veljkovic´, VB; Lakicevic, SH; Stamenkovic, OS; Todorovic, ZB. and Lazic, KL. (2006). Biodiesel Production from Tobacco (*Nicotiana tabacum* L.) Seed Oil with a High Content of Free Fatty Acids. Fuel, Vol. 85,pp 2671–2675.
Vetter, RL; Friederick, DJ. & Huntington, P. (1990). Full Scale Anaerobic Digester and Waste Management System for a 300-Cow Dairy. Proceed. 6th Internl Symp. On Agric and Food Process Wastes ASAE. St.Joseph, MI, USA, pp 236-249.
Voermans, JAM. (1985). Biogas Production on Livestock Farms. Proceed 5th Internl Symp on Agric Wastes. ASAE, St. Joseph, MI, USA, pp 132-37.
Volunteers in Technical Assistance, VITA. (1980). The 3-Cubic Metre Biogas Plant. A Construction Manual,VITA, pp 7.
Wadman, WP; Sluijsmans, CMJ. & De la Cremer, LCN. (1987). Value of Animal Manure: Changes in Perception. In: Animal Manure on Grassland and Fodder Crops. Fertilizer or Waste? H.G. Van der Meer, R.J. Unwin, Martinus Nijhoff Public, Dordrecht. pp 1-6.
Williams, J. & Haq, N. (2002). Global Research on Underutilised Crops: An Assessment of Current Activitiesand Proposals for Enhanced Cooperation. International Centre for Underutilised Crops, Southampton, UK:
Wis, (2009). Anaerobic Digestion Feedstock Classification. (http://www.wisbiorefine.org/ proc/anaerobic.pdf) Accessed May 28, 2009.
World Bank (2008). Biofuels: The Promise and the Risks, in World Development Report 2008. The WorldBank, Washington, DC, USA. pp. 70-71.
Zehnder, AJ; Ingvorsen, K. & Marti, T. (1982). Anaerobic Digestion. In Hughes, DE. et al., ed. Anaerobic Digestion, Elsevier Biomedical Press, Amsterdam.

3

Rheological Characterization

Annika Björn, Paula Segura de La Monja,
Anna Karlsson, Jörgen Ejlertsson and Bo H. Svensson
*Department of Thematic Studies, Water and Environmental Studies,
Linköping University,
Sweden*

1. Introduction

The biogas process has long been a part of our biotechnical solutions for the handling of sewage sludge and waste. However, in many cases the existing process applications need to be optimized to improve the extent of biogas production as a part of the energy supply in a sustainable and viable society. Although the principles are well known, process disturbances and poor substrate utilization in existing biogas plants are common and are in many cases likely linked to changes in the substrate composition.

Changes in substrate composition can be done as a means to obtain a more efficient utilization of existing biogas facilities, which today treating mainly manure or sewage sludge. By bring in more energy rich residues and wastes a co-digestion process with higher biogas potential per m^3 volatile solids (VS) can often be obtained. However, new and changing feedstocks may result in shift in viscosity of the process liquid and, hence, problems with inadequate mixing, break down of stirrers and foaming. These disturbances may seriously affect the degradation efficiency and, hence, also the gas-production per unit organic material digested. In turn, operational malfunctions will cause significant logistic problems and increased operational costs. Changes of the substrate profile for a biogas plant may also infer modifications of the downstream treatment of the digestate.

Together with high digestion efficiency, i.e. maximum methane formation per reactor volume and time, the economy of a biogas plant operation depends on the energy invested to run the process. A main part of the energy consumed during operation of continuous stirred tank reactors (CSTRs) is due to the mixing of the reactor material (Nordberg and Edström, 2005). The shear force needed is dependent on the viscosity of the reactor liquid, where increasing viscosity demands a higher energy input. Active stirring must be implemented in order to bring the microorganisms in contact with the new feedstock, to facilitate the upflow of gas bubbles and to maintain an even temperature distribution in the digester. Up to 90% of biogas CSTR plants use mechanical stirring equipment (Weiland *et al.*, 2010).

In this context the rheological status of the reactor liquid as well as of the residual digestate are important for process mixing design and dimensioning. In addition experiences on rhelogical characterisation of sewage sludge revealing their dependence on the suspended solid concentration and on the characteristics of the organic material as well as on the interactions between particles and molecules in the solution (Foster, 2002). Therefore, this type of characterisation can be important in process monitoring and control.

The aim of this chapter is to briefly introduce the area of rheology and to present important parameters for rheological characterization of biogas reactor fluids. Examples are given from investigations on such parameters for lab-scale reactors digesting different substrates.

2. Rheology

Rheology describes the deformation of a body under the influence of stress. The nature of the deformation depends on the body's material conditions (Goodwin & Hughes, 2000). Ideal solids deform elastically, which means that the solid will deform and then return to its previous state once the force ceases. In this case, the energy needed for deformation will mainly be recovered after the stress terminates. If the same force is applied to ideal fluids, it will make them flow and the energy utilized will disperse within the fluid as heat. Thus, the energy will not be recovered once the forcing stress is terminated (Goodwin & Hughes, 2000).

For fluids a flow curve or rheogram is used to describe rheological properties. These properties may be of importance in anaerobic digestion for the dimensioning of e.g. feeding, pumping and stirring. Rheograms are constructed by plotting shear stress (τ) as a function of the shear rate (γ) (Tixier *et al.*, 2003; Guibad *et al.*, 2005).

The stress applied to a body is defined as the force (F) divided by the area (A) over which this force is acting (Eq. 1). When forces are applied in opposite directions and parallel to the side of the body it is called shear stress (Goodwin & Hughes, 2000). Shear stress (τ; Pa) is one of the main parameters studied in rheology, since it is the force per unit area that a fluid requires to start flowing (Schramm, 2000). The shear rate (γ; s^{-1}) describes the velocity gradient (Eq. 2). Hence, shear rate is the speed of a fluid inside the parallel plates generated when shear stress is applied (Pevere & Guibad, 2005).

$$\tau = F/A = N/m^2 = Pa \quad (1)$$

$$\gamma = dv_x/dy = (m/s)/m \quad (2)$$

2.1 Newtonian fluids

Ideal fluids (e.g. water, methanol, olive oil and glycerol) perform linearly in rheograms, as illustrated for glycerol in figure 1, and are identified as Newtonian fluids. The Newtonian equation (Eq. 3) illustrates the flow behaviour of an ideal liquid (Schramm, 2000), where η is the viscosity (Pa*s). Dynamic viscosity, also called apparent viscosity, describes a fluid's resistance of deformation (Pevere & Guibad, 2005). In terms of rheology it is the relation of shear stress over the shear rate (Eq. 4). For Newtonian fluids the dynamic viscosity maintains a constant value meaning a linear relationship between τ and γ.

$$\tau = \eta * \gamma \quad (3)$$

$$\eta = \tau / \gamma \quad (4)$$

When measuring the dynamic viscosity, the fluid is subjected to a force impact caused by moving a body in the fluid. Resistance to this movement provides a measure of fluid viscosity. The dynamic viscosity can be measured using a rotation rheometer. The device consists of an external fixed cylinder with known radius and an internal cylinder or spindle with known radius and height. The space between the two cylinders is filled with the fluid subjected to dynamic viscosity analysis.

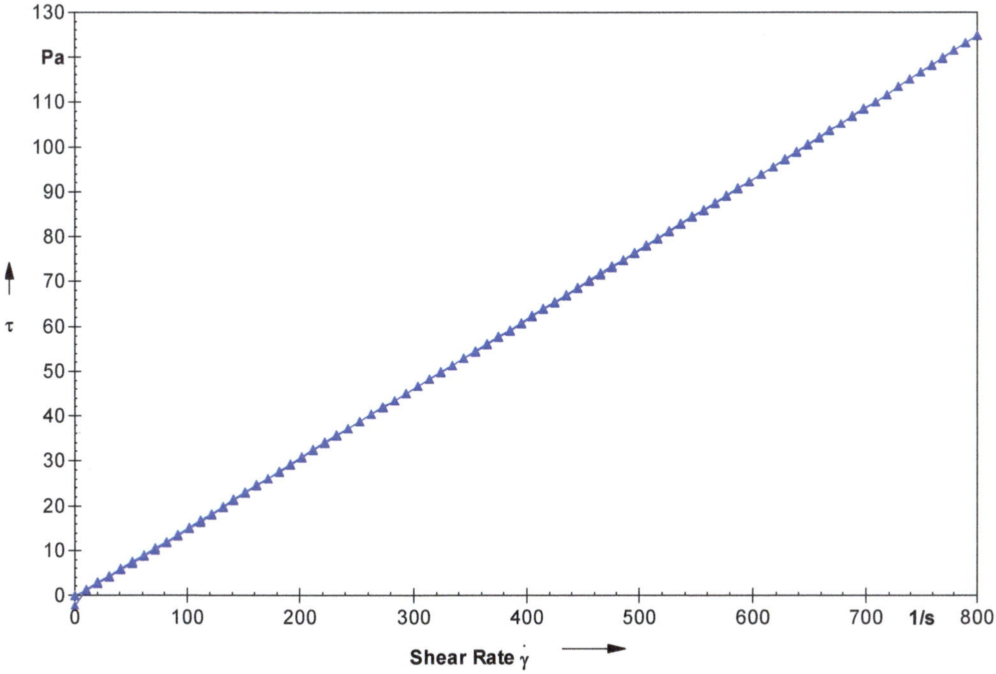

Fig. 1. Rheogram - flow curve of glycerol (▲) at 20 °C with a linear relationship between shear stress (τ; Pa) and shear rate (γ; s^{-1}), representing a Newtonian liquid.

2.2 Limit viscosity

Limit viscosity (η_{lim}) corresponds to the viscosity of a fluid at the maximum dispersion of the aggregates under the effect of the shear rate (Tixier & Guibad, 2003). The limit viscosity is estimated through the rheogram, when the dynamic viscosity becomes linear and constant. This parameter has been shown to be of great value when studying the rheological characteristics of sludge, since it determines the level of influence of important factors such as the total solids fraction (TS; Lotito *et al.*, 1997). TS (%) and volatile solids (VS, % of TS) are parameters measured in the biogas process in order to control the amount of solids that may be transformed to methane. Also, Pevere and Guibad (2005) reported that the limit viscosity was sensitive to the physicochemical characteristics of granular sludge, i.e. it was influenced by changes in the particle size or the zeta potential.

2.3 Dynamic yield stress

Yield stress (τ_0) is defined as the force a fluid must be exposed to in order to start flowing. It reflects the resistance of the fluid structure to deformation or breakdown. Rheograms from rotational viscometer measurements are used as a means to calculate yield stress. It can also be obtained by applying rheological mathematical models (section 2.6; Spinosa & Loito 2003). Yield stress is important to consider when mixing reactor materials, since the yield stress is affecting the physico-chemical characteristics of the fluid and impede flow even at relative low stresses. This might lead to problems like bulking or uneven distribution of material in a reactor (Foster, 2002).

2.4 Static yield stress

The static yield stress (τ_s) is the yield stress measured in an undisturbed fluid while dynamic yield stress is the shear stress a fluid must be exposed to in order to become liquid and start flowing. The fact that both dynamic yield stress and static yield stress sometimes may appear is explained by the existence of two different structures of a fluid. One structure is not receptive to the shear stress and tolerates the dynamic yield stress, while a second structure (a weak gel structure) is built up after the fluid has been resting a certain period of time (Yang *et al.*, 2009). When these two structures merge, a greater resistance to flow is generated translated to the static yield stress.

The formation of the weak gel structure may be a result from chemical interactions among polysaccharides or between proteins and polysaccharides (Yang *et al.*, 2009). The weak gel structure is quite vulnerable and, thus easily interrupted by increasing shear rates.

2.5 Non-Newtonian fluids

Non-Newtonian fluids do not show a linear relationship between shear stress and shear rate. This is due to the complex structure and deformation effects exhibited by the materials involved in such fluids. The non-Newtonian fluids are however diverse and can be characterised as e.g. pseudoplastic, viscoplastic, dilatant and thixotropic fluids (Schramm, 2000).

2.5.1 Pseudoplastic fluids

Pseudoplastic fluids become thinner when the shear rate increases, until the viscosity reaches a plateau of limit viscosity. This behaviour is caused by increasing the shear rate and the elements suspended in the fluid will follow the direction of the current. There will be a deformation of fluid structures involving a breaking of aggregates at a certain shear rate and this will cause a limit in viscosity. For pseudoplastic fluids the viscosity is not affected by the amount of time the shear stress is applied as these fluids are non-memory materials i.e. once the force is applied and the structure is affected, the material will not recover its previous structure (Schramm, 2000). Some examples are corn syrup and ketchup.

2.5.2 Viscoplastic fluids

Viscoplastic fluids, such as e.g. hydrocarbon greases, several asphalts and bitumen, behave as pseudoplastic fluids upon yield stress. They need a predetermined shear stress in order to start flowing. One type of these, the Bingham plastic, requires the shear stress to exceed a minimum yield stress value in order to go from high viscosity to low viscosity. After this change a linear relationship between the shear stress and the shear rate will prevail (Ryan, 2003). Examples of Bingham plastic liquids are blood and some sewage sludge's.

2.5.3 Dilatant fluids

Dilatant fluids become thicker when agitated, i.e. the viscosity increases proportionally with the increase of the shear rate. Like for the pseudoplastic fluids the stress duration has no influence, i.e. when the material is disturbed or the structure destroyed it will not go back to its previous state. Some examples of shear thickening behaviour are honey, cement and ceramic suspensions.

2.5.4 Thixotropic fluids

Thixotropic fluids are generally dispersions, which when they are at rest construct an intermolecular system of forces and turn the fluid into a solid, thus, increasing the viscosity. In order to overcome these forces and make the fluid turn into a liquid and which may flow, an external energy strong enough to break the binding forces is needed. Thus, as above a yield stress is needed. Once the structures are broken, the viscosity is reduced when stirred until it receives its lowest possible value for a constant shear rate (Schramm, 2000). In opposite to pseudoplastic and dilatant fluids, the viscosity of thixotrpic fluids is time dependent: once the stirring has ended and the fluid is at rest, the structure will be rebuilt. This will inform about the fluid possibilities of being reconstructed. Wastewater and sewage sludge can be examples of fluids with thixotropic behaviour (Seyssieq & Ferasse, 2003) as well as paints and soap.

2.6 Rheological mathematical models

There are several rheological mathematical models applied on rheograms in order to transform them to information on fluid rheological behaviour. For non-Newtonian fluids the three models presented below are mostly applied (Seyssiecq & Ferasse, 2003).

2.6.1 Herschel Bulkley model

The Herschel Bulkley model is applied on fluids with a non linear behaviour and yield stress. It is considered as a precise model since its equation has three adjustable parameters, providing data (Pevere & Guibaud, 2006). The Herschel Bulkley model is expressed in equation 5, where τ_0 represents the yield stress.

$$\tau = \tau_0 + K * \gamma^n \qquad (5)$$

The consistency index parameter (K) gives an idea of the viscosity of the fluid. However, to be able to compare K-values for different fluids they should have similar flow behaviour index (n). When the flow behaviour index is close to 1 the fluid´s behaviour tends to pass from a shear thinning to a shear thickening fluid. When n is above 1, the fluid acts as a shear thickening fluid. According to Seyssiecq and Ferasse (2003) equation 5 gives fluid behaviour information as follows:

$\tau_0 = 0$ & $n = 1 \Rightarrow$ Newtonian behaviour
$\tau_0 > 0$ & $n = 1 \Rightarrow$ Bingham plastic behaviour
$\tau_0 = 0$ & $n < 1 \Rightarrow$ Pseudoplastic behaviour
$\tau_0 = 0$ & $n > 1 \Rightarrow$ Dilatant behaviour

2.6.2 Ostwald model

The Ostwald model (Eq. 6), also known as the Power Law model, is applied to shear thinning fluids which do not present a yield stress (Pevere *et al.*, 2006). The n-value in equation 6 gives fluid behaviour information according to:

$$\tau = K * \gamma^{(n-1)} \qquad (6)$$

$n < 1 \Rightarrow$ Pseudoplastic behaviour
$n = 1 \Rightarrow$ Newtonian behaviour
$n > 1 \Rightarrow$ Dilatant behaviour

2.6.3 Bingham model

The Bingham model (Eq. 7) describes the flow curve of a material with a yield stress and a constant viscosity at stresses above the yield stress (i.e. a pseudo-Newtonian fluid behaviour; Seyssiecq & Ferasse, 2003). The yield stress (τ_0) is the shear stress (τ) at shear rate (γ) zero and the viscosity (η) is the slope of the curve at stresses above the yield stress.

$$\tau = \tau_0 + \eta * \gamma \qquad (7)$$

$\tau_0 = 0 \Rightarrow$ Newtonian behaviour
$\tau_0 > 1 \Rightarrow$ Bingham plastic behaviour

3. Rheological characterization of biogas reactor fluids

When considering the rheology for biogas reactors their viscosity is estimated to correspond to a given TS of the reactor fluid. This is mainly based on historically rheological data from sewage sludge with known TS values. However, problems may arise when using these TS relationships for other types of substrates which may impose other rheological characteristics of the reactor fluids. Furthermore, often low consideration is given to possible viscosity changes due to variation in feedstock composition etc.

Shift in the viscosity and elasticity properties of the reactor material related to substrate composition changes can alter the prerequisites for the process regarding mixing (dimension of stirrers, pumps etc. or reactor liquid circulation) and likely also foaming problems (Nordberg & Edström, 2005; Menéndez *et al.*, 2006). It may also call for changes in the post treatment requirements and end use quality of the organic residue e.g. dewatering ability, pumping and spreading on arable land (Baudez & Coussot, 2001). The additions of enzymes can be used to reduce the viscosity of the substrate mixture in the digester significantly and avoid the formation of floating layers (Weiland, 2010; Morgavi *et al.*, 2001). All these factors affect the total economy for a biogas plant.

In this context differences in the rheological characteristics of biogas reactor fluids as depending on substrate composition were analyzed and used as examples in this presentation.

3.1 Rheological measurments

A rotational rheometer RheolabQC coupled with Rheoplus software (Anton Paar) was used for different reactor fluids, which recorded the rheograms´ and allowed subsequent data analysis. The temperature was maintained constant at 37±0.2 °C. The reactor fluid volume used for each measurement was 17 ml. Reactor fluids from mesophilic (37°C) lab-scale reactors (4 L running volume), with a hydraulic retention time (HRT) of 20 days, were sampled.

Five lab-scale reactors (A-E) were sampled before the daily feeding of substrates. All reactors had been running for at least three HRTs prior to sampling. The different substrates treated were slaughter household waste, biosludge from pulp- and paper mill industries, wheat stillage and cereal residues. The TS values ranged between 3.1−3.9 % for four of the reactors while one was at 7.7 % (Table 1).

Reactor	Digested substrate	TS (%)
A	Slaughter house waste	3.9
B	Biosludge from pulp- and paper mill industry 1	3.8
C	Biosludge from pulp- and paper mill industry 2	3.7
D	Wheat stillage	3.0
E	Cereal residues	7.7

Table 1. Fluids from five lab-scale reactors were chosen for rheological measurements. A short description of their TS values and substrates are presented.

Rheological measurements were carried out with a three-step protocol where (1) the shear rate increased linearly from 0 to 800 s^{-1} in 800 sec., (2) maintaining constant shear rate at 800 s^{-1} in 30 sec, (3) decreasing linearly the shear rate from 800 to 0 s^{-1} in 800 sec., according to Björn *et al.* (2010). For each sample three measurements were carried out and performed immediately after sampling or stored at +4 °C pending analysis.

The fluid behaviour was interpreted by the flow- and viscosity curves according to Schramm (2000), and the dynamic viscosity, limit viscosity and yield stress were noticed. The three most common mathematical models for non-Newtonian fluids; Herschel Bulkley model; Ostwald model (Power Law) and Bingham model, were applied in order to transform rheogram data values to the rheological behaviour of the fluids. Flow behaviour index (n) and consistency index (K) were studied.

3.2 Flow and viscosity behaviour characteristics

The flow curves for reactor fluids A-E (Figures 2−3) indicated different flow behaviour according to the definitions by Schramm (2000). A Newtonian behaviour of reactors A and D, fed with slaughter house waste and wheat stillage, respectively, was illustrated where the exerted shear stress was almost proportional to the induced shear rate. However, a small yield stress of 0.2 Pa and 0.3 Pa were detected, indicating a pseudo-Newtonian behaviour.

Fluids from reactor B, receiving biosludge from paper mill industry 1 as substrate, indicated an unusual performance at the beginning of the rheogram with decreasing shear stress, thereafter a linear increase in shear stress. A yield stress of 14 Pa was detected. A space between the curves was noticeable when the shear rate increased and afterwards decreased for reactor B (Fig. 2). This area describes the degree of thixotropy of this fluid, which means that the increase of this area is related to the amount of energy required to breaking down the thixotropic structure. Thus, the flow curves obtained with the three-step protocol indicated a thixotropic behaviour of reactor fluid B.

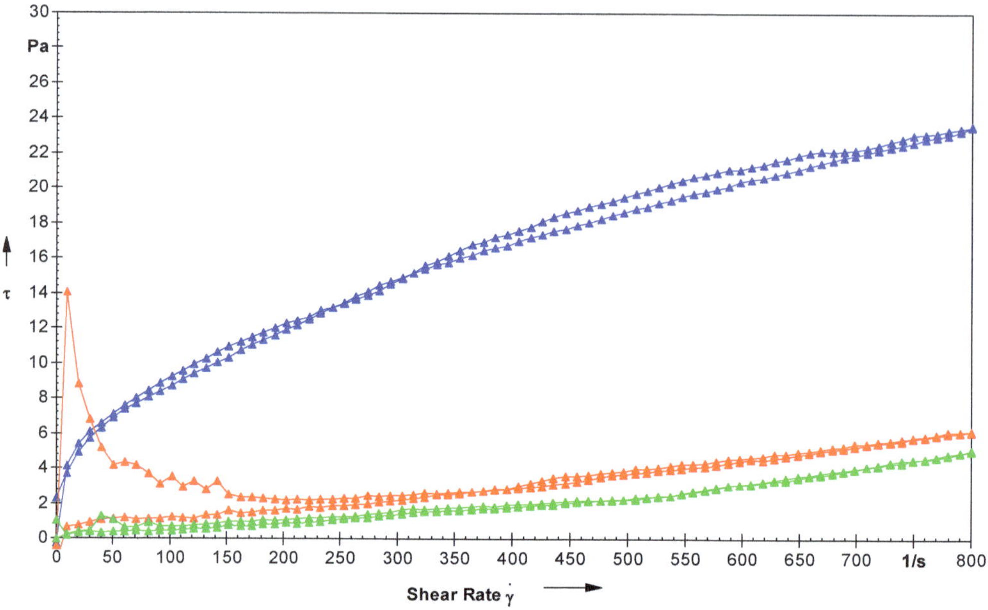

Fig. 2. Rheogram - flow curves illustrating shear stress (τ; Pa) vs shear rate (γ; s^{-1}) for fluids from reactor A (▲), B (▲) and C (▲) with a three-step protocol.

Reactors C and E revealed viscoplastic behaviours, i.e. a pseudoplastic behaviour with yield stress. Reactor C, fed with biosludge from paper mill industry 2, showed a yield stress of 4 Pa (Fig. 2), and reactor E, receiving cereal residues, a yield point of 4.5 Pa (Fig. 3). The yield stress is defined as the force that a fluid must overcome in order to start flowing (Spinosa & Lotito, 2003). Also for reactor E, a small space between the curves was noticeable when the shear rate increased and afterwards decreased (Fig. 3). This area difference might indicate some degree of thixotropy.

Reactor	Flow curve behaviour	Viscosity curve behaviour
A	Newtonian; pseudo-Newtonian	Viscoplastic (pseudo-Newtonian)
B	Thixotropic	Thixotropic
C	Viscoplastic	Viscoplastic
D	Newtonian; pseudo-Newtonian	Pseudoplastic or viscoplastic
E	Viscoplastic	Viscoplastic

Table 2. The flow and viscosity curves for reactor fluids A-E indicated different fluid behaviour according to Schramm (2000).

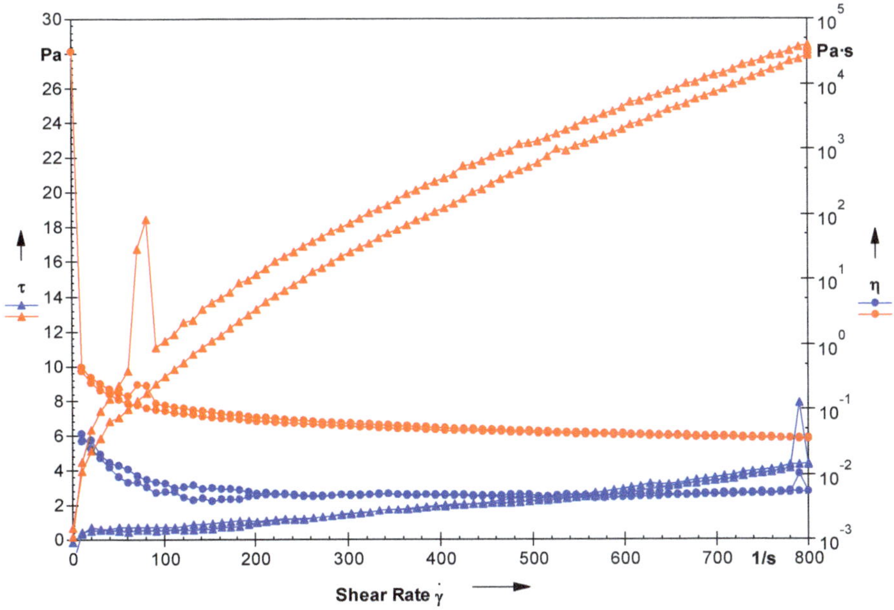

Fig. 3. Rheogram - flow and viscosity curves for reactors D (▲; ●) and E (▲; ●) with a three-step protocol. Flow curves illustrating shear stress (τ; Pa) vs shear rate (γ; s^{-1}) and viscosity curves illustrating dynamic viscosity (η; Pa*s) vs shear stress (γ; s^{-1}).

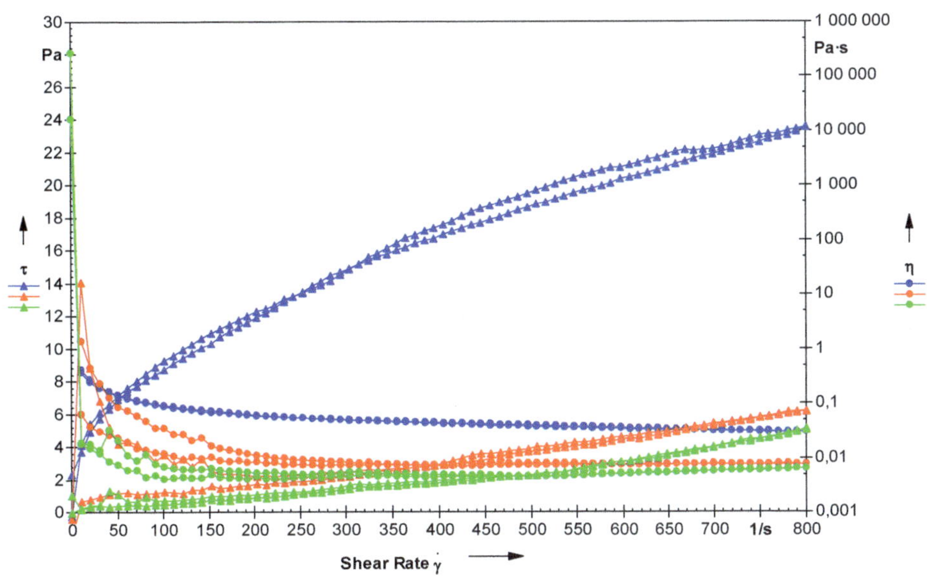

Fig. 4. Rheogram - flow and viscosity curves for reactors A (▲;●), B (▲;●) and C (▲;●) with a three-step protocol. Flow curves illustrating shear stress (τ; Pa) vs shear rate (γ; s^{-1}) and viscosity curves illustrating dynamic viscosity (η; Pa*s) vs shear stress (γ; s^{-1}).

The viscosity curves (Figures 3–4) did almost correspond to the flow curve behaviour for the investigated biogas reactor fluids (Table 2). Using the scheme by Schramm (2000) the viscosity curves for reactor A indicated a viscoplastic liquid and for reactor D a viscoplastic or pseudoplastic liquid. The viscosity initially dropped very quickly for reactor A, specifically indicating Bingham viscoplastic fluids with pseudo-Newtonian behaviour. Generally for reactors A–E, the viscosity decreased with increasing shear rate, until it reached its limit viscosities (Table 3).

Reactor	Treated substrate	TS (%)	Dynamic viscosity (mPa*s)	Limit viscosity (mPa*s)
A	Slaughter house waste	3.9	18	6
B	Biosludge paper mill industry 1	3.8	436	8
C	Biosludge paper mill industry 2	3.7	267	29
D	Wheat stillage	3.0	33	6
E	Cereal residues	7.7	443	36

Table 3. The initial dynamic viscosity and the limit viscosity obtained during interval 1 in the 3-step protocol analysis for each reactor fluid.

The limit viscosities ranged 6–36 mPa*s with the highest value for reactor E (Table 3). However, the limit viscosity was similar for reactor C and E despite a difference in TS (%). The dynamic viscosity ranged 18–443 mPa*s for the reactor fluids (Table 3). The reactors A and D showed lower dynamic viscosity values compared to reactor B, C and E, possibly due to their pseudo-Newtonian behaviour. Also, there was a difference in dynamic viscosity between reactor fluids B and C both receiving biosludge with similar TS (%) but coming from two different paper mill industries in Sweden. Thus, the results demonstrated that similar TS values do not necessarily correspond to similar dynamic or limit viscosity values. This contradicts the results presented by Tixier and Guibad (2003), who reported that an increase in TS for activated sludge corresponded to a higher limit viscosity and higher yield stress. Nor, did biosludge from two different Swedish pulp- and paper mill industries with similar TS give similar viscosity values.

Samples from reactor B showed different rheological behaviour depending on when they were measured. The yield stress and viscosity increased when the reactor fluids had been stored and resting compared to when analyzed immediately after sampling, indicating thixotropic behaviour. Figure 5 illustrates how the B reactor fluid after been resting for 48 hours showed a resistance to flow, known as static yield stress. This value (24 Pa) decreased until it reached the dynamic yield stress (7 Pa), which was the value needed in order to become liquid and start flowing. When the analysis was done right after the first measurement (Second measurement), the fluid had already been stirred, so the two different structures that form the resistance to flow were now mixed and no static yield stress was detected. The static yield stress might be initiated by several factors, e.g. weakness of the fluid structure, low mixed liquid solid suspension (MLSS) concentration, small size of particles and poor dewater ability (Pevere *et al.* 2006).

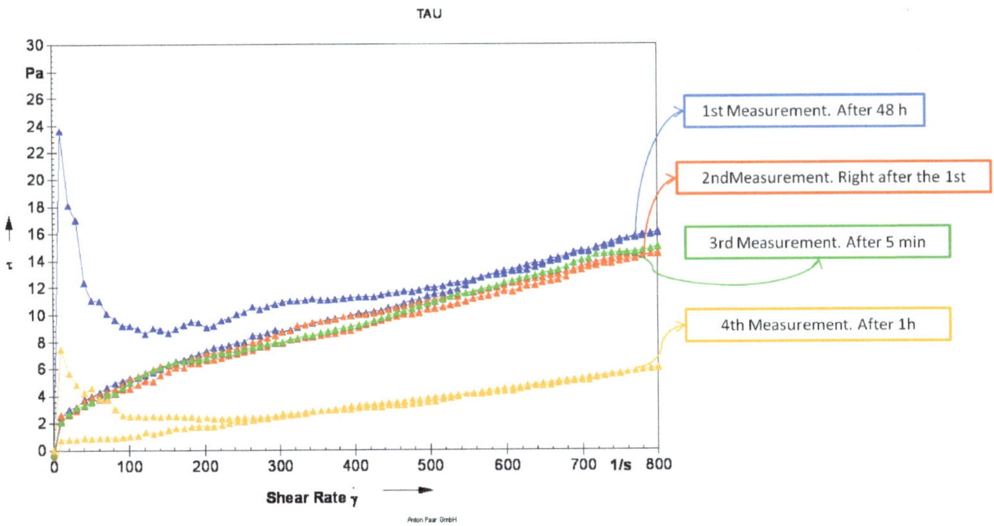

Fig. 5. Rheogram - shear stress (τ; Pa) vs shear rate (γ; s^{-1}) of reactor B. Four measurements of the same sample were made after different sample resting times (1st▲; 2nd▲; 3rd▲; 4th ▲).

3.3 Mathematical modeling

Also, the Herchel-Bulkley model indicated that reactor fluid A performed as a pseudo-Newtonian fluid called Bingham plastic, since the yield stress-value was > 0 (0.24 Pa) and a flow behaviour index of 1.06 (Table 4). Results obtained by the Ostwald and Bingham models confirmed a Bingham plastic behaviour of reactor A. However, since the τ_0-value was almost 0 and the n-value 1 it was also closely performing as a Newtonian fluid which is consistent with the flow curve appearance (Fig. 2). However, when studying the viscosity curve (Fig. 4) the results showed an initial viscosity decrease and then a constant viscosity indicating a pseudo-Newtonian fluid behaviour.

The Herschel-Bulkley and Ostwald models both indicated a pseudoplastic behaviour of reactor D, since the τ_0-value was 0 and $n < 1$ (Table 4). The Bingham model gave a yield stress of 0.33 Pa which did not indicate Newtonian or Bingham plastic behaviour. Thus, the common results for reactor D strongest indicate a pseudoplastic fluid behaviour.

Reactor B was hard to define also when modelling the rheogram data values of figure 4. The regression values were low for all three mathematical models (Table 4). However, the Herschel-Bulkely model had a flow behaviour index $n>1$ indicating that the fluid acted as a shear thickening (dilatant) fluid, but the Ostwald and Bingham models indicated pseudoplastic and Bingham plastic behaviours, respectively. When the static yield stress appeared in the reactor B rheogram (Figures 2 and 4), the flow behaviour index showed shear thickening fluid behaviour ($n=3.4$) and a limit viscosity of 8 mPa*s. This also corresponded to a low consistency value ($5*10^{-10}$). At the static yield stress of 24 Pa (Fig. 5), the flow behaviour index showed shear thickening fluid behaviour (n=1.41) and a limit viscosity of 22 mPa*s. This also corresponded to a low consistency value ($5*10^{-4}$). As soon as the fluid was measured again, n decreased (0.70) showing a pseudoplastic behaviour and K increased (0.11) indicating that the consistency of the reactor material was higher. The limit viscosity was 17 mPa*s. These results showing a time dependency and structure recovery strengthen the arguments for a thixotropic fluid behaviour of reactor B. Once the stirring has ended and the fluid was at rest, the fluid structure starts to rebuild. Therefore, the viscosity become time dependent. This information is important to consider for biogas reactor performance, e.g. when applying semi-continuous mixing.

	Herschel-Bulkley				Ostwald			Bingham	
	τ_0	n	K	R^2	n	K	R^2	τ_0	R^2
A	0.24	1.06	0.003	0.93	0.69	0.35	0.84	0.21	0.92
B	2.57	3.40	$5*10^{-10}$	0.45	0.08	2.28	0.002	1.88	0.12
C	2.89	0.59	0.42	0.99	0.44	1.23	0.99	6.36	0.95
D	0	0.65	0.04	0.88	0.64	0.04	0.87	0.33	0.95
E	2.38	0.49	0.98	0.96	0.39	1.98	0.96	8.31	0.91

Table 4. The results obtained from mathematical modelling of rheogram data of fluids from reactors A–E. τ_0: yield stress (Pa); n: flow behaviour index; K: Consistency index; R^2: regression coefficient.

Also, fluids from reactor C and E were hard to define from modelling of the rheogram data because they gave indications for fluids being between pseudoplastic and Bingham plastic behaviours, i.e. the τ_0-values were >0 (2.89 and 2.38) and n <1 (Table 4).

4. Conclusion

The biogas reactor fluids investigated were behaving viscoplasticly, since they had yield stress and one of them was also thixotropic, due to its partial structure recovering. However, the reactor treating slaughterhouse waste was very close to act as a Newtonian fluid. Also, there was a difference in dynamic- and limit viscosities depending on the substrates used. The results demonstrated that similar TS values did not necessarily correspond to similar flow and viscosity behaviours. Nor, did biosludge from two different Swedish paper mill industries with similar TS show similar viscosity values.

To encounter problems related to involvement of new substrates and/or co-digestions in existing facilities, investigations for possible viscosity changes are needed. Ongoing research will hopefully provide an important basis for predictions of changes in rheology linked to the composition of the organic materials, which are translated in the process. This is important in order to achieve proper designs in relation to possible variation in substrate mixes in conjunction with new constructions, but also to better control material flows in the existing facilities to avoid disturbances in the reactor performance.

5. References

Baudez, J-C. & Coussot, P. (2001). Rheology og aging concentrated, polymeric suspensions: application to pasty sewage sludges. *Jour Rheol,* Vol.45, No.5, pp. 1123-1139.

Björn, A. ; Segura de La Monja, P. ; Karlsson, A. ; Ejlertsson, J. & Svensson, BH. (2010). Differences and shifts in the rheological characteristics of fluids in controlled stirred tank reactors for biogas production. Poster presentation at World Conference in Anaerobic Digestion (AD12), Guadalajara, Mexico October 31rd o November 4th 2010.

Foster, CF. (2002). The rheological and physico-chemical characteristics of sewage sludges. *Enzyme Microb Technol,* Vol.30, pp. 340-345.

Goodwin, JW. & Hughes, RW. (2000). *Rheology for Chemists, an introduction.* Royal Society of Chemist. Cambridge, UK.

Guibad, G.; Tixier, N. & Baudu, M. (2005). Hysteresis area, a rheological parameter used as a tool to assess the ability of filamentous sludges to settle. *Process Biochem,* Vol.40, pp. 2671-2676.

Lotito, V.; Spinosa, L.; Minini, G. & Antonaci, R. (1997). The rheology of sewage sludge at different steps of treatment. *Water Sci Tech,* Vol.36, No.11, pp. 79-85.

Menéndez, M. & Paredes, B. (2006). Rheological behaviour and organoleptic effects of ovalbumin addition in yoghurt mousse production. *Jour Dairy Sci,* Vol.89, pp. 951-962.

Morgavi, DP.; Beauchemin, KA. & Nsereko, LM. (2001). Resistance of feed enzymes to proteolytic inactivation by rumen microorganisms and gastrointestinal proteases. *Jour Anim Sci,* Vol.79, pp. 1621-1630.

Nordberg, Å. & Edström, M. (2005). Co-digestion of energy crops and the source-sorted organic fraction of municipal solid waste. *Water Sci Tech,* Vol.52, No.1-2, pp. 217-222.

Pevere, A. & Guibad, G. (2005). Viscosity evolution of anaerobic granular sludge. *Biochem Engin Jour,* pp. 315-322.

Pevere, Å.; Guibad, G; van Hullenbusch, E.; Lens, P. & Baudu, M. (2006). Viscosity evolution of anaerobic granular sludgee. *Biochem Engin Jour,* Vol.27, pp. 315.322

Schramm, G. (2000). *A practical Approach to Rheology and Rheometry,* (2nd Ed.), Thermo Haake Rheology, Germany

Seyssiecq, I.; Ferasse, J.H. & Roche, N. (2003). State-of-the-art: rheological characterization of waste water treatment sludge. *Biochem Engin Jour,* Vol.16, pp. 41-56.

Spinosa, L., & Lotito, V. (2003). A simple method for evaluating yield stress. *Adv Environ Research,* Vol.7, pp. 655-659.

Tixier, N.; Guibad, G. & Baudu, M. (2003). Determination of some rheological parameters for the characterization of activated sludge. *Bioresource Tech,* Vol.90, pp. 215-220.

Weiland, P. (2010). Biogas production: current state and perspectives. *Appl Microbiol Biotechnol,* Vol.85, pp. 849-860.

Yang, F. ; Bick, A. & Shandalov, S. (2009). Yield stress and rheological characteristics of activated sludge in an airlift membrane bioreactor. *Jour Membrane Sci,* Vol.334, pp. 83-90.

4

Biogas Production from Anaerobic Treatment of Agro-Industrial Wastewater

Jorge del Real Olvera and Alberto Lopez-Lopez
Environmental Technology Unit, Centre of Research and Assistance in Technology and Design of the State of Jalisco, (CIATEJ), Mexico

1. Introduction

Today, globally most energy is provided by burning oil and only a very small percentage is generated by nuclear power plants. The contribution of energy from renewable resources is almost negligible. But this will change in the future with increasing in environmental pollution and fossil fuel depletion, in addition to environmental problems generated by the Fukushima nuclear power plant.

One of the most attractive ways to obtain sources of alternative energy and the pollution control is the recover resource and energy from waste streams through bioconversion processes (Cantrell et al., 2008). In this respect, intensive studies have been conducted in the past few decades and various "green technologies" have been extensively reviewed (Kleerebezemand and Loosdrecht, 2007; Hallenbeck and Ghosh, 2009). For many years, anaerobic digestion has been a prevailing technology for biogas production, in which substrates are converted to methane and other products under a joint effort of several microbial groups in a reaction system (Sterling et al., 2001).

In this context biogas generated by agro-industrial wastewater will play a vital role in future. Biogas is a versatile renewable energy source, which can be used for replacement of fossil fuels in power and heat production, and it can be used also as gaseous vehicle fuel. Methane-rich biogas can replace also natural gas, as a feedstock in the production of chemicals and materials (Shin et al., 2010).

Sustainable development must be the foundation for economic growth in the twenty-first century. It is necessary redirect the efforts toward bioenergy production from renewable material, low-cost and locally available feedstock such as waste and wastewater agro-industrial. This effort will not only alleviate environmental pollution, but also reduce energy insecurity and demand for declining natural resources. The most cost-effective and sustainable approach is to employ a biotechnology option. Anaerobic treatment is a technology that generates renewable bioenergy necessary to replace the energy requirements around the world through the production of methane and hydrogen. However, it has also been employed for production of polyhydroxyalkanoates (PHA), these are linear polyesters generated by bacterial fermentation of sugar or lipids. They are produced by the bacteria to store carbon and energy. More than 150 different monomers can be combined within this family to give materials with extremely different properties. These plastics are biodegradeable and are used in the production of bioplastics (Mu et al., 2006) and other biochemicals.

This chapter intends to bring together the knowledge obtained from different applications of anaerobic technology in the treatment of various kinds of agro-industrial wastewaters to generate biogas. The first part covers essential information on the fundamentals of anaerobic technology, to demonstrate how the anaerobic treatment is able to generate significant volumes of methane-rich biogas. The wastewaters used in this chapter to generate biogas, contribute significantly in the pollution of the water bodies. In this opportunity the wastewater from Tequila vinasses were treated by different microbial consortia with energy purpose. This chapter illustrates the basics concepts of microbiology and biochemistry involved in the wastewater anaerobic treatment. The remainder focuses on various anaerobic reactor configurations and operating conditions used for the treatment of agro-industrial wastewaters different, show some examples with technical viability and the potential benefits that would be obtained by the utilization of the biogas as source of energy to full scale.

2. Historical background

Very old sources indicate that using wastewater and so-called renewable resources for the energy supply it is not new, it was already known before the birth of Christ. Even around 3000 BC the Sumerians practiced the anaerobic cleansing of waste. The Roman scholar Pliny described around 50 BC some glimmering lights appearing underneath the surface of swamps (Lee et al., 2010).

In 1776 Alessandro Volta personally collected biogas from the Lake Como to examine it. His findings showed that the formation of gas depends on a fermentation process and that may form an explosive mixture with air. The English physicist Faraday also performed some experiments with marsh gas and identified hydrocarbons as part of this. Around the year 1800, Dalton, Henry and Davy first described the chemical structure of methane, however the final chemical structure of methane (CH_4), was first elucidated by Avogadro in 1821 (Horiuchi et al., 2002).

In the second half of 19th century, more systematic and scientific in-depth research was started in France to better understand the process of anaerobic fermentation. The objective was simply suppress the bad odor released by wastewater pools. During their investigations, researchers detected some of the microorganisms which today are known to be essential for the fermentation process. It was Béchamp who identified in 1868 that a mixed population of microorganism is required to convert ethanol into methane, since several end products were formed during the fermentation process, depending on the characteristic of substrate (Lee et al., 2010).

In 1876, Herter reported that acetate found in wastewater, stoichiometrically form methane and carbon dioxide in equal amounts. Louis Pasteur tried in 1884 to produce biogas from horse dung collected from Paris roads. Together with his student Gavon he managed to produce 100 L methane from 1 m^3 dung fermented at 35°C. Pasteur claimed that this production rate should be sufficient to cover the energy requirements for the street lighting of Paris. The application of energy from renewable resources started from this time on (Deublein and Steinhauser, 2008).

3. Fundamentals of microbiology and biochemistry in anaerobic digestion

One of the key factors in the success of microbial-mediated processes is an adequate understanding of process microbial, more specifically the study of microscopic organisms involved in wastewater degradation and byproduct formation. The low growth rate, the specific nutrient and trace mineral requirements of methanogens, coupled with their susceptibility to changes in environmental conditions demand meticulous process control for stable operation (Khanal, 2008). The biochemistry mainly involves enzyme-mediated chemical changes (the chemical activities of microorganism), type of substrate (kind wastewater) microorganism can destroy or transform to new compounds, and the step-by-step pathway of degradation (Sachdeva et al., 2000).

3.1 Organics conversion in anaerobic systems

The anaerobic digestive process is a natural biological process in which an interlaced community of bacteria cooperates to obtain a stable and auto-regulated fermentation through assimilation, transformation and decomposition of the residual organic matter present in waste and wastewater into biogas. This is a complex multistep process in terms of chemistry and microbiology, where the organic material is degraded to basic constituents to obtain methane gas under the absence of an electron acceptor such as oxygen. The common metabolic pathway and process microbiology of anaerobic digestion is shown in Fig. 1 (Khanal, 2008).

Generally, the anaerobic digestion process consists of four stages; the first one is called hydrolysis (or liquefaction), it consists in the transformation of complex organic matter such as proteins, carbohydrates and lipids into simple soluble products like sugars, long-chain fatty acids, amino acids and glycerin, this stage is carried out by the action of extracellular enzymes excreted by the fermentative (group 1) (Khanal, 2008).

In the second step, called the acidogenic stage fermentative bacteria use the hydrolysis products to form intermediate compounds like organic acids, including volatile fatty acids (VFA). Theses VFA along with ethanol are converted to acetic acid, hydrogen and carbon dioxide by other group of bacteria known as hydrogen-producing acetogenic bacteria (group 2) (Khanal, 2008).

Organic acids are oxidized partially by bacteria called acetogenic in the third stage, which produce additional quantities of hydrogen and acetic acid. The acetogenesis is regarded as thermodynamically unfavorable unless the hydrogen partial pressure is kept below 10^{-3} atm, pathway efficient removal of hydrogen by the hydrogen-consuming organisms such as hydrogenotrophic methanogens and/or homoacetogens (Zinder, 1988).

Finally, in the fourth stage, both acetic acid and hydrogen are the raw material for the growth of methanogenic bacteria, converting acetic acid and hydrogen to biogas composed mainly of methane, carbon dioxide and hydrogen sulfide (Khanal, 2008).

Acetate, H_2 and CO_2 are the primary substrate for methanogenesis. On chemical oxygen demand (COD) basis about 72% of methane production comes from the decarboxylation of acetate, while the remainder is from CO_2 reduction (McCarty, 1964). The groups of microorganisms involved in the generation of methane from acetate are known as acetotrophic or aceticlastic methanogens (group 3). The remaining methane is generated from H_2 and CO_2 by the hydrogenotrophic methanogens (group 4). Since methane is largely generated from acetate, acetotrophic methanogenesis is the rate-limiting step in anaerobic wastewater treatment. The synthesis of acetate from H_2 and CO_2 by homoacetogens (group 5) has not been widely studied. Mackie and Bryant (1981) reported that acetate synthesis through this pathway accounts for only 1-2% of total acetate formation at 40°C and 3-4% total solids at 60°C in a cattle waste digester.

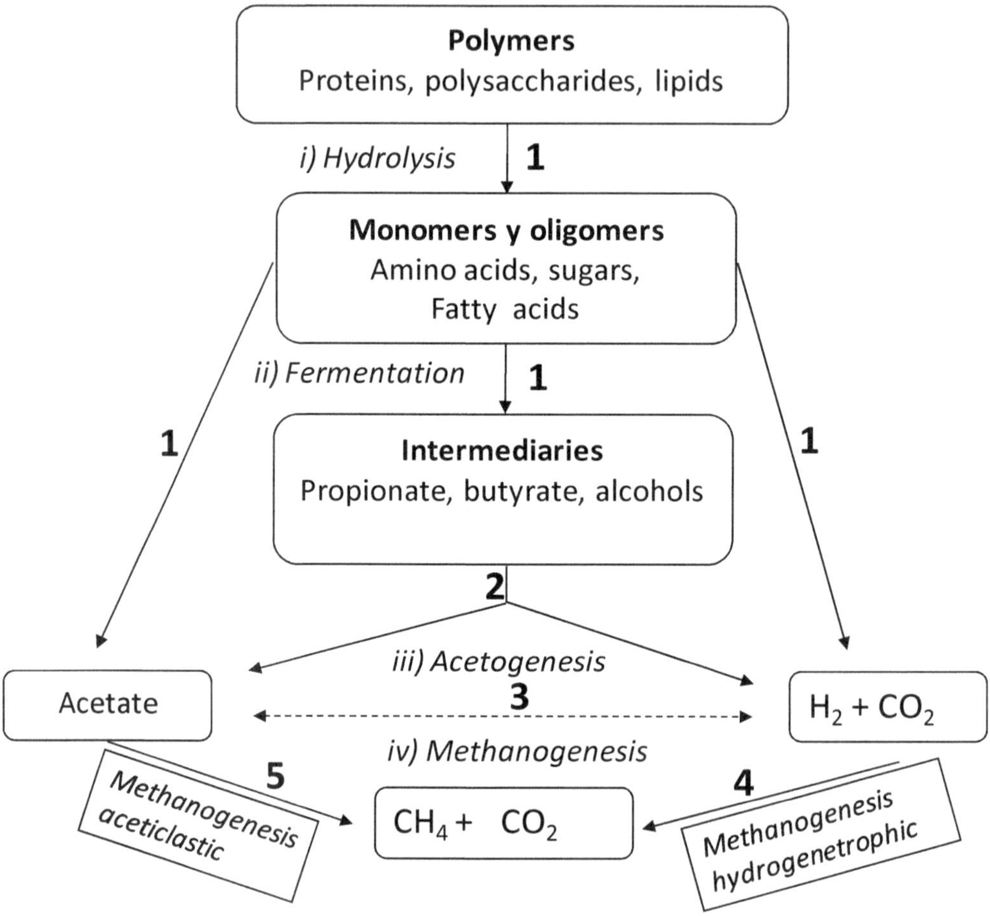

Fig. 1. Steps of anaerobic digestion of complex organic matter (the number indicate the group of bacteria involved in the process).

3.2 Process microbiology

The anaerobic degradation of complex organic matter is carried out by different groups of bacteria as indicated in Fig. 1. These exists a coordinated interaction among these bacteria. All process may fail if one group is inhibited (Khanal, 2008).

a. **Fermentative Bacteria (group 1):** This group of bacteria is responsible for the first stage of anaerobic processes. The anaerobic species belonging to the family of Streptococcaceae and Enterobacteriaceae and the genera of *Bacteroides, Clostridium, Butyrivibrio, Eubacterium, Bifidobacterium* and *Lactobacillus* are most commonly involved in this process (Novaes, 1986).
b. **Hydrogen-Producing Acetogenic Bacteria (group 2):** This group of bacteria metabolizes higher organic acids (propionate, butyrate, H_2, etc.), ethanol and certain aromatic compounds (i.e. benzoate) into acetate, H_2 and CO_2 (Zinder, 1998). The anaerobic oxidation of these compounds is not favorable thermodynamically by hydrogen-producing bacteria in a pure culture, however in a coculture of hydrogen-producing acetogenic bacteria and hydrogen-consuming methanogenic bacteria, these exists a symbiotic relationship between these two groups of bacteria. It is important to point out that during anaerobic treatment of complex wastewater such as vinasses or slaughterhouse, as many as 30% of the electrons is associated with propionate oxidation. Thus, these chemical appears to be more critical than oxidation of other organic acids and solvents (Deublein and Steinhaunser 2008).
c. **Homoacetogens Bacteria (group 3):** Homoacetogenesis has attracted much attention in recent years because of its final product acetate, an important precursor to methane generation. The responsible bacteria are either autotrophs or heterotrophs. The autotrophic homoacetogens utilize a mixture of hydrogen and carbon dioxide, with CO_2 serving as the carbon source for cell synthesis. The heterotrophics homoacetogens, on the other hand, use organic substrate such as formate and methanol as a carbon source while producing acetate as the end product (Eq. 1 to 4) (Khanal, 2008).

$$CO_2 + H_2 \rightarrow CH_3COOH + 2H_2O \quad (1)$$

$$4CO + 2H_2O \rightarrow CH_3COOH + 2CO_2 \quad (2)$$

$$4HCOOH \rightarrow CH_3COOH + 2CO_2 + 2H_2O \quad (3)$$

$$4CH_3OH + 2CO_2 \rightarrow 3CH_3COOH + 2CO_2 \quad (4)$$

Acetobacterium woodii and *Clostridium aceticum* are the two mesophilic homoacetogenic bacteria isolated from sewage sludge (Novaes1986). Homoacetogenic bacteria have a high thermodynamic efficiency; as result there is no accumulation of H_2 and CO_2 during growth on multicarbon compounds (Zeikus 1981).

d. **Metanogenic Bacteria (group 4 and 5):** Methanogens are obligate anaerobes and considered as a rate-limiting specie in anaerobic treatment of wastewater. Abundant methanogens are found in anaerobic environments rich in organic matter such as swamps, marches, ponds, lake and marine sediments, and rumen of cattle. Most methanogens can grow by H_2 as a source of electrons via hydrogenase as shown in the follow reaction (Eq. 5) (Khanal, 2008):

$$4H_2 + CO_2 \rightarrow CH_4 + 2H_2O \quad (5)$$

The source of H_2 is the catabolic product of other bacteria in the system, such as hydrogen-producing fermentative bacteria, especially *Clostridia* (group 1) and hydrogen-producing acetogenic bacteria (group 2). The hydrogenotrophic pathway contributes up to 28% of the methane generation in an anaerobic treatment system. It bears mentioning that there are many H_2-using methanogens that can use formate as a source of electrons for the reduction de CO_2 to methane, as show in reaction (Eq. 6):

$$4HCOO^- + 2H^+ \rightarrow CH_4 + CO_2 + 2HCO_3^- \quad (6)$$

4. Factors affecting the generation of methane

Anaerobic microorganisms, especially methanogens are highly susceptible to changes in environmental conditions. Many researchers evaluate the performance of an anaerobic system based on its methane production rate because methanogenesis is regarded as a rate-limiting step in anaerobic treatment of wastewater. Methanogens are highly vulnerable and extremely low growth rate in an anaerobic treatment system require careful maintenance and monitoring of the environmental conditions. A temperature change in the substrates or substrates concentration can lead to shutdown of gas production (Novaes, 1986).

The microbial metabolism processes are dependent on many parameters, so that for an optimum fermenting process, numerous parameters must be taken into consideration and be controlled. Some of these environmental conditions are shown in the Table 1 (Deublein and Steinhauser, 2008). A brief discussion of the factors more reported in literature is shown follows.

Operation Parameters	Inhibitors
Hydrogen partial pressure	Oxygen (O_2)
Concentration of the microorganisms	Sulfur compounds
Type of substrate	Organic acids (fatty acids and amino acids)
Specific surface of material	Nitrate (NO_3^-)
Disintegration	Ammonium (NH_4^+) and ammonia (NH_3)
Cultivation, mixing and volume load	Heavy Metals
Light and Mixing	Tannins
Temperature	Disinfectants, herbicides and insecticides
Alkalinity and pH	Degree of decomposition of organic matter
Organic Loading Rate (OLR)	Foaming
Nutrients (C/N/P-ratio)	Scum
Trace elements	
Precipitants (calcium carbonate, MAP, apatite)	
Biogas removal	

Table 1. Environmental conditions and inhibitors in the degradation methanogenic (Deublein and Steinhauser, 2008).

4.1 Temperature

It is interesting to note that anaerobic digestion in the natural environments occurred in a wide range of temperatures between 4 °C (lake sediment) to 60 °C (thermophilic digestion process); however, for the industrial practices, the temperature range is limited to 20-55 °C

(Fannin, 1987). In the natural environments, the optimum temperature for the growth of methane forming *archaea* is 5-25 °C for psychrophilic, 30-35 °C, for mesophilic, 50-60 °C, for thermophilic and >65 °C for hyprethermophilic (Tchobanoglous and Burton, 1996).

It is generally understood that higher temperature could produce higher rate of reaction and thus promoting higher application of organic loading rate (OLR) without affecting the organic removal efficiency (Chae et al., 2007; Choorit and Wisarnwan, 2007; Poh and Chong, 2009). Using palm oil mill effluent as the substrate, Choorit and Wisarnwan (2007) demonstrated that when the digester was operated at thermophilic temperature (55 °C), showed higher OLR application than the that of mesophilic (17.01 against 12.25 g COD/ m^3-d) and the methane productivity was also higher (4.66 against 3.73 L/L/d) (Choorit and Wisarnwan, 2007). A similarly study by Chae et al (2007), indicated that the higher temperature of 35 °C led to the highest methane yield as compared to 30 °C and 25 °C although the methane contents only changed slightly.

Using cheese whey, poultry waste and cattle dung as substrates, Desai et al. (1994) showed that when the temperature was increased from 20, 40 and 60 °C, the biogas production and methane percentage increased as well. The digestion rate temperature dependence can be expressed using Arrhenius expression:

$$r_t = r_{30}(1.11)^{(t-30)} \quad (7)$$

where t is temperature in °C, and r_t, r_{30} are digestion rates at temperature t and 30°C, respectively. Based in Eq. 7, the decrease in digestion rate for each 1 °C decreased in temperature below the optimum range is 11%. Similarly, the calculated rate at 25 °C y 5 °C are 59 and 7% respectively, relative to the rate at 30 °C (Dasai et al., 1994).

Although the thermophilic anaerobic process could increase the rate of reaction, the yield of methane that could be achieved over the specified organic amount is the same regardless of the mesophilic or thermophilic conditions. That value is 0.25 kg CH_4/kg COD removed or 0.35 m^3 CH_4/kg COD removed (0 °C, 1 atm) which is derived by balancing the following equation (Eq. 8), taking into account the different operating conditions worked, can be explained that the values obtained for methane production is different in many scientific reports:

$$CH_4 + 2O_2 \rightarrow CO_2 + 2H_2O \quad (8)$$

Although thermophilic condition could result in higher application of organic loading rates and better destruction of pathogens, at the same time it is more sensitive to toxicants and temperature control is more difficult (Gerardi, 2003; Choorit and Wisarnwan, 2007). Furthermore, biomass washout that could lead to volatile fatty acids accumulation and methanogenesis inhibition could also occur if the thermophilic temperature could not be controlled (Poh and Chong, 2009). As a result, in tropical regions mesophilic temperatures are the preferred choice for anaerobic treatment (Yacob et al., 2005, Sulaiman et al., 2009).

4.2 Alkalinity and pH

As far as the anaerobic digestion process is concerned, it is more appropriate to discuss alkalinity and pH together because these parameters are related to each other and very promising to ensure a suitable environment for successful methanogenesis process. Alkalinity is produced in the wastewaters as results of the hydroxides and carbonates of calcium, magnesium, sodium, potassium or ammonia and may also include borates, silicates and phosphates (Tchobanoglous and Burton, 1991). The alkalinity plays an important pH controlling role in the anaerobic treatment process by buffering the acidity derived from the acidogenesis process (Gerardi, 2003; Fannin, 1987).

Methane producing methanogens are known to be strongly affected by pH (Poh and Chong, 2009) and could only survive on a very narrow range of pH (Table 2) (Gerardi, 2003).

Genus	pH Range
Methanosphaera	6.8
Methanothermus	6.5
Methanogenium	7.0
Methanolacinia	6.6-7.2
Methanomicrobium	7.0-7.5
Methanosprillium	7.0-7.5
Methanococcoides	6.5-7.5
Methanohalobium	6.5-6.8
Methanolobus	6.5-6.8
Methanothrix	7.1-7.8
Methanosaeta	7.6

Table 2. The optimum pH range for selected methanogens (Gerardi, 2003; Steinhaus et al.2007, Tabatabaei et al., 2011)

As such, the methanogenic activity will be severely affected once the optimum pH range is not met. Steinhaus and coworker studied the optimum growth conditions of *Methanosaeta concilii* using a portable anaerobic microtank (Steinhaus et al., 2007). They reported an optimum pH level of 7.6 revealing that even little variations on both sides of the optimum pH suppressed the growth of the methanogens. Several studies have also reported reactor failure or underperformance simply due to pH reduction caused by accumulation of high volatile fatty acids in the anaerobic treatment system (Fabián and Gordon, 1999; Poh and Chong, 2009; Tabatabaei et al., 2011).

In a study using synthetic wastewater in the thermophilic temperature, was found that at the pH of above 8.0, the methanogenesis was strongly inhibited and the value recorded for acetotrophic methanogenic test was zero (Visser et al., 1993). When investigating the role of pH in anaerobic degradation test; Fabián and Gordon (1999), found out that the acidification led to the low performance of the anaerobic degradation, however the biodegradation was significantly increased once the wastewater when the pH was adjusted to above 6.5.

4.3 Organic Loading Rate (OLR)

The OLR variation can be derived from either variation in influent chemical oxygen demand (COD) or variation in flow rate with constant COD. An increase in OLR beyond the optimum level is followed by a decrease in the main process parameters such as COD removal, specific methane production. In addition, high amount of suspended solids "known as biomass wash-out" are observed in the effluent, indicating that the reactor suffered a process imbalance and that biomass accumulated in the reactor (Converti et al., 1993; Fezzani and BenCheikh, 2007; Rincón et al., 2008). This could be ascribed to an increase in the concentrations of the VFA with a consequent decrease in pH (Tiwari et al., 2006) or to escalated levels of inhibitory or toxic compounds such as phenols, lignin and others.

Therefore, there is a maximal operational value for this parameter. For instance, Rizzi and coworkers in the year of 2006 reported a decrease in COD removal and specific methane production when OLR was increased from 10 to 15 kg COD/m^3-d. With the OLR increase to 20 kg COD/m^3-d the biomass excess started to wash out, followed by deterioration of the reactor performance. In a different study, stable reactor performance was observed when the OLR increased from 1.5 to 9.2 kg COD/m^3-d with the maximum methane production rate achieved for an OLR of 9.2 kg COD/m^3-d. However, a significant decrease in the pH value (from 7.5 to 5.3) was observed when OLR was further raised to 11.0 kg COD/m^3-d. In addition, the increase in the effluent COD with increased OLR was paralleled to a sharp increase in the effluent total volatile fatty acids (TVFA, g acetic acid/L) by about 400% (Rincón et al., 2008). This indicates that, at higher OLR the effluent total COD and mainly soluble COD is largely composed of the unused volatile acids produced in the reactor due to the inhibition of methanogenesis.

Methanobacteriaceae and *Methanosaeta* were found the main methanogens in a laboratory scale up-flow anaerobic digester treating olive mill wastewater (Rizzi et al., 2006). However, the authors also reported an interesting population shift by OLR variation. At lower OLR i.e. 6 kg COD/m^3-d, hydrogenotrophic Methanobacterium predominated in the reactor but the number of cells/g sludge showed a 1000 fold decrease from 10^{11} to 10^8 when the OLR was increased to 10 kg COD/m^3-d. In contrast, phylotypes belonging to the acetoclastic Methanosaeta were not affected by OLR variation and at 10 kg COD/m^3-d, dominated in the biofilm (10^9 cells/g sludge) (Rizzi et al., 2006).

Olive oil wastewater is characterized by high levels of inhibitory compounds such as tannins, and lipids. As a result, increased OLR leads to higher concentration of these substances and a consequent inhibition of methanogenic cells. However, acetoclastic *Methanosaeta* due to its high affinity for acetate is capable of occupying the deepest and thus more protected niches in the granule or biofilm with low concentrations of substrate (acetate) (Gonzales-Gil et al., 2001). Phylotypes belonging to the genus *Methanosaeta* were also dominant independent of different OLR in other anaerobic digesters (Rincón et al., 2008).

In a different study was investigated the microbial ecology of granules in UASB reactor fed by synthetic wastewater under various OLR. The authors showed that the predominant microbial biomass was *Methanosaeta*. However, increasing the OLR led to a substantial increase of *Methanosarcina* in the granules (Kalyuzhnyi et al., 1996). The increase of *Methanosarcina* in the studied synthetic wastewater (toxin-free) due to increasing OLR is explained by the low affinity of these methanogens for acetate in comparison with *Methanosaeta*. Hence, by increasing OLR and consequent VFA concentration, *Methanosarcina* is favored.

As reviewed earlier, under mesophlic conditions *Methanosaeta* plays a significant role in making cores of sludge granules (Sekiguchi et al., 2001) and thus their ratio seems to control the speed of granulation (Rincón et al., 2008). Higher OLR, result in consequent higher concentration of substrates (i.e. acetate) in the reactor. Morvai and coworkers in 1990 investigate the influence of organic load ranging from 0.5-3.0 g/L on granular sludge development in an acetate-fed system. They argued that in the range of feed acetate levels examined, higher concentrations of acetate caused faster granulation of the sludge bed and, presumably of the microbial population, and resulted in better sludge structure and improved sludge settleability.

Low OLR has been reported to cause acute mass transfer limitation leading to disintegration of the larger granules (Ahn et al., 2002). The disintegration begins at the core of the granules due to substrate limitation with a consequent loss of granules strength and stability. However, this was not in agreement with the studies reported, which low OLR (<1.5 kg COD/m^3-d) did not lead to disintegration of the granules in UASB reactors (Tiwari et al., 2005). This could be ascribed to the different experimental settings and wastewaters used in these studies. Teo and coworker (2000), treat a high iron bearing wastewater in a UASB reactor. Evidence shows that the presence of divalent and trivalent cations ions, such as Fe^{2+} and Fe^{3+}, helps bind negatively charged cells together to form microbial nuclei that promote further granulation.

Tiwari et al. (2006) tried to enhance the granulation process by using natural ionic polymer additives. These may thus reduce the effect of low OLR (i.e. substrate limitation) on the granules and delayed the disintegration. Meanwhile was reported that COD removal rate, the COD specific removal rate (r_s) and methane production rate were not suppressed by increasing OLR when treating wine wastewater and sewage mixture (Converti et al., 1990). That indicated that no inhibition factor related to the organic content of the effluent was present in both wine wastewater and sewage mixture studied.

This was further supported by the cell mass concentration varied very little with increasing the OLR. However as completely noticed by the authors, even at the absence of inhibitory compounds in the initial part, the removal rate increased with the OLR, following a first order kinetic. In the second part, instead the removal rate tended to a constant maximum value, following a zero order kinetic. Afterwards, the removal efficiencies as well as the methane production yield gradually decreased with increasing influent COD due to increasing the OLR, which evidently showed a substrate inhibition occurrence (Converti et al., 1990).

This supports the idea that even at the absence of the inhibitory compounds in the wastewater, increasing influent COD by the means of increasing OLR could lead to substrate inhibition and consequent reduced removal efficiencies. In other study is described the dependence of the removal rate on the OLR by an empirical equation similar to Monod's model (Eq. 9) to compare the degradability of different effluents (Converti et al., 1990):

$$r_s = \frac{r_{s(max)} OLR}{(k + OLR)} \quad (9)$$

where $r_{s(max)}$ (kg COD/kg of vss d) is the maximum value of r_s, and k is a constant which physically is expressed in units of OLR, an increase of k indicates increased treatment ability of the studied effluent. The desired OLR is the function of the favorable effect of OLR on stimulating the growth of methanogens in the bioreactor by providing them with higher substrate concentrations, its reverse effect on elevating the concentration of inhibitory compounds and the buffering capacity of methanogenic community. In the other words, the maximal operational value of OLR is translated into the highest methane production (indicating the highest conversion efficiency of the system) that the buffering capacity of methanogenic community is still capable of compensating for elevated concentrations of inhibitory compounds (Tabatabaei et al., 2011).

4.4 Mixing

There are only a limited number of studies found specifically focused on the effects of mixing on the treatment efficiency and biogas production using various types of agro-industrial wastewater including palm oil mill effluent, wash water of animal waste, lixiviate of municipal waste and fruit and vegetable wastes (Kaparaju et al., 2007; Sulaiman et al., 2009). Adequate mixing is very important in order to achieve successful anaerobic treatment of organic rich wastewater. In another word, it enhances the anaerobic process rate by preventing stratification of substrate, preventing the formation of surface crust, ensuring the remaining of solid particles in suspension, transferring heat throughout the digester, reducing particle size during the digestion process and releasing the biogas from the digester content (Kaparaju et al., 2007; Sulaiman et al., 2009).

Prior to 1950s, anaerobic digesters treating sewage sludge were not equipped with mechanical mixing and thus caused the formation of scum layer at the surface (Fannin, 1987). To overcome this problem, mixing was employed to disrupt scum formation and enhance contact between microorganisms and substrates. It has been reported that the acetate-forming bacteria and methane-forming bacteria are required to be in close contact to achieve continuous degradation of organic materials (Tabatabaei et al., 2011). In addition to the mentioned advantages, mixing also helps to eliminate thermal stratification inside the digesters, maintain digester sludge chemical and physical uniformity, rapid dispersion of metabolic products and toxic materials and prevent deposition of grit (Gerardi, 2003).

4.5 Heavy metals inhibition

Heavy metals are present in various types of wastewater, including agro-industrial wastewater, landfill leachate and cane vinasses (Del Real et al., 2009; Yusof et al., 2009). Although many metals are required in trace amounts to provide sufficient growth to methanogens, the methanogenic activity in anaerobic reactors is strongly affected by excess amounts of heavy metals (Colussi et al., 2009). The toxic effects of metals in biological process is particularly due to the inhibition of enzymes activity as a result of metals binding to the SH group of the enzyme. The inhibitory concentrations of four heavy metals on methane-producing granular sludge that caused 50% reduction in cumulative methane production was found to be 7.5 mg/L of Zn, 27 mg/L of Cr, 35 mg/L of Ni and 36 mg/L of Cd with an order of Zn>Cr>Ni≈Cd (Altas, 2009). Whereas a different study revealed that 50% reduction in methane production occurred at 6.4 mg/L of Cu (II), 4.4 mg/L of Cd(II) and 18.0 mg/L of Cr(VI) with an order of Cd(II)>Cu(II)>Cr(VI) in anaerobic digestion of cattail with rumen culture (Yue et al., 2007).

Yue and coworker in 2007, indicated that metals cause anaerobic system failures when they are in the form of free ions (in its soluble form) and above certain concentrations (Table 3). The differences reported in the metals inhibitory concentration might be due to the several factors including variation in sludge characteristics, chemical form of heavy metals and microbial resistance to metals (Altas, 2009). Various heavy metals presence in wastewater also showed synergistic effects during anaerobic treatment process. For instance, the presence of chromium in the sludge results in higher toxicity of copper (Colussi et al., 2009).

Altas in 2009 showed that low concentrations of metals in the anaerobic reactor can be extremely toxic. Meanwhile, Cantrell and coworkers (2008) indicated that high concentrations of soluble metals have come to completely stop the production of biogas in an anaerobic system. To combat metal toxicity in the anaerobic degradation process, they can be precipitated as sulphate salts and carbonate salts, except iron and chromium.

Substance	**Total Concentration (mg/L)**	**Soluble Concentration (mg/L)**
Cu	200	0.5
Cr (VI)	50 – 70	3
Cr (III)	200 – 260	-
Ni	180 - 420	2
Zn	30	1

Table 3. Concentrations of inorganic compounds inhibitory of anaerobic process (Yue *et al.*, 2007).

5. Proprieties and volumetric compensation of the gases from anaerobic process

The physical and chemical proprieties of biogas, affect the choice of technology used for clean-up and combustion; therefore, knowledge of these proprieties is useful for optimization biogas utilization. Since biogas contains primarily methane and carbon dioxide, this section is focused on their respective physical characteristics (Table 4). Because others components (nitrogen, hydrogen sulfide, traces organics), are present in relatively small quantities are not considered in the table. The magnitude of CH_4 and CO_2 varies greatly and depends on the composition of the organic material digested in the wastewater.

Proprieties	**Methane (CH_4)**	**Carbon dioxide (CO_2)^a**
Molecular weight (g/mol)	16.04	44.01
Specific gravity, (air =1^c)	0.554	1.52
Boiling point, (14.7 psia)	126.0 °C	43.0 °C ^{b}
Freezing point, (14.7 psia)	-182.4 °C	- 56.55 °C
Specific volume (m^3/kg)	1.51	7.03x10^{-3}
Critical temperature (°C)	46.62	31.08
Critical pressure (kPa)	4,640.1	7391.15
Heat capacity (kJ/kg K)	2.26	0.858
Ratio Cp/Cv	1012	
Heat of combustion (kJ/m^3)	377	
Limit of inflammability	5-15% by volume	
Stoichiometry in air ^{c}	0.0947 by volume 0.0581 by mass	

a: pure gas given at 77 °F and 101.3 kPa; b: sublimes; c: Air at 101.3 kPa and 15.54 °C

Table 4. Physical constants of methane and carbon dioxide

The volumetric measurement of biogas must compensate the pressure and temperature differences. The equation 10, illustrate a simple method of gas volume compensation for a saturated gas taking into account the adjustment by pressure and temperature (Salisbury, 1950):

$$V_s = V * 17.626 * \frac{(H-A)}{(459.6+T)} \quad (10)$$

where V is the observed volume, V_s volume at standard conditions (60°F and 30 inches Hg), H is absolute gas pressure (inches Hg), A water vapor pressure (inches Hg), and T temperature of gas (°F). Pure methane at standard temperature and pressure has a lower heating value of 912 BTU/ft^3 (34 kJ/L). Typical biogas of 65% methane has a heating value of approximately 600 BTU/ft^3 (22.36 kJ/L), since only the methane portion will burn, approximate equivalents of biogas to others fuels are presented in the table 5.

Biogas with 65% of methane (1000 L)
600 L of natural gas
25.0 L of propane
22.3 L of Butane
17.79 L of gasoline
16.28 L of diesel

Table 5. Equivalents of biogas others fuels (Palmer, 1981)

6. Reactor types

Many reactor configurations are used for the anaerobic treatment of agro-industrial wastewaters. Among them, the most common types are discussed and illustrated in Fig. 2.

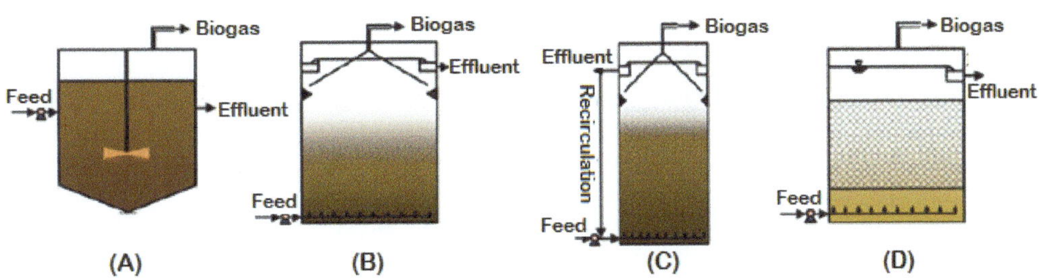

Fig. 2. Most commonly used anaerobic reactors types: (A) Completely mixed anaerobic digester, (B) UASB reactor, (C) AFB reactor, (D) Upflow AF reactor (Ersahin et al., 2011)

6.1 Completely stirred anaerobic digester

The completely stirred anaerobic digester (CSTR) is the basic anaerobic treatment system with an equal hydraulic retention time (HRT) and solids retention time (SRT) in the range of 15-40 days in order to provide sufficient retention time for both operation and process stability. Completely mixed anaerobic digesters without recycle are more suitable for wastes with high solids concentrations (Tchobanoglous et al., 2003). A disadvantage of this system is that a high volumetric loading rate is only obtained with quite concentrated waste streams with a biodegradable COD content between 8,000 and 50,000 mg/L. However, many waste streams are much dilute (Rittmann and McCarty, 2001). Thus, COD loading per unit volume may be very low with the detention times of this system which eliminates the cost advantage of anaerobic treatment technology. Typical the OLR for this digester is between 1-5 kg COD/m^3-d (Tchobanoglous et al., 2003).

6.2 Upflow anaerobic sludge blanket reactor

One of the most notable developments in anaerobic treatment process technology is the upflow anaerobic sludge blanket (UASB) reactor invented by Lettinga and coworkers (Lettinga et al., 1980) with its wide applications in relatively dilute municipal wastewater treatment and over 500 installations in a wide range of industrial wastewater treatment including food-processing, paper and agro-industrial process (Tchobanoglous et al., 2003).

Influent flow distributed at the bottom of the UASB reactor travels in an upflow mode through the sludge blanket and passes out around the edges of a funnel which provides a greater area for the effluent with the reduction in the upflow velocity, enhancement in the solids retention in the reactor and efficiency in the solids separation from the outward flowing wastewater. Granules which naturally form after several weeks of the reactor operation consist primarily of a dense mixed population of bacteria that is responsible for the overall methane fermentation of substrates (Rittmann and McCarty, 2001). Good settleability, low retention times, elimination of the packing material cost, high biomass concentrations (30,000-80,000 mg/L), excellent solids/liquid separation and operation at very high loading rates can be achieved by UASB systems (Speece, 1996). The only limitation of this process is related to the wastewaters having high solid content which prevents the dense granular sludge development (Tchobanoglous et al., 2003). Designed for OLR is typically in the range of 4 to 15 kg COD/m^3-d (Rittmann and McCarty, 2001).

6.3 Fluidized and expanded bed reactors

The anaerobic fluidized bed (AFB) reactor comprises small media, such as sand or granular activated carbon, to which bacteria attach. Good mass transfer resulting from the high flow rate around the particles, less clogging and short-circuiting due to the large pore spaces formed through bed expansion and high specific surface area of the carriers due to their small size make fluidized bed reactors highly efficient. However, difficulty in developing strongly attached biofilm containing the correct blend of methanogens, detachment risks of microorganisms, negative effects of the dilution near the inlet as a result of high recycle rate and high energy costs due to the high recycle rate are the main drawbacks of this system. The expanded granular sludge bed (EGSB) reactor is a modification of the AFB reactor with a difference in the fluid's upward flow velocity. The upflow velocity is not as high as in the fluidized bed which results in partial bed fluidization. (Rittmann and McCarty, 2001). OLR of 10-50 kg kg COD/m^3-d can be applied in AFB reactors (Ozturk, 2007; Ersahin et al., 2011).

6.4 Anaerobic filters

The anaerobic filter (AF) has been widely applied in the beverage, food-processing, pharmaceutical and chemical industries due to its high capability of biosolids retention. In fact clogging by biosolids, influent suspended solids, and precipitated minerals is the main problem for this system. Applications of both upflow and downflow packed bed processes can be observed. Prevention of methanogens found at the lower levels of the reactor from the toxicity of hydrogen sulfide by stripping sulfide in the upper part of the column and solids removal from the top by gas recirculation can easily be achieved in downflow systems in comparison to upflow systems. However, there is a higher risk of losing biosolids to the effluent in the downflow systems. Design OLR is often in the range of 8-16 kg COD/m^3-d which is more than tenfold higher than the design loading rates for aerobic processes (Rittmann and McCarty, 2001).

7. Bioenergy production from different kinds of wastewaters

Methanogenic anaerobic digestion is a classical anaerobic bioconversion process that has been practiced for over a century and used in full-scale facilities worldwide. This is a complicated process that involves a mixture of population of microorganisms and several gasses and liquid products, thus strict process control and product purification are required. Biogas production have been demonstrated in numerous studies with great success like can see in the Table 6 (Gavrilescu, 2005).

Wastewater	Reactor Type	HRT (days)	OLR (kg COD/m^3-d)	COD removal (%)	CH_4 Yield (m^3/kg COD)	Reference
Brewage distillery	UASB	--	16.5-44.0	80	16.5	Shin et al., (1992)
Cane-molasses stillage	AFB	5.6-32	4.65-20	85	0.168	Yeoh (1997)
Cheese whey and dairy	Hybrid reactor	--	10	98	--	Malaspina et al (1996)
Cheese whey and dairy	Hybrid reactor	--	0.97-2.82	91-97	0.28-0.35	Strydom et al (1997)
Cheese whey and dairy	CSTR	2.0	5	90	--	Ince (1998)
Cheese whey and dairy	CSTR	4-7	--	--	0.55	Yilmazer y Yeningüm (1999)
Landfill leachate	AFB	4.7-16	2.41-7.98	>90	---	Lin (1990)

Table 6. Typical performance of anaerobic reactor used for wastewater treatment (Gavrilescu, 2005)

8. Biogas production from agro-industrial wastewaters

8.1 Case vinasses of tequila

Tequila is a Mexican regional alcoholic beverage obtained from the fermentation of sugars from the cooked stems of blue agave (Agave tequilana Weber var. azul). Its production and commercialization is verified and certified by the Mexican Tequila Regulatory Council (CRT) (NOM-006-SCFI-2005, 2006). In 2008 the CRT registered 139 producers and 1,018 brands of Tequila (bottled in Mexico and in foreign countries, CRT 2008). Based on the number of employees, only 7% are large factories and the rest are small and medium factories, with a grand total of around 30,500 direct employees (National Tequila Industry Chamber, CNIT 2009). Therefore, this industry represents an important economic activity for the 180 Mexican municipalities within the *appellation d'origine contrôlée* granted in 1995 for Tequila.

Tequila production has had an important increase from 2004 to 2008, as it is shown in Fig. 3. In 2010 about 187.3 million liters of Tequila (55% Alc. Vol.) has been produced with a projection for annual growth of at least 10% (CNIT 2010); there is also a decrease in production of Tequila between 2000 and 2003, due to the agave crisis (Dalton 2005). Although exhaustive reviews regarding the treatment of different distillery wastewaters are published elsewhere (Satyawali and Balakrishnan 2008; Mohana et al. 2009), it is considered that special attention should be paid to distillery effluents from the Tequila industry due to their complex composition. This section present the potential generation of energy from wastewater treatments to generate biogas from the Tequila industry.

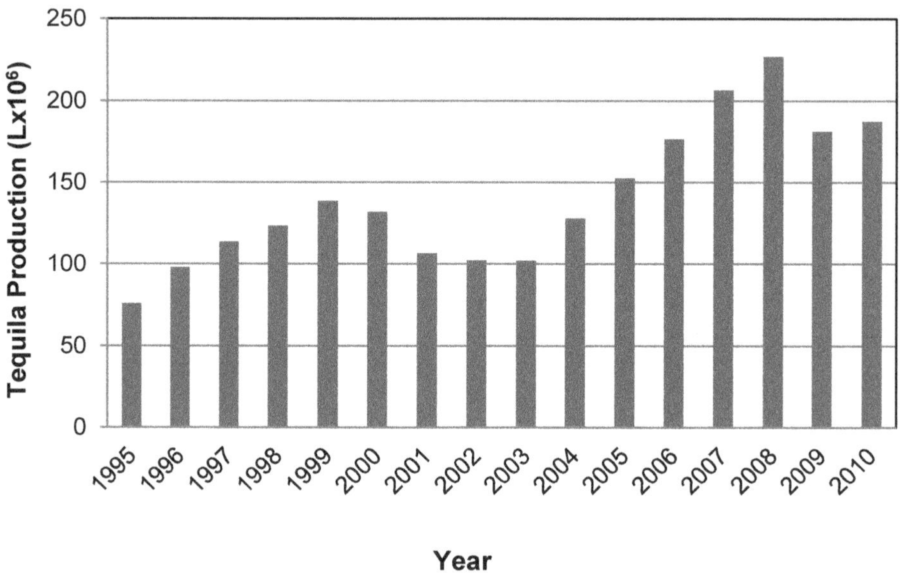

Fig. 3. Dynamics of Tequila production (55% Alc. Vol.). (calculated from CNIT 2010)

The production of Tequila generates large quantities of bagasse and vinasses. Bagasse is a residual solid; it is generated in the elaboration of Tequila and is produced during the extraction of juice from the cooked heads of agave. Vinasses are the liquid residues that are generated and remain in the bottom of the still after the distillation of the must of fermented agave.

Vinasses are dark brown in color, because they contain phenolics (tannic and humic acids), melanoidins that are low and high molecular weight polymers formed as one of the final products of Maillard reaction (Satyawali and Balakrishnan 2008). It is known that for each liter of Tequila produced, 1.4 kg of bagasse and 10–12 L of vinasses are generated. Under this basis of calculation, it is estimated that the production of Tequila in 2010 generated 262.2 million kilograms of bagasse and 1,873.0 million liters of vinasses.

In the majority of the Tequila factories, bagasse is converted into compost, which is also done in the agave plantations. However, approximately 80% of the vinasses are discharged directly into water bodies (rivers, streams, lakes, reservoirs) and municipal sewer systems or directly onto the soil without receiving adequate treatment for discharge. This common practice causes a deterioration of different degrees to the water bodies receiving the discharges due to low pH, high temperature and elevated concentrations of both BOD and COD of these effluents. On the contrary, if the vinasses receive appropriate treatment and management, they can be used as a source of nutrients and organic matter in agricultural activities; they can also be a potential source of renewable energy. A summary of the physicochemical characteristics of the vinasses generated from the process of producing traditional Tequila (100% agave) is shown in Table 7 (Lopez-Lopez, 2010).

Parameter	Value
pH	3.4-4.5
Oils and fats (mg/L)	10-100
Total COD (mg/L)	60,000-100,000
Soluble COD (mg/L)	40,000-80,000
Total BOD (mg/L)	35,000-60,000
Soluble BOD (mg/L)	25,000-50,000
Total solids (mg/L)	25,000-50,000
Total suspended solids (mg/L)	2,000-8,000
Fixed suspended solids (mg/L)	10-500
Volatile suspended solids (mg/L)	1,990-7,500
Total dissolved solids (mg/L)	23,000-42,000
Settleable solids (mL/L)	10-900
Total alkalinity (mg/L)	< 6.00
Total acidity (mg/L)	1,500-6,000
Fixed acidity (mg/L)	1,480-5,800
Volatile acidity (mg/L)	20-200
Ca (mg/L)	200-1,100
Mg (mg/L)	100-300
K (mg/L)	150-650
Phosphates (mg/L)	100-700
Total nitrogen (mg/L)	20-50
NH_4^+-nitrogen (mg/L)	15-40
Organic nitrogen (mg/L)	5.0-10
Total reducing sugars (% w)	0.5-2.0
Direct sugars (% w)	0.4-1.0
Cu (mg/L)	< 3.0
Fe (mg/L)	< 45
Ni (mg/L)	< 0.02
Zn (mg/L)	< 1.0

Table 7. Physicochemical characteristics of Tequila vinasses (Lopez-Lopez 2010)

The anaerobic biological process has been utilized for treating Tequila vinasses on laboratory, pilot and industrial scales due to technical and economical advantages over aerobic processes (Linerio and Guzman 2004; Mendez, et al. 2009). On a laboratory scale, Lopez-Lopez and coworkers (2011), Mendez and coworkers (2009) showed an anaerobic digester capable of removing 90–95% of organic material as COD; generating significant amounts of biogas rich in methane. The most common system found at an industrial level in treating Tequila vinasses is of anaerobic type. Fig.4 shows the amount of energy that can be generated if the entire volume of vinasses is treated.

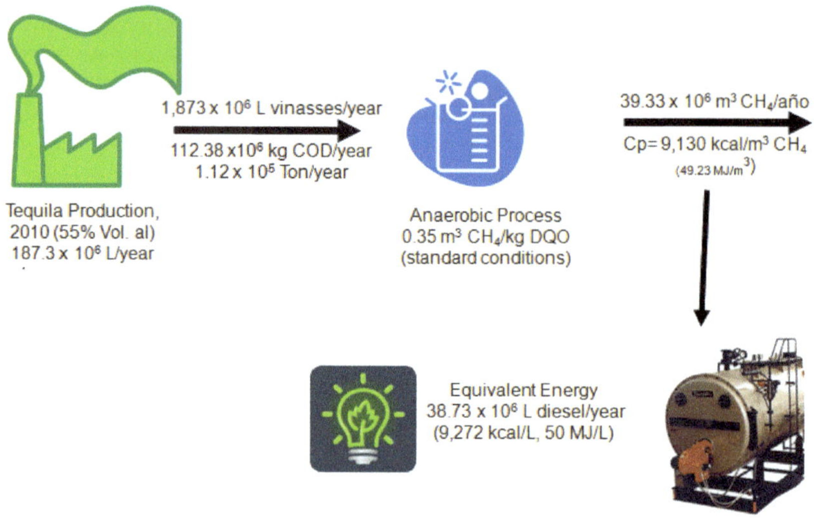

Fig. 4. Production of biogas from Tequila vinasses as a source of energy

8.2 Others cases

In general, all types of wastewater can be used as substrates as long as they contain carbohydrates, proteins, fats, cellulose and hemicelluloses as main components. It is important that the following points are taken into consideration when selecting the wastewater industrial.

The content of organic substance should be appropriate for the microorganisms selected in anaerobic process.

- The high nutritional value of the organic substance for the microorganism, hence the potential for gas formation and should be as high as possible.
- The substrate should be free of pathogens and others organism which would need to be made innocuous prior to the anaerobic process.
- The content of harmful substances and trash should be low to allow the fermentation process to take place smoothly.
- The composition of the fermentation residue should be such that it can be used, e.g. as fertilizer.

In this section some of agro-industrial wastewater employed like organic substrates is shows, because the degree to which the organic substances in the wastewater is decomposed in the bioreactor depends on the origin of the liquid. In the Table 8, is shows the methane production rate from wastewater of different types.

Wastewater type	Reactor Type	HRT (days)	OLR (kg COD/m^3-d)	Temperature (°C)	COD removed (%)	MPR (m^{3}CH$_4$/kg COD)	Ref.
Slaughter-house	Anaerobic filter	0.6-3.0	3.7 -16.5	25	50-81	0.41	[1]*
Slaughter-house	CSTR	20-30	0.2-0.3	37	70-80	0.45	[2]*
Tequila vinasses	UASB	2.0-2.5	2.0-12.0	37	50-85	0.46	[3]*
Cane vinasses	CSTR	20-30	2.5-12.7	35	50-75	0.42	[4]*
Pulping coffee	CSTR	20-30	0.2-0.4	35	60-75	0.37	[5]*

Table 8. Methane production rate from wastewater of different type

In all previous cases, the wastewaters are discharged directly into the body of water, causing several environmental pollution in addition to the loss of the energetic potential contained in the effluents.

9. References

Ahn, Y.H., Song, Y.J., Lee, Y.J. and Park, S. (2002). Physicochemical characterization of UASB sludge with different size distributions. *Environ. Technol.* 23, 889–897.

Altas, L. (2009). Inhibitory effect of heavy metals on methane-producing anaerobic granular sludge. *J. Hazard. Mater.* 162, 1551-1556.

Cantrell, K.B., Ducey, T., Ro, K.S. and Hunt, P.G. (2008). Livestock waste to bioenergy generation opportunities. *Biores.Technol.* 99:7941–7953.

Chae, K.J., Jang, A., Yim, S.K. and Kim, I.S. (2007). The effect of digestion temperature and temperature shock on the biogas yields from the mesophilic anaerobic digestion of swine manure. *Bioresour. Technol.* 99(I), 1–6.

Choorit, W. and Wisarnwan, P. (2007). Effect of temperature on the anaerobic digestion of palm oil mill effluent. *Electron. J . Biotechnol.* 10(3), 376–385.

* [1] López-López, A., Vallejo-Rodríguez, R. and Méndez-Romero, D. C.(2010) Evaluation of a combined anaerobic and aerobic system for the treatment of slaughterhouse wastewater, Environmental Technology, 31: 3, 319 — 326.

[2] Padilla-Gasca E, López-López A, Gallardo-Valdez J (2011) Evaluation of Stability Factors in the Anaerobic Treatment of slaughterhouse Wastewater. J Bioremed Biodegrad 2:114. doi:10.4172/2155-6199.1000114.

[3] López-López A (2011) Evaluación de alternativas de tratamiento para vinazas tequileras, internal project CIATEJ-IVINAZAS2/09/PT. (unpublished data)

[4] Del Real J., Prieto F., Santos E.M., Román A.D., and Gordillo A.J. (2007). Factibility of treat wineyard wastewater through anaerobic biodigestion with from cow's waste. Rev. Téc. Ing. Univ. Zulia. Vol. 30 (2), 128 - 135.

[5] Del Real Olvera J. and Islas Gutiérrez J. (2010). Anaerobic biodegradation of wastewater produced in coffee pulping. Revista Colombiana de Biotecnología. Vol. 12 (2), 230-239.

CNIT. (2010). Informe estadístico 2010. www.tequileros.org.mx Accessed: December 2010

Colussi, I., Cortesi, A., L., Vedova, D.V., Gallo, V. and Robles, F.K.C. (2009). Start-up procedures and analysis of heavy metals inhibition on methanogenic activity in EGSB reactor. *Bioresour. Technol.* 100, 6290-6294.

Converti, A., del Borghi, M. and Ferraiolo, G. (1993). Influence of organic loading rate on the anaerobic treatment of high strength semisynthetic wastewaters in a biological fluidized bed. *Chem. Eng. J.* 52(1), 21-28.

Dalton, R. (2005). Alcohol and science: saving the agave. *Nature.* 438(7071):1070-1071.

Del Real J., Prieto F., Santos E.M., Román A.D. and Gordillo A.J. (2009). Simulación matemática de un digestor anaerobio tipo tanque agitado para el tratamiento de vertidos residuales. *Rev. Int. Contam. Ambient.* 25 (1), 33-41.

Desai, M., Patel, V. and Madamwar, D. (1994). Effect of temperature and retention time on biomethanation of cheese whey-poultry waste-cattle dung. *Environ. Pollut.* 83, 311-315.

Deublein, D. and Steinhauser, A. (2008). Biogas from Waste and Renewable Resources. An Introduction. Ed Wiley-VCH. Germany.

Ersahin, M.E., Ozgun, H., Dereli, R.K. and Ozturk, I. (2011). Anaerobic Treatment of Industrial Effluents: An Overview of Applications. In Waste water- treatment and reutilization. Ed. Fernando Sebastián García, In Tech.

Fabián, R.M. and Gourdon, R. (1999). Effect of baling on the behavior of domestic wastes: laboratory study on the role of pH in biodegradation. *Bioresour. Technol.* 69, 15-22.

Fannin, K.F.(1987). Start-up, Operation, Stability and Control. In Anaerobic Digestion of Biomass, Eds., Chynoweth, D.P. and R. Isaacson., Elsevier Applied Science Publishers Ltd, 171-196.

Fezzani, B. and BenCheikh, R. (2007). Anaerobic co-digestion of olive mill waste water with olive mill solid waste in a tubular digester at a mesophilic temperature. *Bioresour. Technol.* 98, 769-774.

Gavrilescua, M. and Chisti, Y. (2005). Biotechnology—a sustainable alternative for chemical industry. *Biotechno.l Adv.* 23:471-499.

Gerardi, M.H. (2003). The microbiology of anaerobic digesters. John Wiley and Sons Inc., New Jersey, pp. 91-118.

Gonzales-Gil, G., Lens, P.N.L., Van Aelst, A., Van As, H., Versprille, A.I. and Lettinga, G. (2001). Cluster structure of anaerobic aggregates of an expanded granular sludge bed reactor. *Appl. Environ. Microbiol.* 67, 3683-3692.

Hallenbeck, P.C. and Ghosh, D. Advances in fermentative biohydrogen production: the way forward? *Trends. Biotechnol.* 27:287-297.

Horiuchi, J.I., Shimizu, T., Tada, K., Kanno, T., and Kobayashi, M. (2002). Selective production of organic acids in anaerobic acid reactor by pH control. *Bioresour. Technol.* 82: 209-213.

Kalyuzhnyi, S.V., Sklyar, V.I., Davlyatshina, M.A., Parshina, S.N., Simankova, M.V., Kostrikina, N.A. and Nozhevnikova, A.N. (1996). Organic removal and microbiological features of UASB-reactor under various organic loading rates. *Bioresour. Technol.* 55,47-54.

Kaparaju, P., Buendia, I., Ellegaard, L. and Angelidakia I. (2007). Effects of mixing on methane production during thermophilic anaerobic digestion of manure: Lab scale and pilot scale studies. *Bioresour. Technol.* 99, 4919-4928.

Khanal, S. (2008). Anaerobic Biotechnology for Bioenergy Production. Principles and Applications. Ed. Wiley-Blackwell. New Delhi, India.

Kleerebezem, R. and Loosdrecht, M.C.V. (2007). Mixed culture biotechnology for bioenergy production. *Curr. Opin. Biotechnol.* 8:207-212.

Lee, H.S., Vermaas, W.F.J. and Rittmann, B.E. (2010). Biological hydrogen production: prospectives and challenges. *Trends. Biotechnol.* 28:262–271.

Lettinga, G., van Velsen, A. F. M., Hobma, S. W., de Zeeuw, W. and Klapwijk, A. (1980). Use of the upflow sludge blanket (USB) reactor concept for biological wastewater treatment, especially for anaerobic treatment. Biotechnology and Bioengineering, 22 (4), 699-734.

Linerio-Gil, J.E. and Guzmán-Carrillo, A. (2004). Tratamiento de efluentes y aprovechamiento de residuos. In: Gschaedler-Mathis, AC. (ed) Ciencia y tecnología del tequila; Avances y perspectivas. Centro de Investigación y Asistencia en Tecnología y Diseño del Estado de Jalisco (CIATEJ). Guadalajara, México, pp 258–284.

Lopez-Lopez, A. (2009). Evaluación de alternativas de tratamiento para vinazas tequileras. Internal project CIATEJ-PIVINAZAS/08/PT. (unpublished data).

López-López A., Dávila-Vazquez G., León-Becerril E., Villegas-García E., Gallardo-Valdez J. Tequila vinasses: generation and full scale Treatment processes. (2010). Rev. Environ. Sci Biotechnol.9 (1): 109-116

Mackie, R.L. and Bryant M.P. (1981). Metabolic activity of fatty acid oxidizing bacteria and the contribution of acetate, propionate, butyrate and CO_2 to methanogenesis in cattle waste at 40-60 °C. *Appl. Environ. Microbiol.* 40: 1363-1373.

Mccarty, P.L. (1964). Anaerobic waste treatment fundamentals. Public Works 95: 91-94.

Méndez, H.G. Snell, R. Alcaraz, V., González, V. and Pelayo, C. (2009). Anaerobic treatment of Tequila vinasses in a CSTR-type digester. Biodegradation. Published online, doi:10.1007/s10532- 10009-19306-10537.

Mohana, S., Acharya, B.K. and Madamwar, D. (2009). Distilley spent wash: treatment technologies and potential applications. *J. Hazard. Mater.* 163(1):12–25.

Morvai, L., Miháltz, P., Czakó, L., P'terfy, M. and Holló, J. (1990). The influence of organic load on granular sludge development in an acetate-fed system. *Appl. Microbiol. Biotechnol.* 33, 463-468.

Mu, Y., Wang, G. and Yu, H.Q. (2006). Biohydrogen production by mixed culture at various temperatures. *Int. J. Hydrogen Energy.* 31:780–785.

Novaes R.F. (1986). Microbiology of anaerobic digestion. Water Sci. Technol. 18 (12), 1-14.

Ozturk, I. (2007). Anaerobic Treatment and Applications, Water Foundation Press, ISBN: 978- 975-6455-30-2, Istanbul, Turkey (in Turkish).

Poh, P.E. and Chong, M.F. (2009). Development of anaerobic digestion methods for palm oil mill effluent (POME) treatment. *Bioresour. Technol.* 100, 1–9.

Rincón, B., Borja, R., González, J.M., Portillo, M.C. and Saiz-Jiménez, C.S. (2008). Influence of organic loading rate and hydraulic retention time on the performance, stability and microbial communities of one-stage anaerobic digestion of two-phase olive mill solid residue. *Biochem. Eng. J.* 40, 253–261.

Rittmann, B. E. and McCarty, P. L. (2001). Environmental Biotechnology: Principles and Applications. McGraw-Hill, ISBN: 0072345535, New York, United States of America.

Rizzi, A., Zucchi, M., Borin, S., Marzorati, M., Sorlini, C. and Daffonchio, D. (2006). Response of methanogen populations to organic load increase during anaerobic digestion of olive mill wastewater. *J. Chem. Technol. Biot.* 81(9), 1556-1562.

Sachdeva V, Tyagi RD, Valero JR. Production of biopesticides as a novel method of wastewater sludge utilization/disposal. Water Sci Technol 2000;42:211–6.

Satyawali, Y. and Balakrishnan, M. (2008). Wastewater treatment in molasses-based alcohol distilleries for COD and color removal: a review. *J. Environ. Manage.* 86(3):481–497.

Sekiguchi, Y., Takahashi, H., Kamagata, Y., Ohashi, A. and Harada, H. (2001). In situ detection, isolation, and physiological properties of a thin filamentous microorganism abundant in methanogenic granular sludges: a novel isolate affiliated with a clone cluster, the green non-sulfur bacteria, subdivision I. Appl. Environ. Microbiol. 67, 5740-5749.

Shin SG, Han G, Lim J, Lee C, Hwang S. A comprehensive microbial insight into twostage anaerobic digestion of food waste-recycling wastewater. Water Res 2010;44:4838–49.

Speece, R. E. (1996). Anaerobic Biotechnology for Industrial Wastewaters, Archae Press, ISBN: 0- 9650226-0-9, USA.

Steinhaus, B., Garcia, M.L., Shen, A.Q. and Angenent L.T. (2007). A Portable Anaerobic Microtank Reveals Optimum Growth Conditions for the Methanogen *Methanosaeta concilii. Appl. Environ. Microbiol.* 73, 1653–1658.

Sterling JMC, Lacey RE, Engler CR, Ricke SC. Effects of ammonia nitrogen on hydrogen and methane production during anaerobic digestion of dairy cattle manure. Bioresour Technol 2001;77:9-18.

Sulaiman, A., Tabatabaei, M., Hassan, M.A. and Shirai, Y. (2009). The effect of higher sludge

Tabatabaei, M., Sulaiman, A., Nikbakht A.M., Yusof, N. and Najafpour G. (2011). Influential Parameters on Biomethane Generation in Anaerobic Wastewater Treatment Plants. In Alternative Fuel, Ed. Maximino Manzanera, In tech.

Tchobanoglous, G. and Burton, F.L., (1996). Wastewater engineering treatment disposal and reuse. McGraw-Hill International Edition, Singapore, pp. 84–827.

Tchobanoglous, G., Burton, F. L. and Stensel, H. D. (2003). Wastewater Engineering Treatment and Reuse, Metcalf and Eddy, Inc., 4th ed. Revised, Mc-Graw-Hill, ISBN: 0-07- 041878-0, New York, USA.

Teo, K.C., Xu, H.L. and Tay, J.H. (2000). Molecular mechanism of granulation: II. Proton translocating activity. *J. Environ. Eng.* 126, 411–418.

Tiwari, M.K., Guha, S., Harendranath, C.S. and Tripathi, S. (2006). Influence of extrinsic factors on granulation in UASB reactor. *Appl. Microbiol. Biotechnol.* 71(2), 145–154.

Visser, A., Gao, Y. and Letingga, G. (1993). Effects of pH on methanogenesis and sulphate reduction in thermophilic (55 °C) UASB reactors. *Bioresour. Technol.* 44, 113–121.

Yacob, S., Shirai, Y., Hassan, M.A., Wakisaka, M. and Subash, S. (2005). Start-up operation of semi-commercial closed anaerobic digester for palm oil mill effluent treatment. *Process Biochem.* 41, 962–964.

Yue, Z.B., Yu, H.Q. and Wang, Z.L. (2007). Anaerobic digestion of cattail with rumen culture in the presence of heavy metals. *Bioresour. Technol.* 98, 781-786.

Yusof, N., Haraguchi, A., Hassan, M.A., Othman, M.R., Wakisaka, M. and Shirai, Y. (2009). Measuring organic carbon, nutrients and heavy metals in rivers receiving leachate from controlled and uncontrolled municipal solid waste (MSW) landfills. *Waste. Manage.* 29 (10), 2666-2680.

Zeikus J.G. (1981). Microbial intermediary metabolism. In Anaerobic Digestion, edited by Hughes et al., 23-25. Procceding of the 2nd International Symposium on Anaerobic Digestion. Travemunde, Germany.

Zhang, W.; Xie, Q., Rouse, J. D., Qiao,S. and Furukawa, K. (2009). Treatment of high-strength corn steep liquor using cultivated Polyvinyl alcohol gel beads in an anaerobic fluidized-bed reactor. Journal of Bioscience and Bioengineering, 107 (1), 49–53.

Zinder E.R. (1998). Conversion of acetic acid to methane by thermophiles. In Anaerobic Digestion, edited by E.R. Hall and P.N. Hobson, pp. 1-12. Procceding of the 5th International Symposium on Anaerobic Digestion, Bologna, Italy.

5

Influence of Substrate Concentration on the Anaerobic Degradability of Two-Phase Olive Mill Solid Waste: A Kinetic Evaluation

Bárbara Rincón and Rafael Borja
Instituto de la Grasa (CSIC), Avda. Padre García Tejero, Sevilla, Spain

1. Introduction

The evolution of modern technology for olive oil extraction has affected the industrial sector depending directly on the by-products obtained. The traditional three-phase continuous centrifugation process for olive oil extraction was introduced in the 1970s, notably to increase the processing capacity and extraction yield and to reduce labour. This three-phase manufacturing process of olive oil usually yields an oily phase (20%), a solid residue (30%) and an aqueous phase (50%), the latter coming from the water content of the fruit, which is usually defined as vegetation water. Such water, combined with that used to wash and process the olives, make up the so-called "olive mill wastewater" (OMW) and also contains soft tissues from olive pulp and a very stable oil emulsion (Borja et al., 2006). This process generates a total volume of traditional OMW of around 1.25 litres per kg of olives processed. Consequently, the three-phase centrifugation process caused an increase in the average mill size, a decrease in the total number of mills, increased water consumption and increased production of wastewaters.

The OMW composition is not constant either qualitatively or quantitatively and it varies according to cultivation soil, harvesting time, the degree of ripening, olive variety, climatic conditions, the use of pesticides and fertilizers and the duration of aging. The three-phase OMW is characterized by the following special features and components: intensive violet-dark brown to black in colour; specific strong olive oil smell; high degree of organic pollution (chemical oxygen demand -COD- values up to 220 g/L); pH between 3 and 6 (slightly acidic); high electrical conductivity; high content of poly-phenols (0.5-24 g/L) and high content of solid matter (Niaounakis and Halvadakis, 2004).

The annual OMW production of Mediterranean olive-growing countries is estimated to ranging from 7 million to over 30 million m^3. This huge divergence of results can partly be explained by the fact that the production of olives varies from one year to another due to weather conditions and plagues that can affect the olive trees. The average total production amounts approximately to $10\text{-}12 \times 10^6$ m^3 per year and occurs over a brief period of the year (November-March). Spain produced 20% of the OMW of the Mediterranean basin ($2\text{-}3 \times 10^6$

m^3/year) before the implantation of the two-phase extraction process in most of the Spanish olive oil factories, which represented an equivalent pollution of 10-16x10^6 inhabitants in the short milling period (Nioaunakis and Halvadakis, 2004).

The efforts to find a solution to the OMW problem are more than 50 years old (Borja et al., 2006). There are many different types of processes that have been tested: detoxification processes (such as physical, thermal, physicochemical, biological and combination of processes), recyclying and recovery of valuable components, production system modification, etc. However, none of the detoxification techniques on an individual basis allow the problem of disposal of OMW to be solved to a complete and exhaustive extent, effectively and in an ecologically satisfactory way. At the present state of OMW treatment technology, industry has shown little interest in supporting any traditional process (physical, chemical, thermal or biological) on a wide scale. This is because of the high investment and operational costs, the short duration of the production period (3-5 months) and the small size of the olive mills (Borja et al., 2006).

2. The two-phase olive oil manufacturing process

The failure to develop a suitable and economical effluent wastewater treatment technology for OMW has lead manufacturers of technology to develop the "ecological" two-phase process, which delivers oil as the liquid phase and a very wet olive cake (**two-phase olive mill solid waste -OMSW-**) as the solid residue. This technology has attracted special interest where water supplies are restricted and/or aqueous effluent must be reduced (Borja et al., 2006).

In the two-phase process a horizontally mounted centrifuge is used for primary separation of the olive oil fraction from the vegetable solid material and vegetation water. The resultant olive oil is further washed to remove residual impurities before finally being separated from this wash water in a vertical centrifuge. Therefore, the two-phase olive mills produce three identifiable and separate waste streams. These are:

1. The wash waters generated during the initial cleansing of the fruit.
2. The aqueous solid residues generated during the primary centrifugation (two-phase OMSW).
3. The wash waters from the secondary centrifuge generated during the washing and purification of virgin olive oil.

Spain was the first country where the two-phase system was used and from there this new technology was installed around the world. The two-phase decanting reduces the water requirements. Nevertheless it has created a new solid residue, two-phase OMSW, which requires further investigation to find out how it must be handled.

The two-phase olive oil extraction process has several advahtages over the three-phase centrifugation process (Alba et al., 2001; Di Giovacchino et al., 2001 and 2002):

- The construction of the two-phase scroll centrifuge is less complicated and thus is more reliable in operation and less expensive than the three-phase decanter.
- During operation of the three-phase scroll centrifuge the separated oil and water may be remixed; volatile compounds from the vegetation water may cause a sticky deposit on the centrifuge.

- The throughput of the two-phase centrifuge in relation to the oil quantity is higher because no additional water is required to produce the pulp. Energy consumption is also reduced as a result of the lower processing quantity.
- Oil produced by the two-phase centrifuge is of higher quality; in particular, it has higher oxidation stability and better organoleptic characteristics.
- The operating costs are lower. Water utilization in the olive mill decreases considerably.

In addition, the disadvantages of two-phase manufacturing process are:

- The two-phase process, although it produces no olive mill wastewater as such, generates the wash waters derived from the initial cleansing of the fruit and from the purification of virgin olive oil. In addition, it combines the olive vegetation water that is generated with the solid waste to produce a single effluent stream in semi-solid form. This doubles the amount of "solid" waste (OMSW or 'alperujo') requiring disposal, and it cannot be composted or burned without some form of expensive pre-treatment.
- Two-phase OMSW has a moisture content significantly higher than that of traditional cake from three-phase centrifuges. This increased amount of moisture, together with the sugars and fine solids that in the three-phase system were contained in OMW give two-phase OMSW a doughy consistency and makes transport, storage and handling difficult -it can not be piled and must be kept in large ponds.
- Two-phase OMSW is characterized by higher values of the pulp/stone ratio, as well as the greater weight produced.
- This two-phase technology transfers the problem of disposing of the olive-mill waste from the mill to the seed-oil refineries. Two-phase OMSW, prior to oil solvent extraction, must be dried with considerably higher energy requirements than in the three-phase continuous oil production process, making the industrial recovery of the residual oil difficult and expensive.

2.1 The two-phase Olive Mill Solid Waste (OMSW)

The characteristics of two-phase OMSW are obviously very different from the characteristics of olive cake resulting from three-phase centrifuge systems. Two-phase OMSW is a thick sludge that contains pieces of stone and pulp of the olive fruit as well as vegetation water. It has a moisture content in the range of 60-70% while olive cake from a three-phase extraction process has only around 40-45% moisture. It also contains some residual olive oil (2-4%), 2% ash with a 30% potassium content (Alba et al., 2001).

The average composition of the two-phase OMSW is: water (60-70%), lignine (13-15%), cellulose and hemicellulose (18-20%), olive oil retained in the pulp (2.5-3%), mineral solids (2.5%). Among their organic components, the major ingredients are as follows: sugars (3%), volatile fatty acids (C2-C7) (1%), poly-alcohols (0.2%), proteins (1.5%), poly-phenols (0.2%) and other pigments (0.5%) (Borja et al., 2002).

As it can be seen, the two-phase OMSW has a high organic matter concentration giving an elevated polluting load. The high polluting power and large volumes of solid waste generated (around 2 millions of tons per year in Spain) can pose large-scale environmental problems, taking into account the 2000 Spanish olive oil factories, most of them located in the Andalusia Community (Borja et al., 2002).

3. Anaerobic digestion as an alternative for treatment of two-phase OMSW

Anaerobic digestion (AD) is an attractive treatment for this waste of difficult disposal. AD processes transform the organic matter contained in a certain waste in biogas as main product. This process is carried out for different kind of microorganisms which work in a coordinate and interdependent chain until biogas obtaining.

Anaerobic treatment of moderate and high strength wastes with high biodegradable content presents a number of advantages in comparison to the classical aerobic processes: a) quite a high degree of purification with high-organic load feeds can be achieved; b) low nutrient requirements are necessary; c) small quantities of excess sludge are usually produced and finally, d) a combustible biogas is generated. The production of biogas enables the process to generate or recover energy instead of just energy-saving; this can reduce operational costs as compared with other processes such as physical, physico-chemical or biological aerobic treatments (Borja et al., 2006).

Previous works carried out at pilot-scale have shown that most of agro-industrial residues, such as sugar beet pulp, potato pulp, potato thick stillage and brewer´s grains, can be treated anaerobically with an efficient solids stabilisation and energy recovery, if the applied process-type (one or two stages) is selected according to the C:N ratio of the residues. These works demonstrated that at hydraulic retention times (HRT) of between 10 and 20 days, normally, the 50-60% of the organic matter was degraded. The ultimate anaerobic biodegradability was higher and lied between 76% (brewer´s grain) and 88% (potato pulp), which demonstrated that more than 60% of the available energy potential could be used in the industrial processes. The gas production varied between 300 and 500 m^3 biogas per ton of dry matter with a methane content of 60-70%. The undigested solids, which were separated from the effluent of the reactors could be completely stabilised after a short aerobic post-treatment to be used as a soil conditioner (Borja et al., 2006).

A number of kinetic models have been proposed for the process of anaerobic digestion. Early models were based on a single-culture system and used the Monod equation or variations. More recently, several dynamic simulation models have been developed based on a continuous multi-culture system; these correspond to the major bioconversion steps in anaerobic digestion but again make the assumption that culture growth obeys Monod type kinetics. Doubt has been expressed by several investigators on the validity of applying the Monod equation to waste treatment as the specific growth rate is expressed only as a function of the concentration of the limiting substrate in the reactor. The Monod equation contains no term relating to input substrate concentration; this implies that the effluent substrate concentration is independent of the input concentration. Experimental results do not always agree with this implication; for example the anaerobic digestion of dairy manure, beef cattle manure at mesophilic and thermophilic temperatures, rice straw or poultry litter (Borja et al., 2003).

Deviation from the Monod relationship in many digestion systems may be due to their complexity. This complexity has necessitated the use of generalized measures of feed and effluent strength, namely total Chemical Oxygen Demand (COD) and volatile solids (VS), which may not truly reflect the nature of the growth-limiting substrate. Utilizable carbon in the digester is derived from the hydrolysis of polymeric compounds, constituting the waste, by exo-enzymes in the extracellular medium or on the surface /vicinity of the microorganisms: only these hydrolysed, assimilable compounds can be considered as the growth-limiting substrate in terms of the Monod relationship. Extra-cellular hydrolysis is often considered the rate-limiting step in anaerobic digestion of organic wastes (Borja et al., 2003) and for a model to be truly valid this must be taken into account.

Multi-culture system kinetics may be desirable in view of the heterogeneous nature of the microbial population performing the various bioconversion steps involved. However, the kinetic models based on this premise necessarily involve a number of kinetic equations and coefficients making them highly complex, as shown by the reported models (Borja et al., 2003). Complexity does not necessarily equate to accuracy and there is still a strong case in favour of a simpler kinetic treatment based on a single culture system. Methanogenesis is particularly suited to this approach as there is a strong holistic characteristic in the process. Various cultures and bioconversion steps in digestion are interdependent and the whole process has certain self-regulatory characteristics within the process limits.

Kincannon and Stover (1982) proposed a widely used mathematical model to determine the kinetic constants for immobilized systems and high-rate reactors. In this model the substrate utilization rate is expressed as a function of the organic loading rate by monomolecular kinetics for biofilm reactors such as rotating biological contactors and biological filters (Kapdan and Erten, 2007). A special feature of the modified Stover-Kincannon model is the utilization of the concept of organic loading rate as the major parameter to describe the kinetics of an anaerobic filter in terms of organic matter removal and methane production (Büyükkamaci and Filibeli, 2002; Kapdan and Erten, 2007).

The modified Stover-Kincannon model allows to calculate the maximum substrate utilization rate by the microorganisms (R_{max}) and the saturation constant (K_B) in anaerobic digestion processes (Yu et al., 1998). Therefore, this model allows determining the effluent substrate concentration for a known volume of reactor and an initial concentration of the substrate. The modified Stover-Kincannon model has been used for different substrates and reactor configurations: anaerobic hybrid reactors treating petrochemical waste (Jafarzadeh et al., 2009), anaerobic treatment of synthetic saline wastewater by *Halanaerobium lacusrosei* (Kapdan and Erten, 2007), anaerobic digestion of soybean wastewaters (Yu et al., 1998) and molasses (Büyükkamaci and Filibeli, 2002) in a filter and in a hybrid reactor, respectively.

The aim of the present study was focused on the AD of two-phase OMSW at two different influent substrate concentrations and on the determination of kinetics constants of the system using the above-mentioned modified Stover-Kincannon model.

4. Materials and methods

4.1 Equipment

An anaerobic reactor with a working volume of 1 litre equipped with magnetic stirring and placed in a thermostatic chamber at 35 °C was used. The reactor had an upper settling zone designed to minimize loss of the biomass responsible for the process. The reactor was fed daily by means of an external feeder and liquid effluent removed daily through a hydraulic seal, comprising 25 cm liquid column, designed to prevent air from entering the reactor and biogas from leaving. This reactor has been described in detail elsewhere (Martín et al., 1991).

The methane volume produced in the process was measured using a 5 litre Mariotte reservoir fitted to the reactor. A tightly closed bubbler containing a NaOH solution (3 M) to collect the CO_2 produced in the process was intercalated between the two elements. The methane produced displaced a given volume of water from the reservoir, allowing ready determination of the gas (Martín et al., 1991).

4.2 Inoculum

The reactor was inoculated with methanogenically active biomass from a laboratory-scale anaerobic reactor processing olive mill wastewater. The composition and features of the biomass used were: pH, 7.2; total solids (TS), 60.3 g/L; mineral solids (MS), 19.3 g/L; volatile solids (VS), 41.0 g/L; total suspended solids (TSS), 59.9 g/L; mineral suspended solids (MSS), 18.8 g/L; volatile suspended solids (VSS), 41.1 g/L.

4.3 Two-phase Olive Mill Solid Waste (OMSW)

The OMSW used for the experiments was collected from a two-phase technology mill. The OMSW was derived from olives with a high ripening index (6.5) and an intense purple colour. Before use the small stone pieces were removed by sieving the OMSW through a 3.15 and 2.00 mm sieve. Two influent substrate concentrations were used for the experiments: 35 g COD/L (OMSW 1) and 150 g COD/L (OMSW 2). These concentrations were obtained by dilution of the collected waste. The features and composition of these two-phase OMSWs are summarised in Table 1.

	Units	**OMSW 1**	**OMSW 2**
pH	*	5.6	5.8
COD	g O_2/L	35	150
SCOD	g O_2/L	15	67
TVFA	g acetic acid/L	0.70	2.90
Alkalinity	g $CaCO_3$/L	0.74	2.20
TS	g/L	40.2	165.3
MS	g/L	5.6	21.1
VS	g/ L	34.6	144.2
TSS	g/ L	35.2	142.2
MSS	g/ L	4.1	15.7
VSS	g/ L	31.1	126.5
Total phenolic compounds	g caffeic acid/L	0.61	2.44

COD: total chemical oxygen demand; SCOD: soluble chemical oxygen demand; TVFA: total volatile fatty acids (as acetic acid); Alkalinity (as $CaCO_3$). Values are averages of five determinations; there was virtually no variation (less than 3 %) between analyses.

Table 1. Composition and features of the OMSWs.

4.4 Experimental procedure

The anaerobic reactor was initially charged with 300 mL of distilled water, 500 mL of the inoculum and 200 mL of a nutrient-trace element solution. The composition of this nutrient-trace element solution is given in detail elsewhere (Borja et al., 2001).

The start-up of the reactor involved stepped increases in COD loading using an influent substrate concentration of 17.2 g COD/L. During this period the organic loading rate (OLR) was gradually increased from 0.25 to 0.50 g COD/(L d) between 1 and 15 d, 0.75 g COD/(L d) between 16 and 30 d, 1.00 g COD/(L d) between 31 and 45 d and finally 1.25 g COD/(L d) between 46 and 60 d.

After the preliminary step, the reactor was fed in series of semicontinuous experiments using OLRs of 0.9, 1.2, 1.4, 1.7, 2.1, 2.8, 3.5, 4.1 L COD/(L d) for the OWSW1, which correspond to hydraulic retention times (HRTs) of 40.0, 28.6, 25.0, 20.0, 16.6, 12.5, 10.0 and 8.3 d, respectively. After these experiments with OMSW 1 five different OLRs were assessed for the OMSW 2, 3.0, 6.0, 9.05, 12.0 and 15.0 g COD/(L d), these OLRs corresponded to HRTs of 50.0, 25.0, 16.6, 12.5 and 10.0 d, respectively.

Once steady-state conditions were achieved at each feed flow-rate, the daily volume of methane produced, and total and soluble COD, pH, total volatile fatty acids (TVFA) and volatile solids (VS) of the different effluents were determined. The samples were collected and analysed for at least 5 consecutive days. The steady-state value of a given parameter was taken as the average of these consecutive measurements for that parameter when the deviations between the observed values were less than 3% in all cases. Each experiment had a duration of 2-3 times the corresponding HRT.

The organic loadings applied in this work were increased in a stepwise fashion in order to minimise the transient impact on the reactor that might be induced by a sudden increase in loadings.

4.5 Chemical analyses

The following parameters were determined: total and soluble COD, pH, total solids, mineral solids, volatile solids, total suspended solids, mineral suspended solids, volatile suspended solids, total volatile fatty acids (TVFA), alkalinity and total phenolic compounds. All analyses were carried out according to the recommendations of the Standard Methods of American Public Health Association (APHA, 1989).

In each steady-state experiment, samples were collected and the above parameters analysed. The pH and gas volume were determined daily, whilst the remaining parameters were measured at least five times per week on five different samples taken on different days to ensure that representative data were obtained.

5. Results and discussion

5.1 Influence of substrate concentration and OLR on the COD removal efficiency and operational parameters

The anaerobic degradability studies were carried out using two different two-phase OMSWs with COD concentrations of 35 g COD/L (OMSW 1) and 150 g COD/L (OMSW 2). The experiments were performed using progressive influent substrate concentrations, those corresponding to the OMSW 1 being the first ones and those corresponding to the OMSW 2 carried out at the end of the study.

Tables 2 and 3 summarize the steady-state operating results including HRT, OLR, methane production rates (r_{CH4}), total and soluble CODs, VS, TVFA, alkalinity and TVFA/alkalinity ratio for the OMSW 1 and OMSW 2, respectively (Borja et al., 2002).

Figure 1 shows the variation of the COD removal efficiency with the OLR for the two OMSWs used.

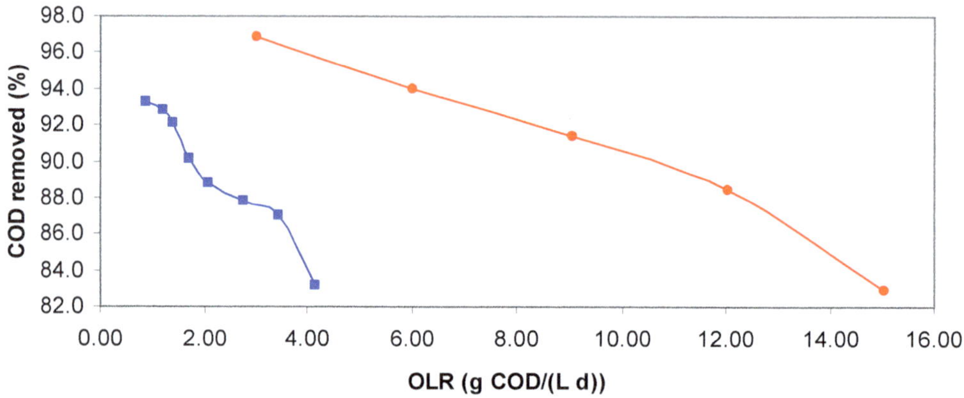

Fig. 1. Variation of the percentage of COD removed with the OLR for the two OMSWs used (■: OMSW 1; ●: OMSW 2).

OLR (g COD/(L d))	0.86	1.21	1.38	1.72	2.08	2.76	3.45	4.14
HRT (d)	40.0	28.6	25.0	20.0	16.6	12.5	10.0	8.3
pH	7.9	7.8	8.0	7.9	8.0	7.9	7.8	7.1
* r_{CH4} (L CH_4/(L d))	0.24	0.34	0.38	0.47	0.56	0.73	0.91	0.85
COD (g/L)	2.30	2.50	2.74	3.40	3.85	4.20	4.50	5.80
Soluble COD	0.72	1.20	1.40	1.65	1.90	2.15	2.35	3.80
VS (g/L)	1.70	1.88	2.07	2.40	2.75	3.10	3.40	4.50
TVFA (g acetic acid/L)	0.105	0.155	0.180	0.205	0.215	0.260	0.310	0.495
Alkalinity (g $CaCO_3$/L)	1.950	1.850	1.715	1.690	1.640	1.690	1.670	1.410
TVFA/Alkalinity	0.04	0.07	0.09	0.10	0.11	0.13	0.15	0.29

Values are the averages of 5 determinations taken over 5 days after the steady-state conditions had been reached. The differences between the observed values were less than 3 % in all cases. (* r_{CH4}: methane production rates)

Table 2. Steady-state results under different experimental conditions for the OMSW 1 with a COD of 35 g/L.

OLR (g COD/(L d))	3.00	6.01	9.05	12.02	15.03
HRT (d)	50.0	25.0	16.6	12.5	10.0
pH	7.2	7.0	7.0	6.9	6.5
* r_{CH4} (L CH_4/(L d))	0.59	1.13	1.64	2.12	2.05
COD (g/L)	4.80	9.05	12.95	17.50	25.70
Soluble COD	3.05	6.00	8.25	11.30	15.05
VS (g/L)	3.60	6.80	9.70	13.10	19.30
TVFA (g acetic acid/L)	0.56	0.81	1.08	1.25	1.57
Alkalinity (g $CaCO_3$/L)	1.98	1.90	1.81	1.70	1.32
TVFA/Alkalinity	0.23	0.35	0.40	0.61	0.95

Values are the averages of 5 determinations taken over 5 days after the steady-state conditions had been reached. The differences between the observed values were less than 3 % in all cases. (* r_{CH4}: methane production rates)

Table 3. Steady-state results under different experimental conditions for the OMSW 2 with a COD of 150 g/L.

As can be seen in Figure 1 the percentage of COD removed decreased with increased OLR for the two influent substrate concentrations studied. The percentage of COD removal decreased from 93.3% to 83.2% when OLR increased from 0.86 to 4.14 g COD/(L d) for the most diluted substrate (OMSW 1). For the most concentrated influent (OMSW 2) OLRs were varied from 3.00 to 15.03 g COD/(L d) and COD removal efficiencies higher than 88% were obtained at an OLR of 12.02 g COD/(L d). Even under a higher OLR of 15.03 g COD/(L d), corresponding to an HRT of 10 days, COD removal was 82.9%.

The total effluent CODs of the anaerobic reactor increased with increased OLR for the two influent substrate concentrations studied, as summarized in Tables 2 and 3. Such an increase in the effluent COD was paralleled by a similar increase in the effluent total volatile fatty acids (TVFA). This seems to indicate that, at higher OLR, the effluent total COD and mainly soluble COD is largely composed of the unused volatile acids produced in the reactor.

Given that the buffering capacity of the experimental system was found to be at favourable levels with excessive total alkalinity present at virtually all loadings, the efficiency of the process and the rate of methanogenesis was not very affected. The experimental data obtained in this work indicate that a total alkalinity of about 1.7 g/L as $CaCO_3$ is sufficient to prevent the pH from dropping to below 7.0 at an OLR of 9.05 g COD/(L d) for the most concentrated substrate used (OMSW 2).

The pH in the reactor was always higher that 7.0 for all the HRTs and OLRs studied corresponding to the most diluted OMSW studied. In addition, pH values equal or higher than 6.9 were observed for OLRs lower than 12.02 g COD/(L d) and HRTs higher than 12.5 d when the most concentrated influent was processed, with pH of 7.2 as a maximum value achieved. This high stability can be attributed to carbonate/bicarbonate buffering. This is produced by the generation of CO_2 in the digestion process which is not completely removed from the reactor as gas. Buffering in anaerobic digestion is normally due to bicarbonate, as carbonate is, generally, negligible if compared to the bicarbonate (carbonate/bicarbonate ratio is equal to 0.01 for pH 8.2) (Speece, 1983). The buffering guards against possible acidification of the reactor giving a pH of the same order as the optimal for methanogenic bacteria (Wheatley, 1990).

The TVFA/Alkalinity ratio can be used as a measure of process stability (Wheatley, 1990): when this ratio is less than 0.3-0.4 the process is considered to be operating favourably without acidification risk. As was observed in Tables 2 and 3 the ratio values were lower than the suggested limit value for OLRs lower than 9.05 g COD/(L d) in the experiments corresponding to the highest influent substrate concentrations studied (OMSW 2). For this substrate, between HRTs of 50.0 and 16.6 days, the TVFA/Alkalinity ratio was always lower than the above-mentioned failure limit and the TVFA values were always lower than 1,08 g/L (as acetic acid). However, at a HRT of 10.0 days, a considerable increase of the TVFA/Alkalinity ratio until a value of 0.95 was observed in the reactor, which was mainly due to a considerable increase in the TVFA concentration (1.57 g/L as acetic acid) with simultaneous decrease in alkalinity (1.32 g/L, as $CaCO_3$).

5.2 Influence of substrate concentration on the methane production rates and methane yield coefficients

The volumetric methane production rates as a function of OLR are illustrated in Figure 2. As can be seen the volume of methane produced per day increased linearly with increased OLR up to OLR values of 3.45 and 12.02 g COD/(L d) for the influents OMSW 1 and OMSW 2, respectively. After these values a slight decrease was observed in the cases studied over the different ranges tested. Apparently, the activity of methanogenic bacteria was not impaired up to OLR values of 12.02 g COD/(L d) for the most concentrated influent (OMSW 2) used because of the appropriate stability and adequate buffering capacities provided in the experimental system. Nevertheless, the methane production rate decreased slightly from 2.12 to 2.05 L CH_4/(L d) when the OLR was increased from 12.02 to 15.03 g COD/(L d).

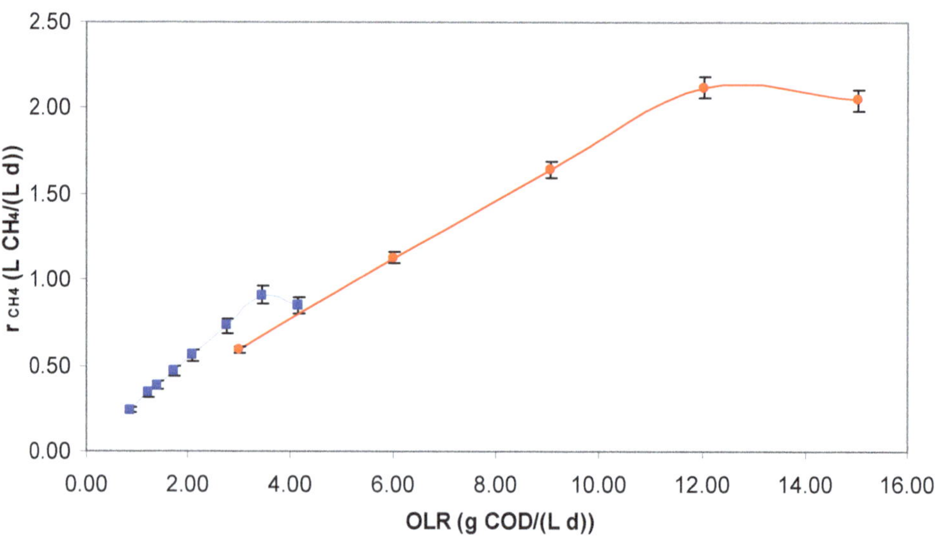

Fig. 2. Variation of the methane production rate, r_{CH4}, with the OLR (g COD/(L d)) of the reactor for the two OMSWs used as influents (■: OMSW 1; •: OMSW 2).

This decrease in the methane production at the highest OLR values might be attributed to an inhibition of the methanogenic bacteria at high OLR values, which caused an increase in effluent TVFA contents and TVFA/Alkalinity ratio, as can be seen in Table 3. Specifically, TVFA content increased from 1.25 to 1.57 g/L (as acetic acid) when the OLR was increased from 12.02 to 15.03 g COD/(L d).

The experimental data listed in Tables 2 and 3 were used to determine the methane yield coefficient, Y_p. As the volume of gas produced per day, r_{CH4}, is assumed to be proportional to the amount of substrate consumed, then:

$$r_{CH4} = Y_p q \, (S_0 - S) \tag{1}$$

where S_0 and S are the substrate concentrations (expressed as g COD/L) at the digester inlet and effluent, respectively, and q is the feed flow-rate. By plotting Eq (1) in the form r_{CH4} against $q\ (S_0 - S)$ (Figure 3), the following values of the methane yield coefficients with their 95% confidence limits were obtained for the two substrate concentrations used: 0.300 (± 0.001) and 0.200 (± 0.006) L methane STP/g COD removed when the OMSW 1 and OMSW 2, respectively, were processed. These values agree with the data reported in the literature for anaerobic treatment of food industry wastewaters (Borja et al., 1995; Maqueda et al., 1998; Martín et al., 1993). Taking into account that, theoretically, 0.35 L of methane is produced per gram of COD removed when the starting compound is glucose (Wheatley, 1990), the effectiveness of the anaerobic process in converting OMSW into methane at mesophilic temperature is demonstrated.

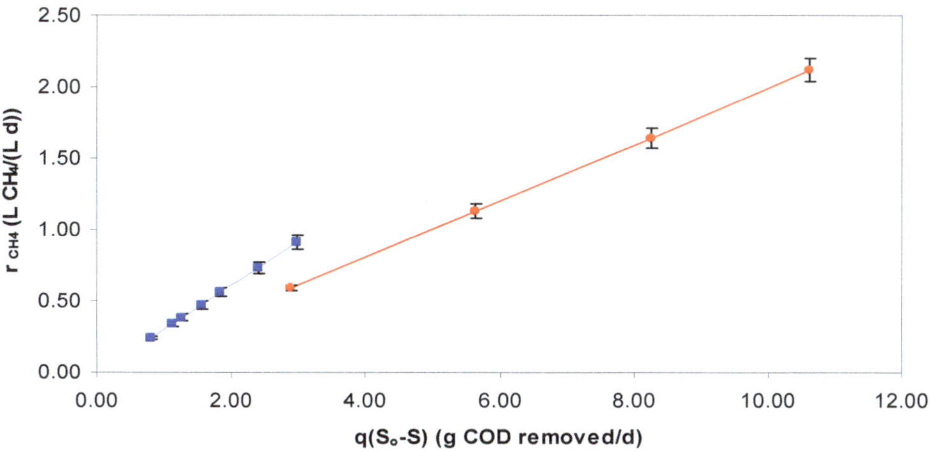

Fig. 3. Variation of the volume of methane produced per day, r_{CH4}, as a function of the product of the differences of substrate concentrations at the reactor inlet (S_0 in g COD/L) and outlet (S in g COD/L) and the feed flow-rate (q in L/day) for the two OMSWs used as influent. (■: OMSW 1; •: OMSW 2).

5.3 Kinetic evaluation

Since the early 1980's, Stover and Kincannon have proposed a design concept of total organic loading rate and established a kinetic model for biofilm reactors. In this model the substrate utilization rate is expressed as a function of the organic loading rate by monomolecular kinetics for biofilm reactors such as rotating biological contactors and biological filters (Yu et al., 1998). This kinetic model can be used to describe carbonaceous removal in terms of BOD (biochemical oxygen demand), COD (chemical oxygen demand) and TOC (total organic carbon) as well as for nitrification.

The original Stover-Kincannon model (Kincannon and Stover, 1982) (Equation 2) was initially proposed for rotating biological contactor (RBC) systems and can be expressed by the following equation:

$$dS/dt = [R_{max} (qS_o/A)]/[K_B+(qS_o/A)] \tag{2}$$

where: A is the disc surface area where the active biomass is attached; S is the substrate concentration in the reactor (in COD units) for a time (t); S_o is the initial substrate concentration; q is the flow rate; R_{max} is the maximum removal rate constant and K_B is the saturation value constant (in g COD/(L d)).

In the modified Stover-Kincannon model the substrate utilization rate is expressed as function of the organic loading rate as follows (Yu et al., 1998).

$$dS/dt = [R_{max} (qS_o/V)]/[K_B+(qS_o/V)] \tag{3}$$

where V is the volume of the anaerobic reactor. The term dS/dt is defined for a steady-state relationship for different authors as:

$$dS/dt = q (S_o - S)/V \tag{4}$$

Linearization of equation (3) gives:

$$V/[q (S_o - S)]= [K_B V/(R_{max} q S_o)]+[1/R_{max}] \tag{5}$$

In continuously stirred tank reactors the hydraulic retention time (HRT) can be defined as: $HRT=V/q$, so equation (5) can be written as follows:

$$(HRT)/(S_o - S)= [K_B (HRT)/(R_{max} S_o)] + [1/R_{max}] \tag{6}$$

According to this model a plot of $(HRT)/(S_o - S)$ versus HRT should give a straight line of intercept $[1/R_{max}]$ and slope equal to $K_B /(R_{max} S_o)$.

As can be seen in Figure 4 the experimental data fitted to a straight line with R^2= 0.9992 for OMSW 1 and R^2= 0.9999 for OMSW 2. The maximum removal rate constant (R_{max}) increased from 26.6 to 83.3 g COD/(L d) when the OMSW concentration changed from 35 to 150 g COD/L, indicating a good adaptation of the initial inoculum to the OMSW treated and to increasing concentrations of organic matter fed. The saturation value constants (K_B) were 27.7 g COD/(L d) and 82.7 g COD/(L d) for OMSW 1 and OMSW 2, respectively. The values of R_{max} and K_B obtained for the concentrated OMSW were similar to those obtained by other authors for the anaerobic digestion of soybean wastewaters (Yu et al., 1998) and molasses (Büyükkamaci & Filibeli, 2002). Stover and Campana (2003) have shown that in the model R_{max} is reduced by refractory organics and toxicity. Moreover, the refractory compounds change K_B significantly from R_{max}. These affirmations are in agreement with the data obtained in these experiments, where the higher organic concentration of OMSW 2 gave R_{max} values higher than for OMSW 1.

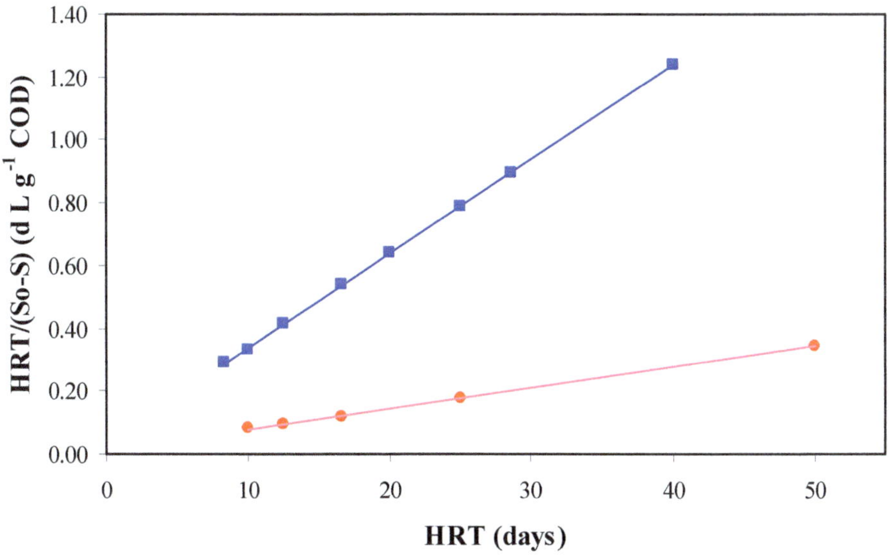

Fig. 4. Determination of the kinetic parameters using the modified Stover-Kincannon model for the two-phase OMSW 1 and OMSW 2. (■: OMSW 1; ●: OMSW 2).

6. Conclusions

The kinetic constants obtained define the bio-treatability of the two-phase olive mill solid waste. The values obtained for R_{max} and K_B were similar to those obtained for other substrates of high organic content. The increase in the maximum methane removal rate for the most concentrated two-phase olive mill solid waste used demonstrated the good adaptation of the bacterial inoculum used to the increase in the substrate concentration. This adaptation allowed the microorganisms to work with high stability even with high organic matter concentrations in the fed substrate. These results can be used to estimate the treatment efficiency of industrial-scale reactors working with similar operational conditions.

7. References

Alba, J.; Hidalgo, F.; Ruiz, M.A.; Martínez, F.; Moyano, M.J.; Borja, R.; Graciani, E. & Ruiz, M.V. (2001). Elaboración de aceite de oliva virgen. In: *El Cultivo del Olivo*, D. Barranco, R. Fernández-Escobar, L. Rallo (Ed.), 551-588, Mundi-Prensa, Madrid, Spain, ISBN 84-7114-983-4.

APHA (American Public Health Association). (1989). Standard Methods for the Examination of Water and Wastewater, 17th ed., APHA, Washington DC, USA, ISBN 087553161X.

Borja, R.; Banks, C.J. & Wang, Z. (1995). Effect of organic loading rate on anaerobic treatment of slaughterhouse wastewater in a fluidised-bed reactor. *Bioresource Technology*, Vol. 52, No. 2, pp. 157-162, ISSN 0960-8524.

Borja, R.; González, E.; Raposo, F.; Millán, F. & Martín, A. (2001). Performance evaluation of a mesophilic anaerobic fluidized bed reactor treating wastewater derived from the production of proteins from extracted sunflower flour. *Bioresource Technology*, Vol. 76, No. 1, pp. 45-52, ISSN 0960-8524.

Borja, R.; Rincón, B.; Raposo, F.; Alba, J. & Martín, A. (2002). A study of anaerobic digestibility of two-phases olive mill solid waste (OMSW) at mesophilic temperature. *Process Biochemistry,* Vol. 38, No. 5, pp. 733-742, ISSN 0032-9592.

Borja, R.; Martín, A.; Rincón, B. & Raposo, F. (2003). Kinetics for substrate utilization and methane production during the mesophilic anaerobic digestion of two-phases olive pomace (TPOP). *Journal of Agricultural and Food Chemistry,* Vol. 51, No. 11, pp. 3390-3395, ISSN 0021-8561.

Borja, R.; Rincón, B. & Raposo, F. (2006). Review: Anaerobic biodegradation of two-phase olive mill solid wastes and liquid effluents: Kinetic studies and process performance. *Journal of Chemical Technology and Biotechnology,* Vol. 81, No. 9, pp. 1450-1462, ISSN 0268-2575.

Büyükkamaci, N. & Filibeli, A. (2002). Concentrated wastewater treatment studies using an anaerobic hybrid reactor. *Process Biochemistry,* Vol. 38, No. 5, pp. 771-775, ISSN 0032-9592.

Di Giovacchino, L.; Costantini, N.; Serraiocco, A.; Surrichio, G. & Basti C. (2001). Natural antioxidants and volatile compounds of virgin olive oils obtained by the two or three-phase centrifugal decanters. *European Journal of Lipid Science and Technology,* Vol. 103, pp. 279-285, ISSN 1438-7697.

Di Giovacchino, L.; Sestili, S. & Di Vencenzo, D. (2002). Influence of olive processing on virgin olive quality. *European Journal of Lipid Science and Technology,* Vol. 104, pp. 587-601, ISSN 1438-7697.

Jafarzadeh, M.T.; Mehrdadi, N. & Hashemian, S.J. (2009). Kinetic constants of anaerobic hybrid reactor treating petrochemical wastes. *Asian Journal of Chemistry,* Vol. 21, No. 3, pp. 1672-1684, ISSN 0970-7077.

Kapdan, I.K. & Erten, B. (2007). Anaerobic treatment of saline wastewater by *Halanaerobium lacusrosei*. *Process Biochemistry,* Vol. 42, No. 3, pp. 449-453, ISSN 1359-5113.

Kincannon, D.F.; Stover, E. L. (1982). Design methodology for fixed-film reactors-RBCs and biological towers. Civil Engineering for Practicing and Design Engineers 2: 107.

Maqueda, C.; Pérez-Rodríguez, J.L. & Lebrato, J. (1998). An evaluation of clay minerals as support materials in anaerobic digesters. *Environmental Technology,* Vol. 19, No. 8, pp. 811-819, ISSN 0959-3330.

Martín, A.; Borja, R.; García, I. & Fiestas, J.A. (1991). Kinetics of methane production from olive mill wastewater. *Process Biochemistry,* Vol. 26, No. 2, pp. 101-107, ISSN 0032-9592.

Martín, A.; Borja, R. & Chica, A. (1993). Kinetic study of an anaerobic fluidized-bed system used for the purification of fermented olive mill wastewater. *Journal of Chemical Technology and Biotechnology,* Vol. 56, No. 2, pp. 155-162, ISSN 1097-4660.

Niaounakis, M. & Halvadakis, C.P. (2004). *Olive-Mill Waste Management: Literature review and Patent Survey.* Typothito - George Dardanos Publications, ISBN 0-08-044851-8, Athens, Greece.

Speece, R.E. (1983). Anaerobic biotechnology for industrial wastewater treatment. *Environmental Science and Technology,* Vol. 17, pp. 416-427, ISSN 0013-936X.

Stover, E. L.; Campana, C.K. (2003). Effluents from food processing/On-site Processing of waste, In: *Encyclopedia of Food Sciences and Nutrition.* Caballero, B., Trugo, L.C., Finglas, P.M. Editors, pp. 1975-1976, ISBN:0-12-227055-X, 978-0-12-227055-0.

Wheatley, A. (1990). *Anaerobic digestion: a waste treatment technology,* published for SCI, ISBN 1-85166-526-9, London, UK, Elsevier Applied Science.

Yu, H.; Wilson, F. & Tay, J.H. (1998). Kinetic analysis of an anaerobic filter treating soybean wastewater. *Water Research,* Vol. 32, No. 11, pp. 3341-3352, ISSN 0043-1354.

6

Biogas Production and Cleanup by Biofiltration for a Potential Use as an Alternative Energy Source

Elvia Ines Garcia-Peña,
Alberto Nakauma-Gonzalez and Paola Zarate-Segura
Bioprocesses Department, Unidad Profesional Interdisciplinaria de Biotecnología, IPN, Mexico City, Mexico

1. Introduction

As many countries have taken advantage of the richness of crude oil, fossil fuels have become the main energy source, and human activities have become entirely dependent on petroleum products. However, this is not sustainable because of the huge environmental cost of harvesting and utilizing vast amounts of fossil fuels (Fairley, 2011). Therefore, the need for alternative fuels has become critical, especially for a new generation of advanced biofuels that can maximize petroleum (crude oil) displacement and minimize the side effects of burning fossil fuels. The primary objective is then to produce biofuels from corn stalks or other 'cellulosic plants' (or even from municipal garbage) and jet fuels from dedicated energy crops such as the fast-growing *Camelina sativa* (Fairley, 2011). The challenges are then to develop the agriculture for these plants and improve their utilization at an industrial scale. In this way, net reductions in petroleum use and greenhouse-gas emissions will be long-lasting and ethical. Bridging this gap will require continued investment, research, government regulations and development of technology. The International Energy Agency (IEA) has recommend the maximized use of farm, forestry and municipal wastes as well as increased cultivation of dedicated energy crops away from lands that provide carbon sequestration and other critical environmental services. One way to develop biofuels along an environmental friendly path is to draft a set of standards and practices that biofuels producers must comply with, either voluntarily or by mandate (Fairley, 2011).

In large cities, such as Mexico City with a population of more than 20 million, concerns about waste disposal and the use of alternative energy sources has steadily increased. This population produces a tremendous amount of solid waste, more than 12,000 tons per day. On the other hand, to provide sufficient food for this population, many markets are distributed throughout the city. The central market for food distribution in Mexico City, Central de Abasto (CEDA), is the second largest market in the world, receiving 25,000 tons of food products and producing 895 tons of organic solid waste each day (84% of the total solid waste produced is organic waste, 50% of that is from fruits and vegetables).

Fruit and vegetable waste (FVW) is produced in large quantities in markets in many large cities (Mata-Alvarez et al., 1992; Misi and Forster, 2002; Bouallagui et al., 2003; Bouallagui et al., 2005). The application of an anaerobic digestion process for simultaneous waste treatment and renewable energy production from the organic fraction of these residues could therefore be of great interest (Bouallagui et al., 2005). The high biodegradability of FVW promotes the rapid production of volatile fatty acids (VFAs), resulting in a rapid decrease in pH, which in turn could inhibit methanogenic activity (Bouallagui et al., 2003; Bouallagui et al., 2009). A strategy to avoid the acidification of the system is the addition of cosubstrates. Data obtained during the codigestion of FVW and other substrates resulted in the design of an efficient digestion process, improving methane yields through the positive synergistic effects of the mixed materials exhibiting complementary characteristics and the supply of missing nutrients from the cosubstrate (Agdag and Sponza, 2005, Habiba et al. 2009, Bouallagui et al. 2009). In a recently published study (Garcia-Peña et al., 2011), a 30-Liter anaerobic digestion reactor operated with a mixture of FVW:MR (meat residues) (75:25) had a stable CH_4 production percentage of 53 ± 2 % and a sustained pH of 6.9 ± 0.5 (naturally regulated) in a co-digestion process. The adequate and sustained performance and stable CH_4 production were a result of an appropriate buffering capacity and highly stable operation of the experimental system. However, the biogas produced during this anaerobic process needs to be cleaned before use by eliminating a relatively high content of other compounds as CO_2 and H_2S.

Biogas consists of approximately 60–70% (v/v) methane (CH_4), 30–40% (v/v) carbon dioxide (CO_2), 1-2% (v/v) nitrogen (N_2), 1000–3000 ppmv H_2S, 20-30 ppmv of VFAs and 10–30 ppmv of ammonia (NH_3), depending on the organic substrate used during the anaerobic process (Angelidaki et al., 2003). Hydrogen sulfide (H_2S) is one of the most commonly reported reduced sulfur compounds, and represents up to 2% (v/v) . However, this H_2S concentration can be higher when a rich protein feedstock is used. H_2S elimination is thus required because it reduces the life span of combustion engines by corrosion, forms SO_2 upon combustion and is a malodorous and toxic compound (Angelidaki et al., 2003, Pride, 2002). Malodorous gases include mainly H_2S (around 3000 ppmv) and some volatile fatty acids (VFAs).

Reducing CO_2 and H_2S content will significantly improve the quality of biogas. There have been many technologies developed for the separation of CO_2 from gas streams, including absorption by chemical solvents, physical absorption, cryogenic separation, membrane separation and CO_2 fixation by biological or chemical methods (Abatzoglou and Boivin , 2009, Granite and O'Brien, 2005). These techniques are of significant industrial importance and are generally applied during natural gas sweetening and in the removal of CO_2 from flue gases of power plants.

H_2S is currently removed using chemical, physical or biological methods. The most commonly used method is chemical absorption by selective amines, such as diglycolamine, monoethanolamine and methyldiethanolamine, but also by absorption into aqueous solutions, physical absorption on solid adsorbents or conversion to low-solubility metal sulfides (Horikawa et al., 2004, Osorio and Torres, 2009). Water scrubbing systems are also frequently used because of their simplicity and low cost (Kapdi et al., 2005, Rasi et al., 2008). Their use allows the production of high quality CH_4 enriched gas from biogas by chemical absorption where a packed bed column and a bubble column are normally used to provide liquid/gas contact (Krumdieck et al., 2008). However, the main drawbacks of these chemical technologies are the high energy requirement, the stability and selectivity of the chemicals used, the high cost of the chemicals and their regeneration, the negative environmental impacts from liquid wastes, the large equipment size requirements and the high equipment corrosion rate (Tippayawong and Thanompngchart, 2010, Fortuny et al., 2008).

Biological treatments are cost effective and environmentally friendly processes (Shareenfdeen et al., 2003, Ng et al., 2004, Maestre et al., 2010). Biofiltration is one of the most promising clean technologies for reducing emissions of malodorous gases and other pollutants into the atmosphere (van Groenestijn and Hesselink, 1993, van Groenestijn and Kraakman, 2005). This technology has been proven to effectively control reduced sulfur compounds in diluted gas streams (Yang et al., 1994, Smet et al., 1998, Ergas et al., 1995, Chung et al., 1996, Devinny et al., 1999; Gabriel and Deshusses, 2003, Kim and Deshusses, 2005). However, the elimination of H_2S from fuel gases requires systems that can handle high loads of pollutants for extended periods of time (Maestre et al., 2010). Surprisingly, there is still a limited number of reports on the removal of high concentrations of H_2S (>1000 ppmv) using biofilters, biotrickling filters and bioscrubbers. On the other hand, two processes have been effectively applied for the removal of high concentrations of H_2S from biogas or fuel gas in industrial processes: the Thiopaq process (Paques, The Netherlands) and the Biopuric process (Biothane, USA). The first one is a chemical process that uses a conventional caustic scrubber and an expanded bed bioreactor for the recovery of spent caustic and elemental sulfur generation. The Biopuric process combines a chemical scrubber with a subsequent biological treatment.

Although H_2S treatment for industrial processes has already been applied through the above-mentioned commercial systems, there is a need for the development of alternative and sustainable biological processes. Regarding the development of biofiltration and/or biotrickling filter systems to eliminate high H_2S concentrations, Rattanapan et al., 2009, compared the elimination of 200 to 4000 ppmv of H_2S in two biofiltration systems. One of the biofilters was a sulfide oxidizing bacterium immobilized on Granular Activated Carbon (GAC) (biofilter A) and the other was GAC without cell immobilization (biofilter B). The results showed that in the GAC system, the H_2S was autocatalytically oxidized when it absorbed into the CAG, reaching a removal percentage of 85%. The removal was enhanced to over 98% (even at a concentration as high as 4000 ppmv) through the biological activity in biofilter A. In this last system, the maximum elimination capacity was approximately 125 gH_2S/m^3GAC h. In addition, Fortuny et al., 2008, reported the performance of a biotrickling filter system for treating high concentrations of H_2S in simulated biogas using a single reactor. Two laboratory-scale biotrickling filters filled with different packing materials were evaluated, the inlet H_2S concentration ranged from 900 to 12000 ppmv. During long-term operation, a removal percentage of 90% was determined with an extremely high H_2S concentration (6000 ppmv). Maximum elimination capacities of 280 and 250 g H_2S/m^3 h were obtained at empty-bed residence times of 167 and 180 s, respectively. During this study, the main end products of the biological oxidation of H_2S were sulfate and elemental sulfur; the final percentage of these products varied as a function of the ratio of O_2/H_2S supplied (v/v). At a value of 5.3, corresponding to an inlet H_2S concentration of 3000, the main product was sulfate (60-70%), whereas at the higher H_2S concentration of 6000 ppmv, the sulfate recovery decreased to 20-30%. Elemental sulfur production varied inversely with the O_2/H_2S supplied (v/v), it was low at a ratio of 5.3 and increased up to 68-78% as the ratio decreased.

In a biofiltration system, a gas stream is passed through a packed bed on which pollutant-degrading organisms are immobilized as biofilms. Biotrickling filters use the same principle, but an additional liquid phase will flow through the reactor. In both systems, the microorganisms in the biofilms transform the absorbed H_2S by metabolic activity into elemental sulfur or sulfate depending on the amount of available oxygen. Oxygen is thus the key parameter that controls the level of oxidation. Sulfur production (Eq. 1) results from the partial oxidation of sulfide instead of complete oxidation to sulfate (Eq. 2) when oxygen is limited, as is shown in Equations 1 and 2 (Kennes and Veiga, 2001).

$$H_2S + 0.5O_2 \rightarrow S_0 + H_2O \quad (1)$$

$$H_2S + 2O_2 \cdot SO_4^{-2} + 2H+ \quad (2)$$

As the performance of a biofiltration system depends on the microbial community present in the reactor, the determination of the microorganism and the microbial activity responsible for the behavior of the process is very important. However, there is still a lack of understanding of the structure and dynamics of microbial communities and the physiological role of the main microbial population as well as the correlation between the global performance of the system with the metabolic activities of the microorganisms involved in the process. This knowledge could allow control of the reactor behavior and the design of enhanced processes to eliminate high concentrations of H_2S in the gas phase because the performance of the process depends on the robustness of the microbial communities (Maestre et al., 2010).

Some authors have characterized microbial population diversity present in different gas phase reactors by analysis of biomarkers such as phospholipid fatty acids (Webster et al., 1997), molecular techniques such as fluorescent *in situ* hybridization (FISH) (Moller et al., 1996), cloning and sequencing of ribosomal RNA genes (Roy et al., 2003), terminal restriction fragment length polymorphism (Maestre et al., 2009) and denaturing gradient gel electrophoresis (Borin et al., 2006). There are only a few studies in the literature that focused on determining the microbial diversity of microorganisms capable of removing reduced sulfur compounds in biofilters or gas phase bioreactors using molecular biological approaches. Ding et al., 2006, reported the changes in the microbial diversity of a biofilter-treating methanol and H_2S. In this study, the biofilter's initial microbial community had a high diversity, but after the biofiltration system was fed with H_2S, the microbial diversity decreased to adapt to the low pH and use H_2S as an energy source. Maestre et al., 2010 studied and described the bacterial composition of a lab-scale biotrickling filter (BTF) treating high loads of H_2S using 16S rRNA gene clone libraries. The authors reported the diversity, the community structure and the changes in the microbial population on days 42 and 189 of reactor operation. The main changes in microbial diversity were observed at the beginning of the process and again when steady state operation was reached (i.e., neutral pH and at an inlet H_2S concentration of 2000 ppmv). At steady state, the major sequences associated with SOB included *Thiothrix* spp., *Thiobacillus* spp., and *Sulfurimonas denitrificans*. Additionally, FISH analysis was used to determine the spatial distribution of sulfur-oxidizing bacteria (SOB) along the length of the reactor under pseudo-steady state operation. The aerobic species were found to be predominantly along the system, but some facultative anaerobes were also found. The anaerobic microorganisms were associated with higher H_2S concentrations (inlet) with lower oxygen availability. The distribution of a microbial community was associated with changes in the dissolved oxygen (DO) concentration, and the accumulation of elemental sulfur and the pH (Maestre et al., 2010). Recently, Omri et al., 2011 studied the microbial community structure of the three layers (bottom, middle and top) of a biofilter using the polymerase chain reaction-single strand conformation polymorphism (PCR-SSCP) analysis. The results obtained showed a high microbial diversity for bacteria, with the relative diversity of the bacterial community represented by the number of peaks in the profiles. Significant differences were observed between the microbial communities of the three layers of the biofilter. The Simpsons diversity index was used to determine the microbial diversity in the system, and the results indicated that the bottom and middle layers exhibited high diversity (1/D of 13.6 and 10.8, respectively). However, the microbial distribution in the top layer (1/D=8.75) was associated with the vertical gradient of the substrate, as higher H_2S concentrations near the inlet allowed the growth of sulfur-oxidizing bacteria and low pH provided a favorable environment for the oxidation of H_2S. The predominant bacteria in samples of the operation were found to be *Pseudomonas* sp, *Moraxellacea, Acinetobacter* and *Exiguobacterium* belonging to the phyla Pseudomonadaceae, gamma-Proteobacteria and Firmicutes.

In the present chapter, the data obtained for the potential use of FVW and meat residues for methane production will be presented. The results demonstrating how a codigestion process of FVW and MR enhanced methane production by increasing the C/N ratio and controlling the natural pH in a 30L reactor will also be analyzed and discussed. At different stages of the start up of the anaerobic digestion system, methane production increased from 14 to 50% as a result of the use of a protein rich feedstock (MR). However, the H_2S concentration also increased in the biogas stream under these conditions. Due to the increased H_2S content, and considering that this compound does not allow for the efficient use of methane as fuel, a biofiltration system was evaluated in the elimination of H_2S. The results obtained for the elimination of H_2S and VFAs (average concentrations of 1500 ppmv and less than 10 ppmv, respectively) in the gas stream from an anaerobic process by a biofiltration system will then be presented. The microbial population in the biofilter when operating at steady state conditions is also presented and discussed.

2. Materials and methods

2.1 Microbial culture

Microbial culture in the biofiltration system was adapted to 10% (v/v) acetic acid and 40 mM $Na_2SO_3.5H_2O$ in mineral media and used for VFA and H_2S degradation experiments, respectively. In situ immobilization in the biofilter was facilitated by recirculating the solution containing the microbial culture previously adapted (Ramirez-Saenz et al., 2009).

2.2 Experimental setup

2.2.1 Anaerobic digestion system

The anaerobic digestion process was evaluated in an anaerobic digestion reactor (ADR) consisting of a stainless steel tubular reactor with a total volume of 30 L into which 20 L of a (50:50) mixture of FVW and meat residues (MR) was initially packed. After the initial period of operation (20 days), the ADR was inoculated with cow manure (10% v/v) to enrich the methanogenic population. The reactor was stirred by recirculating the FVW twice a day. The pH was initially set at 7 by the addition of a 0.8 N NaOH solution. Later in the process, the pH was naturally regulated by the metabolic intermediates produced during digestion. The bioreactor was kept at room temperature and operated in a fed-batch mode. To avoid inhibition due to metabolic products and to ensure a sufficient supply of organic matter, 2.5Kg of different compositions of fresh feedstock mixtures were fed periodically (approximately every 12-15 days), and an equal volume of exhausted sludge was removed.

2.2.2 Biofiltration system

A lava rock biofilter was used to evaluate the degradation of H_2S from the AD gas stream. The experimental setup for the biofilter used in this study was previously described (Ramirez-Saenz et al 2009). The gas stream was humidified and fed in the top of the biofilter using a mass flow controller. Sample ports were located in the output and input of the gas stream. For H_2S degradation experiments, the biofiltration system was fed at the top with an air-diluted gas stream originated from the ADR, as previously reported (Ramirez-Saenz et al., 2009). Periodic water additions (once a week) were used to control moisture loss and to avoid SO_4^{-2} accumulation. Recirculation was provided at a flux of 0.5 L/min over 1 h. All experiments were conducted at room temperature (20-25°C).

2.2.3 H_2S elimination tests in the biofilter

Different dilutions of the biogas stream produced in the AD at an initial concentration of 3000 ppmv of H_2S were prepared by mixing the biogas with humidified air. Two empty-bed residence times (EBRT), 31 and 85 s, were chosen for the performance of the reactor during the H_2S and VFA biodegradation tests. Increasing mass loading rates from 99 g/m^3h to 400 g/m^3h (corresponding to 850 and 3000 ppmv H_2S) were used for evaluating H_2S removal at both EBRT. Gas samples of the inlet and outlet ports of the biofilter were periodically collected and diluted in 10 L Tedlar bags before taking measurements to determine the H_2S and VFA consumption in the biofiltration system. Details of the analysis conditions were previously reported in Ramirez-Saenz et al., 2009.

2.3 Analytical methods

The fruit and vegetable waste samples were analyzed for total solids (TS) and volatile solids (VS) according to the standard methods of the American Public Health Association (APHA, 2005).

Biogas production in the anaerobic digester was periodically measured using a water displacement setup in which the biogas was passed through a 5% NaOH solution (Anaerobic Lab Work, 1992). Biogas samples were taken periodically from the gas collection lines prior to the water displacement setup, and the gas composition was analyzed using a gas chromatograph (GowMac Series 550, Bethlehem, PA) equipped with a thermal conductivity detector. A CTR1-packed column (Alltech Co., Deerfield, IL) was used for the analysis. The analysis conditions were the same as those reported previously (Garcia-Peña et al., 2009). VFA samples were analyzed in a gas chromatograph (Buck Scientific, East Norwalk, CT) as previously reported (Garcia-Peña et al., 2009).

Biofilter samples were analyzed for H_2S consumption by measuring H2S concentrations of the inlet and outlet of the biofilter using a gas analyzer (Testo 350XL, Clean Air Engineering, Inc., Pittsburgh, PA).

2.4 Microbial characterization and identification in the biofiltration system

DNA samples of the microbial consortium along different lengths of the biofilter were extracted using an Easy-DNATM Kit (Invitrogen, USA) following the manufacturer's instructions. The 16S rRNA gene was amplified using universal bacterial primers. Polymerase chain reaction (PCR) was conducted as previously reported (Garcia-Peña et al., 2011). The PCR amplification products were purified using a QIAquick PCR Purification Kit (Qiagen, UK). The PCR-amplified DNA products were separated by DGGE on 8% polyacrylamide gels with a linear gradient of 5-30% denaturant (100% denaturant was 40% [v/v] formamide plus 42% [w/v] urea) using the DCodeTM System (Bio-Rad, Hercules, CA, USA). After purification, the PCR products were sent to a sequencing service (Macrogen Inc., Korea). The nucleotide sequences of each 16S rRNA gene were aligned using the Clustal X software (Higgins et al., 1996). Identification of the sequences was made after performing BLAST searches of the NCBI database.

2.5 Data analyses

The data obtained for the microbial population by DGGE were further analyzed by Jaccard's (3) and Sorensen-Dice´s (4) indexes. Similarity indices are frequently used to study the coexistence of species or the similarity of sampling sites. A matrix of similarity coefficients, between either species or locations, may be used to analyze changes in microbial populations over time or at different locations (Real and Vargas, 1996). Jaccard's index is one of the most useful and widely used indices to determine similarity between binary samples. Jaccard's index may be expressed as follows:

$$J = \frac{n_{AB}}{n_A + n_B - n_{AB}} \tag{3}$$

where n_{AB} is the number of bands in both samples A and B, n_A is the number of bands present in sample A, and n_B is the number of bands present in sample B.

Sorensen-Dice´s index, also known as Sørensen's similarity coefficient, is also used to compare the similarity of two samples. It can also be applied to the presence/absence of data. The index is described by the expression, the terms of the equations are the same as describe above:

$$S_D = \frac{2n_{AB}}{n_A + n_B} \tag{4}$$

3. Results

3.1 Methane production from FVW

Evaluation of the feasibility of producing methane from fruit and vegetable waste (FVW) was performed in two different stages: an initial stage with different conditions for high biogas and methane yields and VS removal evaluated in batch experiments. The evaluation of the anaerobic digestion process in a fed-batch mode and the feasibility of the long operation time under stable conditions were determined in a second stage of the project.

The design and efficient operation of an anaerobic digestion system can be determined by establishing the physical and chemical characteristics of the organic waste and the culture conditions, which have an effect on biogas production and process stability of the system. The residue mixture used as feedstock was composed of products most frequently and consistently sold in the market (i.e., excluding seasonal products). The FVW mixture (equal proportions of each residue, w/w) showed a total solid (TS) content of approximately 73-100 g/Kg waste (approximately 10%), a pH of 4, and a moisture content of 90% (Garcia-Peña et al., 2011). The FVW had a higher soluble carbohydrate content (only contained fruits and vegetables and did not include a source of protein), a high moisture content, and could be considered a highly degradable substrate. These properties make it an ideal candidate for CH_4 production.

Since the FVW is highly degradable, large amounts of volatile fatty acids (VFA) are produced and a rapid acidification of the system could inhibit the biological activity of the methanogens. Some reports have demonstrated that pH control is one of the most important parameters in achieving high biogas production (Mata-Alvarez 1992, Bouallagui et al., 2005). Additionally, as mentioned above, the FVW contains no nitrogen source. Some experiments were also performed to evaluate the effect of adding a nitrogen source, considering that an adequate C/N ratio is necessary to enhance the anaerobic digestion process. Different conditions were thus tested to optimize biogas and methane production using FVW, including buffered and nitrogen supplemented systems with and without inoculation.

Figure 1 shows the data obtained for TS removal, biogas production (m^3/kgVS) and final pH under the evaluated conditions. As an overall result, the inoculation, pH control and nitrogen source addition all had positive effects on VS reduction and biogas yield. Higher degradation percentages of approximately 86 % of the initial total volatile solid (tVS) were obtained in the inoculated and pH-regulated system (IpH) as well as in the inoculated, pH-regulated and nitrogen-added system (IpHN) in a 28-day culture. Higher VS removal was correlated with higher biogas production. The highest biogas production (0.42 m^3/kgVS) was obtained in the inoculated system with pH control and nitrogen addition (IpHN), reaching a VS removal percentage of 86%.

Lower biogas productions of 0.15 m^3/kgVS and 0.08 m^3/kgVS were measured in the systems without inoculum (WIpH and WIpHN) and had correspondingly low VS removals of 42 and 34%, respectively.

The results suggested a correlation of pH with biogas productivity, with higher productivity occurring at pH values close to the optimum pH of 7. Methane production (approximately 45-53+0.5% v/v in the biogas mixture) only occurred in the inoculated systems, with the highest methane percentage (53 %, v/v) observed for the IpHN system.

Biogas and CH_4 production were in the range (0.16-0.47 m^3/kgVS) of those reported by other authors for anaerobic digestion processes using FVW as a feedstock (Rajeshwari et al., 1998, Alvarez 2004, Mata-Alvarez et al., 1992; Boullagui et al., 2003). In the inoculated systems, both nitrogen addition and pH control had positive influences on biogas production. Therefore, these conditions should be used to produce the maximum amount of methane from FVW.

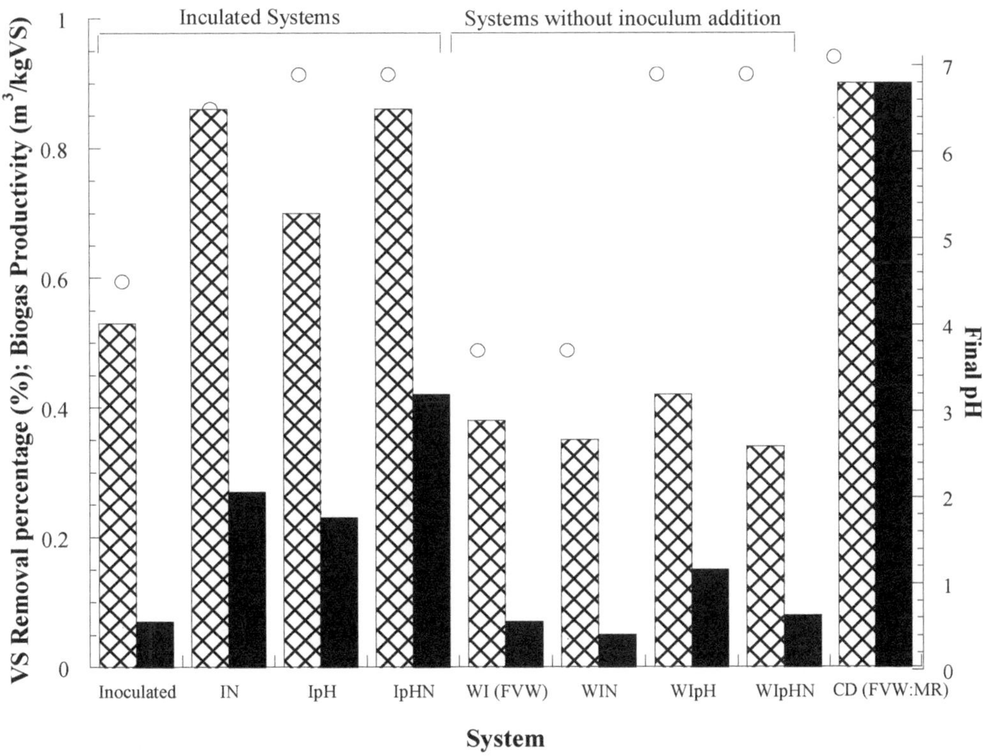

Fig. 1. VS removal percentage (%/100); Biogas productivity in $m^3/kgVS$; (○) Final pH obtained for different conditions evaluated in the batch systems. I, FVW inoculated with cow manure (10%); IN, FVW inoculated and supplemented with NH_4Cl as a nitrogen source; IpH FVW inoculated and salts added (buffer) to control pH; IpHN, FVW inoculated, buffering salts, and NH_4Cl added; WI, FVW without inoculation (Control); WIN, FVW and NH_4Cl; WIpH, FVW and buffering salts; and WIpHN, FVW buffering salts and NH_4. Modified from Garcia-Peña et al., 2011.

Similar conditions were obtained by the codigestion of a mixture of FVW and meat residues (obtained from meat packaging operations at the same market). The meat residues (MR) provide a high nitrogen concentration, and protein hydrolysis could result in natural pH control due to NH_4 production. For the codigestion system of FVW and MR, the highest biogas yield of 0.9 $m^3/kgVS$ (methane yield of 0.45 $m^3/KgVS$) was observed, reaching an organic matter degradation of 93% (Figure 1).

The feasibility of an anaerobic digestion process using FVW and MR was then evaluated in a 30 L ADR system. To determine the effect of MR addition on biogas and methane production, experiments were carried out using different MR proportions. The biogas productivity and methane percentages obtained under different conditions after 130 days of operation are presented in Table 1.

	Time of operation (days)	Biogas production (m^3/kgVS)	VS removal (%)	Methane (%)[b]	Methane production (m^3/kgTS)
Start-up	0-19	1.03	89	0	0
Inoculation	20-43	0.64	81	16	0.10
75:25	43-55	0.5	70	28	0.14
50:50	55-64	0.4	73	30	0.12
100:0	64-76	0.24	80	14	0.033
50:50	76-83	0.2	75	30	0.06
75:25	83-130	0.25[a]	78	53	0.135
Current operation		0.4	65	63	0.252

Table 1. Biogas production, VS removal and methane production during the start-up of the ADS. a average, b biogas was composed mainly of CO_2 and CH_4, with the remaining biogas percentage (v/v) accounted for by CO_2.

In the first stage, no methane was observed in the biogas effluent, thus biogas production resulted from the hydrolysis of the easily degradable components of the feedstock. After 20 days of operation, the anaerobic digestion reactor (ADR) was inoculated with (13 %, v/v cow manure) and a new, strong biological activity was observed. In this period, the CH_4 content in the biogas was 16 % (v/v) due to the initial activity of the methanogenic population introduced into the ADR with the inoculum. The methane percentage increased from 16 to 30% (v/v) as the proportion of MR (50:50) increased. When steady state was reached (after 80 days of operation) with a 75:25 mixture of FVW and MR, the CH_4 percentage was stable at 53 ± 2 %, and the pH was stable at 6.9 ± 0.5 (naturally regulated during this last stage of the process).

For the next stage (on the 70th day of operation), a 50:50 mixture of FVW:MR was added and the CH_4 percentage recovered to 30%. Once stable operation was achieved after 83 days, the biogas production showed a constant value of approximately 0.25 m^3/kgVS and a methane percentage of 53%, corresponding to a methane production of 0.135 m^3/kgVS and a VS removal of 78%. The reactor was regularly fed with a 75:25 mixture of FVW:MR. Under these conditions, the CH_4 percentage was stable at 53 ± 2 %, and the pH was stable at 6.9 ± 0.5.

An appropriate buffering capacity and a highly stable experimental system were observed with Organic Loading Rates (OLRs) in the range of 2.4 to 2.7 g COD/L day (Hydraulic Retention Times (HRT) in the range of 15–20 days). The natural pH regulation during the stable operation of the ADS was a result of NH_3 release from protein hydrolysis. The results could also be explained as an effect of the alkalinity in the system. At normal percentages of CO_2 in the digester gas, between 25 and 45%, a total alkalinity of at least 500 to 900 mg/L is required to keep the pH above 6.5. Higher CO_2 partial pressure makes alkalinity requirements larger (Rittmann and McCarty, 2001). When the meat residues were introduced into the ADS, the alkalinity started to increase considering that the moles of bicarbonate alkalinity was equal to the moles of NH_4 according to the stoichiometric equation for methanogenesis of an organic mixture (carbohydrates and proteins) (Garcia-Peña et al 2011). Additionally, at stable and at long operation times the CO_2 percentage was lower than those obtained during the start-up of the ADS. The required alkalinity was 3445 mg/L, while an alkalinity as $CaCO_3$ of 4804.6 mg/L was calculated under the experimental conditions (i.e., 70% removal efficiency and an initial substrate concentration of 50 gCOD) using the stoichiometric equation for an organic mixture of carbohydrates and protein (50:50) (Garcia-Peña et al, 2011). This total alkalinity value was high enough to avoid a possible acidification of the Anaerobic Digestion System (ADS), and the high buffering capacity allowed stable operation without external control. The increase in methane production after the 80th day of operation could have resulted from an increase in the methanogenic population and its adaptation to the operating conditions of the ADS. The high VS removal, the increased methane yield, and the natural pH control during the stable period of the ADS was due to an adequate ratio of nutrients and the availability of proteins for new cell synthesis (Garcia-Peña et al 2011).

3.2 H_2S elimination from the biogas stream using a biofiltration system

A characterization of the three potential biofilter packing materials was performed. The highest water retention capacity (WRC) was found in vermiculite (65%), while the WRC for lava rock was 15%. Although vermiculite showed a higher WRC, lava rock favored water irrigation, which ensured that the desired moisture level was maintained and avoided sulfate accumulation at the same time.

The acetic, propionic, butyric and valeric acid could also be present in the biogas stream, their assimilation was determined in batch experiments. The biodegradation of different loadings of acetic and propionic acids as individual substrates were also evaluated in the lava rock biofilter as it as previously described elsewhere (Ramirez-Saenz et al., 2009).

As the MR proportion increased in the ADS, the H_2S concentration in the biogas stream also increased. The biodegradation of H_2S was determined in the lava rock biofilter under two different empty-bed residence times (EBRT). Results for H_2S elimination capacity as a function of H_2S inlet loading in the lava rock biofilter, operated at 85 sec and 31 sec EBRT, are depicted in Figure 2. As shown in Figure 2A, at an EBRT of 85 sec, the relationship between the inlet loading and the elimination capacity was linear, and the critical H_2S elimination capacity defined by Devinny et al., 1999 (i.e., deviation from the 100% removal capacity) was not yet reached at an inlet loading of 144 g/m^3h. Under these operation conditions, the removal efficiency of H_2S for loadings between 36 and 144 g/m^3h was always above 98 %. Furthermore, the EC reached a maximum of 142 g/m^3h when the H_2S loading was 144 g/m^3h.

For an EBRT of 31 sec (Figure 2B), the H_2S elimination capacity was found to be linear with respect to H_2S inlet loading up to 200 g/m^3h (100% removal efficiency). A higher inlet loading of 300 g/m^3h reduced the removal efficiency in the system to 85 %. An inlet loading of 400 g/m^3h (corresponding to 3000 ppmv) caused the removal efficiency to drop to 75%, which suggested inhibition of biological activity and/or insufficient mass transfer. In this case, the critical H_2S EC was 200 g/m^3h, whereas a maximum H_2S EC value of 232 g/m^3h was achieved in the biofilter. At the same time, however, the removal of VFAs present in the gaseous stream (approximately 10 ppmv) reached 99%.

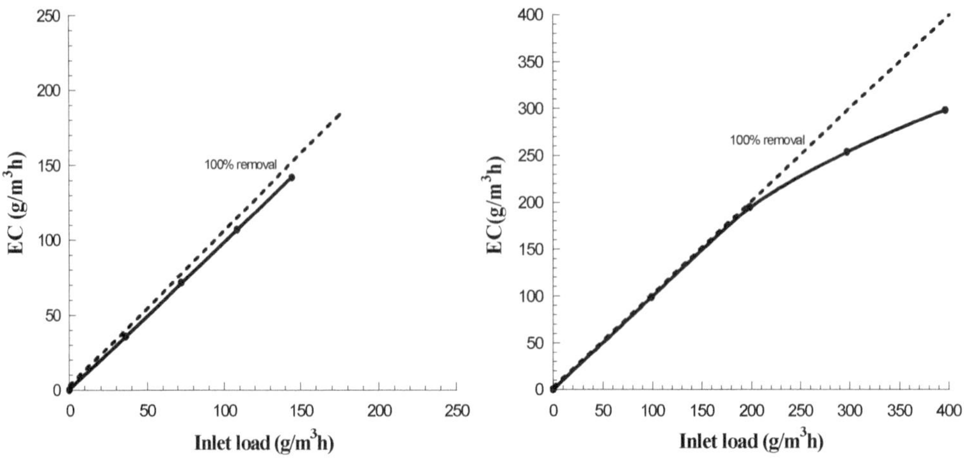

Fig. 2. H2S elimination capacities as a function of H2S inlet loadings in a biofilter operated at A) 85 sec EBRT and B) 31 sec EBRT. The points and solid lines represent the experimental data, and the dashed line (--) is the 100% removal line measured at both EBRTs.

For long operation times, the biofilter nearly eliminated all the H_2S from the biogas stream. The H_2S concentrations of the AD gas stream were previously diluted to maintain an inlet concentration of 1500 ppmv and to allow complete elimination (99% removal efficiency), but the high removal efficiency was maintained over 90 days and complete biodegradation of VFAs was also observed. Fifty days after the start-up period, a technical failure in the AD system blocked the feeding of the biofilter, no data was obtained during that time. After operation conditions were restored, an inlet H_2S concentration was maintained at approximately 1500 ppmv from day 103 to 194 at an EBRT of 31 sec. Under these conditions, a removal efficiency of 95% was maintained for 90 days. Higher concentrations, around 3000 ppmv, caused a drop in the biofilter efficiency to 50% (Ramirez-Saenz et al., 2009). The biofilter was fed with the ADS gas stream every two weeks, which corresponded to the HRT of the AD system.

According to the stoichiometry of aerobic biological H_2S oxidation (Eq. 1 and 2) and the sulfate determinations obtained between days 103 and 194 of the operational period of the biofilter, 51 to 60% of the H_2S was completely oxidized to sulfate. These data are correlated with those reported by Fortuny et al., 2008 with respect to the H_2S conversion to sulfate. The elimination of these compounds allowed the potential use of the biogas while maintaining the methane (CH_4) content throughout the process.

3.3 Microbial community characterization

Samples from three different positions of the biofiltration system were taken to evaluate the spatial distribution of the microbial population. Figure 3 shows a picture of the biofiltration system and the positions where the samples were collected. The samples at the top of the reactor "a" correspond to the inlet of the biogas stream mixed with different air fluxes, samples "b" and "c" correspond to the middle part of the reactor and sample "d" was located at the bottom of the biofilter (outlet of the biogas stream).

Fig. 3. Biofiltration system coupled to an anaerobic digestion system. Arrows indicate the position of sampling used for DGGE analysis.

Additionally, the changes in the bacterial community were also determined by taking samples during long-term operation of the biofiltration system (Samples from 1a to 9a). The analysis was performed by a DDGE system using 16S rRNA as a bacteria-specific target for PCR amplification. Figure 4 shows an example of denaturing gradient gel DDGE (15% to 60%) from samples of different times of cultivation compared with the initial bacterial community. In summary, around 13 bands for the total bacterial community were systematically detected over long-term operation of the biofiltration system. (Samples 1a to 3a correspond to days 5, 10 and 20th of operation. Sample 4a was obtained at day 45th of operation, when the inlet load was increased to 3000 ppmv. Samples from 5a to 8a were obtained in days 90, 110 130 and 150 of operation, respectively.) In view of the total bacterial community, the bands remained constant until variations in intensity appeared. In lane 4a, lower intensity bands revealed a weakening pattern, which suggest a decrease in certain types of bacteria, when the H_2S concentration increased from 1500 ppm to 3000 ppm. Both the decrease in removal efficiency and the decrease in the microbial population could be explained by the toxicity of the extremely high H_2S concentration. This factor was assumed to be responsible for the disappearance of some of the microbial species. Increased intensity in some bands in the gel (boxes A and B) demonstrated intensifications of specific band patterns. These data suggested the eventual dominance of H_2S-oxidizing bacteria (SOB) and bacteria able to consume VFA.

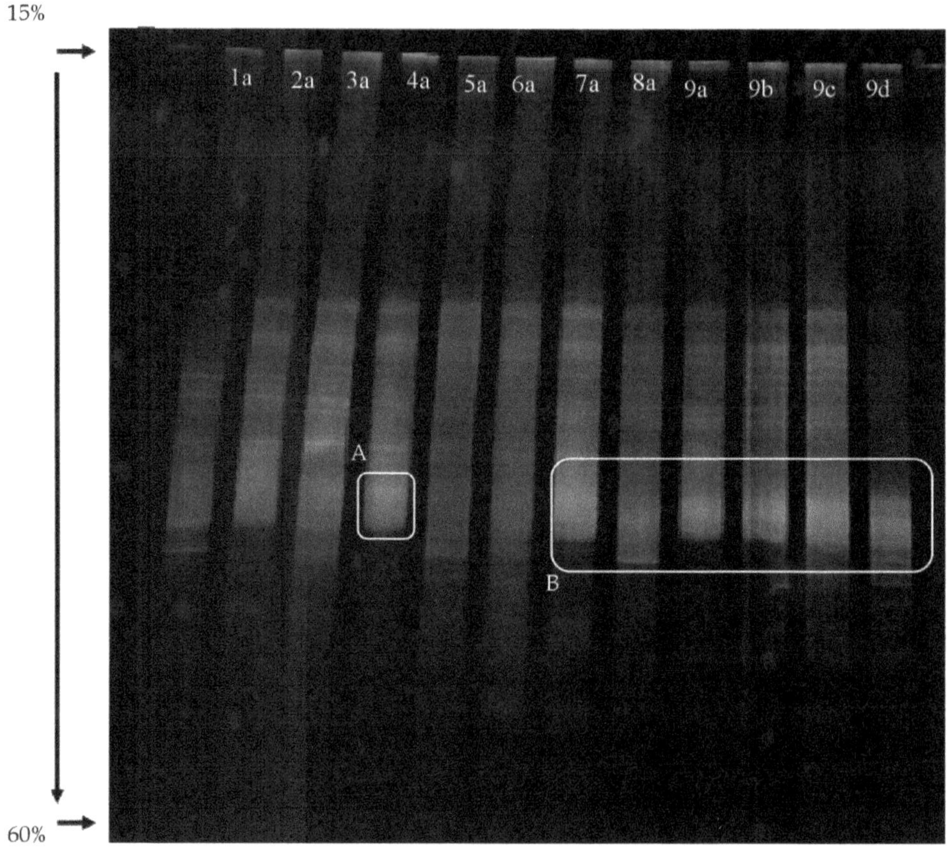

Fig. 4. Polyacrylamide denaturing gradient gel (15–60%) with DGGE profiles of 16 S rRNA gene fragments of the samples taken from different operation times (Lanes from 1a to 8a) and locations of the biofilter (Lanes from 9a, inlet to 9d, outlet), (6-h run, 200 V, 60 °C).

To compare bacterial community between different samples and to determine possible changes in composition, the presence or absence of a band in a DDGE gel was analyzed using a binary system. A 0 value was assigned when the band was absent (i.e., different band is considered a different microorganism) and 1 when the band in two or more samples was present (i.e., same microorganism) at similar positions in the gel. Jaccard's index and the Sorensen-Dice index could then be calculated. Table 2 shows the matrix constructed using the DDGE gel containing different band patterns obtained at different times of operation (lanes 1a to 9a) and at different lengths along the biofilter (9a, 9b, 9c and 9d). Nineteen different bands (arbitrarily named A to S) were found in the samples analyzed by gradient DDGE.

Once the number of bands that were similar or different between the two samples was determined, the similarity of the different samples was determined by calculating the Jaccard and Sorensen-Dice indexes. Two different aspects were analyzed: the similarity of the samples during the time of cultivation (lanes 1a to 9a) and the similarity at a different position in the reactor (lane 9a compared to 9b, 9c and 9d).

Band/Lane	1a	2a	3a	4a	5a	6a	7a	8a	9a	2b	2c	2d
A	1	1	1	1	1	1	1	1	1	1	1	1
B	1	1	1	1	1	1	1	1	1	1	1	1
C	0	0	1	0	1	1	1	0	0	0	0	0
D	0	1	1	1	1	1	1	1	1	1	1	0
E	0	1	1	1	0	1	1	1	1	1	1	0
F	1	1	1	1	1	1	1	1	1	1	1	1
G	1	1	1	1	1	1	1	1	1	1	1	1
H	1	1	1	1	1	1	1	1	1	1	1	0
I	1	1	1	1	1	1	1	1	1	0	1	0
J	1	1	1	0	1	1	1	1	1	1	1	0
K	1	1	1	1	0	0	1	1	1	1	1	0
L	1	1	0	1	0	0	0	1	0	0	0	0
M	1	0	1	0	0	0	0	0	0	0	0	0
N	1	0	1	0	1	1	0	1	0	0	0	1
O	1	0	1	0	1	1	1	0	0	0	0	1
P	0	1	1	0	1	1	1	1	0	0	0	1
Q	0	0	0	0	0	1	1	0	0	0	1	1
R	1	1	1	0	1	1	1	0	0	1	1	1
S	0	0	0	0	0	0	0	0	0	0	0	1
Total	13	13	16	10	13	15	15	13	10	10	12	10

Table 2. Matrix constructed using the DDGE gel containing different band patterns. Lanes 1a-8a correspond to samples at 8 different times of cultivation. Lanes 9a to 9d were samples at different lengths of the biofiltration system.

Regarding time of cultivation, the bands A, B and E to J constantly appeared in the microbial community, suggesting little change in the microbial populations during the operation of the biofilter. The most similar microbial communities were found in lanes 6a and 7a, with a Jaccard index of 0.875 and a Sorensen-Dice index of 0.933, which corresponded to the steady state of the biofiltration system at an average H_2S inlet concentration of 1500 ppmv and a removal efficiency of 95%. In contrast, the least similarity was found between lanes 4a and 5a, 6a and 7a (Jaccar´s indexes of 0.438, 0.47 and 0.56, respectively). In lane 4a, the microbial community sample was exposed to an increased H_2S concentration of 3000 ppmv. These data differed from those found by Maestre et al., 2010. These authors reported a wide phylogenetic diversity and showed that the initial populations became more specific, being the SOB the dominant community.

The similarity between the microbial communities along the biofilter was also calculated. For this purpose, the bands in lane 9a were compared with the bands in lanes 9b, 9c and 9d. Significant differences in the microbial population were observed at different lengths along the biofilter. The highest divergence was found between lanes 9d and 9b, with a Jaccard index of 0.333 and a Sorensen-Dice index of 0.5. These data could be partially explained by the H_2S concentration gradient: a higher concentration at the inlet of the biofilter and a lower concentration at the outlet (sample 9d). The accumulation of metabolic products could also explain the divergence. The highest similarity was found in samples of lanes 9b and 9c, which corresponded to the middle of the reactor, where apparently the environmental conditions were more homogeneous (Jaccard index of 0.833 and Sorensen-Dice index of 0.909). These data are in agreement with the results obtained by Maestre et al., 2010 and Omri et al., 2011 about the divergence in microbial populations along the reactor.

Sequence analysis of DNA extracted from single bands representing specific species were then used as an approach for further community characterization. Sequence analyses of bands (Table 3) revealed the predominant bacteria in the biofiltration system. The structure of the bacterial community sequenced was associated with microbial activity in the system as a function of the pollutant eliminated in the biofiltration system.

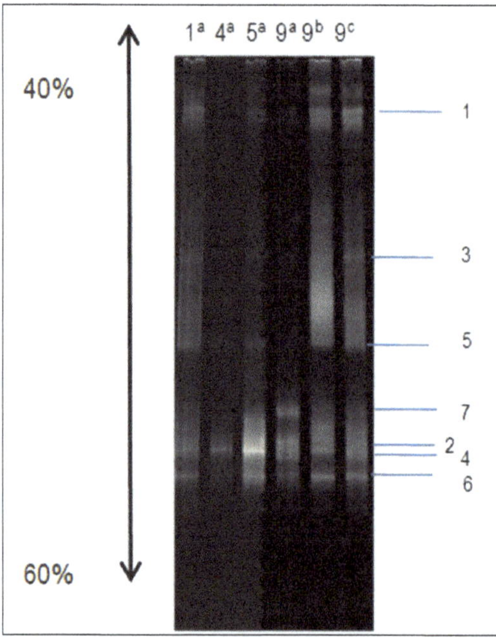

Band No.	Closest relative	Identity (%)
1	*Agromyces mediolanus*	100
2	*Arcobacter butzleri*	99%
3	*Bacillus cereus*	98%
4	*Bosea thiooxidans*	94%
5	*Butirivibrio fibrisolvens*	99%
6	*Thiobacillus sp.*	100%
7	*Uncultured bacteria*	98%

Table 3. Sequence analysis and species identification of the major (7) DGGE bands for the biofilter samples.

Band sequencing results showed that the dominant members of SOB consisted of *Bosea thiooxidans* and *Thiobacillus* sp (Table 3). Das et al., 1996 reported that *Bosea thiooxidans* was a new gram-negative bacterium isolated from agricultural soil and capable of oxidizing reduced inorganic sulfur compounds. Data showed that this microorganism was strictly aerobic. Experiments conducted to evaluate thiosulfate oxidation showed that the growth yield varied with the concentration of this compound; the greatest growth was observed at a concentration of 5 g/L. Under these conditions, conversion of thiosulfate to sulfate was stoichiometric, and the pH of the medium decreased from 8.0 to 6.6. The distance matrix phylogenetic tree based on the level of difference between *Bosea thiooxidans* and 19 reference strains of the alpha subclass of the Proteobacteria indicated that strain BI-42 belonged to a new lineage located between the methylotrophs, the genus *Beijerinckia*, and the *Rhodopseudomonas palustris* group. No close relationship was found between the strain and other sulfur-oxidizing bacteria, such as *Thiobacillus acidophilus* and *Acidiphilium* species (Das et al., 1996). Microorganisms utilize sulfur compounds for the biosynthesis of cellular material or transform these compounds as part of a respiratory energy-generating process. Most of the known sulfur-oxidizing bacteria belong to the genera *Thiobacillus, Thiothrix, Beggiatoa, Thiomicrospira, Achromatium, Desulfovibrio, Desulfomonas, Desulfococcus,* and *Desulfuromonas* . Furthermore, members of the genus *Thiobacillus* have been studied extensively to increase understanding of the coupling of oxidation of reduced inorganic sulfur compounds to energy biosynthesis and assimilation of carbon dioxide.

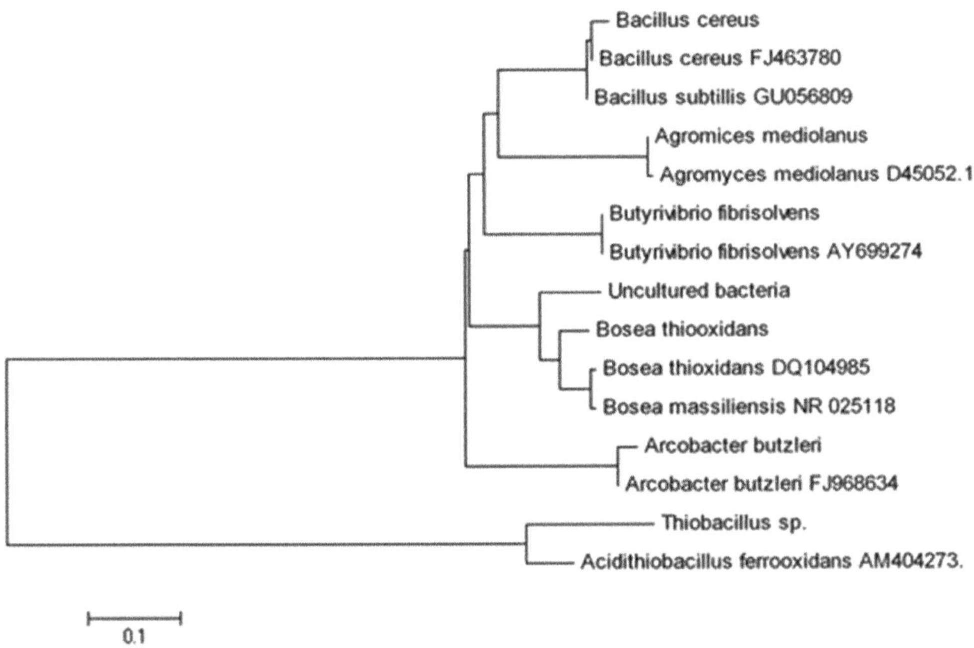

Fig. 5. Neighbor-joining tree of partial 16S rRNA sequences (approximately 750 bp) recovered by denaturing gradient gel electrophoresis (DGGE) bands in the biofilter. The bar indicates 1% sequence variation.

The presence of *Arcobacter butzleri,* belonging to the Phylum Proteobacteria (e-Proteobacteria), could be associated with VFA degradation (these compounds were introduced into the biofilter through the biogas stream). This microorganism is able to grow under both aerobic and anaerobic conditions over a wide temperature range (15–42 °C). However, optimal growth occurs under microaerobic conditions (3–10% O_2). *Arcobacter butzleri* was also recently found as a member of the microbial population in a microbial fuel cell (MFC) used to produce electricity from synthetic domestic wastewater that contained a mixture of VFAs as electron donors (Freguia et al., 2010). Similar results were obtained by Nien et al., 2011, where an *Arcobacter butzleri* strain, ED-1, was also determined to be part of the microbial community of a MFC fed with acetate. Although aerobic species were predominant because the metabolic activity determined (sulfate as the main product), the DGGE showed that profile some facultative anaerobes were as part of the microbial population, which could be related to the trophic properties of the community, and the different substrates in the biogas stream (H_2S and VFAs).

Some of the species found in the present study agreed with those previously reported in the literature for biofiltration systems used in the removal of reduced sulfur compounds. For example, Ding et al., 2006 studied a packed compost biofilter for the treatment of a mixture of H_2S and methanol using 16S rRNA sequencing analysis. The authors established that the microbial community was composed of strains of *Thiobacillus, Sulfobacillus,* and *Alicyclobacillus hesperidensis*. In a biofilter packed with compost, activated carbon and sludge used for the removal of H_2S, Chung, 2007 determined a microbial population composed of *Pseudomonas citronellolis, P. fluorescens, P. putida, S. capitis, Bacillus subtilis* and *Paracoccus denitrificans*. In a recently published work (Omri et al., 2011), it was reported that most bacteria in the operation samples were of the genera *Pseudomonas* sp., *Moraxellacea, Acinetobacter* and *Exiguobacterium,* which belong to the phyla Pseudomonadaceae, gamma-Proteobacteria and Firmicutes.

A neighbor-joining tree (Fig 5.) of partial 16S rRNA sequences (approximately 750 bp) was constructed in MEGA4 (Tamura et al., 2007) by considering sequences obtained and comparing them with others in the data bank.

4. Conclusion

The feasibility of CH_4 was demonstrated. High VS removal, the increased methane yield, and the natural pH control during the stable period of the ADS was obtained by codigestion of VFW and MR, due to an adequate ratio of nutrients and the availability of proteins for new cell synthesis. However, the increasing MR concentration in the ADS increased the H_2S concentration in the gas stream. The elimination of H_2S and VFAs by a biofiltration system was successfully determined, reaching high removal efficiencies of both compounds (95% and 99%, respectively). This approach could allow the potential use of the biogas maintaining the methane (CH_4) content throughout the process. The microbial population characterization of the bioifltration system showed that dominant members of SOB were *Bosea thiooxidans* and *Thiobacillus* sp. Some facultative anaerobes were also determined in the system, which could be explained by the composition of the biogas stream and the conditions at different length of the reactor.

5. Acknowledgment

This work was supported through funding provided by the CONACYT grant 60976 and Instituto Politécnico Nacional, grant SIP 20113067.

6. References

Abatzoglou, N. & Boivin, S. (2009). A review of biogas purification processes. *Biofuels, Bioproducts and Biorefining*, 3, 42-71.

Agdag, O.N. & Sponza, D.T. (2005). Co-digestion of industrial sludge with municipal solid wastes in anaerobic simulated landfilling reactors. *Process Biochemistry*, 40, 1871-1879.

Alvarez, R. (2004). Produccion anaerobica de biogas, aprovechamiento de los residuos del proceso anaerobico. Technical report. Research project.

Anaerobic Lab Work (1992). International course on anaerobic waste water treatment. Wageningen University and IHE Delft. The Netherlands.

Angelidaki, I., Ellegaard, L., Ahring B.K. (2003). Applications of the anaerobic digestion process. *Advances in Biochemical Engineering/Biotechnology*, 82, 1-33.

APHA (2005). Standard methods for the examination of water and wastewater. American Public Health Association, Washington, DC.

Borin, S., Marzorati, M., Brusetti, L., Zilli, M., Cherif, H., Hassen, A., Converti, A., Sorlini, C., Daffonchio, D. (2006). Microbial succession in a compost-packed biofilter treating benzene-contaminated air. *Biodegradation*, 17, 79–89.

Bouallagui, H., BenCheikh, R., Marouani, L., Hamdi, M. (2003). Mesophilic biogas production from fruit and vegetable waste in tubular digester. *Bioresource Technology*, 86, 85-90.

Bouallagui, H., Lahdheb, H., Ben Romdan, E., Rachdi, B., Hamdi, M. (2009). Improvement of fruit and vegetable waste anaerobic digestion performance and stability with co-substrates addition. *Journal of Environmental Management*, 90, 1844-1849.

Bouallagui, H., Touhami, Y., Ben Cheikh, R., Hamdia, M. (2005). Bioreactor performance in anaerobic digestion of fruit and vegetable wastes: Review. *Process Biochemistry*, 40, 989-995.

Chung, Y.C., Huang, C., Tseng, C.P. (1996). Operation and optimization of *Thiobacillus thioparus* CH_{11}. Biofilter for hydrogen sulphide removal. *Journal of Biotechnology*, 52, 31-39.

Chung,Y.C. (2007(. Evaluation of gas removal and bacterial community diversity in a biofilter developed to treat composting exhaust gases. *Journal of Hazardous Materials*, 144, 377–385.

Das, S.K., Mishra, A.K., Tindall, B.J., Rainey, F.A., Stackebrandt, E. (1996). Oxidation of thiosulfate by a new bacterium, Bosea thiooxidans (strain BI-42) gen. nov., sp. nov.: Analysis of phylogeny based on chemotaxonomy and 16s Ribosomal DNA Sequencing. *International Journal of Systematic Bacteriology*, 46, 981-987.

Devinny, J.S., Deshusses, M., Webster, T. (1999). Biofiltration for air pollution control. Lewis Publishers. USA.

Ding, Y., Das, K. C., Whitman, W. B., Kastner, J. R. (2006). Enhanced biofiltration of hydrogen sulfide in the presence of methanol and resultant bacterial diversity. *American Society of Agricultural and Biological Engineers*, ASABE. 49, 2051- 2059.

Ergas, S.J., Schroeder, E.D., Chang, D.P.Y., Morton, R.L. (1995). Control of volatile organic compounds emissions using a compost biofilter, *Water Environmental Resource*, 67, 816-821.

Fairley, P. (2011). Introduction: Next generation biofuels. *Nature*, 474, S2–S5 23. doi:10.1038/474S02a

Fortuny, M. Baeza, J.A., Gamisans, X., Casas, C., Lafuente, J., Deshusses, M.A, Gabriel, D. (2008). Biological weetening of energy gases mimics in biotrickling filters. *Chemosphere*, 71, 10–17.

Freguia, S., The, E.H., Boon, N., Leung, K.M., Keller, J., Rabaey, K. (2010). Microbial fuel cells operating on mixed fatty acids. *Bioresource Technology* 101, 1233–1238.

Gabriel, D. & Deshusses, M.A. 2003. Retrofitting existing chemical scrubbers to biotrickling filters for H_2S emission control. *Proceedings of the National Academy of Sciences USA* 100, 6308–6312.

Garcia-Peña E.I., Parameswaran P., Miceli J., Canul-Chan M., Krajmalnik R. (2011). Anaerobic digestion process from vegetable and fruit residues; Process and microbial ecology studies. In press *Bioresource Technology* (doi:10.1016/j.biortech.2011.07.068).

García-Peña E.I., Ramirez D., Guerrero-Barajas C., Arriaga-Hurtado L.G. (2009). Semi-continuos biohydrogen production as an approach to generate electricity. *Bioresource Technology*, 100, 6369-6377.

Granite, E.J. & O'Brien T. (2005). Review of novel methods for carbon dioxide separation from flue and fuel gases. *Fuel Process Technology*, 86, 1423-1434.

Habiba, L., Bouallagui, H., Hamdi, M. (2009). Improvement of activated sludge stabilization and filterability during anaerobic digestion by fruit and vegetable waste addition. *Bioresource Technology*, 100, 1555-1560.

Higgins, D.G., Thompson, J.D., Gibson, T.J. (1996). Using CLUSTAL for multiple sequence alignments. *Methods in Enzymology*, 266, 383-400.

Horikawa, M.S., Rossi, F., Gimenes, M.L., Costa, C.M.M., da Silva, M.J.C. (2004). Chemical absorption of H_2S for biogas purification. *Brazilian Journal of Chemical Engineering*, 21, 415-422.

Kapdi, S.S., Vijay, V.K., Rajesh, S.K., Prasad, R. (2005). Biogas scrubbing, compression and storage: perspective and prospectus in Indian context. *Renewable Energy* 30, 1195-1202.

Kennes, C. & Veiga, M.C. (2001). Bioreactors for Waste Treatment, Kluwer Academic Publisher.

Kim, S. & Deshusses, M.A. (2005). Understanding the limits of H2S degrading biotrickling filters using a differential biotrickling filter. *Chemical Engineering Journal*, 113, 119–126.

Krumdieck, S., Wallace, J., Curnow, O. (2008). Compact low energy CO_2 management using amine solution in a packed bubble column. *Chemical Engineering Journal*, 135, 3-9.

Kuenen, J. G. (1989). The colourless sulfur bacteria, p. 1834-1837. In J. T. Staley, M. P. Bryant, N. Pfennig, and J. G. Holt (ed.), Bergey's manual of systematic bacteriology, vol. 3. Williams and Wilkins, Baltimore.

Maestre, J. P., R. Rovira, F.J. Álvarez-Hornos, M. Fortuny, J. Lafuente, X. Gamisans, D. Gabriel. (2010). Bacterial community analysis of a gas-phase biotrickling filter for biogas mimics desulfurization through the rRNA approach. 2010. *Chemosphere* 80, 872–880.

Maestre, J.P., Rovira, R., Gamisans, X., Kinney, K.A., Kirisits, M.J., Lafuente, J., Gabriel, D., 2009. Characterization of the bacterial community in a biotrickling filter treating high loads of H2S by molecular biology tools. *Water Science and Technology*, 59, 1331–1337.

Mata-Alvarez, J., Cecchi, F., Llabrés, P., Pavan, P., (1992). Anaerobic digestion of the Barcelona central food market organic wastes: plant design and feasibility study. *Bioresource Technology*, 42, 33-42.

Misi, S.N., Forster, C.F., (2002). Semi-continuous anaerobic co-digestion of agro-waste. *Environmental Technology*, 23, 445–51.35.

Moller, S., Pedersen, A.R., Poulsen, L.K., Arvin, E., Molin, S. (1996). Activity and threedimensional distribution of toluene-degrading *Pseudomonas putida* in a multispecies biofilm assessed by quantitative in situ hybridization and scanning confocal laser microscopy. *Applied and Environmental Microbiology*, 62, 4632–4640.

Ng, Y.L., Yan, R., Chen, X.G., Geng, A.L., Gould, W.D., Liang, D.T., Koe, L.C.C. (2004). Use of activated carbon as a support medium for H_2S biofiltration and effect of bacterial inmobilization on available pore surface. *Applied Microbiology and Biotechnology*, 66, 259-265.

Nien, P.Ch., Lee, C.Y., Ho, K.Ch., Adav, S.S., Liu, L., Wang, A., Ren, N., Lee, D.J. (2011). Power overshoot in two-chambered microbial fuel cell (MFC). *Bioresource Technology*, 102, 4742–4746.

Omri, I., Bouallagui, H, Aouidi, F., Godon, J.-J., Hamdi, M. (2011). H_2S gas biological removal efficiency and bacterial community diversity in biofilter treating wastewater odor. In press *Bioresource Technology*, (doi:10.1016/j.biortech. 2011.05.094).

Osorio, J. & Torres, J.C. (2009). Biogas purification from anaerobic digestion in a wastewater treatment plant for biofuel production. *Renewable Energy*, 34, 2164-2171.

Pride C. 2002. ATADs, Odors and Biofilters, Florida Water Res. J. , 18-26

Rajeshwari, K.V., Panth, D.C, Lata, K., Kishore, V.V.N. (1998). Studies on biomethanation of vegetable market waste. *Biogas Forum*, 3, 4-11.

Ramirez-Saenz, D., Zarate-Segura, P.B., Guerrero-Barajas, C., García-Peña, E.I. (2009). H_2S and volatile fatty acids elimination by biofiltration: clean-up process for biogas potential use. *Journal of Hazardous Materials*, 163, 1272–1281.

Rasi, S., Lantela, J., Veijanen, A., Rintala, J. (2008). Landfill gas upgrading with countercurrent water wash. *Waste Management*, 28, 1528-1534.

Rattanapan, C., Boonsawang, P., Kantachote, D. (2009). Removal of H_2S in down-flow GAC biofiltration using sulfide oxidizing bacteria from concentrated latex wastewater. *Bioresource Technology*, 100, 125–130

Real R. & Vargas, J.M. (1996). The probabilistic basis of Jaccard's index of similarity. *Systems Biology*, 45, 380-385.

Rittmann, B.E. & McCarty, P.L. (2001). Environmental biotechnology: Principles and applications. New York: McGraw-Hill.

Roy, S., Gendron, J., Delhomenie, M.C., Bibeau, L., Heitz, M., Brzezinski, R. (2003). *Pseudomonas putida* as the dominant toluene-degrading bacterial species during air decontamination by biofiltration. *Applied Microbiology and Biotechnology*, 61, 366–373.

Shareenfdeen, Z., Herner, B., Webb, D. Wilson, S. (2003). Biofiltration eliminates nuisance chemical odors from industrial air streams. *Journal of Industrial Microbiology and Biotechnology*, 30, 168 – 174

Smet, E., Lens, P., van Langenhove, H. (1998). Treatment of waste gases contaminated with odorous sulfur compounds. *Critical Reviews in Environmental Science and Technology*, 28, 89-117.

Suzuki, I. & Silver, M. (1965). The initial product and properties of the sulfur oxidizing enzyme of thiobacilli. *Biochimica et Biophysica Acta*, 122, 22- 33.

Tippayawong, N. & Thanompongchart, P. (2010). Biogas quality upgrade by simultaneous removal of CO_2 and H_2S in a packed column reactor. *Energy*, 35, 4531-4535.

van Groenestijn, J. & Hesselink, P. (1993). Biotechniques for air pollution control, *Biodegradation*, 4, 283-301.

van Groenestijn, J.W. & Kraakman, N.J.R. (2005). Recent developments in biological waste gas purification in Europe. Review. *Chemical Engineering Journal*, 113, 85-91.

Webster, T.S., Devinny, J.S., Torres, E.M., Basrai, S.S. (1997). Microbial ecosystems in compost and granular activated carbon biofilters. *Biotechnology and Bioengineering*, 53, 296-303.

Widdel F. & Pfenning, N. (1984). Dissimilatory sulfate- or sulfur-reducing bacteria, In N. R. Krieg and J. G. Holt (ed.), Bergey's manual of systematic bacteriology, vol. 1. Williams and Wilkins, Baltimore. p. 663-679.

Yang, Y. and Allen, E.R. (1994) Biofiltration control of hydrogen sulphide. 2. Kinetics, biofilter performance and maintenance. *Journal of the Air & Waste Management Association*, 44 1315-1321.

7

Gas Quality Parameter Computation in Intermeshed Networks

Peter Hass
IPSOS Industrial Consulting GmbH, Berlin, Germany

1. Introduction

The increasing number of biogas plants which is favored now as a part of the energy concept of the German Government and the European Union has major impact on traditional gas distribution and transmission systems. In addition, synthetic methane gas or hydrogen injections must be considered in the near future which will originate from wind power generation (conversion of excess capacities). The main aspect of this change is the resulting calorific value which may be subject to changes in a short time which must therefore be measured, calculated and permanently surveyed.

This chapter describes the basics of gas mixing and the various situations which may be encountered and must be handled in the transportation or distribution process. There are some limitations which must be considered for industrial consumers and power plants. Measurements and simulations are required to survey and control the process of gas distribution and finally generate figures for accounting and billing. Some typical examples are presented to give an insight into real situations and projects.

The problems and limitations of the gas distribution process in heterogeneous networks and biogas injections are discussed with respect to the IT-structure and organizational environment. The final benefit that can be achieved is an individual calorific value for each consumer in the grid enabling a fair billing despite the variations of many gas injections from many sources.

2. Current and future gas injections of biogas vs. gas demand

2.1 Biogas to the grid

When the biogas finally had been produced, treated und conditioned it will either be fed into a nearby gas pipeline system or grid or it will be burned and transformed into electric power which is fed into the electric grid. In the following chapters we review and discuss the aspects of the gas grid, only.

Normally, the gas grid used by biogas plants will be a low or medium pressure operated distribution network. Certain conditions may require that the biogas is compressed to a higher level and being fed into a high pressure transportation network.

Typical locations of biogas plants and the major cities the corresponding gas transportation and distribution network are shown in figure 1.

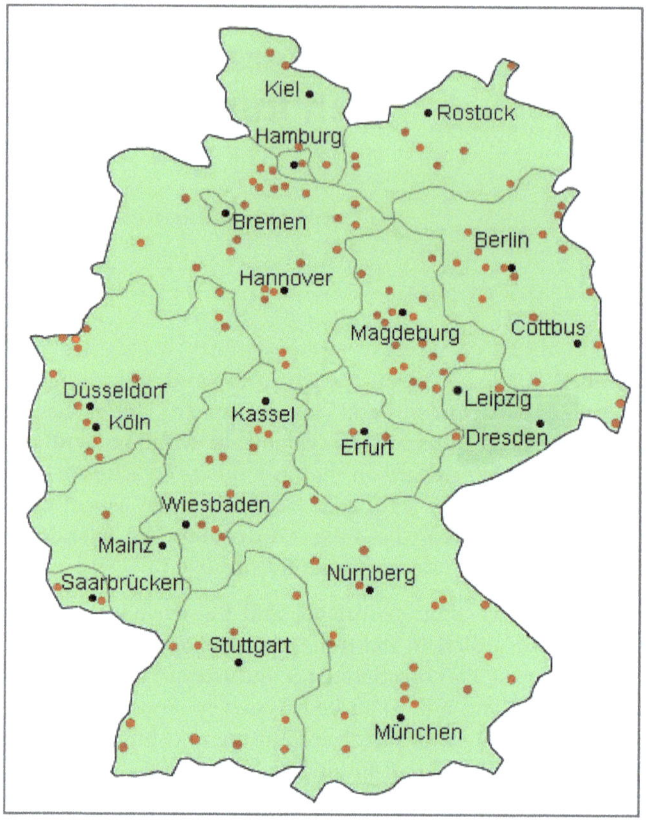

Fig. 1. Biogas plant locations in Germany 2011 (source: DENA)

2.2 The location paradox

Typical biogas plants have preferred locations in rural areas, where the renewable material grows and the transportation is short and easy. In opposition to the rural places the areas of higher gas demand or actual consumption are located in or near urban areas. So, in most cases the biogas must be transported in pipelines over a longer distance spreading also over a wider network area before the biogas is consumed totally. Depending on the network structure and the total consumption figures there exist mixing areas of biogas and natural gas or pure biogas.

2.3 Adverse production-load situations (summer-winter)

As gas is used - at least in middle Europe and Germany - for heating to a high percentage there will be a big difference in consumption figures between winter and summer (see figure 2).

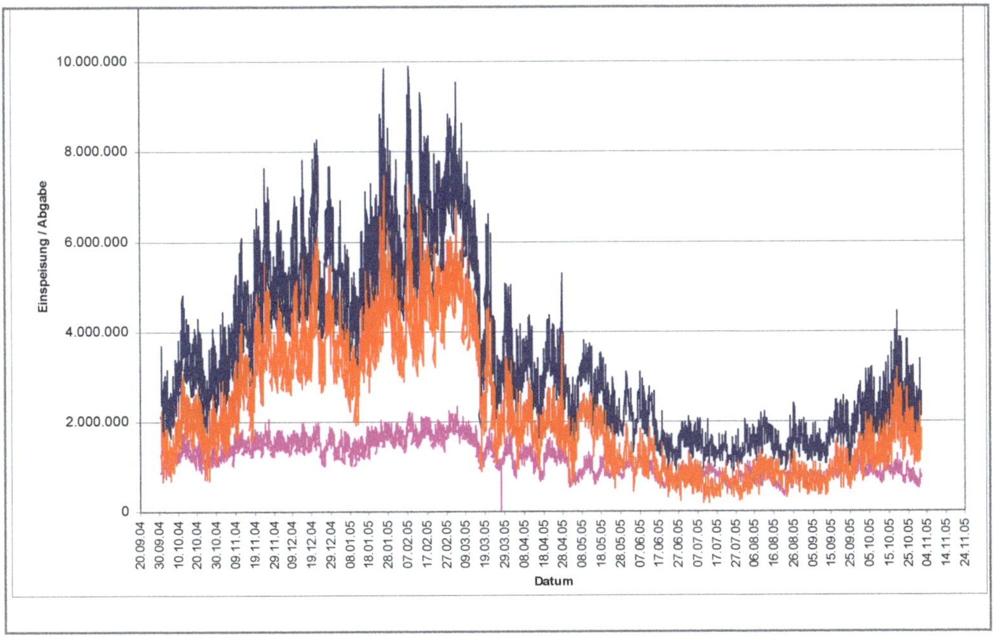

Fig. 2. Total load/feed-in (blue), industrial (RLM) and small consumer load (red) vs. time (hours) of a big city

In certain areas the biogas feed-in in the network is in wintertime only a small percentage and the area of influence is therefore small, too; but it is large in summertime. This fact principally leads to problems in pipeline connection, operation, constant gas quality delivery and fair billing (see below, Operational Aspects). In the future - when the number of plants and/or biogas production will increase - we will expect a considerable higher impact on network operation and surveillance tasks.

3. Basic methods of the gas mixing process

3.1 Gas parameters, gas quality figures (G 260, G 685)

The gas used in the networks for the final customer has to fulfill quality and composition requirements. According to the standard defined by DVGW G 260 working sheet two main types of natural gas (gas families) are distinguished which stem from different sources and production locations:

- H-Gas, high calorific value (Russian source, typically)
- L-Gas, low calorific value (North Sea source, mainly)

Aside from the composition of the gas, for technical reasons the values of calorific value and Wobbe-Index are important characteristics. A typical range of these values is used in practice and will be permanently measured and surveyed:

Value	Shorthand	Unit	Group L	Group H
Wobbe-Index Nominal Value	$W_{S,n}$	kWh/m³	10.5 ... 13.0 12.4	12.8 ... 15.7 15.0
Calorific Value	$H_{S,n}$	kWh/m³	8.4 ... 11.0	10.7 ... 13.1
Relative Density	d_n	-	0.55 ... 0.75	0.55 ... 0.75

Table 1. Essential gas parameters

The calorific value is generally used for the billing, as the final consumer/customer must receive his bill with the energy value included, meaning in the unit of kWh in a period (i.e. a year, a month). The energy value yields from multiplication of accumulated flow and calorific value, e.g. 3000 m^3/a * 10.1 kWh/m^3 equal to 30300 kWh/a.

3.2 Basic equations for mixing

Gas mixing occurs when streams of different gas qualities are united to a single stream. In pipeline systems this means that different gas sources meet in a T-type or Y-type of pipeline. A simple example of two streams of volume (V) with two calorific values (H) is given below (see figure 3). The resultant value depends on the product of volumes or flow (Q) and the amount of each calorific value according to:

$$H_{1,2} = \frac{V_1 * H_1 + V_2 * H_2}{V_1 + V_2}$$

$$H_s = \frac{\sum_i (V_i * H_i)}{\sum_i V_i} = \frac{\sum_i (Q_i * H_i)}{\sum_i Q_i}$$

$H_{1,2}$, H_s = resulting calorific value
V_1 , V_2 = volume of stream #1, #2
H_1 , H_2 = calorific value of stream #1, #2
Q_i = flow of stream i (= dV/dt)

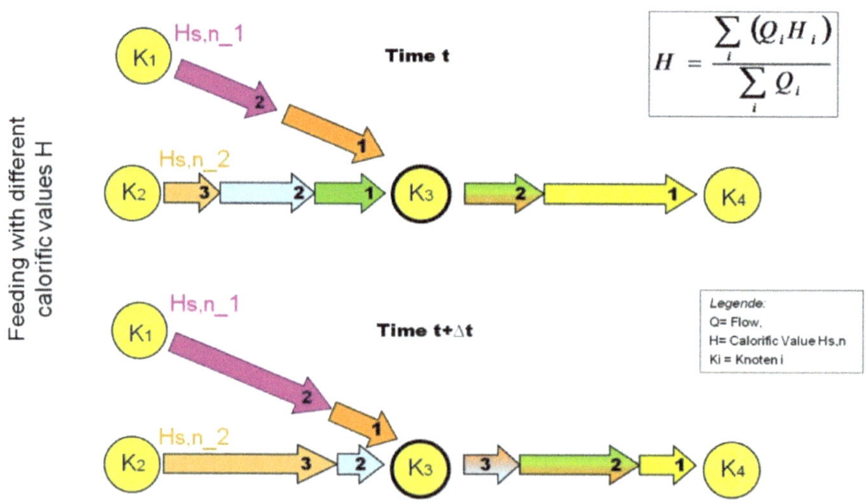

Fig. 3. Mixing and dynamic tracking of gas parameters at point K3

The mixing of the flows and thus the resulting value is not perfect in the vicinity of the mixing point. This is due to the pipe dimensions which may be large or different and the flow characteristic: laminar or turbulent. The mixing process is better and faster if the flow is turbulent. If the flow is laminar mixing may take a long way and time as even a layering effect may occur. In order to speed up mixing in such cases static mixer pipes, which have small obstacles inside to provoke little turbulence, will be built in the line.

3.3 Laminar vs. turbulent flow

Streams of gas and fluids show a typical behaviour when the flow changes from laminar state to the turbulent state (see figure 4). The areas of the flow type are described by the Reynold number (Re). Beyond Re= 2300 the flow switches suddenly from laminar to turbulent. But he transition spreads over a range and is dependent on physical and technical parameters. In pipeline systems the main parameters are pipeline inner diameter and roughness value which determine the flow type inside. Generally spoken, smaller diameters and higher roughness values lead to turbulent flow (but also to a higher pressure loss which is not really wanted).

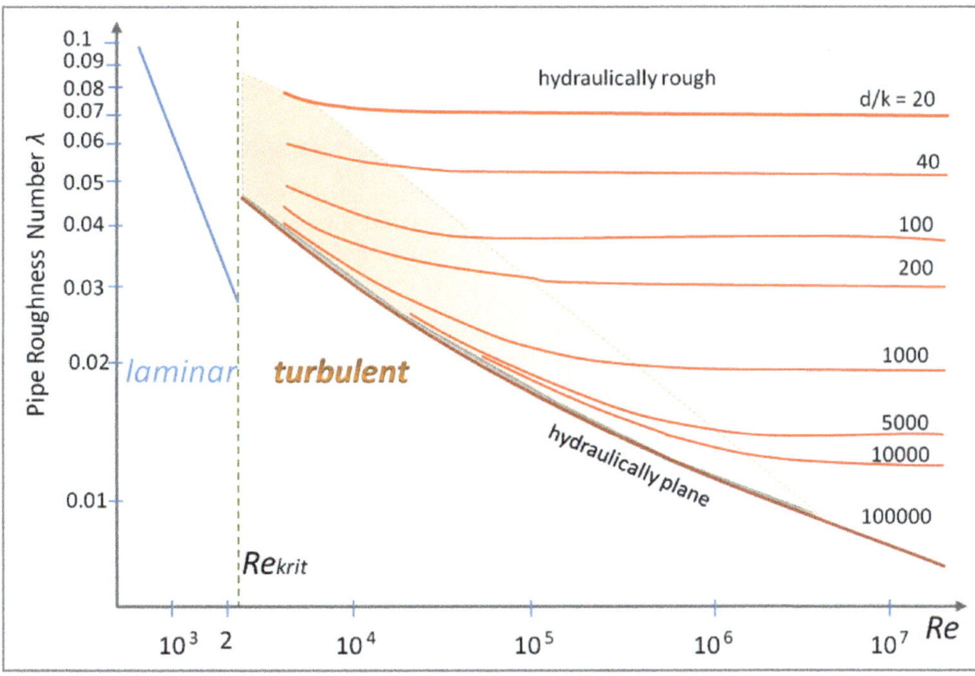

(*Re* is defined as: $Re = v\,d / \upsilon$, (v = velocity, d = pipe diameter, υ = dynamic viscosity))

Fig. 4. Laminar and turbulent flow in Reynold number vs. pipe roughness number (source /1/)

3.4 Gas conditioning, process optimization

Gas conditioning is required in biogas plants in order to provide gas of near or equal quality of natural gas when it is fed into the network. Many consumers, especially industrial consumers need gas with stable calorific value and/or Wobbe-Index. The allowed tolerance in Germany/Europe is ± 2 %, only.

By means of propane gas which has a higher calorific value (28.1 kWh/m³, approx.) the biogas will be mixed (conditioned) by special equipment to an appropriate final calorific value. The final value is selected and continuously controlled by the network operator depending on gas type (H or L), network structure, flow situation and consumer requirements.

The conditioning of gas is a high extra cost for the network operator (operational cost and propane, especially). An optimization of gas conditioning means that propane gas usage and cost should be minimized. This can be achieved in the conditioning equipment by selecting and setting the set point of the final calorific value to an acceptable low value. But this value must still be compatible with the calorific values in the surrounding network which - of course - must be known. Appropriate measurements, the mixing equations and simulation help to solve the optimization and set point problem, even on-line.

In gas network the gas flow may have a number feeder points - including different gas qualities - and in addition the intermeshed network structure contains potential mixing points at every pipe crossing or branch. Normally, insight of the gas flow and calorific value at certain interesting points in the intermeshed network is only possible by many measurements (e.g. gas chromatography) and/or network computation and simulation.

4. Measurement and data requirements for network computation

Any computation of a gas network is based on a model for the network structure and equipment accompanied by many data for the place and time of interest. These are data of materials, parameters of the medium, physical states, flow data (input and output) and controlling equipment. The sources of these data are shown schematically in table 2.

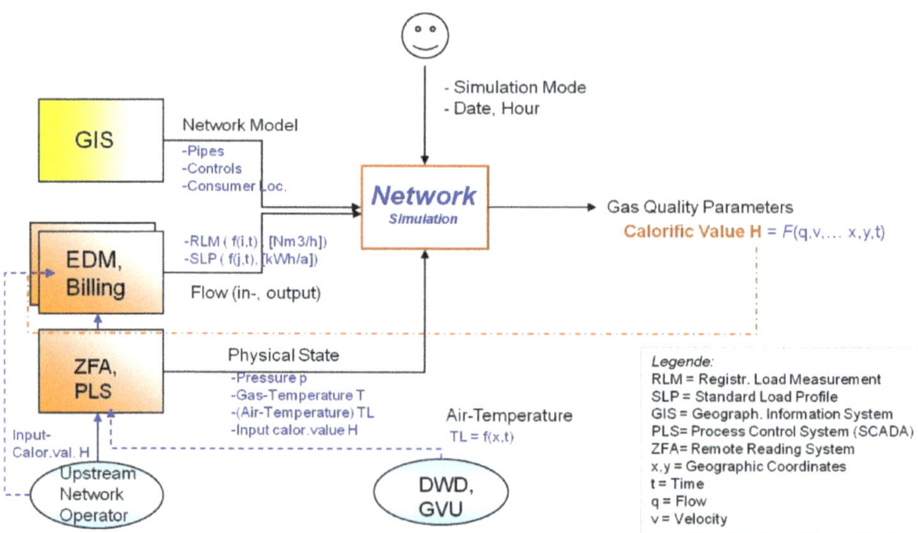

Fig. 5. Schematic data flow of reconstruction simulation for gas parameters/calorific value and feed back

4.1 Data types, accuracy, positioning, time scale/acquisition cycles

The data types and sources are summarized, collected and described in the following table to give a short overview:

Data Type	Data Detail	Source	Accuracy	Time Scale
Pipe	Inner diameter, Length, Roughness, Material class, ...	Geogr. Information System (GIS)	0.1 mm 1 m 0.01 mm	actual as possible
Node	Geographic (schematic) coordinates, node type (branch, ...)	GIS	1 m	
Control equipment	Valve (open/closed) Regulator (pressure/flow control operation mode), (max./min. flow, pressure)	GIS SCADA		hour, minute
Medium	Gas density, Gas temperature, Law of pressure loss	PGC		
Physical State	Pressure Flow (intake)	SCADA	0.001 bar 0.1 - 1 m^3	per hour per hour
Consumer	Flow (output), Type: RLM, SLP	SCADA, EDM	0.1 m^3	per hour, per month, per year
Other/ Derived	Outside air temperature, Consumption history/forecast, planned intake flow		°C (1 m^3)	per hour

Table 2. Data types required for control and simulation

The network model must be as actual as possible and should be updated whenever changes occur in reality. Very sensible with regard to the computation result is the information of actual or historical valve positions (open/closed); its tracing is indispensible, because wrong position will cause wrong/deviating results. The size of the network model may extend from some 1000 pipes up to 700 000 pipes (transport system to large distribution system including the transport system). The pressure range of larger networks may go down in several levels from 100 (84) bar to 0.020 bar finally at distribution level.

The data sources of different systems and their type of data which are necessary to build up a network model for simulation is shown schematically below (see figure 5). When computing the calorific value for each node (geographic position x,y) over time (t) the resulting value will be used to support the billing process.

4.2 Control system, full and sparse measurement coverage

In reality the gas networks are operated and surveyed by control systems (SCADA) consisting of a control center and remote control stations including remote control and data transmission. In general, a lot of data is acquired transmitted and stored but there is not always a full coverage for each point in the network. One can rely on:

- Input points: flow, pressure, gas quality
- Output points: flow of consumers (registered continuously)
- Intermediate points: flow, pressure (sparsely)

Intermediate points in the network are sparsely equipped or positioned, strongly depending on operational needs. Transport networks have a more detailed data view than distribution networks.

4.3 Consumer data (home, standard load profiles; Continuous registration)

Consumer data, especially small "home" consumers, are read once a year, only. When executing computations with small consumers their hourly values are deduced from yearly readings using standardized methods (standard load profiles, SLP). In all computations - when SLP's are involved - it must be kept in mind that this method influences the accuracy of the computing results when a short time period is considered. Opposite to the former small consumers the big consumers are measured and registered continuously; so computations can be made very accurate even in short periods.

5. Operational aspects

5.1 Smooth operation goal

Network operators favor smooth operating conditions for a number of reasons:

- Avoid sudden pressure changes (could generate shock waves, higher gas velocity)
- Avoid bigger and many flow changes (leads to pressure changes , regulator instability and wear out)
- Deliver/provide constant gas quality (i.e. colorific value), (operate within allowed limits).

In reality more or less big changes are likely to occur each hour (minute) due to changes of the consumption, scheduled feed-in according to delivery contracts, natural variation of gas quality of gas sources/production fields.

5.2 Local biogas flooding

In wintertime gas consumption is high and normally biogas is a smaller amount of the total consumption causing normally no transportation problems. But in summertime when the gas consumption is very low some areas face the situation that biogas production is higher than the total consumption in a distribution (sub-) network. In effect, biogas floods the network but could not be consumed, even if the regulating devices would have been changed to different settings to retain natural gas flow. So the exceeding or all biogas must be transported to other areas via additional pipes which have to be provided by the network operator.

5.3 Reverse feeding to high pressure trunk lines

So called "Reverse Feeding" to the transportation network is required to solve the problem of excess biogas production in a local area. An extra pipeline leads the biogas that has been previously compressed to an appropriate point (nearest one) of the transportation network. The level of compression depends on the pressure level of the transportation system (occasionally this can be up to 80 bar).

5.4 Odorization and deodorization

If biogas is fed into a distribution network it must be odorized before. Odorization adds the typical alarming and disgusting smell to the gas that warns human beings in case of leakage. If in the situation that excess biogas exists in a network area odorized biogas must be deodorized before entering the transportation network that has no odorized gas (odorization will be added downstream at the distribution level, only).

6. Network modeling, simulation

Building a network model is a complex task. It starts first with data extraction from the geographical information system (GIS) via a special interface. This will include and deliver all pipe data, node geographic coordinates and equipment forming the basic network structure. Control equipment such as valves and controlling regulators are read or derived from GIS data; regulator devices often must be connected manually or corrected afterwards.

Next the inputs and outputs of the network have to be introduced. Aside from feeder points or underground gas storage the outputs - better said the outflows – are modeled by consumers. Each consumer (up to hundreds of thousands) has his individual set of data, static or dynamic. Static data are used for long term planning for a certain scenario, dynamic data is used for short term planning (i.e. some days). When the simulation finally starts the correct pressure data - at least for the most important points - must be given to the simulator. Consumer data, valve position and pressure data must be taken from different IT-systems: Energy Data Management (EDM) system and Process Control system (SCADA).

6.1 Transport networks

Transport networks are designed to transport gas over longer distance. They are equipped with remote control which transmits all relevant data to the control centre. From the point of view of modeling these networks have an excellent information base for a moderate number of pipes and nodes (measurement points) making modeling straightforward and easy. The network structure of a transport system tends to be sparsely intermeshed.

6.2 Distribution networks

Distribution networks are designed to transport gas over shorter distance, e.g. within a city. They are equipped also with remote control, but only important data is transmitted to the control center. The amount of data handled may be subject to changes in the future when for each customer Smart Metering and on upper level Smart Grid will be introduced. From the point of view of modeling the distribution networks have an acceptable information base for all pipes but moderate number of measurement points making modeling an intensive work. The network structure of a distribution system tends to be strongly intermeshed. Distribution networks may have also a smaller trunk transportation system at a higher pressure level (e.g. 25, 16, 10 or 4 bar) while most of the pipes in the final distribution area are operated at 0.022 to 0.8 bar depending on the required flows.

6.3 Influence of intermeshing

Intermeshing is a basic concept in pipeline/network planning: it provides intrinsic redundancy for gas delivery in case of trouble/break at single points of pipeline. It supports continuous operation and pressure of the system; as it is most important to keep the whole pipeline system under pressure all the time (if the pressure would drop to zero, oxygen could enter the pipeline system exposing some areas or the system to the risk of explosion). Exaggerated use of intermeshing lead to higher investment cost in pipes and decrease a cost-efficient network structure in principle (besides, an item of passionate discussions among planners). Intermeshed networks need, because of their complexity, simulation support to get detailed insight into physical state and variables of the pipeline system.

6.4 Online- and offline-simulation

An advanced feature of simulation or operating mode is online-simulation. Here the simulation is coupled with a SCADA system and is executed corresponding to the cycle of the data acquisition (minutes to hours). It is a good tool to watch and control the network by additional detailed information almost in real time. This type of simulation is very demanding as it requires correct and complete data all the time what must be carefully prepared. Offline-simulation is the normal application it can be executed when needed - at arbitrary time.

6.5 Static vs. dynamic simulation

Static simulation means to execute a simulation for a fixed time; it is therefore good and often used for checking a set of scenarios or planning alternatives. Dynamic simulation reflects the changes of variables over time. This can be applied for historical data or for the future (based on forecasts). In case of reconstruction and tracking of certain values in the network (e.g. gas quality, calorific value) dynamic simulation is required. The typical time scale for simulations are hours which correspond well to most of the measurement cycles which are read from the field devices.

6.6 Tracking of gas quality parameters in the network

Online-simulation is used for tracking of gas quality parameter in the network. This method saves measuring equipment (gas chromatographs) in the field giving a much more detailed view of the way, value and distribution the gas quality parameters (see figure 6). Most often it will be applied for tracking the calorific value for all nodes on an hourly base (if used with historic data it will be called a reconstruction run). If the computed results shall be applied for billing purposes, then an official acknowledgement/permit of the technical authority (board of weights and measures) is required and a permanent surveyance system has to be installed.

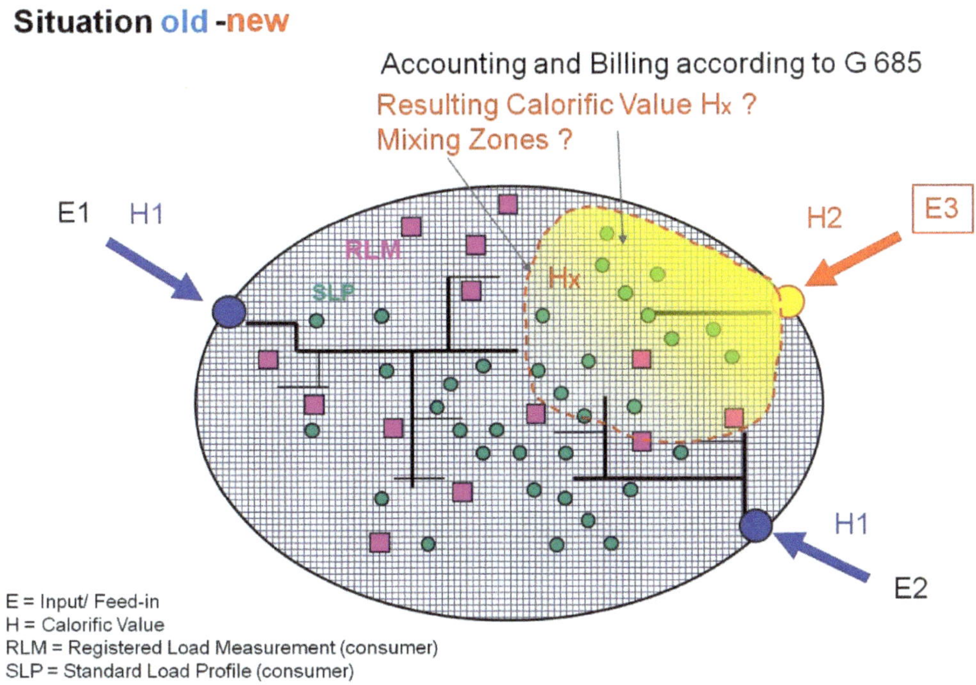

Fig. 6. Schematic view of areas of influence from different input of calorific values at feeding points

6.7 Examples

The following example demonstrates the results of a simulation which tracked the calorific value with historic data for 24 hours (reconstruction simulation) in a smaller city with a medium sized distribution network with some trunk lines (see figure 7).

There are six feeder points in total; in the north and the west are equal calorific values, in the south there is one point with different calorific value. (The different calorific values are made visible by different colors on the pipe segments, arrows indicate flow directions). In the middle of the network there is a mixing area (blue and pink) while near the southern feeder point the initial value dominates (dark yellow); the eastern branch of the grid shows a moderate mixed value (red). The small diagrams aside the network lines show the variable flow at the feeder points.

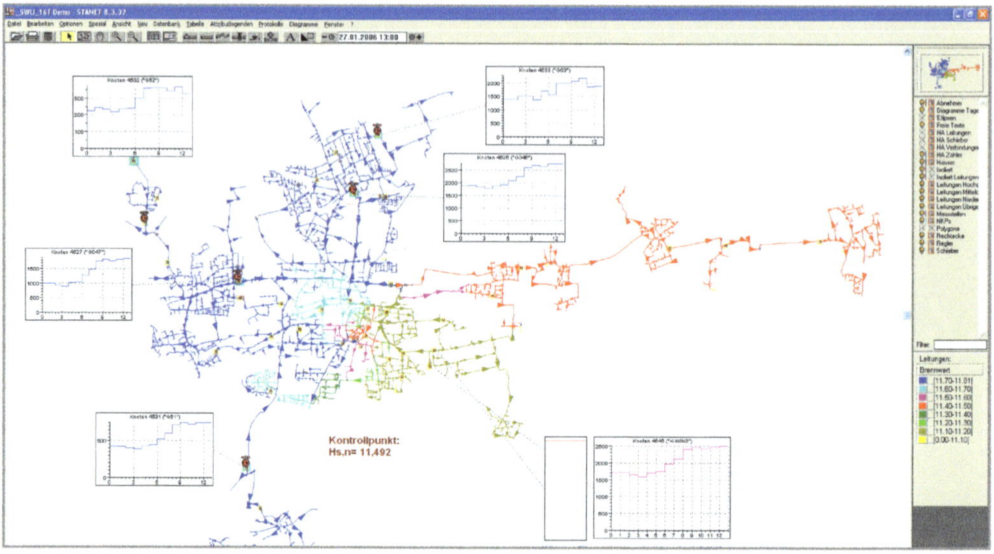

Fig. 7. Result of calorific value tracking in a distribution network; distinction by color scale (courtesy of STANET)

Figure 8 shows a detailed view of the distribution of the calorific value in the middle area of the network, figure 9 demonstrates the area of influence from the different feeder points in the same network by different colors in the background.

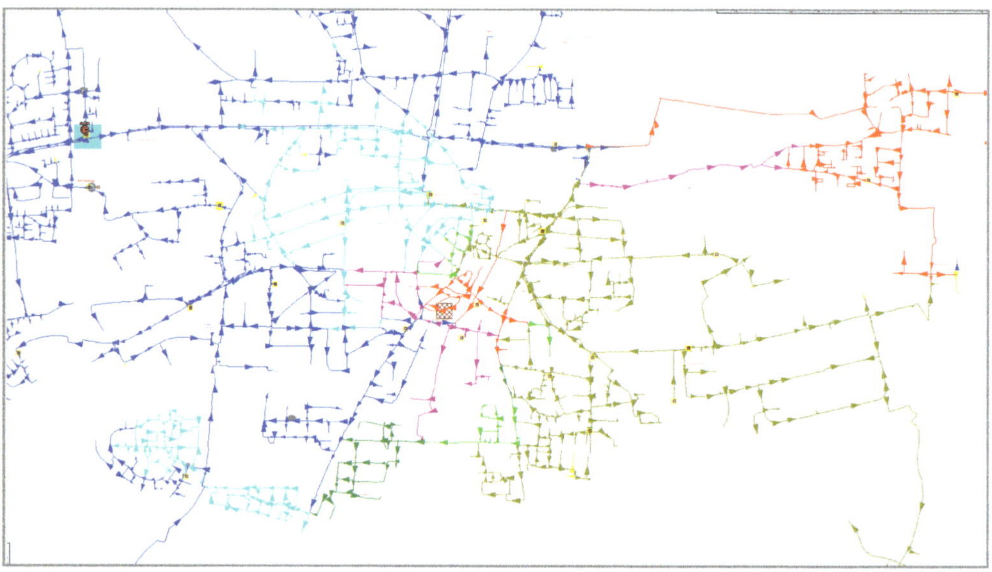

Fig. 8. Detailed view of calorific value distribution in the inner city area network (distinction by color)

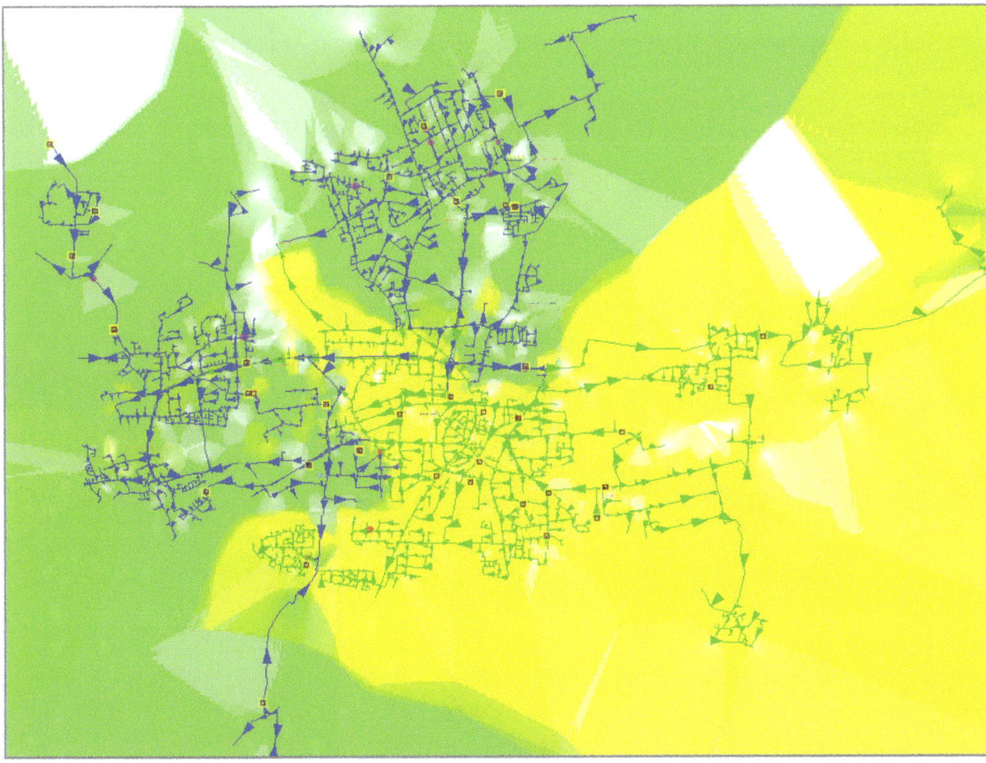

Fig. 9. Areas of influence of the feeder points (calorific value distribution distinct by colored borders in the background: yellow/green)

The next example shows the result of the tracking of the calorific value for high pressure network (see figure 10). Here two biogas plants feed into the network at different points; the three feeder points from other upstream transportation systems (not shown here) are equipped with process gas chromatographs ensuring complete information of the incoming gas qualities. The pipelines are color coded according to their calorific value. As the calorific value changes with time there are different values which are transported downstream in the pipes (visible in simulation only, not in this snapshot at a distinct time). Apart from the transport of different calorific values the areas of influence are discernable (e.g. blue colored pipes).

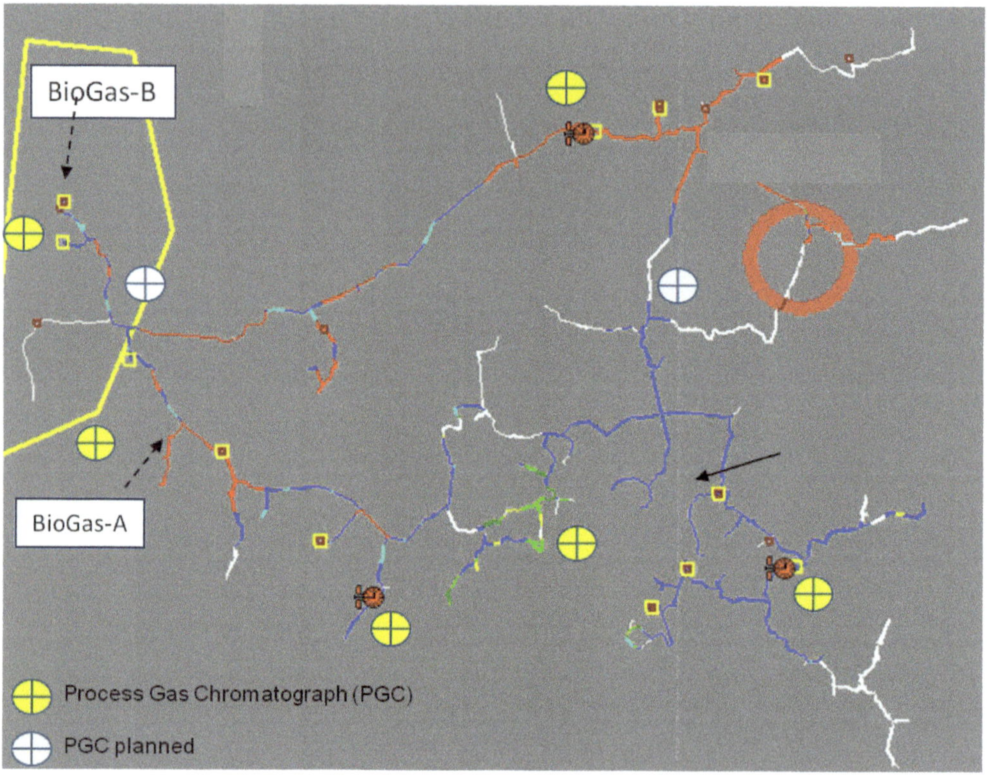

Fig. 10. Tracking oft he calorific value in a high pressure transport network and biogas feedings (A, B)

7. Tools, programs

7.1 Features of available simulation programs, short overview

Simulation programs are available as stand-alone versions or as integrated versions in a SCADA system. The best known and well reputated simulation programs for gas in Europe/Germany are summarized below highlighting some features:

Program Name	Producer	Features	Remarks
GANPRODA	PSI AG, Berlin	Integrated in SCADA; (GANESI based); online + offline simulation; tracking/reconstruction of calorific value	Medium: gas; static + dynamic computation
SIMONE	Liwacom, Essen	Stand-alone version; online + offline simulation; tracking/reconstruction of calorific value	Medium: gas; static + dynamic computation
STANET	Fischer-Uhrig Engineering, Berlin	Stand-alone version; online + offline simulation; tracking/reconstruction of calorific value, quality parameters	Medium: gas, water, steam, electricity; static + dynamic computation
Stoner SPS	Advantica/ German Lloyd, Hamburg	Stand-alone network calculation	

Table 3. Programs for simulation and gas quality parameter tracking

7.2 IT-systems requirements

The simulation programs can be executed already on a powerful PC; for big networks which have more than 100,000 pipes a powerful server type of computer with fast and large storage capacity is recommended. Computing time for static simulation ranges from seconds to few minutes (10,000 pipes about 20 s, 200,000 pipes about 150 s); dynamic simulations will take time according to the length of the period to be simulated.

8. Shortcomings, limitations

Unfortunately there are some limitations which should be known and considered when one will work and plan with simulation results:

8.1 Calibration aspects for network model and verification

In order to verify the computing results the network should be calibrated first. This means in the initial phase the simulation results must be compared with and validated against measurements taken in the field, e.g. pressure at selected (better many) control points. In most cases the network points which have appropriate measurements are sparsely installed. Calibration is a project that needs an extra effort and maybe temporary installation of additional measurements. If calibration is omitted then the computed/simulation results could be uncertain at a (small) percentage in a specific area (of few measurement points). Pipeline roughness value is another value that is often not really known but estimated, only. Instead, an "integral" roughness value is used for the pipes which averages individual values and some small scale effects on flow (e.g. sharp bendings).

8.2 Consumer behavior and data acquisition cycles

The small consumer's individual daily behavior is not known exactly; only an aggregation of all consumers can be calculated. This is due to the fact that consumption values are normally read and collected once a year. For computing purposes a yearly value has to be deduced to a daily or even hourly value by standard load profiles which have implicit uncertainties, of course. One day this shortcoming will be overcome stepwise when Smart Metering will be widely introduced and used.

8.3 Accuracy, cost-efficiency

Many measurements which are useful from the technical or simulation point of view may not be useful for economic reasons. Additional measurements cost extra money: equipment, erection, survey and maintenance, etc. So, in fact any computation will have to cope with less data points and less accuracy than desired or theoretically possible.

8.4 Technical board acceptance procedure

Introduction and use of quality parameter tracking systems (for calorific value) require an individual technical acceptance procedure from the Technical Board and a special operating permit. The required accuracy of the computed result must be better than ± 2% in the network area or ± 0.8% from total measuring range. In addition, a number of permanently recorded control measurements are requested to prove the correctness and integrity of the used data for the Technical Board authority.

9. Conclusion, benefits

Calorific value tracking is a most useful simulation tool in order to derive detailed information all over the network for every node and not only at distinct points. This method replaces many costly measurements in the network. Applying historical data to this tool helps to reconstruct all gas parameters and physical states of the past. Using the forward-looking method it enables even optimization of the gas conditioning for all biogas injections into the network. Aside from existing limitations the precision of the computed calorific value is high. Based on these data billing can be made fair as each consumer could get to know his energy consumption the best way possible.

10. References

[1] Cerbe, G.; Grundlagen der Gastechnik, Hanser 2004

[2] Hass, P.; Röttger, S.; Gasqualitätsberechnung in vermaschten Endverteilnetzen, gwf Gas/Ergas 9-2009, p. 512 , Oldenbourg Verlag

[3] PTB, Technische Richtlinie G14, Einspeisungen von Biogas in das Erdgasnetz, 11/2007.

[4] Wernekinck, U. et al., Ermittlung des Brennwertes von Erdgasen - Gasmessung und Abrechnung, 4. Aufl. Vulkan Verlag Essen 2009.

[5] Schley,P.; Schenk, J.; Hielscher, A.; Brennwertverfolgung in Verteilnetzen, gwf Gas/Erdgas, 9/2011, p. 553-556, Oldenbourg Verlag.

[6] STANET User Manual, Fischer-Uhrig Engineering, Berlin 2011.

8

Economic and Ecological Potential Assessment for Biogas Production Based on Intercrops

Nora Niemetz[1], Karl-Heinz Kettl[1],
Manfred Szerencsits[2] and Michael Narodoslawsky[1]
[1]*Graz University of Technology,*
[2]*Ökocluster,*
Austria

1. Introduction

Biogas production is discussed controversially, because biogas plants with substantial production capacity and considerable demand for feedstock were built in recent years. As a consequence, in most cases corn becomes the dominating crop in the surrounding and the competition on arable land is intensified. Therefore biogas production is blamed to raise environmental risks (e. g. erosion, nitrate leaching, etc.). Furthermore it is still discussed, that a significant increase of biogas production could threaten the security of food supply. The way out of this dilemma is simply straight forward but also challenging: to use preferably biogenous feedstock for biogas production which is not in competition with food or feed production (e. g. intercrops, manure, feedstock from unused grassland, agro-wastes, etc.). However, the use of intercrops for biogas production is not that attractive since current biogas technology from harvest up to the digestion is optimized for corn. Additionally current reimbursement schemes do neither take the physiological advantages and higher competitiveness of corn into account nor compensate lower yield potentials of intercrops which are growing in late summer or early spring. Higher feed-in tariffs for biogas from intercrop feedstock, as they are provided for the use of manure in smaller biogas systems, would not only be justified, as shown below, but also stimulating. Beyond that, the plant species used as intercrops as well as the agronomic measures and machinery used for their growing seem to provide lots of opportunities for optimization to increase achievable yields. Moreover, adaptations of biogas production systems, as discussed in this chapter, facilitate biogas production from intercrops.

Further advantages of intercrops growing are that they contribute to a better soil quality as well as humus content and reduce the risk of nitrous oxide emissions. Simultaneously intercrops allow a decrease of the amount of chemical fertilizer input, because the risk of nitrate leaching is reduced and if leguminosae are integrated in intercrop-mixtures, atmospheric nitrogen is fixed. This is important, because conventional agriculture for food and feed production utilizes considerable amounts of mineral fertilizers. Due to the fact that the production of mineral nitrogen fertilizers is based on fossil resources, it makes economically and ecologically sense to reduce the fertilizers demand.

In the case study, a spa town in Upper Austria, the set-up of the supply chain is seen as key parameter. An important issue in this case are more decentralized networks for biogas production. This can be achieved e.g. with several separated decentralized biogas fermenters which are linked by biogas pipelines to a centralized combined heat and power plant.

2. Methodologies

Process Network Synthesis (PNS) was used as a tool for economic decisions to get an optimal technology solution for biogas production with particular consideration of feedstock which is not in competition with food or feed production. Ecological evaluation of the resulting optimal PNS solution through footprint calculation was based on the Sustainable Process Index (SPI). These calculations are based on the data, which was gathered in three field tests, and the practical experiences, that were gained in the growing and harvesting of intercrops on more than 50 hectares of arable land. Besides the determination of dry matter yields of different kinds of intercrops and intercrop mixtures the effects on ground water, soil and nutrient management were investigated in the field experiments with time-domain-reflectometry, soil water and mineral nitrogen content measurement. Additionally, the potential biogas production was measured by means of biogas fermenter lab scale experiments.

2.1 Process Network Synthesis (PNS)

Process Network Synthesis (PNS) (Friedler et. al., 1995) uses the p-graph method and works through energy and material flows. Available raw materials are turned into feasible products and services, while in- and outputs are unequivocally given by each implemented technology. Time dependencies like resource availability (e.g. harvesting of renewable resources) as well as product or service demand (e.g. varying heat demand for district heating over the year) are part of the optimization.

The necessary input for this optimization includes mass and energy balances, investment and operating costs for the technologies considered, costs for resources and utilities, prices for products and services as well as constraints regarding resource supply and product/service demand. For the case study all data were provided from project partners and are specific for the considered region. First the so called maximum structure is generated linking resources with demands. From this starting point the optimization is carried out resulting in an optimum solution structure representing the most economical network.

2.2 Sustainable Process Index (SPI)

Sustainable Process Index (SPI) was developed by Krotscheck and Narodoslawsky in the year 1995 and is part of the ecological footprint family. The SPI represents as a result the area which is required to embed all human activities needed to supply products or services into the ecosphere, following strict sustainability criteria. Based on life cycle input (LCI) data from a life cycle assessment (LCA) study, SPI can be used to cover the life cycle impact assessment (LCIA) part. LCA studies are standardized and described by the ISO norm 14040 (ISO, 2006). Within the methodology there are seven impact categories defined which are indicated by different colors:

- Area for area
- Area for non-renewable resources
- Area for renewable resources
- Area for fossil carbon
- Area for emissions to water
- Area for emissions to soil
- Area for emissions to air

A high footprint is equal to a high environmental impact!

The freeware tool SPIonExcel (Sandholzer et. al., 2005) was used to calculate the ecological footprint (Graz University of Technology, n.d.) This offers the possibility to measure not only the economical performance of the PNS scenarios.

To assess the sustainability of biogas production from intercrops it is necessary to consider the whole crop rotation and the effects of intercrop on main crops. A direct comparison of biogas feedstock from main crops (e. g. corn) and intercrops is not possible, because inter crops grow with lower temperatures and less hours of sunshine. Therefore one of the systems compared, was corn as main crop, commonly cultivated with plow, and an intercrop cultivated with conservation tillage and harvested with a chopper for biogas production. It was assumed, that biogas was processed to natural gas quality. In the second system with intercrops corn was cultivated with conservation tillage whereas the intercrop was grown with direct drilling and harvested with a self-loading trailer instead of a chopper. Since a late harvest of a winter intercrop with high yields would reduce corn yields, an early harvest with an average intercrop yield of only 4 tons dry matter was assumed. In the reference system corn was grown without intercrop and the biogas produced in the intercrop systems was substituted by natural gas. The yield of the main crop corn was equal in all systems (15 tons dry matter of the whole plants per hectare for silage).

	common intercrop system		**improved intercrop system**		**reference system without intercrop**
position in crop rotation	main crop	intercrop	main crop	intercrop	main crop
tillage	plow	conservation tillage	conservation tillage	direct drilling	plow
harvest	chopper	chopper	chopper	self-loading trailer	chopper

Table 1. Systems compared with the Sustainable Process Index (SPI)

3. Intercrops

In temperate climate zones, allowing only the cultivation of one main crop per year, intercrops are planted after the harvest of the main crops (e.g. wheat, corn or triticale) or as undersown crops, while the main crop is still growing. Summer intercrops are harvested in

September or October as long as the trafficability of fields is sufficient. Achievable yields of summer intercrops are higher, the earlier main crops are harvested and intercrops are sown. The variety of plant species, suitable for biogas production from summer intercrops is very high and reaches from different kinds of millet, over grainlegumes, clover, sun flowers to cruciferae or other plants, adequate for regional conditions and the specific crop rotation of the fields. If cultivated as undersown crops, the variety of usable plant species (e. g. specific types of clover and grass) is restricted to those, not growing too fast and capable to resist a long period with shadow from the main crops.

Winter intercrops (e. g. feeding rye, triticale, different types of clover or rape) are sown in autumn and reaped before the cultivation of summer main crops (e. g. corn or soybean). The later winter intercrops are harvested, the higher are the achievable intercrop yields but the higher is also the risk of diminishing yields of the main crop. For example, output cuts of corn may be higher than additional yields of the intercrop, if intercrops are harvested in the middle of May or later. Therefore, the harvest of the intercrop at exactly the right moment with immediate subsequent cultivation of the main crop is crucial for the overall outcome of this type of crop rotation.

Dry matter yields, achievable with intercrops, vary to a higher extent than those of main crops, because they grow at the edges of the growing season and have less opportunities to compensate unfavourable conditions for growing. Furthermore, there are only a few farmers with experience and appropriate machinery for cultivation and harvesting of intercrops for biogas production at present.

Dry matter yields of summer intercrops in own field experiments in the years 2009 and 2010 averaged out at about 3 tons per hectare. After early cultivation with adequate machinery yields achieved 5 tons and more in some cases. However, intercrops did not achieve yields worthy for harvest in other cases, because of late harvest of main crops in the middle of august in connection with high precipitation and low temperatures in august and September. Under these conditions undersown summer intercrops (e. g. red clover under wheat and spelt) were advantageous and reached yields of almost 5 tons in the middle of September.

The yields of winter intercrops depend mainly on the time of harvest and the average temperature in March and April. If harvested at the end of April or the beginning of May, yields of about 4 tons dry matter were achieved with feeding rye or mixtures of rye or triticale with winter pea or rape. Yields of the following corn were equal or at maximum 10 percent lower than corn without preceding intercrop, if the intercrop was sufficiently manured with biogas digestate. A comparison with average yields found by other authors is compiled in Table 2.

	summer intercrops	winter intercrops
	dry matter yields in tons per hectare	
Own experiments	3	4 (without reduction of corn yields)
Neff, 2007	5	
Aigner/Sticksel/Hartmann, 2008	3	4,9 (middle of April)7,5 (5. Mai)
Laurenz, 2009	4,5	6 (with a reduction of corn yield of 2,5)
Koch, 2009	5	

Table 2. Average yields of summer and winter intercrops

Methane yields per hectare, achievable with winter intercrops, average out at about 1100 cubic meter with a methane content per kg organic dry matter of 310 liter. The methane yields of summer intercrops are lower and achieved 800 cubic meter per hectare in average. The methane content amounts in average 290 liter methane per kg organic dry matter. Therefore, between 4 and 6 hectare of intercrops are required to substitute one hectare of corn as biogas feedstock. This may seem little at the first glance. Considering the fact, that only rates of 10 or 20 percent of arable land should be used for biogas production at maximum, if the security of food supply should not be threatened, it becomes a considerable dimension, since intercrops for biogas production may be cultivated on 60 up to 90 percent of the arable land, if crop rotations are designed accordingly. Therefore the overall biogas potential of intercrops is comparable with the potential of corn.

However, the realization of these potentials requires adaptations of farmers' conditions for biogas production, as current reimbursement schemes and common technical equipment for tillage, drilling, harvest and biogas production make the use of intercrops profitable, only if farmers also apply for agro-environmental payments. Since these payments are only available in certain countries and are not guaranteed for the same period as biogas plants have to be operated, the risk for specific investments is considerable. To stimulate biogas production from intercrops, the physiological advantages and higher competitiveness of corn should be taken into account in the design of reimbursement schemes and tariffs should compensate lower yield potentials of intercrops. Higher feed-in tariffs for biogas from intercrop feedstock, as they are already provided for the use of manure in smaller biogas systems, would also encourage the optimization of agronomic practices (e. g. plant species used as intercrops, tillage, drilling) and technical equipment. In this way, the amount and reliability of intercrop yields would be increased additionally.

3.1 Ecological evaluation of intercrops

Based on input data for the production of main crops with and without intercrops several ecological footprints were calculated. Corn silage as main crop has a yield of 15 ton per hectare (dry matter) and 4 t (dry matter) per hectare of intercrop. SPI calculation includes

		common intercrop system	improved intercrops system	common intercrop system	improved intercrops system	conventional
		intercrop	intercrop	main crop	main crop	main crop (no intercrops combination)
	LCI input data	workings hours per ton (dry matter)				
machinery input	Tractor (<45 kW), light workload	0.40	0.23	0.04	0.04	0.04
	Tractor (<45 kW), normal workload	0.18	0.18	0.00	0.00	0.00
	Tractor (<70 kW), normal workload	0.88	0.44	0.55	0.52	0.55
	Tractor (<70 kW), heavy workload	0.00	0.00	0.13	0.00	0.13
	Tractor (70-110 kW), light workload	0.24	0.24	0.00	0.00	0.00
	Tractor (70-110 kW), normal workload	0.36	0.24	0.20	0.28	0.20
		kg per ton (dry matter)				
fertilizer	Application of N-Fertiliser	9.33				12.67
	Application of P-Fertiliser	1.57				1.57
	Application of K-Fertilisation	9.29				9.29
	Application of Ca-Fertiliser	8.43				8.43
		g per ton (dry matter)				
pesticides	Herbicide Phenmediapham	0.00	0.00	61.56	61.56	61.56
	Herbicide Terbuthylazin SP	0.00	0.00	108.05	108.05	108.05
	Herbicide Pyridate SP	0.00	0.00	6.91	6.91	6.91

Table 3. LCI data

machinery working hours, fertilizers, pesticides, agricultural area, and nitrogen fixation by leguminosae and seeds. Input data for the footprint calculation is listed in Table 3 which is derived from (KTBL, n.d.).

In terms of nitrogen fertilizer demand the use of leguminosae in intercrop mixtures reduces the demand of mineral nitrogen fertilizer through nitrogen fixation. Based on these data the ecological footprint results are listed in Table 4.

	SPI results [m^2 / t (dry matter)]		
	common intercrop system	improved intercrops system	conventional
main crop	27,217.8	26,374.6	31,528.6
intercrop	13,988.1	9,250.2	-------

Table 4. LCIA results

These footprints are per ton dry matter of intercrop or main crop. In general the lower machinery input for reduced tillage results in an accordingly lower footprint which points out the advantage of this method. This effect becomes more important as the yield of the crop decreases. The yields of intercrops are inevitably lower than of main crops, because of lower temperatures and less sunshine hours. Therefore, the footprint of intercrops sown with direct drilling and harvested with self-loading trailer is 34 % lower than of intercrops grown with conservation tillage and harvested with chopper. The amount of fertilizer for the main crops can be reduced with leguminosae intercrops. For this reason the footprint of the main crop in the reference system is higher than in the first system with intercrops with common tillage. If the effect of reduced nitrogen leaching or nitrous oxide emissions would be considered in the SPI-calculation, the difference would become even bigger.

For an overall assessment of the three systems, biogas produced in the systems with intercrops was processed to natural gas quality and substituted with natural gas in the system without intercrop. With processing the average methane content of biogas from about 60 % is increased to 96 % CH_4. Of course, biogas from intercrops can also be used in combined heat and power plants (CHP). Its processing is only obligatory for the comparison with natural gas. Although the footprint per ton dry matter of intercrops, even if they are sown with direct drilling, is bigger than the footprint of main crops, it is much smaller than the footprint of natural gas, it may substitute.

Table 5 illustrates this overall balance per hectare of agriculture area. Biogas purification SPI relies on life cycle data from ecoinvent database (Ecoinvent, n.d.). This balance can be seen as a rough estimation of the footprint reduction potential, if not only agriculture but also natural gas consumption is considered.

Table 5 points out an advantage for intercrop cultivation with direct seeding and harvesting with self-loading trailer in comparison with intercrops grown with conservation tillage and harvest with chopper. The footprint of intercrops used for green fertilizing to increase soil quality, was not calculated in detail. Nevertheless it can be assumed that the footprint is worse than the footprints of intercrops for biogas production, because the efforts for drilling are the same and instead of harvesting energy is needed for their incorporation into the soil.

For natural gas the SPI value is 540.4 m^2/Nm^3. Although further biogas purification is needed the whole balance points out a footprint reduction potential of 39 - 42 %.

	with intercrops		conventional
	common intercrop system	improved intercrops system	
CH_4 yield [m^3 / t (dry matter)]	1,200	1,200	
overall purified biogas [m^3/ha]	4,800	4,800	
intercrop SPI [m^2/ha]	408,266	395,619	472,929
maincrop SPI [m^2/ha]	55,952	37,001	0
provision of natural gas [m^2 /ha]	0	0	648,480
biogas fermentation process (electricity, heat) [m^2 / ha]	21,074	21,074	
biogas purification [m^2 / ha]	193,500	193,500	0
SPI [m^2 / ha]	**678,793**	**647,194**	**1,121,409**

Table 5. Energy balance per hectare

4. PNS optimization

A case study, as part of the so called Syn-Energy[1] project, was carried out in a spa town in Upper Austria wherein the set-up of the supply chain was seen as one of the key parameters. Beside detailed analyses of intercrops (e.g. biogas content, yields) a main focus was to find a network in respect of a higher degree of decentralization for biogas production. This can be achieved e.g. with several separated decentralized fermenters that are linked by biogas pipelines to a single combined heat and power plant. The specific data for intercrops were used to carry out the evaluations. Of note was to show how intercrops can affect networks from an ecological and economical point of view.

4.1 Case study

Figure 1 shows three potential decentralized locations for biogas production. As there is a spa town located in the considered region it was not possible to contemplate a fourth, central location for a fermenter as it would infringe with the touristy activity there. There is already an existing district heating network in town that should be extended. The heat needed could be either generated by a centrally placed CHP with biogas transported via pipelines or heat produced with decentralized CHPs could be used for fermenter heating and/or transported via long -distance heat pipelines to the town. In the first case, with central CHP, fermenter heating is provided by wood chip furnace.

The fermentation could work with different feedstock types to find out the most lucrative way of using intercrops, manure, grass silage and corn silage. Corn as additional feedstock was taken into consideration for economic reasons, because it is favored under current economic conditions. For the optimization it was assumed that proportional to the availability of manure biomass in an amount of 34 % intercrops, 18 % grass silage and 16 % corn silage (referring to fresh weight) per livestock unit can be supplied. As there are several

[1] Syn-Energy „Klima- und Wasserschutz durch synergetische Biomassenutzung - Biogas aus Zwischenfrüchten, Rest- und Abfallstoffen ohne Verschärfung der Flächenkonkurrenz"; programme responsibility: Klima- und Energiefonds; programme management: Österreichische Forschungsförderungsgesellschaft mbH (FFG), report not published yet

farmers in and around the considered region eight provider groups (1-8 according to Table 6 and black bordered providers in Figure 1) were defined. The substrate costs were the same for each group.

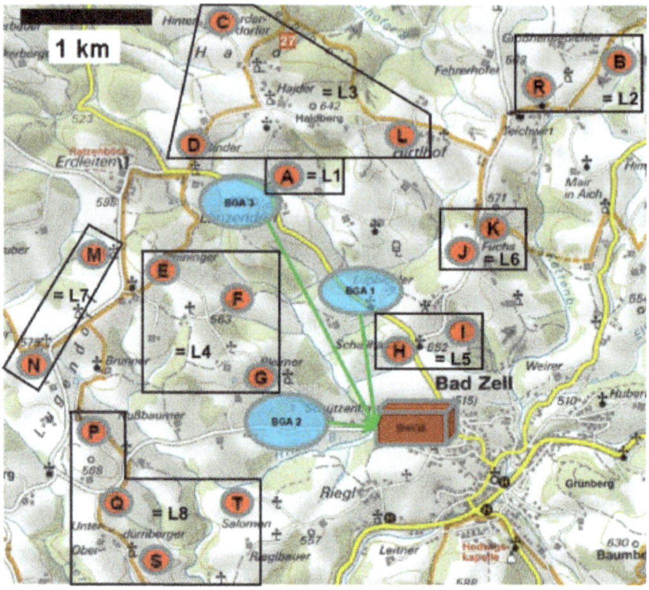

Fig. 1. Substrate providers (A-T) and possible fermenter locations (BGA1-3)

Provider Group	Distances in km to		
	Location 1	Location 2	Location 3
1 (A)	1.6	3.4	0
2 (B, R)	3.3	4.7	4
3 (C, D, L)	2.7	4.6	1.2
4 (E, F, G)	1.9	1.4	3.3
5 (H, I)	0.3	2.1	2.1
6 (J, K)	1.5	2.9	3
7 (M, N)	3.1	3	2.4
8 (P, Q, S, T)	3.8	1.9	3.7

Table 6. Transport distances for substrate provision

The providers differed in the amount of available resources as well as in the distance to each possible fermenter location, which directly correlates with transport distances and costs. Transport costs included fix costs for loading and unloading and variable costs depending on the distance (including unloaded runs). For solid substrates fixed costs of 2 €/t fresh weight were taken into account. Similarly, the conversion was made for the variable costs, which were assumed with 0.49 €/km. Fixed transport costs for manure were defined with 20 €/t dry mass with variable costs of 5 €/t dry mass per kilometer. For grass and corn silage a storage was taken into account. As it is not possible to bring the investment costs down to one number because they are highly depending on the local basic conditions a fix investment of 150,000 € for a silage storage was taken into account. As soon as a location is chosen by the PNS a storage has to be included there. Two locations mean two times investment costs to store the silage that is used for biogas production.

Transportation of heat and biogas could be achieved via pipeline networks. Network energy demands as well as losses caused by transporting were included. Regarding heat it was assumed that the total produced heat amount could be used for district heating. As location 1 and 3 are in one line to the spa town one biogas pipeline could be used for both locations to transport biogas to the central CHP. Therefore no additional costs arise for a biogas pipeline from location 1, if location 3, which is farther away, supplies the center with biogas.

Because of different transport distances the PNS could decide which provision group and amount of substrate should be used to get the most economical optimum solution. The fermentation could run with various substrate feeds. Dependent on them fermenter sizes, costs and exposure times differed. Seven different fermenters were part of the PNS to find the most lucrative way of substrate input. The feeds are shown in Table 7.

Feed [%]	Manure	Inter-crops	Grass silage	Corn silage
1	30	0	0	70
2	30	70	0	0
3	50	50	0	0
4	50	20	10	20
5	75	0	0	25
6	75	25	0	0
7	75	15	10	0

Table 7. Substrate feeds for fermentation

In Table 8 the substrate parameters are described. The optimization was based on two different cost situations (maximum and minimum) concerning substrate provision.

* decided by project partners	Manure	Corn silage	Intercrops	Grass silage
Dry Mass Content [%]	9	33	24	30
Substrate Costs* min. [€/t DM]	5	65	50	50
Substrate Costs* max. [€/t DM]	10	110	80	80
CH_4-output [m^3/ t DM]	200	340	300	300

Table 8. Substrate parameters and costs in € per ton dry matter and cubic meter methane per ton dry matter

Figure 2 shows the so called maximum structure for the PNS optimization, which includes all input and output materials with energy and material flows with economic parameters like investment or operating costs and prices. For the optimization three fermenter sizes (up to a capacity that serves a 250 kW_{el} CHP) were available for biogas production. Four combined heat and power plant capacities (up to 500 kW_{el}) were involved in the maximum structure. The fermenters could be heated by decentralized CHPs or with a wood chip furnace on site in case the biogas is transported to a central CHP.

The biomass furnace that could be a choice to provide fermenter heating was not implemented as separate technology in PNS' maximum structure, but a price of 5 ct/kWh heat was assumed (Wagner, 2008). Produced electricity could be fed into electricity providers' grid, thus benefiting from feed-in tariffs according to Austrian's Eco-Electricity Act (RIS, n.d.).

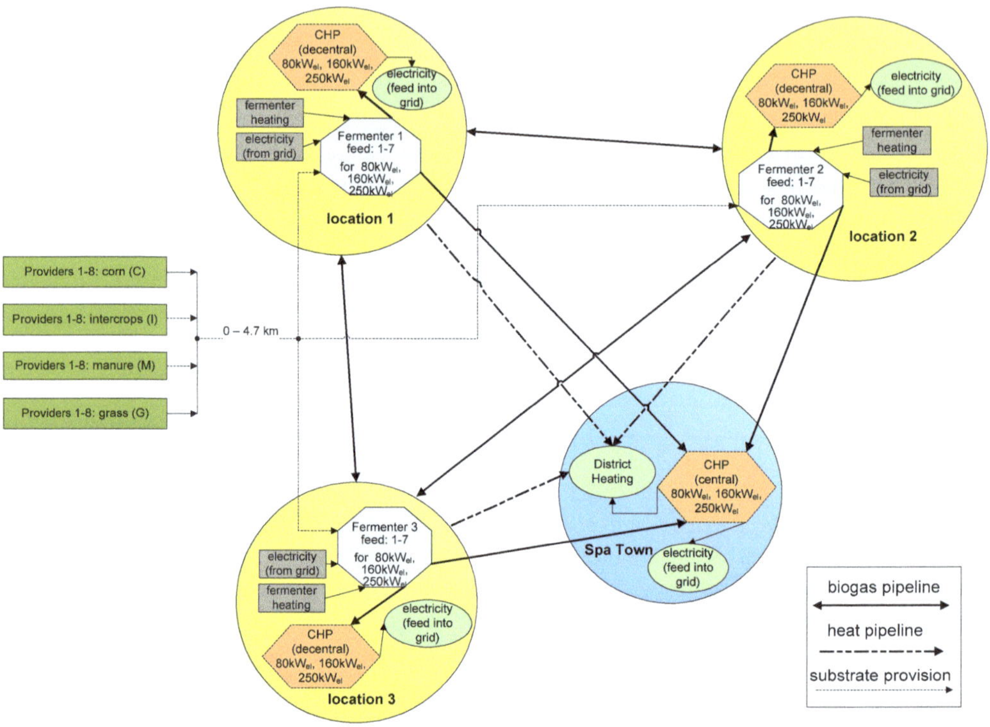

Fig. 2. Maximum structure for PNS Optimization

4.2 PNS optimum solution

The PNS optimization shows that the technology network providing the most benefit for the region includes two different locations (1 and 3) for biogas generation. At location 3 biogas is produced with substrate feed 4, a mixture consisting of manure, intercrops, grass and corn silage. The fermenter runs 7.800 full load hours and is able to provide a 250 kW_{el} CHP with biogas. At location 1 the set up includes a fermenter with same capacity but different load. Substrate mixture 7 is used for biogas production which contains manure, intercrops and grass silage. Both fermenters are heated with a biomass furnace on site. All provider groups can supply the fermenters with at least one substrate. The optimal technology network includes two central 250 kW_{el} CHPs supplied via biogas pipelines with biogas from both locations. For the pipeline coming from location 1 no additional costs have to be incurred because the pipeline would be part of the routing from location 3 to the center. The produced heat covers the central heat demand for a price of 2.25 ct/kWh. The electricity is fed into the grid and feed-in tariffs of 20.5 ct/kWh can be gained. Figure 3 depicts the optimum structure for a situation with maximum substrate costs as listed in Table 8.

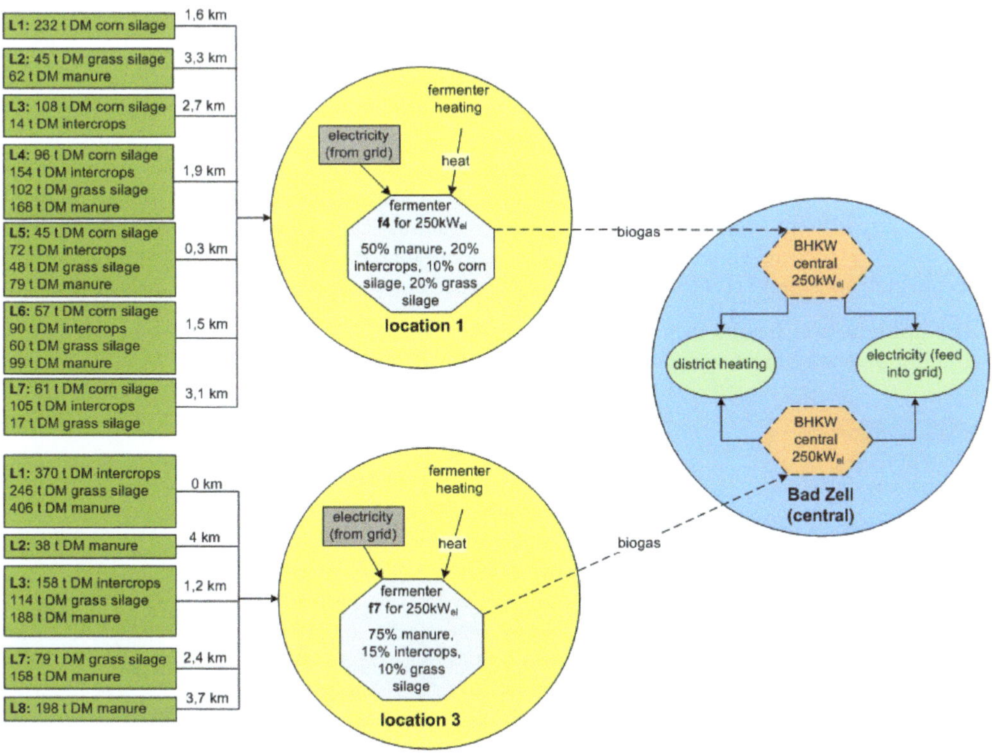

Fig. 3. Optimum structure of a technology network generated with PNS

With this technology network and 15 years payout period a total annual profit of around 196,350 € can be achieved (interest rates are not included). The total material costs including electricity consumed from the grid and costs for fermenter heating add up to approx. 438,000 €/yr with additionally 60,300 € per year for transportation. The total investment costs for this solution would be around 2,895,000 € including district heating and biogas network as well as the costs for fermenters and CHPs.

With minimal substrate costs (see Table 8) there is no change in the optimal structure, but the revenue is higher commensurate to the lower substrate costs (one-third reduction). The revenue for the structure with minimal substrate costs excluding interest accounts for a yearly amount of about 280,400 €.

4.3 Scenarios

To prove plausibility of the optimum PNS structure two scenarios were carried out, both for minimum as well as for maximum substrate cost situations. In the first case the maximum structure was reduced by taking away corn availability. With that only five substrate mixtures could be used for biogas production. The second scenario was set up to get an idea how feed-in tariffs can influence the outcome of an optimization. Therefore it was not allowed that a network set-up results e.g. in two 250 kW_{el} CHPs if a 500 kW_{el} instead could be taken.

4.3.1 Scenario I – No corn silage

As already mentioned in the beginning corn is currently a dominating substrate for biogas production. To show the potential of intercrops no corn is available in this scenario. Not to lose the comparability the amount of corn was compensated with an additional availability of intercrops. The calculation was based on the CH_4-outputs and adds up to additionally 904 t intercrops. With that 2,170 t/yr intercrops, about 1.7 times more than in the basic maximum structure shown in Figure 2, are available in the maximum structure of this scenario. Under these conditions PNS could choose between five different substrate feeds.

The optimization results in a technology network including two locations using the whole amount of available intercrops as shown in Figure 4.

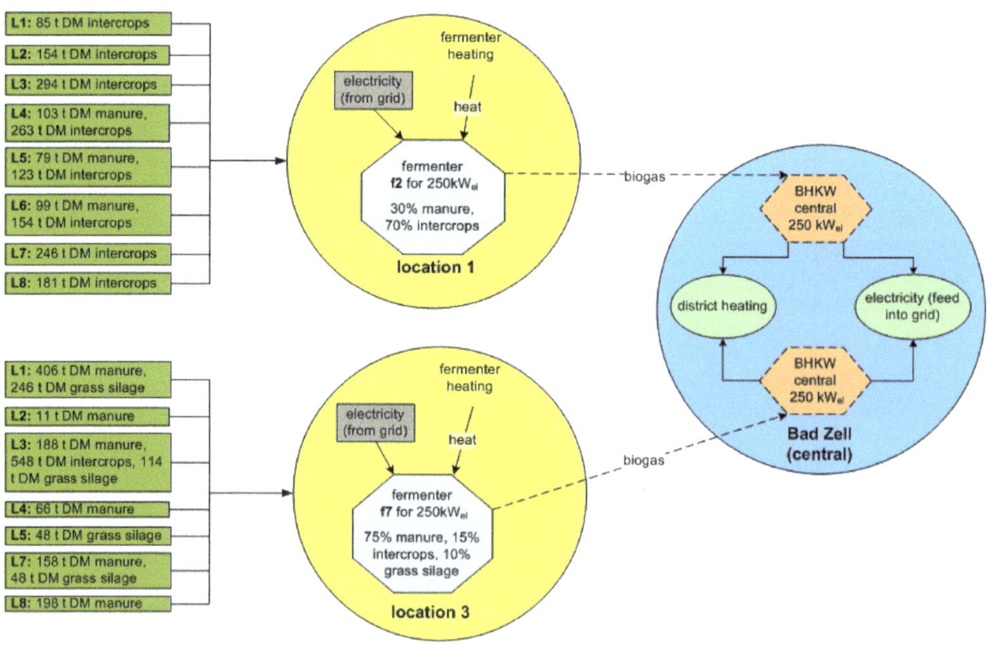

Fig. 4. PNS optimum structure for scenario 1 without corn silage availability

At location 3 a fermenter processing substrate feed 7 with a capacity to produce biogas to supply a 250 kW_{el} CHP runs 7,800 full load hours a year. A second fermenter placed on location 1 and with same efficiency is supplied with substrate feed 2 consisting of 70 % intercrops and 30 % manure. It turned out that with this structure the outcome has yearly revenue of approx. 208,000 €. Compared to the optimum structure it is higher, but the basic conditions are different. Therefore this solution did not come up in the optimization of the maximum structure in the beginning. But it clearly shows that intercrops have a great potential to produce electricity and heat within a highly profitable biogas network without being in competition with food or feed production. But the precondition would be that in the case study a higher amount of intercrops is available as feedstock.

4.3.2 Scenario II – 500 kW_{el} CHP unit

Operating a 500 kW_{el} CHP goes along with reduced feed-in tariffs of 20 €/MWh according to Austrian's Eco-Electricity Act. The positive effect of lower investment and operating costs for larger capacities is therefore narrowed by less revenue for produced electricity. If is forbidden to use two CHPs with same capacity at one location in the maximum structure to gain higher feed-in tariffs the next larger CHP capacity has to be taken although this would

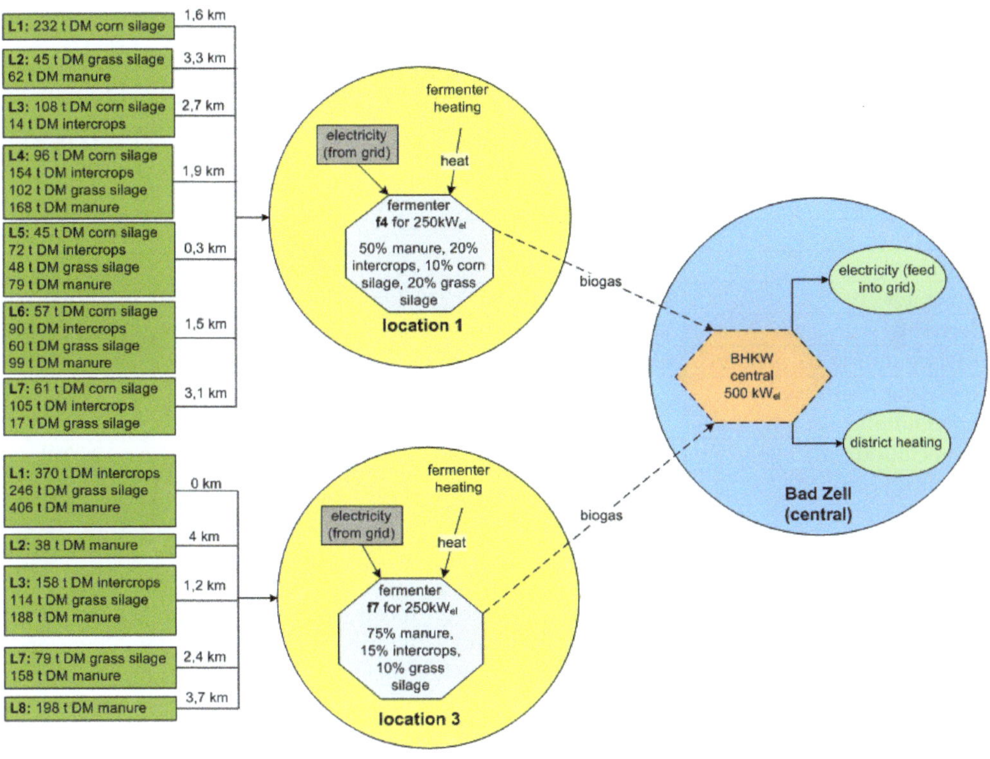

Fig. 5. PNS optimum structure with a central 500 kW_{el} CHP

go along with shortened revenue. With this precondition the optimization of the maximum structure presented in Figure 2 but with only one central 500 kW_{el} CHP unit whereas the rest of the optimum structure (Figure 3) stays the same.

The revenue is narrowed but not as much as it was in scenario 1. To use a 500 kW_{el} central CHP would cause a revenue reduction of yearly 50,000 € within a payout period of 15 years.

4.3.3 Comparison of PNS' optimum solution and the scenarios

Table 9 overviews the results of the three optimizations described before.

	Optimum Structure		Scenario 1		Scenario 2	
Substrate costs	max.	min.	max.	min.	max.	min.
Investment costs [€]						
Total investment costs	2,894,519	2,894,519	2,894,519	2,894,519	2,824,519	2,824,519
Products [MWh / yr] and Revenues [€/yr]						
Total produced electricity	3,826	3,826	3,900	3,900	3,826	3,826
Total produced heat	4,591	4,591	4,680	4,680	4,591	4,591
Revenue for electricity fed in (205 € / MWh)	784,281	784,281	799,500	799,500	707,766	707,766
Revenue for district heating (22,5 € / MWh)	103,296	103,296	105,300	105,300	103,296	103,296
Total revenue [€/yr]	887,576	887,576	904,800	904,800	811,062	811,062
Operating Costs [€/yr]						
Fermentation	114,423	114,423	116,090	116,090	114,423	114,423
CHPs	75,556	75,556	75,556	75,556	51,346	51,346
Transport	60,286	60,286	64,121	64,121	60,286	60,286
Substrates	213,561	129,488	213,400	131,740	213,561	129,488
Electricity	34,432	34,432	35,100	35,100	34,432	34,432
Total operating costs [€/yr]	498,258	414,185	504,267	422,607	474,048	389,975
Operating result without depreciation	389,319	473,392	400,534	482,194	337,015	421,088
Depreciation for 15 years*	192,968	192,968	192,968	192,968	188,301	188,301
Operating result with depreciation*	196,351	280,424	207,566	289,226	148,714	232,787

Table 9. PNS results summary

It turned out that the profitability of a fermenter on location 2 is lower than on the other locations. It was never preferred in any optimum structure. The other locations have one advantage – the shared usage of biogas pipelines whereas low additional costs for location 1 have to be born. There are never heating pipelines from the different locations to the center considered in the optimum technology networks. Just the biogas is transported; heat is produced centrally and distributed within a district heating network, although additional biomass furnaces are required. In scenario 1 the missing corn silage availability was compensated by a higher amount of intercrops, referring to the CH_4 content, and it shows the best revenue, because of higher plant utilization and higher revenue for electricity and heat production. Although in the optimal scenario the amount of corn relating to the total feedstock was not even 17 % of the total (dry matter) the compensation for corn with intercrops results in higher revenue. For more corn that intercrops compensate in the input the impact would be even higher. Therefore it is obvious that intercrops can be a profitable feedstock to run a biogas plant. For the case study the availability of intercrops would have to be raised as described before which would lead to the best technology network for the region.

The system has two limiting factors; on the one hand the distances between the fermenter locations and the feedstock providers accompanying different transport costs and on the other hand the limited resource availability. It could be shown that it is not lucrative to run a central CHP with higher capacity (500 kW_{el}) as feed-in tariffs are lower and less revenue can be gained. Nevertheless, from the point of view of sustainability, it would be preferable to substitute two smaller CHPs with a bigger one. An adaptation of reimbursement schemes to the solutions presented is recommended.

5. SPI evaluation

Based on the economic results of the PNS optimization and previous SPI evaluation of different intercrops, a footprint for the PNS results was calculated. The evaluation includes every substrate, transport, net electricity and infrastructure for fermenters and CHP units. SPIonExcel already provides a huge database of LCIA datasets which can be used for modeling the scenarios. In case of intercrops substrate the SPI value for conservation tillage + self-loading trailer from Table 4 was used.

SPI evaluation results					
		electricity		heat	
	overall SPI [km²]	production [MWh / a]	SPI [m² / MWh]	production [MWh / a]	SPI [m² / MWh]
Optimum solution	93.08	3,825	21,503	4,591	2,360
Scenario 1 - No corn	89.32	3,900	20,236	4,680	2,221
Scenario 2 - 500KW_{el} BHKW	91.51	3,825	20,876	4,591	2,539

Table 10. LCIA results based on PNS scenarios

The overall footprint points out the environmental impact for one year of production. In case of the optimum solution it would need 93.08 km^2 of area which has to be reserved to embed the production sustainably into nature. The overall footprint is shared between both products according the amount of output and the price per MWh (electricity: 205 €/MWh; heat: 22.5 €/MWh). Price allocation of the footprint leads to a higher footprint for the higher valued product.

Scenario 1 has a benefit from the ecological point of view and almost equal revenue according to Table 9. For scenario 2 there is only a slightly difference to the optimum solution because of two small CHP units instead one.

Main impact categories are in every case 'fossil carbon', 'emissions to water' and 'air'. This mainly derives from the utilization of net electricity which contributes around 45 % to the whole footprint. Main contribution to this categories stemming from net electricity and machinery input in agriculture which are still mainly fossil based. This is also the main optimization potential for a further decrease of the footprint.

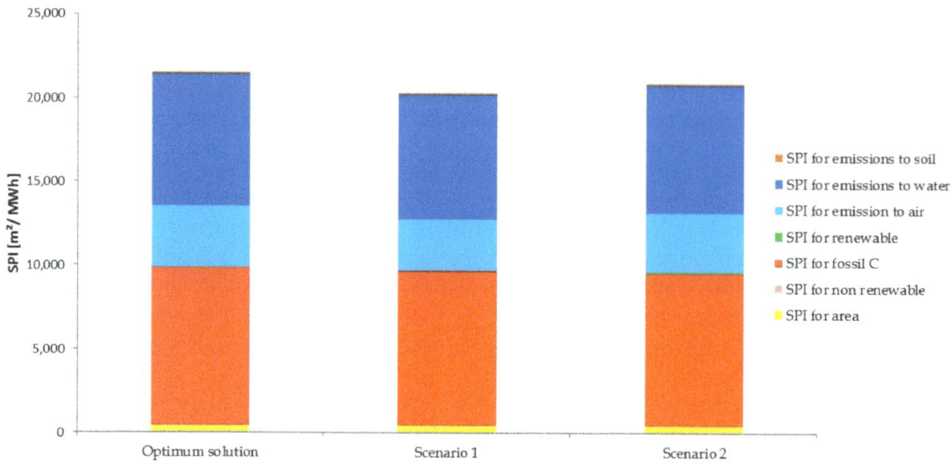

Fig. 6. SPI category comparison

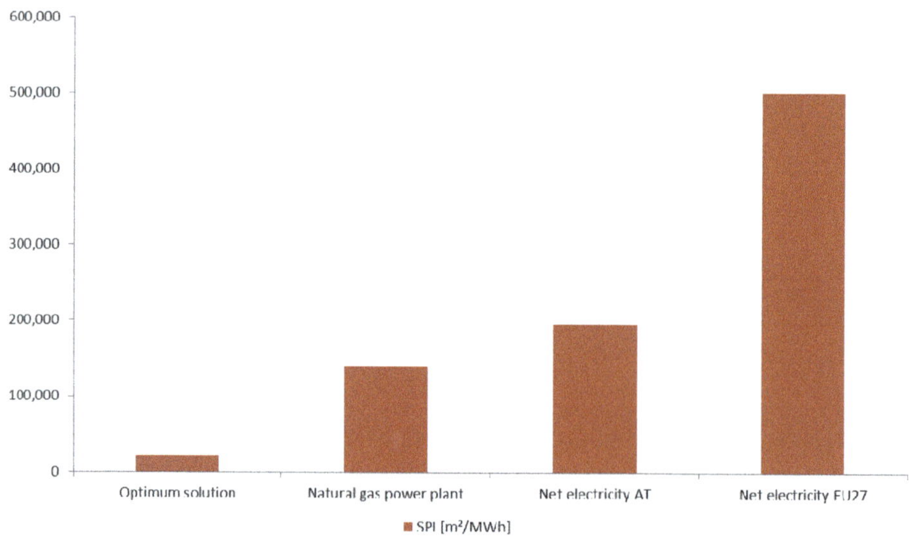

Fig. 7. Comparison of electricity production

Compared to other electricity provision system the optimum solution from the PNS has an ecological benefit in footprint ranging from 61 to 96 % which is pointed out in Figure 7. Although the footprint of the optimum solution could be optimized by using the produced electricity for itself and not selling to the grid (which has economic reasons because of high feed-in tariffs) the ecological benefit compared to other sources is obvious. Every contribution to a greener net infects simultaneously all net participants.

6. Conclusion

The three pillar principle of sustainability serves as conceptual framework to conclude this study. Not only economic and ecological factors are important to implement innovative structures. Often we forget about the social component, the third pillar of sustainability. Not to do so farmers' opinion about intercrops where taken into account. It turned out that intercrops production also abuts on farmers' psychological barriers and the need of intensive cooperation among farmers in the surrounding of a biogas plant. In conjunction with economic risk and high investments, determining farm management for at least 15 years it becomes obvious, that well-considered decisions are to be made. Therefore, it is not astonishing that farmers hesitate, if economic benefits do not clearly compensate social an managerial risks of biogas production from intercrops. Furthermore, the situation that biogas production from corn is favorable regarding practicability in comparison to biogas production from intercrops, reduces farmers motivation to decide for the latter. But even the growing and harvesting of intercrops requires additional work and the strict time frame to cultivate fields, the risk of soil compaction through harvest and potential lower yields of main crops after winter intercrops are counter-arguments to cooperate with farmers already running biogas plants. Higher feed-in tariffs for biogas from intercrops seem to be inevitable and sensitization of decision makers and farmers is needed to emphasize that the planting of intercrops holds many advantages and that intercrops reduce the ecological footprint decisively. Although a higher energy input for agricultural machines is required because of the additional workload for intercrops. In summary the energy balance per hectare including biogas production points out a benefit. In times of green taxes a reduction of CO_2 emissions can diminish production costs. More biogas output per hectare raises the income beside minimized mineral fertilizer demand reduces costs and lowers the ecological footprint. Furthermore, biogas production from intercrops contributes to a reduction of nitrate leaching and nitrous oxide emissions from agriculture. With the transport optimization in-between the network the ecological footprint decreases caused by intelligent fermenter set-up going along with less transport kilometers and fuel demand. A farmer association running an optimal network described before lowers the investment risk and ensures continuous operation and stable substrate availability. On the other hand an association has the potential to strengthen the community and the social cohesion of regions. Some of the advantages mentioned before effect the regional value added positively. On closer examination it could be shown that intercrops can play an important role in sustainable agriculture for the future by running a social and ecological acceptable network and still being lucrative for the operators and the region. Finally biogas production from intercrops does not affect the security of food supply. On the contrary it may even increase productivity in the case of stockless organic farming.

7. Acknowledgment

The research presented here was carried out under the project "Syn-Energy" funded by the Austrian Climate and Energy Fund and carried out within the program "NEUE ENERGIEN 2020" (grant Number 819034).

8. References

Aigner, A., Sticksel, E., Hartmann, S. (2008) Derzeitige Einschätzung von Zwischenfrüchten als Substrat zur Biogasgewinnung, In: LfL Bayern, Date of access: 16.5.2008, Available from:

http://www.lfl.bayern.de/ipz/pflanzenbau_biogas/30222/zwischenfrucht_fuer_biogas_ii.pdf

Ecoinvent, n.d., Swiss Centre for Life Cycle Inventories, Switzerland

Friedler, F., Varga, JB., Fan, LT. (1995). Decision-mapping: a tool for consistent and complete decisions in process synthesis. *Chemical Engineering Science*,50,1755-1768

Graz University of Technology. n.d., Date of access: 03.08.2011, Available from: http://spionexcel.tugraz.at

ISO 14040. (2006). Environmental management -- Life cycle assessment -- Principles and framework, International Organization for Standardization, Geneva

Koch, H. (2009) Zwischenfrüchte nutzen und Ertrag steigern. In: arge kompost & biogas. Engelsberg-Wiesmohl: abel-retec. Date of access: 4.5.2010, Available from: http://www.abel-retec.de/content/files/Zwischenfruchtanbau,fuer,Biogasanlagen.pdf

Krotscheck, C., Narodoslawsky, M. (1996). The Sustainable Process Index - A new Dimension in Ecological Evaluation. *Ecological Engineering,* 6/4,pp. 241-258

KTBL. n.d., Kuratorium für Technik und Bauwesen in der Landwirtschaft, Date of access: 03.08.2011, Available from: http://www.ktbl.de/index.php?id=792

Laurenz, L. (2009) Haupt-, Zweit-, Zwischenfrucht- Alternativen zum Mais - Praxiserfahrungen, In: Vortrag auf der 10. NRW-Energiepflanzentagung am 26.3.2010 im Landwirtschaftszentrum Haus Düsse, Landwirtschaftskammer Nordrhein-Westfalen. Date of access: 31.8.2010, Available from: http://www. duesse.de/znr/pdfs/2009/2009-03-26-biogas-03.pdf

Laurenz, L. (2010) Alternativen/Ergänzungen zu Mais, In: Vortrag auf der NRW-Energiepflanzentagung: Biogas - Alternativen zu Mais am 26.5.2010 im Landwirtschaftszentrum Haus Düsse, Landwirtschaftskammer Nordrhein-Westfalen. Date of access: 31.8.2010, Available from: http://www.landwirtschaftskammer.de/duesse/znr/pdfs/2010/2010-05-26-energiepflanzen-03.pdf

Neff, R. (2007). Biogaspotential einiger landwirtschaftlicher Kulturen - Versuchsergebnisse, In: BioEnTa - BioEnergieTagung 2007 - Tagungsband. pp 28-34; Witzenhausen: Kompetenzzentrum HessenRohstoffe (HeRo) e. V.; Witzenhausen-Institut für Abfall, Umwelt und Energie GmbH.

RIS. (2011). Ökostromgesetz. Date of access: 03.08.2011, Available from: http://www.ris.bka.gv.at/GeltendeFassung.wxe?Abfrage=Bundesnormen&Gesetzesnummer=20002168

Sandholzer, D., Niederl, A., Narodoslawsky, M. (2005) SPIonExcel - fast and easy calculation of the Sustainable Process Index via computer, *Chemical Engineering Transactions*, 7, (2), 443-446.

Wagner, R. (2008) Ökonomischer und energetischer Vergleich von Biogasleitungen mit Warmwasserleitungen zur besseren Wärmeverwertung. Available via C.A.R.M.E.N, [http://www.carmen-ev.de/dt/hintergrund/biogas/vergl_ biogasleitungen.pdf]. Last accessed August 2011

9

Production of Biogas from Sludge Waste and Organic Fraction of Municipal Solid Waste

Derbal Kerroum, Bencheikh-LeHocine Mossaab and Meniai Abdesslam Hassen
University Mentouri Constantine, Algeria

1. Introduction

The pollution of water, air and soil by municipal, industrial and agricultural wastes is a major concern of public autorithies who imperatively have to encourage the development of effective and non expensive treatment technologies. Although it is not recent, the process based on the anaerobic digestion (bio-methanisation) for the treatment of the waste organic fraction, is getting very attractive from the environmental and the economical points of view. It consists of a biological degradation of the organic matter, under anaerobic conditions, where a biogas, mainly methane (CH_4) is evolved, and hence providing a renewable source of energy which may be used in the production of electricity and heat.

Generally various types of residual sludges and solid wastes are generated by human activities. They are composed of organic matter which may or may not be easily biodegradeable, inorganic matter, inert soluble and non soluble matter, toxic matter, etc. In order to treat these solid wastes, it is first required to characterise them and second to choose a treatment mode depending on their types and their possible final destinations. According to the physical state, one may distinguish solid wastes (dehydrated sludges, domestic wastes, etc.), liquid wastes (effluents from food, fresh liquid sludges, etc.) and suspensions (sludges from water treatment plant). Classification in terms of the sources may be as follows:

- Biomass and organic wastes: these are potential biodegrable materials since they are made of natural organic molecules which may be inserted into the biogeochemical of the matter, particularly the carbone cycle. Industrial wastes can be concerned since they have non negligeable organic matter concentrations.
- Agriculture and food wastes such as manure and other wastes from breeding. The treatment of these wastes in treatment plants generate sludges rich in organic matter and hence are potentially biodegradable , requiring adapted methods.
- Municipal wastes which include domestic wastes and other types depending on the mode and nature of the collection: from small and moderate industrial units, public spaces, etc. They represent a good fraction of fermentable ready to undergo a biological treatment.
- Sludges from municipal wastewaters treatment plants: these are the main wastes produced by a treatment plant and are mainly constitued of dead bacteria and

mineralised organic matter. The sludge characterisation is essential for the choice of the most adequate treatment method as well as for the prediction of each treatment stage performance. Generally distinction is made between primary sludges which are recovered by simple waste waters decantation, and are of high concentrations in mineral and organic matter, and the biological or secondary sludges resulting from a biological treatment of waters. These latter have different compositions, depending on the nature of the degraded substrate, the operation load of the biological reactor and the eventual stabilising treatment.

For the treatment of the different pollution types, vvarious techniques and processes of different chemical, biological and physico-chemical natures as well as a coupling of the last two, are developed. The treatment and the final elimination consist of a sequence of unit operations with a great number of possible options among which the best one is to be chosen, taking into account the upstream (nature, characteristics, and waste quantities) and downstream (local possibilities of final eliminations) constraints as well as the cost.

The present study is more concerned by the biodegradable organic solid wastes which are characterised by a high organic matter concentration, recommanding a biological treatment.

One of technologies to carry out the treatment of the organic fraction of this organic waste is anaerobic digestion (bio-methanization, this process is presented with more details in the next sections of this chapter), which consists of a biological degradation in an anaerobic phase of the organic matter into biogas with a high methane percentage. This technology is becoming essential in the reduction of organic waste volume and the production of biogas, a renewable source of energy. It can be used in a variety of ways, with a heating value of approximately 600 -800 Btu/ft and a quality that can be used to generate electricity, used as fuel for a boiler, space heater, for refrigeration equipment, or as a cooking and lighting fuel.

2. Anaerobic digestion process

2.1 Anaerobic digestion historical

The use of anaerobic digestion for the treatment and the stabilization of solid waste is not new. It had been used in the 19th century. In rural parts of China and India, simple reactor constructions were used a long time ago to treat the manure and agricultural wastes in order to recover energy for cooking and lighting (Gijzen, 2002). In 1860s in France (McCarty, 2001), the anaerobic digestion of sludge waste was obtained from wastewater treatment plant, on a large scale, by means of an advanced technology. Furthermore, at the end of 1980s, co-digestion processes treating a mixture of different types of waste, were introduced (Ahring, 2003). Today, anaerobic digestion is one of the most environmentally friendly and suitable treatment methods for of solid organic waste. This technology is widely applied for bio-energy production, because of the increasing request for renewable energy. A consequence of the increasing implementation of this technology is the necessity to determine the ultimate biogas potential for several solid substrates (Angelidaki & al., 20096).

2.2 Anaerobic digestion principle

Among the various techniques of stabilization, anaerobic digestion, or methanisation, is the most interesting one. Indeed, according to Suh and Roussaux (Suh & Roussaux, 2002), it is the least aggressive treatment, towards to the environment. The anaerobic micro-organisms use organic pollution (biodegradable organic matter) as substrate to produce biogas which can be exploited according to several forms. Thus, anaerobic digestion allows a reduction of the dry matter from approximately 50% (OTV, 1997) and the production of a biogas, mainly methane (55-70%) and carbon dioxide (25-40%), with traces of hydrogen and of H_2S, (Mata-Alvares, 2003). Methane can be developed in the form of energy (boiler producing of heat or electricity). At the same time the anaerobic micro-organisms consume little energy, which involves a limited production of muds limited (3 to 20 times lower than an aerobic treatment), (Bitton, 1994). Indeed, the micro-organisms use only approximately 10 to 15 % of the energy of the substrate for their growth (Trably, 2002 and Moletta, 1993), the remaining being used for the production of biogas. Finally, anaerobic digestion allows a reduction of the pathogenic micro-organisms.

Anaerobic digestion consists of sludge fermentation, under strict anaerobic conditions. It is made up of four stages: hydrolysis, acidogenesis, acetogenesis and the methanogenesis. To achieve an anaerobic digestion, it is necessary that the reaction kinetics for the consumed or produced component is balanced. The general diagram of anaerobic digestion is presented on Figure 1 (Edeline, 1997).

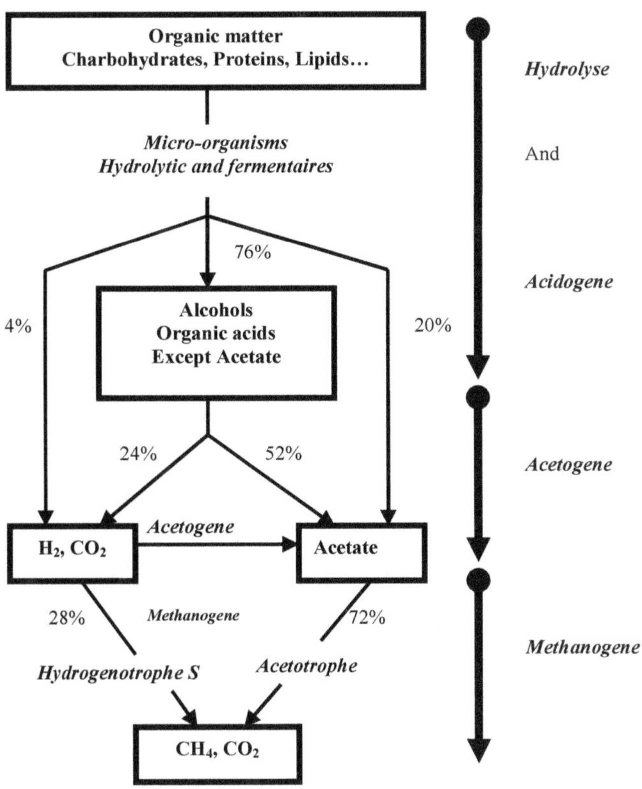

Fig. 1. Diagram of trophic chain of the methanogene and its various stages (Edeline, 1997)

2.3 Anaerobic digestion steps

Anaerobic digestion is a biological process, which is used for the treatment and valorisation of organic waste. Generally It goes through the four steps, as mentionned above, and which are hydrolysis, acidogenesis, acetogenesis and methanogenesis. In the case of co-digestion biodegradable solid waste is added at the head of the process. A preliminary stage of disintegration of the substrate, which is in general a nonbiological step for the transformation of the complex polysaccharide, lipids and proteins, is considered (Thiele, 1991).

2.3.1 Hydrolysis

The hydrolysis is an extracellular process in which complex particulate organic substances (proteins, polysaccharides, lipids, cellulose... etc) are broken up into simple soluble compounds (acid amino, simple, acid sugars fatty, glycérol... etc). It is a significant stage before the process of fermentation, because the fermentative bacteria cannot absorb complex organic polymers directly in their cells. The hydrolytic enzymes include the cellulase, the cellobiase, the xylanase and amylase for the decomposition of sugar polysaccharides, the protease for the degradation of the protein in amino acids, and lipase for the degradation of the glycerol lipids and the fatty acids with long chain (LCFA) (Batstone & al., 2002 and Kaseng & al., 1992).

2.3.2 Acidogenesis

The acidogenic step consists of a degradation of produced components from the hydrolysis step, by the action of acidogenic and fermentative bacteria. It leads to the formation of a mixture of: organic acids, volatile fatty acid (VFA), alcohols, hydrogen, carbon dioxide, ammonium, etc.

Examples of the various products of the fermentation of glucose are shown in the following Table 1:

Products	Reactions
Acetate	$C_6H_{12}O_6 + 2H_2O \rightarrow 2CH_3COOH + 2CO_2 + 4H_2$
Propionate + Acetate	$3C_6H_{12}O_6 \rightarrow 4CH_3CH_2COOH + 2CH_3COOH + 2CO_2 + 2H_2O$
Butyrate	$C_6H_{12}O_6 \rightarrow CH_3CH_2CH_2COOH + 2CO_2 + 2H_2$
Lactate	$C_6H_{12}O_6 \rightarrow 2CH_3CHOHCOOH$
Ethanol	$C_6H_{12}O_6 \rightarrow 2CH_3CH_2OH + 2CO_2$

Table 1. Exemples de la fermentation du glucose (Dolfing, 1988; Angelidaki & Ellegaard, 2002 ; Rodriguez, 2006)

The dominant route depends on several factors such as the concentration in substrate, pH and hydrogen concentration (Balk et al., 2002). Under a very high organic load, the lactic acid production becomes significant (Mattiasson, 2004). With low pH (lower than 5) the production of ethanol is high, whereas with lower pH (lower than 4) there is a strong production of the volatile fatty acids (VFA) (Ren & al., 1997).

However, the partial hydrogen pressure has a great influence on the fermentation route where a low value encourages the fermentation to acetate and hydrogen is favoured (Thauer, 1977).

2.3.3 Acetogenesis

The acetogenic step allows the transformation of the acids, resulting from acidonenic step to acetate, and carbon dioxide, by the action of the acetogenic bacteria. This operation is carried out by different types of bacteria.

2.3.4 Methanogenesis

The mehanogenic step consists of the transformation of acetate, hydrogen and carbon dioxide into methane. For that, there are two main system routes:

1. Aceticlastic methanogens : acetate + $H_2 \Leftrightarrow CO_2 + CH_4$
2. Hydrogenotrophic methanogens: $CO_2 + 4\,H_2 \Leftrightarrow 2\,H_2O + CH_4$

There are other minor routes which have a low importance. In the anaerobic digesters, approximately 60 to 70% of methane are produced by the Aceticlastic methanogens routes (Oles, 1997).

The growth of methanogens bacteria is slow: 3 days in 35°C (Schink, 1997). As they are the most sensitive micro-organisms of the ecosystem, they govern the total kenetics of the process (Ramsay & Pullammanappallil, 2001). Moreover, they are sensitive to the presence of inhibitors such as VFA.

During the methanogenic phase, the products of fermentation such as acetate and H_2 / CO_2 are converted into CH_4 and CO_2 by methanogenic bacteria. Methanogenes bacteria can grow directly on H_2 / CO_2, acetate and all other compounds with only one carbon such as formate, methanol and the methylamine (Puñal & al., 2003).

The methanogenic step is influenced by the operating conditions of the digester, such as temperature, hydraulic loading rate, organic loading rate, and the influent substrate composition (McHugh & al., 2003).

3. Types of digesters and applications

The conventional anaerobic digesters operate as semi continuous, continuous or closed. The operations in semi continuous or continuous are preferable because the maximum growth rate can be obtained by controlling the effluent rate. In the closed system, a balance cannot be obtained while the concentrations of the components in the digester change with time (Karakashev & al., 2005).

The choice of the type of digester used is related to treated waste characteristics. Solid waste and sludge are mainly treated in digester with continuous flow (CSTRs), whereas soluble organic waste is treated by a use of biofilm systems such as the anaerobic filters and fluidized bed digesters with ascending flow (UASB) Smith & al., 2005).

In the systems of biofilm the biomass is maintained in the aggregates of the biofilm/granule where the solid retention time (SRT) is much higher than the hydraulic retention time

(HRT). The advantage is that the digester can operate with a high flow and can tolerate higher concentrations of toxic species than in (CSTR) systems. The biofilm system operates normally in a continuous mode with an (HRT) lower than 5 days. The systems can operate in a wide range of temperature and psychrophils conditions (3°C) up to the extra-thermophiles conditions (80°C). For the anaerobic treatment of soluble organic waste the systems of UASB at high rate are used.

In CSTR systems, the biomass is suspended in the main liquid and will be removed as well as the effluent so that the solid retention time (SRT) is equal to the hydraulic retention time (HRT). This makes it necessary to operate at a high hydraulic retention time (HRT) , between 10 and 20 days, to avoid the scrubbing of the methanogens which have a long time of growth.

4. Factors affecting the anaerobic digestion process stability

The factors affecting the production of biogas are mainly based on the operating conditions of the digester, such as pH and temperature which influence directly the micro-organisms. The perturbations in effluent (including the concentration of substrate and its composition in toxic compounds and inhibitors) can also affect the volume and the quality of the produced biogas. Sometimes, the toxic compounds are not present at the beginning in the effluent waste, but they are produced inside the digester starting from degradation of substrate (example: VFA and ammonia).

4.1 Substrate

The type and the composition of the substrate determine directly the quality of the produced biogas. In anaerobic process the substrate is often measured in term of chemical oxygen demand (COD) or of total volatile solids (TVS). It is significant to distinguish between the degradable and the inert fraction, because a considerable fraction of the COD in effluent is inert (Nielsen, 2006). The waste which contains a high percentage of water has a weak methane yield by COD or VS.

Organic waste contains various compounds: mainly saccharides (which are divided into two fractions, easily and slowly degradables), lipids (easily degradable), proteins (easily degradable), VFA (easily degradable), as well as others compounds (Moosbrugger & al., 1993). The production of methane is generally in the range from 100 to 400 L CH_4 / kg VS.

On the other hand, the majority of organic waste contain a high fraction of the substrate easily degradable, which gives a high production of methane and VFA. It is thus significant to control the organic and hydraulic loading according to the capacity of the digester when the process functions are at low charge that gives also a low production rate of biogas. Although this can prevent the rupture of the process, it is not very ecomical because the capacity of the process is not completly used. The increase in the charge gives more biogas but also there is the risk of the overload, with, as a consequence, the accumulation of the VFA. The high concentration of VFA decreases the pH and making them more toxic for methanogens bacteria.

Sufficient nutriments are also impotant for microbial growth. The macro nutriments such as carbon, hydrogen, nitrogen and oxygen are the main components of the cells in the biomass, with others like sulphur, the phosphorus, the potassium, the calcium, the magnesium and the iron which are required (McMahon & al., 2001). The majority of the nutriments can be inhibiting if they are present at high concentrations.

4.2 Temperature

Anaerobic digestion can be applied in a wide range of temperature, into psychrophilous (< 20 °C) (Vavilin & Angelidaki, 2005), in mesophile (25-40 °C), thermophile (45-60 °C) (Angelidaki & al., 2005), and even in extra thermophile conditions (> 60 °C) (Liu, 2003). The temperature has a direct effect on the physicochemical properties of all the components in the digester and affects also the thermodynamics and the kinetics of the biological processes. The temperature determines if a specific reaction is favorable.

The increase in the temperature has the following advantages:

1. Increase the solubility of the organic compounds which makes them more accessible to the micro-organisms.
2. Increase the chemical and Biological reaction rates and hence accelerates the conversion process, therefore the reactor can be smaller and can operate with a shorter hydraulic retention time (HRT)
3. Improve several physicochemical properties like diffusivity of the soluble substrate, the increase in the rate of transfer of liquid towards gas due to the low solubility of the gas, reduction in the liquid viscosity which makes decreased the energy of agitation necessary and also improves separation liquid-solid separation of the biomass.
4. Particularly increase the death rate of the pathogenic bacteria, which decreases necessary time for the reduction of pathogenic bacteria (Hansson, 2002).
5. Moreover, the reactions of oxidations of organic acid become more energetic at high temperature, which is advantageous for the degradation of fatty acid to long chain fatty acid, and other intermediaries (Chynoweth & al., 1994). Nevertheless, the high temperature can have a certain negative effect. The increase in the temperature decreases the pKa of ammonia, thus increasing the free fraction of ammonia (NH_3) which is an inhibitor of the micro-organisms. Moreover, the rise in the temperature increases the pKa of VFA, which increases its not dissociated fraction, particularly with low pH (4-5), as in the acidogenic reactor (Chynoweth & al., 1994). This makes the thermophilic process more sensitive to inhibition. However, because of the multiplicity of advantages at high temperatures, the thermophilic operation is popular in the anaerobic applications where the ammonia inhibition is not the first consideration.

4.3 pH and buffer value

The level of pH has an effect on the enzymatic activity in the micro-organisms, since each enzyme is in activity only in one specific range of pH, and it has its maximum activity with its optimal pH (Ahring, 1994). A stable pH indicates system equilibrium and digester stability. A falling pH decrease can point toward acid accumulation and digester instability. Gas production is the only parameter that shows digester instability faster than pH. The range of acceptable pH for the bacteria participating in digestion is from 5.5 to 8.5, though the closer to neutral, the greater the chance that the methanogenic bacteria will function

(Golueke, 2002). Most methanogens function in a pH range between 6.7 and 7.4, and optimally between 7.0 and 7.2. The greatest potential for a digester failure is a result of acid accumulation. This would occur if the amount of volatile solids loaded into the digester as fresh waste increased sharply. Maintaining pH is especially delicate in the start-up because fresh waste must undergo acid forming stages before any methane forming can begin, which will lower the pH. To raise the pH during the early stages, operators must add a buffer to the system, such as calcium carbonate or lime.

4.4 Intensity of mixture

Several studies proved that the intensity of mixture in an CSTR digester has an effect on the process inhibition and the re-establishment of the organic overload (Hill & Bolte, 1989). Other researchers (Hill, 1990) studied the accumulation of acetate and propionate in a CSTR digester which treats municipal solid waste and the biosolides with an aggressive starting and an organic overload. They noted that while acetate was consumed thereafter, propionate persisted in the whole system and it started to decrease only after the reducing of mixture intensity. They also noted that a digester with a reduced mixture can tolerate a higher organic load than the digester with an intensive mixture.

4.5 Composed toxic/inhibiting

The inhibiting compounds are one or the other present already in the substrate or product during degradation. The majority of the inhibitors are formed during the degradation of the substrate, such as VFA, LCVA, ammonia and sulphide. Some inhibitors are present already in substrate, such as the heavy LCVA, and metals.

The VFA is the main intermediate in anaerobic digestion, and it accumulates under the action of the non balance of the process. With low pH, the VFA becomes more toxic, due to the increase of the non dissociated fraction.

Ammonia comes mainly from the degradation of protein. A study on 18 central biogas stations in Denmark, proved that ammonia was significatif factor affecting the stability of the process (Hawkes & al., 1994). A concentration about 2 gN/l of ammonia will have no inhibiting effect on acetoclastic methanogens (Hill & Holmberg, 1988). However, the activity of methanogens is decreased during the increase in ammonia concentration, and total inhibition is reached for a concentration of 10 gN/l.

5. Control parameters of the process of biogas

The control of the anaerobic digesters is necessary to ensure a good operating of the digester. Since anaerobic digestion is a complex process implying several groups of micro-organisms which are sensitive to several factors of operation, it is significant to be able to detect the non balance of the process at the beginning to take an action in time to prevent its failure. As with other biological processes, anaerobic digestion can be controlled by measuring the substrate convertion (COD or removed VS), the accumulation of intermediaries (VFA, pH, alkalinity, H_2, CO), the formation of product (gas production rate, CH_4, CO_2).

5.1 Methane and carbon dioxide

Biogas is composed mainly of CH_4 and CO_2. The ratio CH_4 to CO_2 is normally stable in the digester and any change may be due to the process instability. However, the ratio also depends on the composition of the substrate, the temperature, the pH and the pressure (Hickey & Switzenbaum, 1991). Since the dissolution of CO_2 strongly depends on the pH, the fluctuation of the pH can also change the gas composition. A better indicator is thus the production of methane, rather than its composition in the gas (Anderson & Yang, 1992).

The production of methane combines the production of biogas with the measurement of percentage of methane. The production rate of methane (L-CH_4/days) was used successfully like an on line indicator to control a CSTR digester (Feitkenhauer & al., 2002).

5.2 PH

The pH is relatively easy to measure, and is often the only parameter of the liquid phase which is measured on line. The change of the pH can be an indicator, for the stability of anerobic digestion process. Since the micro-organisms can grow at only one specific pH range. The effluent pH can also affect the pH in the digester. The use of the pH as an indicator is normally based on the fact that a decrease of the pH corresponds to the accumulation of VFA. Some anaerobic systems apply the control of the pH where an acid or a base is added to ensure the suitable pH for the microbial growth.

5.3 Alkalinity

Alkalinity is a better alternative than the pH to indicate the accumulation of VFA, because the increase in VFA will directly consume alkalinity before the great change of pH. However, it is proved that the total alkalinity (TA) measured by the titration of the sample with pH 4.3 is not very sensitive because of the combination of result of VFA and bicarbonate to the TA (Hill & Bolte, 1989). Partial Alkalinity (PA) or bicarbonate alkalinity measured by titration of sample in pH 5.75 has an empirical correlation to the VFA accumulation (Wang & al., 2005). However, one does not observe this during the VFA accumulation at the time of the ammonia overload, because this latter increases the alkalinity of the system (Wang & al., 2005).

5.4 Volatile fatty acids

The accumulation of the volatile fatty acids (VFA) during the non balance of the process reflects directly an uncoupling kinetic between the acid producers and consumers (Hickey & al., 1989). The concentration of VFA was suggested for the control and the monitoring of the anaerobic digester (Hill & Bolte, 1989). The VFA is generally measured by gas chromatography (GC) with the use of a detector with ionization of flame (FID), to obtain the individual VFA, or by titration which gives the concentration of total VFA, and which is less expensive and is largely used at the commercial biogas plants. Several methods of titration for the determination of total VFA were proposed, for example a simple titration (Delbès, 2000), a titration at 5-point, and a titration at 8-point.

However, several studies specified that the individual VFA can provide more significant information concerning an early failure of the process the failure of process (Nielsen, 2006).

5.5 Organic matter reduction

There are many industrial applications in which the principal goal of anaerobic digestion is the organic treatment of waste instead of the production of gas. On this subject, the elimination of the difference between the organic matter contained before and after treatment, is a significant parameter that it is necessary to control. This is measured in term of Total solid (TS), volatile solid (VS), total organic carbon (TOC), COD or BOD (Boe & al., 2005). These parameters are appropriate for the control of the anaerobic digestion applied to several types of waste.

5.6 Carbon monoxide

The carbon monoxide is a possible intermediate in the metabolic route of the acetogens and the methanogens (Moletta, 1993); Carbon monoxide was found in a great quantity during toxic inhibition by heavy metals (Liu & al., 2003). According to Moletta (Moletta, 2002) the presence of carbon monoxide is directly related to the acetate concentration, and conversely related to that of methane (Batstone & al., 2002).

N.B: there are other process control parameters of the production of biogas during anaerobic digestion, but they do not find any wide application in practice. However, the hydrogen gas is controlled in the gas phase and its measure in the liquid phase enables the identification of the existing different types of bacterial populations which may influence the process of the anaerobic digestion.

6. Physicochemical conditions necessary to anaerobic digestion

The anaerobic digestion can be carried out only under the following conditions:

- absence of oxygen, nitrates or sulphates (Degrémont, 2004).
- pH close to neutrality: optimum 6,8 - 7,5 (Moletta, 2002)
- concentration in volatile fatty acid (VFA) lower than 2 - 3 g/l (McCarty, 2001).
- a partial hydrogen pressure: 10 - 20 Pa to the maximum (Trably, 2002)
- a potential of oxydoreduction lower than -300 mV (Suh & Roussaux, 2002)
- absence of inhibiting elements: agent chlorinated, antibiotic,...
- an optimal stable temperature for micro-organisms (Bitton, 1994)

7. Advantages and disadvantages of anaerobic digestion

The advantages of anaerobic digestion are:

- A reduction of about 50% of the dry waste;
- A production of a Biogas which may undergo beneficiation in the form of energy (heating, cogeneration of electricity);
- A reduction of the number of pathogenic micro-organisms;
- An agronomic interest, related to a significant phosphate and ammoniacal nitrogen concentration (NH_4^+ (PO_4^{3}) due to the lysis of the organic matter (Münch & Greenfield, 1998);
- Request lower energy compared with aerobic processes ;
- the possibility of treating high organic loads: from 2 to more than 80 kg of COD per cubic meter of digester and per days, with a treatment rate from 80 to 98%

However, it has also certain disadvantages:

- High sensitivity to the toxic compounds (Schnurer & al., 1999);
- A slower degradation compared with aerobic processes (Bitton, 1994);
- Significant capital costs ;
- The growth kinetics of bacteria is low, the pretraitement kinetics is also low and the time needed for the treatment is relatively long;
- the microbial populations are sensitive to the disturbances, in particular with oxygen and with heavy metals (OTV, 1997);
- the treatment by anaerobic digestion is often insufficient to directly reject the effluents in the natural environment: an aerobic postprocessing is necessary to complete the elimination of carbon and possibly nitrogen and phosphorus.

8. Effect of temperature and substrate composition on biogas production

As mentioned previously, the temperature and substrate compostion have a high effect on biogas production. Their effects are confirmed by the study published by several researchers. As an example the results, obtained by Derbal et al. (Derbal & al., 2011) can be cited.

The obtained values of different parameters of the (co)-digestion experiments under mesophilic and thermophilic conditions are presented (see figure 2, 3, 4, 5, 6). It should be noted that the volume of the mesophilic co-digester which is 2000 m^3 is 20 times larger than pilot scale digesters 500 l. Therefore, the absolute biogas volume produced is different from the other cases and no comparison can be made. However, a comparison for different parameters is presented as follows:

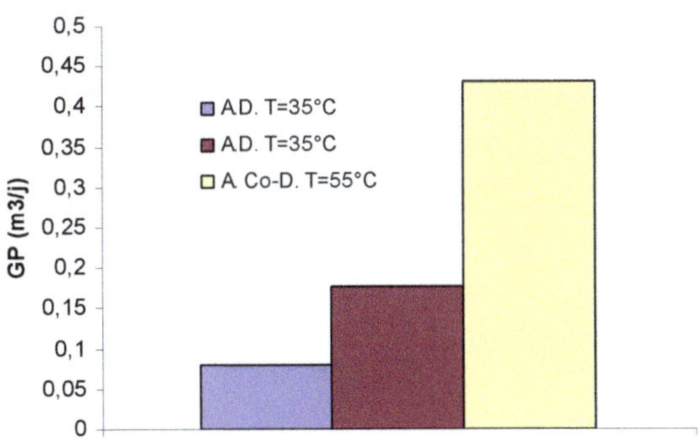

Fig. 2. Comparison of gas production (GP)

Figure 2 represents daily average biogas production values for the four studied cases where the thermophilic co-digestion shows the best results. Eventhough temperature has a certain effect on biogas production, adding solid waste is a contributing factor to this production. In fact, solid wastes contain a high percentage of organic matter.

The use of gas production rate GPR as a comparison parameter led us to include the data from the industrial scale digester. The results shown on Figure 3 confirm that the combined effect of temperature and solid waste addition is positive and considerable. Moreover, thermophilic co-digestion presents the best GPR results wich are confirmed by values of

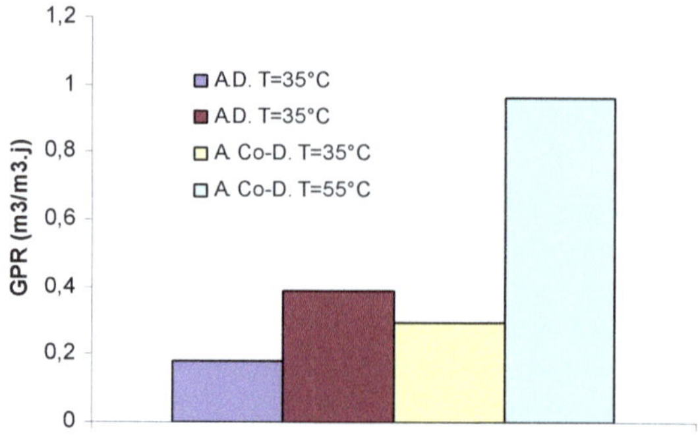

Fig. 3. Comparison of gas production rate (GPR)

SGP of Figure 4. SGP is in relation with the biodegradability of the substrate and with anaerobic process reaction. SGP increased from 0.14 to 0.33 for digestion of wastewater sludge alone when temperature increased from mesophilic conditions (35°C) to thermophilic ones (55°C), whereas for a wastewater sludge mixed with solid waste this parameter increased from 0.31 to 0.51. Adding solid waste under mesophilic comditions results in an increase of SGP from 0.14 to 0.31, whereas under thermophilic conditions SGP increased from 0.33 to 0.51. The combined effect of increasing temperature from mesophilic conditions to thermophilic ones and adding solid waste to wastewater sludge increased SGP from 0.14 to 0.51.

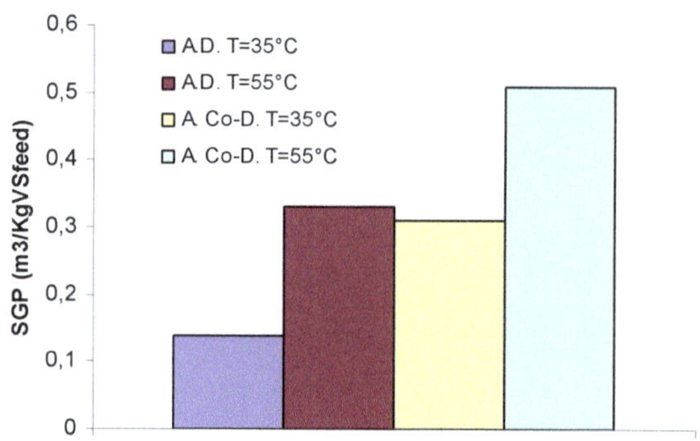

Fig. 4. Comparison of specific gas production (SGP)

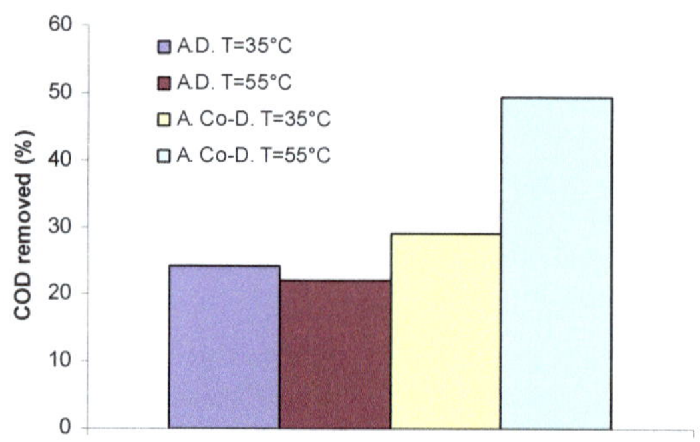

Fig. 5. Comparison of chemical oxygen demand removed (COD)

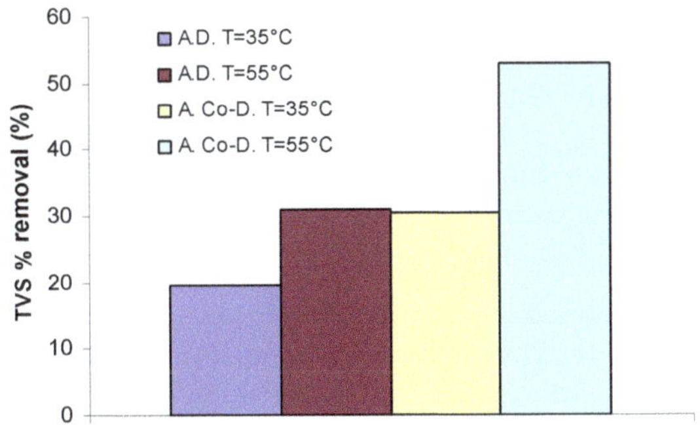

Fig. 6. Comparison of total volatile suspended removal (TVS)

Figure 5 presents the comparison of COD removal in the different studied cases. It increased from 24% for wastewater sludge alone under mesophilic conditions to 49.35% for wastewater sludge mixed with solid waste under thermophilic conditions. Moreover, for TVS thermophilic, co-digestion presents the best removal rate, 52.93%, as shown in Table 5. As a treatment system, anaerobic co-digestion under thermophilic conditions presents the best removal rates as well as specific gas production. It should be noted that changing working conditions from mesophilic to thermophilic ones increases anaerobic kinetic rates and as such the treatment capacity of a known volume will be increased as well. Adding solid waste contributes to the increase of biodegradable organic matter in the substrate (Figure 6).

As a conclusion of this sludy, the obtained results show that thermophilic co-digestion gives the best results. Although the temperature has an effect on the biogas production, it remains however quite relative compared to the effect of solid waste. These results confirm that the combined effect of the temperature and solid waste improves considerably the biogas production rate (GPR). The moving from mesophilic to thermophilic conditions, for waste sludge alone makes GPR pass from 0.18 to 0.39 m^3/m^{3}*d and for the waste sludge mixed with solid waste from 0.29 to 0.96 m^3/m^{3}*d.

The analysis of produced biogas showed that the percentage of biomethane is very high 60.37 and 64.44 for the digestion of sludge waste in mesophilic and thermophic phases, respectively and 65.8 and 60.61 for the co-digestion of solids waste with sludge waste in mesophilic and thermophilic cases, respectively.

9. Modeling of anaerobic digestion process

Due to the importance of anaerobic digestion as a treatment process, different dynamic models exist, such as the AM2 which was developed jointly by researchers of the INRA of Narbonne and the INRIA of Sophia-Antipolis in 2001 (Olivier et al., 2001). It is based on experimental results obtained on the fixed bed reactor of the INRA of Narbonne. This model is made of two steps: acidogenesis and methanogenesis corresponding to acido-acetogens and methanogens bacteria populations, respectively. As a more recent and elaborate model, the ADM1, was developed by an IWA group (Batstone et al., 2002). Its main feature is the consideration of the principal steps of anaerobic digestion process that are, respectively, substrate disintegration (non biological step), hydrolysis, acidogenesis, acetogenesis and finally the methanogenesis with seven different bacteria groups.

Since its development in 2002 and up to now, the ADM1 has been tested and used on different substrates where a great number of research works are reported in the literature. As examples, one can cite (Blumensaat and Keller, 2005) who modified the initial version of ADM1 for the simulation of a dynamic behaviour of a pilot scale digester using sludge, in order to ensure a faultless model implementation. They obtained accurate results for the cases of low to medium loading rates. However the accuracy showed a decline with the increase of the loading rate.

Wayne and Parker, 2005) considered the application of the ADM1 to a variety of anaerobic digestion configurations where the results showed, in most of the considered cases, that the model was able to reproduce the trends of the experimental results.

(Feng et al., 2006) found that the ADM1 is not sensitive to the distribution ratio of carbohydrates, proteins and lipids, whereas the fraction of short chain fatty acids (SCFA) in the influent is rather more important.

Consequently, the great capabilities of ADM1 in modelling different types of substrates and calculations have been the motivating factor to use it in the present work to evaluate the performances of a co-digestion process for the treatment of organic municipal solid waste and waste activated sludge in the above mentioned 2000 m^3 reactor working at a temperature of 37°C.

As mentioned above the ADM1 (Anaerobic Digestion Model No. 1) was developed by the IWA group (Batstone et al., 2002) with the objective to build a full mathematical model based intimately on the phenomenological model in order to simulate, at best, anaerobic reactors. It includes, as a first step, the disintegration of solid complexes (non biological step) into carbohydrates, lipids, proteins and inert material (soluble and particulate inert). The second step is the hydrolysis process of the disintegration products under an enzymatic action to produce sugars, amino acids and long chain fatty acids (LCFA), successively. Then, amino acids and sugars are fermented to produce VFA, hydrogen and carbon dioxide (acidogenesis). Then LCFA, proprionic acid, butyric acid and valeric acid are anaerobically oxided into acetate, carbon dioxide and hydrogen (acetogenesis). Finally, methane can be produced through two paths: the first one is based on acetate whereas the second one is through the reduction of carbon dioxide by molecular hydrogen. The organic species and molecular hydrogen, in this model, are expressed as COD (Chemical Oxygen Demand), whereas inorganic nitrogen and inorganic carbon species are expressed through their molecular concentrations.

Extensions and modifications were brought to ADM1 to enlarge its prediction capabilities by, taking into account other factors such as, for instance, the sulfato-reductors or the degradation of certain substrates (Wolfsberger and Halubar, 2006) and (Batstone and Keller, 2003). Moreover, Usama Zaher (Usama, 2005) considered the toxic effects of cyanide as an inhibition process for acetate.

9.1 Application of ADM1 model for simulation of organic solid waste

Concerning simulations of TCOD and SCOD, and TVFA, after estimation of substrate disintegration and hydrolysis parameters, it can be noticed that the simulated results are in good agreement with the experimental ones as shown on figure 7 and 8 respectively.

Parameters	Middle	Minimum	Maximum	Stand. Dev.	Num. samples
		Sludge			
pH	7.3	6.7	7.9	0.34	36
NH_4^+ (mg N/l)	3.9	1	13	4	24
TKN (mg N/l)	43.1	31.2	49.9	8.23	16
COD (mg COD/l)	670.7	596.8	748	44.49	16
P_{tot} (mg P/g TS)	603.4	241.7	770.6	149.73	10
TS (g/l)	35.6	26.6	47.5	4.67	36
TVS (g/l)	23.1	17.2	31.1	3.05	36
TVS sludge (%ST)	64.8	58.3	80.9	4.35	36
Flow (m^3/day)	0.019	0.019	0.019	0.00	45
		Waste			
TKN (mg N/l)	33.3	21.9	53.5	8.30	13
TCOD (mg COD/l)	996.2	829.7	1124.4	78.26	16
P_{tot} (mg P/g ST)	831.1	183.3	1540.9	411.99	11
TS (g/l)	160.2	72	269.9	56.42	38
TVS (g/l)	141.6	61.5	245.5	51.07	38
TVS (%TS)	89.4	73.7	94.7	4.28	38
Flow (m^3/day)	0.0032	0.0023	0.0036	0.00	45
		Waste mixed with sludge			
TKN (mg N/l)	41,7	29,9	50,4	-	-
TCOD (mg COD/l)	717,6	630,3	802,2	-	-
P_{tot} (mg P/g ST)	636,2	233,3	881,5	-	-
TS (g/l)	53,5	33,1	79,5	-	-
TVS (g/l)	40,2	23,6	62,0	-	-
TVS (%TS)	68,3	60,5	82,9	-	-
Waste (m^3/day)	0.0032	0.0023	0.0036	-	-

Table 2. Influent characteristics

Parameters	Middle	Minimum	Maximum	Stand. Dev.	Num. samp
pH	7.84	7.6	8.1	0.10	44
NH_4^+ (mg N/l)	1022.1	900	1140	70	24
TKN (mg N/l)	37.8	28.7	49.1	5.45	9
TCOD(kg COD/m^3)	22.2	18.3	24.7	1.92	16
SCOD (kg COD/m^3)	4,6	2	7	2.07	5
P_{tot} (mg P/g ST)	752.2	383	1080.8	181.22	12
TS (g/l)	33.1	26.3	52.3	5.01	40
TVS (g/l)	18.9	15.5	26.8	2.18	40
TVS (% TS)	57.2	50	64.3	3.82	40
VFA (mg COD/l)	50.7	7.0	110.3	26.47	36
TA at pH 6 (mg $CaCO_3$/l)	2466.7	2181.5	2911	186.67	44
TA at pH 4 (mg $CaCO_3$/l)	4005.5	3806.4	4356	135.07	44
effluent flow (m^3/day)	0.0225	0.0225	0.0225	0.00	45

Table 3. Effluent characteristics

Parameters	Middle	Minimum	Maximum	Stand. Dev.	Num. samp
Biogas volume (m^3/day)	0.431	0.153	0.728	0.16	31
SGP (m^3 biog/kg TVS)	0.51	0.26	1.06	0.16	29
GPR (m^3 biogas/m^3 day)	0.96	0.34	1.62	0.35	31
% CH_4 (%)	60.6	55	65	2.22	40
% CO_2 (%)	39.4	35	45	2.22	40
Volume of CH_4 (m^3/day)	0.3	0.09	0.44	0.10	31
Volume of CO_2 (m^3/days)	0.17	0.06	0.28	0.06	31
H_2S (ppm)	440	200	1044	204.91	31

Table 4. Characteristics of biogas production

Kinetic parameters	Names	Units	Initial values used in ADM1	Initial values	Estimated values
K_{dis}	Disintegration constant	Day^{-1}	0.5[b]	0.7	0.7
$K_{hyd.Ch}$	Carbohydrate hydrolysis constant	Day^{-1}	10[b]	1.25[a]	1.0
$K_{hyd.Pr}$	Proteins hydrolysis constant	Day^{-1}	10[b]	0.5[a]	0.7
$K_{hyd.Li}$	Lipids hydrolysis constant	Day^{-1}	10[b]	0.4[a]	1.0

[a] Middle values obtained from (Mata-Alvarez , 2003)
[b] Values obtained from (Batstone and Keller, 2003)

Table 5. Initial and estimated values of kinetic parameters

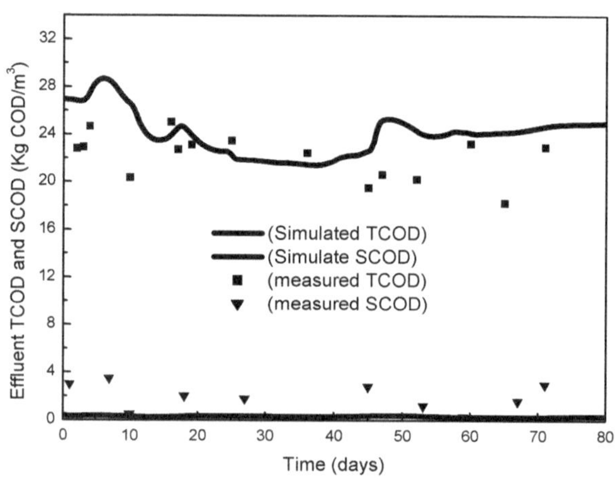

Fig. 7. Comparison between the simulation and the experimental total and souble COD

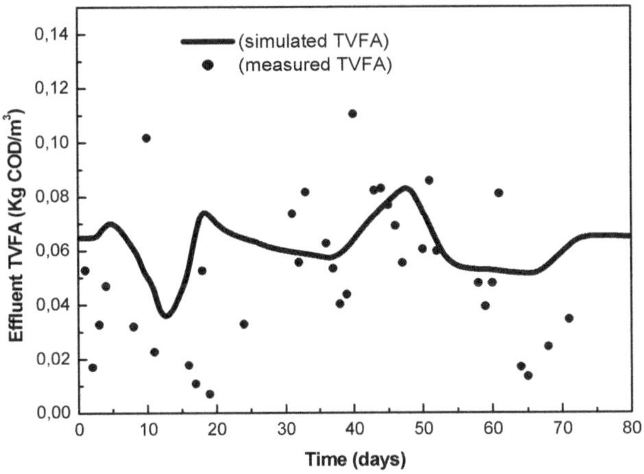

Fig. 8. Comparison between the simulation and the experimental TVFA

However the simulated results of SCOD are somehow underestimated in comparison with the experimental ones. This may be explained by the fact that the substrate distribution between proteins, carbohydrates and lipids was not measured but default model values were adopted for this parameter.

The simulated TVFA results show good digester stability and are in good agreement with the experimental ones as well.

Figure 9 shows variation of the experimental and simulated results of total biogas volume produced, which depends on the nature, the composition and the biodegradability of solids. In this case, mass loading fluctuate as shown by the ORL, it should be underlined that the main objective of these experiments was to increase the ORL to the practical limits in order to treat a maximum quantity of solid waste however it was difficult to maintain it constant. Consequently these variations condition the tendency of biogas volume produced variation. The limitations of ADM1 imply that the simulated biogas production follows an average course; therefore, the simulated data overlaps partially the experimental values.

Figure 10 shows the experimental and simulated results of biogas production. The biogas is composed principally of methane and carbon dioxide and a small percentage of hydrogen. It can be noticed that the simulated results are in good agreement with the experimental ones. A similar remark concerning the average course of the curve can be held as well. Moreover, they show a good stability in the operating of the reactor

To have a clear picture of what is happening within the system, inorganic carbon (IC) and inorganic nitrogen (IN) as well as pH, were represented on the same graph as presented in Figure 11.

Since pH is approximately equal to 8, IC represents alkalinity. Any variation in alkalinity is due to neutralisation of VFA, if accumulated. Furthermore, alkalinity or IC is more sensitive to VFA accumulation than pH and therefore more reliable.

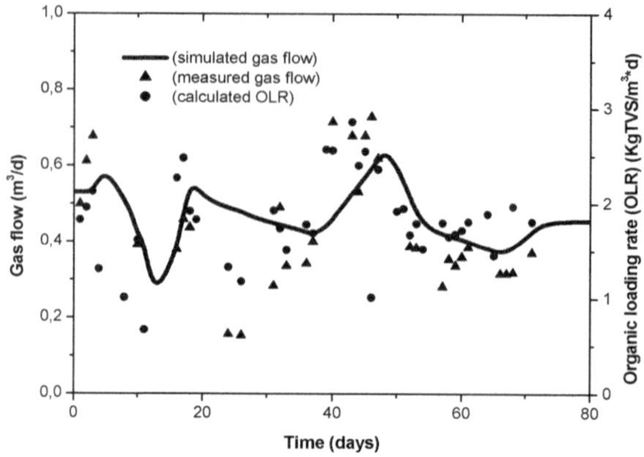

Fig. 9. Comparison between the simulation and the experimental biogas production rate and the variation of the organic loading rate (OLR) with time

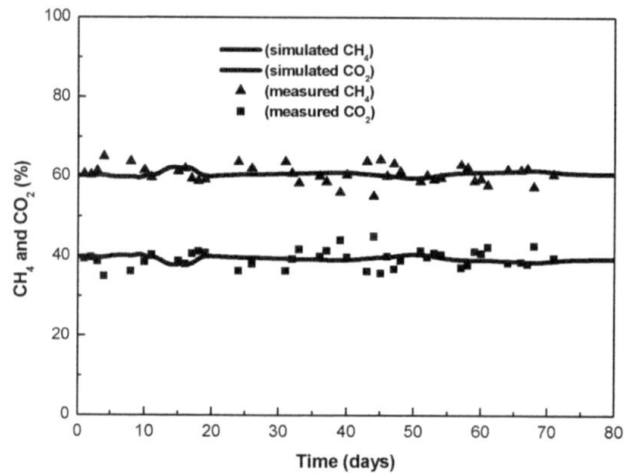

Fig. 10. Comparison between the simulation and the experimental % of CO_2 and CH_4

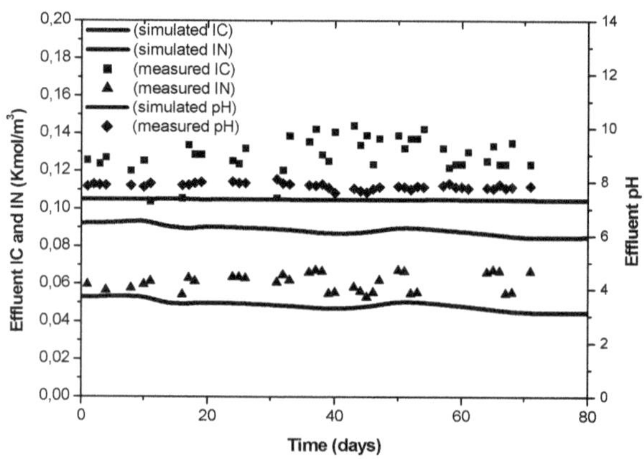

Fig. 11. Comparison between the simulation and the experimental IC, IN and pH

It is noted that in this study, the simulated results show an acceptable fit for Total chemical oxygene demand (COD), biogas volume and composition, pH and inorganic nitrogen (IN). However, for inorganic carbon (IC), the simulated results do not show a good fit. It was confirmed that IC or bicarbonate alkalinity is a very sensitive parameter to volatile fatty acids (VFA) accumulation, compared to pH variations and hence it can be used as a monitoring parameter.

10. References

Ahring, B. K. (1994). Status of science and application of thermophilic anaerobic digestion, Water Science and Technology, 30, (12), 241-249.

Ahring, B.K. (2003). Perspectives for anaerobic digestion, In Advances in biochemical engineering/biotechnology - Biomethanation I, Springer. Berlin, pp. 1-30.

Anderson, G. K. and Yang, G. (1992). Determination of bicarbonate and total volatile acid concentration in anaerobic digesters using a simple titration, Water Environment Research, 64, (1), 53-59.

Angelidaki I, Alves M, Bolzonella D, Borzacconi L, Campos J.L, Guwy A.J, Kalyuzhnyi S, Jenicek P and Van Lier J.B. (2009). Defining the biomethane potential (BMP) of solid organic wastes and energy crops: a proposed protocol for batch assays, Water science and technology 2009; 59(5):927-934.

Angelidaki, I. and Ahring, B. K. (1993). Effect of the clay mineral bentonite on ammonia inhibition of anaerobic thermophilic reactors degrading animal waste, Biodegradation, **3**, 409-414.

Angelidaki, I. and Ellegaard, L. (2002). Anaerobic digestion in Denmark: Past, present and future, p.129-138, In: Anaerobic digestion for sustainability in waste (water) treatment and reuse. Proceedings of 7th FAO/SREN-Workshop, 19-22 Mai Moscow, Russia.

Angelidaki, I., Boe, K. and Ellegaard, L. (2005). Effect of operating conditions and reactor configuration on efficiency of full-scale biogas plants, Water Science and Technology, 52, 189-194.

Balk, M., Weijma, J. & Stams, A.J.M. (2002). Thermotoga lettingae sp nov., a novel thermophilic, methanol-degrading bacterium isolated from a thermophilic anaerobic reactor, International Journal of Systematic and Evolutionary Microbiology 52, 1361-1368.

Batstone D. J., Keller J.,(2003). Industrial applications of the IWA anaerobic digestion Model No. 1 (ADM1), Water Sci. Technol., 47, (12), 199-206.

Batstone D. J.,. Keller J.,. Angelidaki I., Kalyuzhnyi S. V., Pavlostathins S. G., Rozzi, Sanders W. T. M., Siegrist H., Vavilin V. A., (2002). Anaerobic Digestion Model No. 1, International Water Association (IWA), Publishing, London, UK.

Batstone, D. J., Keller, J., Angelidaki, I., Kalyuzhny, S. V., Pavlostathis, S. G., Rozzi, A., Sanders, W. T. M., Siegrist, H. and Vavilin, V. A. (2002). The IWA Anaerobic digestion model no 1. (ADM1), Water Science and Technology, 45, (10), 65-73.

Bendixen, H. J. (1994). Safeguards against pathogens in Danish biogas plants, Water Science and Technology, 30, (12), 171-180.

Bitton, G. (1994). Wastewater microbiology, Wiley-Liss Editor, New York, USA.

Blumensaat F., Keller J., (2005). Modelling of two-stage anaerobic digestion using the IWA anaerobic digestion model No. 1 (ADM1), Water Res., 39, 171-183.

Boe, K., Batstone, D.J. and Angelidaki, I. (2005). Optimisation of serial CSTR biogas reactors using modeling by ADM1, In: The First International Workshop on the IWA Anaerobic Digestion Model No.1 (ADM1), 2-4 September, Lyngby, Denmark. Proceedings, pp. 219-221. International Water Association, London, UK.

Chynoweth, D. P., Svoronos, S. A., Lyberatos, G., Harman, J. L., Pullammanappallil, P., Owens, J. M. and Peck, M. J. (1994). Real-time expert system control of anaerobic digestion, Water Science and Technology, 30, (12), 21-29.

Degrémont. (2004). *Mémento technique de l'eau, 9eme Edition, Tome 1 et 2*, Editions Lavoisier, Paris, France

Delbès, C. (2000). Diversité et dynamique structurales et fonctionnelles de la communauté microbienne d'un digesteur anaérobie : Approche moléculaire à partir des ADNr et ARNr 16s, Université Claude Bernard, Lyon I, p.170.

Derbal, K., Bencheikh-lehocine, M., Meniai A-H. (2011). Pilot study of biogas production from organic solid waste in thermophilic phase, Journal of Science Academy Transactions on Renewable Systems Engineering and Technology (SATRESET), Vol. 1, N° 3, 93-96.

Dolfing, J. (1988). Acetogenesis. p.417-442, In: Zehnder, A. J. B. (ed.) Biology of Anaerobic Microorganisms, John Wiley & Sons, New York.

Edeline, F. (1997). L'épuration biologique des eaux. Théorie & technologie des réacteurs, CEDEBOC Editeur, Paris, France.

Feitkenhauer, H., Sachs, J. V. and Meyer, U. (2002). On-line titration of volatile fatty acids for the process control of anaerobic digestion plants, Water Research, 36, 212-218.

Feng, Y., Behrendt, J., Wendland, C., Otterpohl, R., (2006). Parameters analysis and discussion of the anaerobic digestion model No.1 (ADM1), Water Science & Technology, 54, (4), 139–147.

Gijzen, H.J. (2002). Anaerobic digestion for sustainable development: A natural approach, Water Science and Technology 45, 321-328.

Hansson, M., Nordberg, Å., Sundh, I. and Mathisen, B. (2002). Early warning of disturbances in a laboratory-scale MSW biogas process, Water Science and Technology, 45, (10), 255-260.

Hawkes, F. R., Guwy, A. J., Hawkes, D. L. and Rozzi, A. G. (1994). On-line monitoring of anaerobic digestion: Application of a device for continuous measurement of bicarbonate alkalinity, Water Science and Technology, 30, (12), 1-10.

Hickey, R. F. and Switzenbaum, M. S. (1991). Thermodynamics of Volatile Fatty-Acid Accumulation in Anaerobic Digesters Subject to Increases in Hydraulic and Organic Loading, Research Journal of the Water Pollution Control Federation, 63, (2), 141-144.

Hickey, R. F., Vanderwielen, J. and Switzenbaum, M. S. (1989). The effect of heavy metals on methane production and hydrogen and carbon monoxide levels during batch anaerobic sludge digestion, Water Research, 23, (2), 207-218.

Hill, D. T. (1990). Alkalinity measurements in anaerobic digestion systems as influenced by organic acid level and endpoint pH, Transactions of the ASAE, 33, (5), 1717-1719.

Hill, D. T. and Bolte, J. P. (1989). Digester stress as related to iso-butyric and iso-valeric acids, Biological Wastes, 8, 33-37.

Hill, D. T. and Holmberg, R. D. (1988). Long chain volatile fatty acid relationships in anaerobic digestion of swine waste, Biological Wastes, 23, (3), 195-214.

Karakashev, D., Batstone, D. J. and Angelidaki, I. (2005). Influence of environmental conditions on methanogenic compositions in anaerobic biogas reactors, Applied and Environmental Microbiology, 71, 331-338.

Kaseng, K., Ibrahim, K., Paneerselvam, S. V. and Hassan, R. S. (1992). Extracellular enzyme and acidogen profiles of a saboratory-scale two-phase anaerobic digestion system, Process Biochemistry, 27, 43-47.

Liu, J. (2003). Instrumentation, Control and Automation in Anaerobic Digestion, Ph.D. dissertation, Department of Biotechnology, Lund University, Sweden.

Mata-Alvares, J. (2003). Biomethanization of the organic fraction of municipal solid wastes, IWA publishing, UK, pp 42.

Mathiot, S., Escoffier, Y., Ehlinger, F., Couderc, J. P., Leyris, J. P. and Moletta, R. (1992). Control parameter variations in an anaerobic fluidised bed reactor subject to organic shockloads, Water Science and Technology, 25, (7), 93-101.

Mattiasson, B. (2004). Anaerobic digestion generates fatty acids, Industrial Bioprocessing, 26, (6), 8-9.

McCarty, P.L. (2001). The development of anaerobic treatment and its future, Water Science and Technology 44, 149-156.

McHugh, S., Carton, M., Mahony, T. and O'Flaherty, V. (2003). Methanogenic population structure in a variety of anaerobic bioreactors, FEMS Microbial Ecology, 219, (2), 297-304.

McMahon, K. D., Stroot, P. G., Mackie, R. I. and Raskin, L. (2001). Anaerobic codigestion of municipal solid waste and biosolids under various mixing conditioins- II: Microbial population dynamics" Water Research, 35, (7), 1817-1827.

Moletta, R. (1993). La digestion anaérobie: du plus petit au plus grand, Biofutur, janvier, pp. 16-25.

Moletta, R. (2002). Procédés biologiques anaérobies, Dans Gestion des problèmes environnementaux dans les industries agroalimentaires, Technique et documentation - Editions Lavoisier, Paris, France.

Moosbrugger, R. E., Wentzel, M. C., Ekama, G. A. and Marais, G. v. R. (1993). A 5 pH point titration method for determining the carbonate and SCFA weak acid/bases in anaerobic systems, Water Science and Technology, 28, (2), 237-245.

Münch, E. V. and Greenfield, P. F. (1998). Estimating VFA concentrations in prefermenters by measuring pH, Water Research, 32, (8), 2431-2441.

Nielsen, H. B. (2006). Control parameters for understanding and preventing process imbalances in biogas plants: Emphasis on VFA dynamics, Ph.D. dissertation, BioCentrum-DTU, Technical University of Denmark.

Oles, J., Dichtl, N. and Niehoff, H. (1997). Full scale experience of two stage thermophilic/mesophilic sludge digestion, Water Science and Technology, 36, (6-7), 449-456.

Olivier Bernard, Zakaria Hadj-Sadok, Denis Dochain, Antoine Genovisi, Jean-Philipe Steyer, (2001). Dynamical model development and parameter identification for an anaerobic wastewater treatment process, Biotechnology Bioenergy, 75 (4), 424-438.

OTV. (1997). Traiter et valoriser les boues. Ouvrage collectif n°2, Editions Lavoisier, Cachan.

Puñal, A., Palazzotto, L., Bouvier, J. C., Conte, T. and Steyer, J.-P. (2003). Automatic control of volatile fatty acids in anaerobic digestion using a fuzzy logic based approach, Water Science and Technology, 48, (6), 103-110.

Ramsay, I. R. and Pullammanappallil, P. (2001). Protein degradation during anaerobic wastewater treatment: Derivation of stoichiometry, Biodegradation, 12, (4), 247-257.

Ren, N., Wang, B. and Huang, J. H. (1997). Ethanol-type fermentation from carbohydrate in high rate acidogenic reactor, Biotechnology and Bioengineering, 54, (5), 428-433.

Rodriguez, J., Kleerebezem, R., Lema, J. M. and Van Loosdrecht, M. C. M. (2006). Modeling product formation in anaerobic mixed culture fermentations, Biotechnology and Bioengineering,93, (3), 592-606. 45.

Schink, B. (1997). Energetics of syntrophic cooperation in methanogenic degradation, Microbiology and Molecular Biology Reviews, 61, (2), 262-280.

Schnurer, A., Schink, B. and Svensson, B. H. (1996). Clostridium ultunense sp. nov., a Mesophilic bacterium oxidizing acetate in syntrophic association with a hydrogenotrophic methanogenic bacterium, International Journal of Systematic Bacteriology, 46, (4), 1145-1152.

Schnurer, A., Zellner, G. and Svensson, B. H. (1999). Mesophilic syntrophic acetate oxidation during methane formation in biogas reactors, FEMS Microbial Ecology, 29, (3), 249-261.

Smith, S. R., Lang, N. L., Cheung, K. H. M. and Spanoudaki, K. (2005). Factors controlling pathogen destruction during anaerobic digestion of biowastes, Waste Management, 25, (4), 417-425.

Suh, Y. J. et Roussaux, P. (2002). An LCA of alternative wastewater sludge treatment scenarios", Resources, Conservation and Recycling, 35, 191-200.

Thauer, R. K., Jungermann, K. and Decker, K. (1977). Energy conservation in chemotrophic anaerobic bacteria" Bacteriological Reviews, 41, (1), 100-180.

Thiele, J. H. (1991). Mixed-Culture Interactions in Methanogenesis, In Mixed-Culture in Biotechnology, McGraw-Hill, Inc., New York, USA.

Trably, E. (2002). Etude et optimisation de la biodégradation d'Hydrocarbures Aromatiques Polycycliques (HAPs) et de Polychlorobiphényls (PCBs) au cours de la digestion anaérobie et aérobie de boues urbaines contaminées, Génie des procédés, Université Montpellier II, Montpellier, 280 p.

Usama El-Sayed Zaher, (2005). Modelling and monitoring the anaerobic digestion process in view of optimisation and smooth operation of WWTP's, PhD Thesis in applied biological science: Environmental Technology, Universitait Gent, Faculty of Bioscience Engineering, academic year 2004-2005.

Vavilin, V. A. and Angelidaki, I. (2005). Anaerobic degradation of solid material: Importance of initiation centers for methanogenesis, mixing intensity, and 2D distributed model, Biotechnology and Bioengineering, 89, (1), 113-122.

Wang, J. Y., Zhang, H., Stabnikova, O. and Tay, J. H. (2005). Comparison of lab-scale and pilot-scale hybrid anaerobic solid-liquid systems operated in batch and semi-continuous modes, Process Biochemistry, 40, (11), 3580-3586.

Wayne J. Parker, (2005). Application of the ADM1 model to advanced anaerobic digestion, Bioresource Technology, 96, 1832-1842.

Wolfsberger A. & P. Holubar, (2006). WP7 Biokinetic Data, Modelling and Control 2nd Year CROPGEN meeting, Vienne.

10

Feasibility of Bioenergy Production from Ultrafiltration Whey Permeate Using the UASB Reactors

Marta Kisielewska
University of Warmia and Mazury in Olsztyn, Poland

1. Introduction

Cheese whey is a by-product generated during cheese manufacturing. The disposal of whey is problematic because of its high COD (Chemical Oxygen Demand) (about 50,000 mg L^{-1} - 80,000 mg L^{-1}), low solids content (5% DM), low bicarbonate alkalinity and its tendency to get acidified very rapidly (Aktaş et al., 2006; González Siso, 1996; Venetsaneas et al., 2009). In 2008, Poland produced almost 1123 thousand tonnes of whey (Agricultural Market Agency [ARR], 2009). Traditionally, cheese whey has been used to feed animals, but redistribution of whey to farmers is very expensive. Moreover, lactose intolerance of farm animals also limits the use of whey in feeding (de Glutz, 2009). Since large quantities of whey are produced (about 9 kg of whey in the production of 1 kg cheese) (Zafar & Owais, 2006), there is an increasing concern as how it can be efficiently and cost-efficiently processed without adversely effecting the environment.

Proteins from cheese whey have a high nutritional value. For this reason cheese manufacturers have explored the possibilities of valorisation of whey. They recover proteins by membrane ultrafiltration (UF) process (Silveira et al., 2005). This method of separation has the main advantage - in does not denature proteins, so they save their original nutritional value (de Glutz, 2009). The residual protein-free material is called whey permeate. Permeate streams have very high COD value (about 50,000 - 70,000 mg L^{-1}) (own studies), which represents an important environmental problem, similarly to whey. The chemical and biological instability of the UF whey permeate resulting in difficulties and high cost in its transport and storage. Proper management of this liquid is important due to strict legislation and economic reasons. Because of those there is a strong need to efficiently treat UF whey permeate.

UF whey permeate is composed mainly of lactose. Lactose concentration is about 50,000 mg L^{-1}, so more than 90% of COD is due to lactose (de Glutz, 2009). Moreover, valuable compounds (proteins, vitamins) can be found in its composition. Since UF whey permeate contains significant quantities of lactose, the way to use this waste product could be as a substrate for fermentation to produce biofuels.

Nowadays, the most widely produced biofuels are ethanol and biogas (methane). Alcohol fuels are produced by fermentation of sugars derived from corn, sugar beet, sugarcane, potatoes, wheat, followed by distillation and drying. The production of bioethanol from corn or sugarcane is a mature technology. For example, in Brazil there are 448 bioethanol production units installed and according to a report of the Brazilian Ministry of Mines and Energy, ethanol production was 25 billion liters in 2008 (Soccol et al., 2010). Biogas is produced by anaerobic digestion of organic materials by anaerobic microorganisms. It can be used to produce thermal energy (heating), electricity, or if compressed - it can be used in vehicles. The current operation of biogas plants is relatively large in Europe, especially in Germany. According to Pöschl et al. (2010), the estimated biogas production potential in Germany is 417 PJ per year and 80% of which derived from agricultural resources, including farm waste (96.5 PJ per year), crop residues (13.7 PJ per year), and dedicated energy crops (236 PJ per year).

More recently, hydrogen is playing more important role as a fuel used for heating, lighting and as a motor fuel. The main advantage of hydrogen as a future alternative energy carrier is the absence of polluting emissions when combusted, results in pure water. Today, most hydrogen gas is obtained from fossil fuels which generate greenhouse gas (GHG) that contribute to global warming. The biological hydrogen production is an attractive method because it can be produced from renewable raw materials such as organic wastes. Wastewater from food processing industries show great potential for economical production of hydrogen (Van Ginkel et al., 2005), but today no strategies for industrial-scale productions have been found.

The ability to produce biofuels from low-cost biomass such as agricultural waste and by-products (including for example crop residues, sugar cane waste, wood, grass and wastewater from food processing industries) will be the key to making them competitive with other fuels, for example gasoline. Only biofuels derived from waste products show low environmental effects, such as reduction of GHG emission, small land demand and damage the environment. As a result, since UF whey permeate disposal represent a real problem for the dairy industry, biofuels production offers an ideal alternative to its valorization (de Glutz, 2009; Silveira et al., 2005).

The objectives of this work were to study the applicability of fermentation processes for the production of biogas (methane), fuel bioethanol and biohydrogen in Upflow Anaerobic Sludge Blanket (UASB) reactors fed with raw UF whey permeate. To optimize and enhance the biofuels production, the different processes were used (ultrasonic stimulation of microbial cells, anaerobic steel corrosion process) and the different operational parameters (pH, hydraulic retention times - HRTs regimes, organic loading rates - OLRs) were applied.

2. Biogas production

Biogas is a gas with 50-70% of methane (CH_4) and 50-30% of carbon dioxide (CO_2) content produced by the anaerobic decomposition of organic matter. It can be produced from a wide variety of available waste organic materials, including sewage sludge, animal manure, municipal/industrial organic waste, parts from ethanol production, crop residues, specially grown energy crops and more. Methane, the combustible component of biogas, mainly determines the properties of biogas. A m^3 of biogas produced from food industrial organic wastes has on average a methane content of about 55% and therefore a calorific value of about 6.5 kWh (Angelidaki et al., 2003). Nowadays, the world markets for biogas are booming. Advances in biotechnology, molecular science and microbiology contributed to enhancements in biogas yields production (more high tech resulting in over 70% plant efficiency), which led to the development of commercial biogas plants (Yadvika et al., 2004). As a result, biogas competes with petroleum-based fuels in terms of performance, cost, and additional benefits such as reducing GHG emissions. Currently Europe dominates in biogas production (Prochnow et al., 2009). Germany, the biogas market leader, runs about 5000 biogas plants in 2009, covering more than 1% of the electrical energy consumption from biogas (Meyer-Aurich et al., 2012). However, the trend in producing biogas is also catching up fast in countries like Japan, Australia, New Zealand, USA, China and India.

Innovations are still necessary to support research and development in the field of renewable energy. The main research area is closely related to renewable biomass feedstock. Consequently, the objectives of this work were: (1) to investigate anaerobic biogas potential from UF whey permeate, (2) to evaluate if steel elements could enhance the performance of UASB reactors treating UF whey permeate (COD removal efficiency, phosphorus removal), and (3) to study the influence of steel elements on the biogas production rate and methane content in biogas.

2.1 Materials and methods

2.1.1 Fermentation medium and experimental system

Two identical Plexiglas laboratory-scale UASB reactors (R_0 and R_{Fe}) with a working volume of 2.05 L each, one packed with spiral elements made of steel wire with an iron content of 48% (Fig.1; Table 1), were run in parallel at a constant mesophilic temperature of 35°C ± 1°C throughout a 219-day period. Four running stages were identified in term of OLR applied (Table 1). The OLR was increased stepwise from the initial 2.0 kg COD m^{-3} d^{-1} to finally 12.0 kg m^{-3} d^{-1}. The reactors were operated for 25 - 66 days to ensure the reactors reached steady states at each stages (the steady-state conditions were evidenced when the standard deviations of COD removal efficiencies were within 3%). After the steady-state conditions were achieved, the OLR was increased to the next level. HRT at all stages was 48 h. The pH in the reactors was controlled at the level of 7.0 ±0.05 with 2 M NaOH.

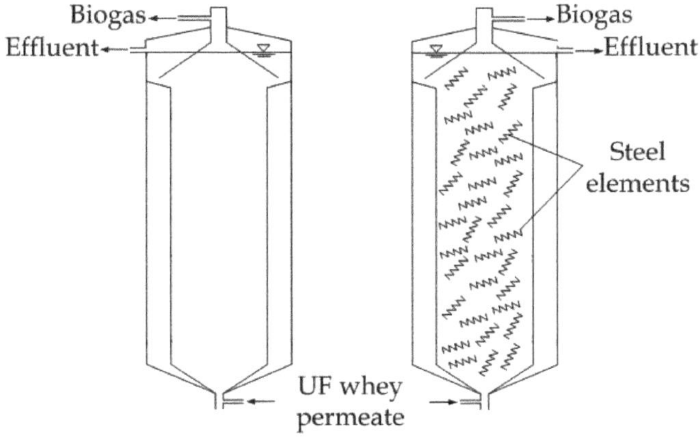

Fig. 1. Schematic of the laboratory-scale anaerobic treatment system

	Stage 1	Stage 2	Stage 3	Stage 4
Operation period (d)	45 – 69	70 – 105	106 – 153	154 – 219
OLR (kg COD m^{-3} d^{-1})	2.0	4.0	7.0	12.0
HRT (h)	24	24	24	24
Contact surface of the packing medium (m^2)	0.00175	0.00175	0.00175	0.00175
Packing volume of steel elements (L)	0.0087	0.0087	0.0087	0.0087

Table 1. Operation regimes for the parallel UASB reactors with and without steel elements

Both UASB reactors were fed with UF whey permeate from the manufacture of dairy products in Nowy Dwór Gdański, Poland. The characteristics of the wastewater used in this study is shown in Table 2. It was received from the factory once a week, was stored at -20°C and was thawed before used. Prior to being fed into the reactor, the substrate was diluted with tap water in accordance with the required organic loading rate (OLR) to obtain wastewater COD concentrations in the average range of 4 – 24 g COD L^{-1}. Diluting UF whey permeate was maintained at a temperature 4°C until used. The reactors were not supplemented with trace elements.

Parameters	Range of values
Total COD (g L^{-1})	52 - 55
Lactose (g L^{-1})	48 - 53
TP (g L^{-1})	0.58 - 0.62
Phosphate (g L^{-1})	0.49 - 0.54
pH (when fresh at 20°C)	4.9 – 5.4
Total iron (mg L^{-1})	0.26 – 0.35

Table 2. Chemical characteristics of wastewater used

The seeding inoculum was taken from a laboratory mesophilic reactor treating synthetic dairy wastewater. Each UASB reactor was seeded up to biomass content of 70 g total suspended solids - TSS L^{-1} at a ratio of 20% (by volume). During startup the reactors were operated at an OLR of 1.0 kg COD m^{-3} d^{-1} and at a HRT of 48 h for 44 days. During the reactors operation, biogas production and composition (CH_4 and CO_2), total COD, total phosphorus – TP, phosphate, soluble iron concentration and pH in the effluent were measured three times a week. After the operation time of 219 days, sludge samples from both UASB reactors were collected for the determination of TSS content, TP and total iron contents.

2.1.2 Analytical methods

All monitored parameters were analyzed according to the Standard Methods for the Examination of Water and Wastewater (PN-74/C-04578.03; PN-90/C-04586.04; PN-EN 1189:2000; PN-75/C-04616.01; PN-67/A-86430; PN-EN ISO 6878:2006; PN-EN 13346:2002). The measurement of the pH was done using an Hanna Instruments pehameter model HI 9107. Biogas production from the UASB reactors was recorded by a water displacement meter, while biogas composition was analyzed by an electronic analyzer (LMSxi/G4.18, Gas Data Ltd.). All selected reactors performance parameters were analyzed with Fisher F-tests using Statistica 7.1 software (Statsoft Inc.). Differences were considered statistically significant if the 95% confidence interval of the mean of the parameters did not overlap.

2.2 Results and discussion

The COD removal efficiency, TP removal efficiency, biogas production and composition were markedly influenced by using steel elements as an additional medium in the UASB reaction chamber.

During Stage 1, both UASB reactors reached the steady-state after 25 days of operation. No statistically significant differences ($p>0.05$) were observed between UASB reactor with steel elements (R_{Fe}) and UASB reactor without steel elements (R_0) in term of the average COD removal efficiency and biogas production rate (Fig. 2; 3). Nevertheless, R_{Fe} indicated higher ($p<0.05$) removal efficiency in phosphate (86.2%) and TP (81.2%) than R_0 in which the analyzed values were 1.8% and 22.8%, respectively (Fig. 3). CH_4 content in biogas produced in R_{Fe} was as high as 67.1% which was higher by 11.9% than in R_0 ($p<0.05$). In Stage 2 and 3 both UASB reactors demonstrated a stable work, but statistically significant differences in the values of all the monitoring parameters between R_0 and R_{Fe} were noticed ($p<0.05$). The duration of each stage were 36 and 48 days, respectively. The average TP removal efficiency and phosphate removal efficiency in R_{Fe} were higher by 77.7% and 83.7%, respectively than in R_0 during Stage 2, and 68.1% and 73.9%, respectively during Stage 3 (Fig. 3). During Stage 2 and 3 high COD removal efficiencies (95.6%, 94.8%, respectively) were remained in R_{Fe}, in contrast to that of 84.2% in Stage 2 and 80.1% in Stage 3 in R_0 (Fig. 3). The average CH_4 content in biogas of 78.0% and biogas production of 2.59 $m^3m^{-3}d^{-1}$ in R_{Fe}, in contrast to that of 60.8% and 0.92 $m^3m^{-3}d^{-1}$, respectively in R_0 ($p<0.05$), were observed during Stage 2. In Stage 3, biogas production increased by 1.12 $m^3m^{-3}d^{-1}$ in R_0 and 1.2 $m^3m^{-3}d^{-1}$ in R_{Fe}, but it was still significantly higher in R_{Fe} (3.79 $m^3m^{-3}d^{-1}$) than in R_0 (2.04 $m^3m^{-3}d^{-1}$), $p<0.05$ (Fig. 2). Moreover in that stage, the highest methane content in biogas of 79.8% in R_{Fe} and 68.1% in R_0 were achieved (Fig. 2). During the last stage it was found the highest biogas production rate in R_{Fe} of 4.01 $m^3m^{-3}d^{-1}$, while 1.86 $m^3m^{-3}d^{-1}$ in R_0 was observed ($p<0.05$). The average

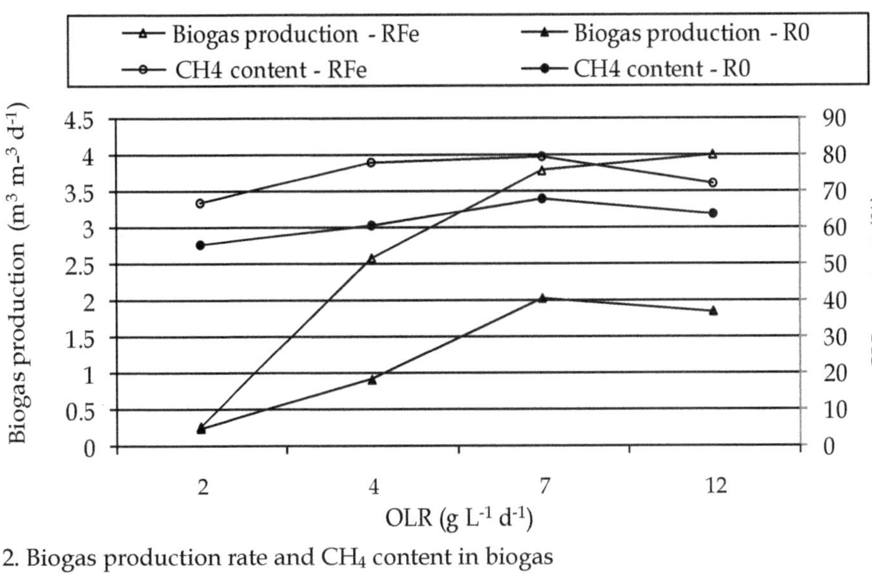

Fig. 2. Biogas production rate and CH_4 content in biogas

CH_4 content in biogas decreased to 72.3% in R_{Fe}, in contrast to that of 64% in R_0, and the differences between R_0 and R_{Fe} were statistically significant ($p<0.05$) (Fig. 2). It was found decrease in TP removal efficiency in R_{Fe} and R_0 to 72.2 and 10.1%($p<0.05$), respectively. According to this, the phosphate removal efficiencies decreased, too (Fig. 3). COD removal efficiency was lower than in Stage 3 and achieved 88.8% in R_{Fe} and 71.8% in R_0, $p<0.05$ (Fig. 3).

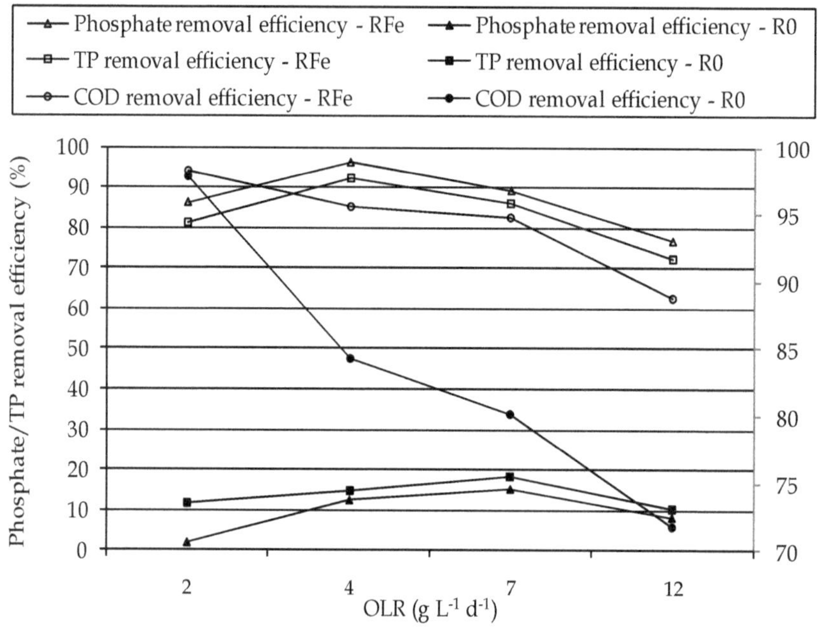

Fig. 3. COD, TP and phosphate removal efficiencies

The study demonstrated that the COD removal efficiency was markedly influenced by using steel elements as an additional medium of the UASB reactor. Iron ions generated from the steel elements must have acted as coagulants and were involved in the removal of suspended organic matter. After the operation time of 219 days, sludge samples from both UASB reactors were collected for the determination of TSS, which was higher by 52.1% in R_{Fe} than in R_0. Moreover, ferrous ions in wastewater could react to form hydroxides which were the sorption areas for suspended organic matter. Additional sorption areas were made by steel elements surface. Enhancement of COD removal efficiency by zero-valent iron processes were reported by Jeon et al. (2003) and Lai et al. (2007). Vlyssides et al. (2009) showed that the addition of ferrous ions in the form of ferrous chloride solution (2% w/v) induced a stable and excellent COD removal efficiency from synthetic milk wastewater, regardless of the increasing in OLR. When the OLR was as high as 10 g COD L^{-1} d^{-1}, the COD removal efficiency of 98% was achieved.

Anaerobic steel media corrosion significantly improved TP and phosphate removal from UF whey permeate, but the removal efficiency was affected by the duration of experiment because of deterioration of steel media. The concentration of TP decreased as the phosphate reacted with ferrous iron to probably form insoluble vivianite precipitated in the reaction chamber. It can be confirmed by significant increasing of TSS (by 52.1% in R_{Fe}) and the accumulative iron ions and phosphorus content detected in the anaerobic granular sludge in R_{Fe} at the end of the experimental period. The TP and total iron percentage in the dry matter was 0.314 and 0.0981, respectively, in R_{Fe} and 0.019 and 0.0129, respectively, in R_0. This results confirmed the anaerobic microbial corrosion occurred in R_{Fe}. Choung & Jeon (2000) and Jeon et al. (2003) obtained similar trends for domestic wastewater treatment under anaerobic conditions. Moreover, the colour of anaerobic sludge granules from R_{Fe} was black, while from R_0 was grey with white conglomerates (Fig. 4). It indicates that the presence of iron determine the colour of granules. The black colour of granules is due to the formation of large amounts of iron sulphide precipitate (Vlyssides et al., 2009). It was seen that the granule diameter in the sludge bed in R_{Fe} was smaller than in R_0. It was different from the data reported by Vlyssides et al. (2009), who observed a considerable increase of 40% in the mean granule diameter resulted in iron accumulation in granules.

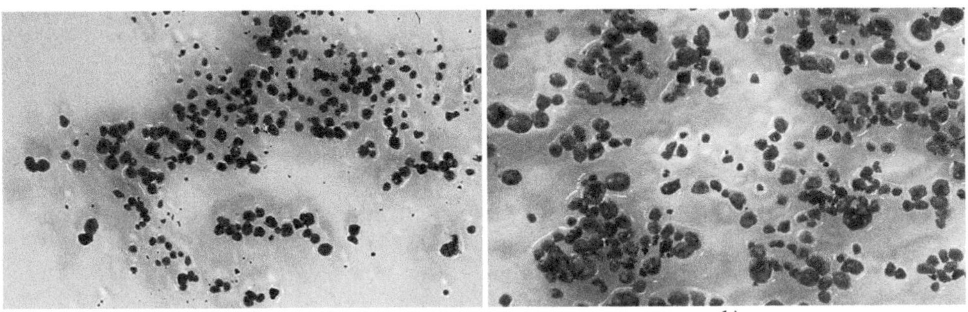

a) b)

Fig. 4. The photography of granular sludge in UASB reactor a) without steel elements, b) packed with steel elements

During the experimental period high iron concentrations in the R_{Fe} effluent were observed. During Stage 1, the highest content of iron was noticed (20.1 mg L^{-1}) and it was consequently decreased to 19.2, 15.8, 14.2 mg L^{-1} in Stage 2, 3, 4, respectively. The decrease of the total iron in the effluent from the UASB reactor packed with steel elements can indicate the formation of a protective layer on the steel surface. According to Volkland et al. (2001) under certain conditions the vivianite could act as a corrosion-inhibiting layer. Moreover, biofilm-forming bacteria can protect steel from corrosion. With a dense suspension of microorganisms (> 10^9 cells mL^{-1}) they can protect the steel surface by forming a corrosion-inhibiting layer in consequence of bacterial adsorption and adhesion (Volkland et al., 2001; Yu et al., 2000). Microbial corrosion and the formation of iron precipitates deteriorate the reactive media of steel elements (Karri et al., 2005). It could explain the gradual decrease in phosphorus and TP removal with the duration of the experiment.

Biogas production rate and CH_4 content in biogas were higher in R_{Fe} than in R_0 in all stages (except Stage 1 where the differences in biogas production between R_{Fe} and R_0 were not statistically significant). According to Karri et al. (2005) zero valent iron was an electron donor for methanogenesis. It suggested that microbial corrosion of steel elements supported methanogenesis which contributed to the more CH_4 and biogas production in R_{Fe}. Iron may play an important role in granulation phenomena and was found to be a component of essential enzymes that carry out numerous anaerobic reactions (Vlyssides et al., 2009; Yu et al., 2000). The conversion of COD to biogas components and bacterial growth may be limited at iron deficient concentrations. However, the accumulation of iron ions may decrease the specific activity of the bacterial groups, including methanogens (Yu et al., 2000). It was reported that high Fe^{2+} concentration in the anaerobic sludge granules led to decrease of the specific activity of biomass due to the presence of a large amount of minerals deposited within the granules, a significant decrease in the water content in granules, and the possible toxicity of high-concentration Fe^{2+} accumulated inside the granules (Yu et al., 2000). During the experiment, biogas production rate was not decreased from Stage 1 to 4, which could indicate that the activity of methanogenic bacteria was not inhibited by anaerobic steel corrosion process. The maximum value for biogas rate was 8.22 L d^{-1} in R_{Fe} and 4.2 L d^{-1} in R_0. Najafpour et al. (2008) achieved the biogas production of 3.6 L d^{-1} for HRT of 48 h with the methane content of 76% from UF whey permeate. Venetsaneas et al. (2009) achieved about 1 L CH_4 d^{-1} and 68% v/v methane content in biogas in the two-stage process for cheese whey fermentation.

2.3 Conclusions

On the basis of this study, it is expected that the UASB reactors packed with steel elements may be applicable to treat UF whey permeate to produce biogas with high CH_4 content. The COD removal efficiency, biogas productivity and CH_4 content in biogas were enhanced by 11.4 - 17.0% ($p<0.05$), 1.67 - 2.15 m^3 m^3 d^{-1} ($p<0.05$), 8.3 - 17.2% ($p<0.05$), respectively, in the UASB reactor packed with steel elements compared to the control reactor performances. In this work, the maximum biogas production rate was 8.22 L d^{-1} in the reactor containing additional iron medium in contrast to about 4.2 L d^{-1} in the control reactor. Total phosphorus removal efficiency obtained in R_{Fe} was higher by 58.4 - 77.7% than in R_0 ($p<0.05$). High iron concentration in the anaerobic granular sludge was not contributed to inhibit the activity of methanogenic bacteria. It should be pointed that during anaerobic corrosion process a protective layer on the steel surface can be formated to decrease phosphorus removal efficiency.

3. Bioethanol production

Bioethanol is an alcohol made by fermenting the rich sugar components of biomass which is seen as a good fuel alternative. The use of bioethanol as a biofuel has very important advantage – it is generally CO_2 neutral. This is achieved because in the growing phase of the biomass plants, CO_2 is absorbed and then released in the same volume during combustion of the fuel (Stephenson et al., 2010). This creates an obvious advantage over fossil fuels which only emit CO_2 as well as other poisonous gasses. Bioethanol can be used as a fuel for transport in its pure form, but it is usually used as a gasoline additive to increase its octane rating and improve vehicle efficiency (Balat & Balat, 2009).

Nowadays, the bioethanol market has continued to grow rapidly, for example, from about 46 billion L of ethanol produced worldwide in 2007 to the expected value of 100 billion L in 2015 (Balat & Balat, 2009; Sarkar et al., 2012). The USA is the world leader in the production of bioethanol with 48 billion L in 2009 (Muthaiyan & Ricke, 2010), followed by Brazil with 27,0 billion L in 2009 (Soccol et al., 2010) which determined 62% of the worldwide production (Sarkar et al., 2012). In the USA, bioethanol is mainly used as a 10% petrol additive (E10 is the standard petrol fuel, in 2011 introduced E15). In Brazil, it is offered both as a pure fuel (E100) and is blended with conventional petrol with a content of 20 to 25% (E20, E25). In Europe, with the adoption of the Biofuel Directive 2003/30/EC in 2003, the framework conditions were especially created for European bioethanol production. Today France is a leading producer of bioethanol, then Germany, Spain, Sweden and Dutch are the significant producers in Europe (Gnansounou, 2010). Current large scale production of fuel ethanol is mainly based on sugarcane (Brasil), corn (the USA), sugar beet and wheat (Europe), (Balat & Balat, 2009). The recent rise in the prices of food ethanol biomass has shifted in focus towards a possibility of deriving fuel ethanol from any type of biomass, especially cellulosic biomass (corn or wheat straw, sugarcane bagasse, wood, grass) and food waste biomass (organic waste and wastewater from food processing industries) (Sarkar et al., 2012; Soccol et al., 2010).

According to the literature, cheese whey could be a suitable substrate for bioethanol production (Kourkoutas et al., 2002; Zafar & Owais, 2006). Lewandowska & Kujawski (2007) used a solution of dried UF whey permeate as a substrate for semi-continuous ethanol fermentation. Silveira et al. (2005) fermented the solution of UF whey permeate in batch cultures. Ghaly & El-Taweel (1997) developed a kinetic model for continuous ethanol fermentation from lactose. Moreover, in 2008 there were two industrial scale whey-ethanol plants in the United States which produced 8 million gallons of fuel ethanol per year (Ling, 2008). In New Zealand there were whey-ethanol plants with an annual production of about 5 million gallons of ethanol (Ling, 2008). Industrial-scale plants producing bioethanol form whey permeate are operated in Ireland (de Glutz, 2009).

There are many reports of potential applications of yeast strains in ethanol production from UF whey permeate streams, but most of them focused on *Kluyveromyces sp.* due to its ability to directly ferment lactose (Kourkoutas et al., 2005; Ozmıhcı & Kargi, 2008; Silveira et al., 2005;). These yeasts generally suffer from low conversion yields (0.4 kg ethanol kg^{-1} lactose) and are very sensitive to product (ethanol) inhibition at concentrations as low as 20 g L^{-1} (de Glutz, 2009). An alternative is to employ indirect fermentation yeasts, such as *Saccharomyces cerevisiae*, which show considerably better ethanol fermentation performance (0.520 kg ethanol kg^{-1} lactose) and much higher alcohol tolerance (100 - 120 g L^{-1}) (Coté et al., 2004; de Glutz, 2009). The disadvantage of using *S. cerevisiae* is the inability to directly ferment lactose. It can be solved by genetic manipulation of yeasts or facilitate the process with a simultaneous lactose hydrolysis, for example by co-immobilization of yeast cells with the enzyme (Coté et al., 2004; Guimarães et al., 2008). Moreover, higher ethanol production could be achieved by application of different stimulation processes, improving biological activity of yeasts. Many researchers have found that ultrasonic stimulation has the function of promoting the activity of enzyme, cell growth and cell membrane permeability (Chisti, 2003; Liu et al., 2003a; Liu et al., 2007; Schläfer et al., 2000). However, application of ultrasonic irradiation at improper intensity or period has destructive impact on cells by disrupting the cell membranes and deactivating biological molecules such as enzymes or DNA (Liu et al., 2007).

The objectives of the studies were: (1) to investigate bioethanol production from UF whey permeate in continuous fermentation in UASB reactors by *K. marxianus* 499, (2) to evaluate the effects of low intensity ultrasound (20 kHz, 1 W L^{-1}) for ethanol production from UF whey permeate by *S. cerevisiae* B4.

3.1 Bioethanol production by *Kluyveromyces marxianus*

3.1.1 Materials and methods

3.1.1.1 Microorganisms

Kluyveromyces marxianus 499 obtained from Institute of Agricultural and Food Biotechnology Warsaw, Poland, in lyophilized form was used in all experiments. The yeast strain was cultivated on plates prepared with Wort Agar growth media from Merck Company Darmstadt (Germany) with the addition of 3% lactose using an incubator shaker under sterile conditions at pH 4.5 and a temperature 25°C for 48 h. The yeast was aseptically transferred from the plates into 300 ml cultivation flasks containing 100 ml of Wort Agar medium from Merck Company Darmstadt (Germany) supplemented with 3% lactose, and cultivated at 25°C for 24 h on a rotatory shaker. The yeast culture was immobilized and suspended in 2% w/v sodium alginate and then added drop-wise to 1.5% w/v $CaCl_2$ solution. The $CaCl_2$ was decanted. The beads were used for inoculation of experimental reactors.

3.1.1.2 Fermentation medium and experimental system

A solution of dried permeate from UF whey permeate from the Dairy Plant in Wolsztyn, Poland, was used as a substrate in this study. The solution was prepared by dissolving dried permeate in distilled warm water to obtain 50 g L^{-1} lactose concentration in wastewater, while the initial COD was 56 g L^{-1}.

Fermentation process was carried out in three UASB reactors with an active volume of 5 L. There were the gas-liquid-solid (G-L-S) separators on the top of each reactor. Whey permeate solution was pumped continuously to the bottom part of the reaction tank by means of the peristaltic pump. The necessary mixing was achieved through the upward wastewater flow and a stirrer operated at 40 rpm. The reactors were water-jacketed and operated at a constant temperature of 25°C ± 1°C. The pH of mixed liquid in the reactors was controlled automatically at pH 4.76 – 4.86 with 2 M NaOH.

For start-up of continuous culture, 1 L of the beads culture medium were grown at 25°C for 24 h in a 2 L Erlenmeyer flask filled with 0.5 L of UF whey permeate after heat sterilization (120°C, 20 min). The concentration of lactose in whey permeate was 50 g L^{-1}. Mixing was achieved by stirring with a magnetic stirrer at 200 rpm. The cell suspension was then aseptically transferred to each UASB reactor which was kept in batch operation for 24 h before switching on the continuous feeding. The reactors were operated at the HRTs of 12, 24 and 48 h. At each HRT the reactors were operated till they had reached the steady-state (the steady-state conditions were evidenced when the standard deviations of the ethanol concentrations and lactose concentrations in the effluent distillate were within 3%).

3.1.1.3 Analytical methods

Lactose concentrations and ethanol concentrations in the effluent distillate were determined according to Standard Methods (PN-67/A-86430; PN-A-79528-3:2007). The biomass concentration of yeast (dry matter) was calculated according to Standard Methods (P-78/C-04541). The samples were analyzed in triplicate and results were reproducible within 3% standard deviation.

3.1.2 Results and discussion

The effects of HRTs on the lactose concentration in the effluent distillate and percent lactose consumption are shown in Fig. 5. When the HRT was 12 h, the average lactose concentration in the effluent distillate was as high as 25 g L^{-1} and the average lactose utilization efficiency was only 50%. Increasing the HRT from 12 to 24 h increased the average yield of lactose utilization to 85%. Further increase in the HRT from 24 to 48 h resulted in the highest lactose utilization of 95%. Similar results obtained Ghaly & El-Taweel (1997). They observed 98% lactose utilization for continuous fermentation from cheese whey with 50 g L^{-1} initial lactose concentration at the HRT of 42 h using the yeast strain of *Candida pseudotropicalis*. Kargi & Ozmihci (2006) reported complete fermentation of lactose (35 g L^{-1} initial lactose concentration) in cheese whey powder (CWP) solution using the yeast strain of *K. marxianus* at HRT of 48 h. Zafar & Owais (2006) obtained about 86% lactose utilization from crude whey within 22 h by *K. marxianus*.

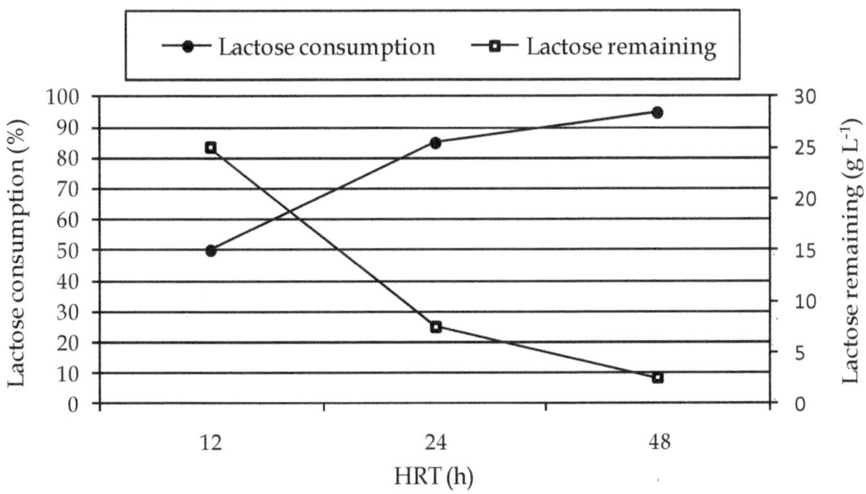

Fig. 5. Effects of HRT on the lactose concentration in the effluent and percent lactose consumption

According to Ghaly & El-Taweel (1997) lower lactose fermentation efficiency under low HRT could be attributed to the cell washout phenomenon and the low cell numbers in the reactor chamber. To remove this problem, during this experiment, the reactors were provided with G-L-S separator and the immobilization of yeast culture was done. The immobilization process made ethanol production more efficient compared to the free system and prolonged the activity of yeast cells (Kourkoutas et al. 2004), which is especially important in continuous fermentation processes. Moreover, the application of immobilization process reduced the risk of microbial cells infection when the yeasts were cultivated on the fermentation medium that was not sterilized before use (Lewandowska & Kujawski, 2007).

Fig. 6 shows the variations of daily ethanol production and ethanol concentration with the initial lactose concentration of 50 g L^{-1}. The maximum daily ethanol production of 8.61 g L^{-1} d^{-1} was obtained at the HRT of 24 h. Increasing in the HRT to 48 h decreased daily ethanol production to the average value of 7.73 g L^{-1} d^{-1} in spite of the fact that alcohol concentration increased from 8.61 to 15.45 g L^{-1}. When the HRT was 12 h, the average daily ethanol production was 4.46 g L^{-1} d^{-1}, while the average ethanol concentration was as low as 2.24 g L^{-1}. The results were similar to the ones obtained by Kourkoutas et al. (2002). The ethanol productivity was 7.0 and 8.0 g L^{-1} d^{-1} at the HRT of 25 and 20 h respectively, using whey as a substrate fermentation and immobilized cells of *K. marxianus*.

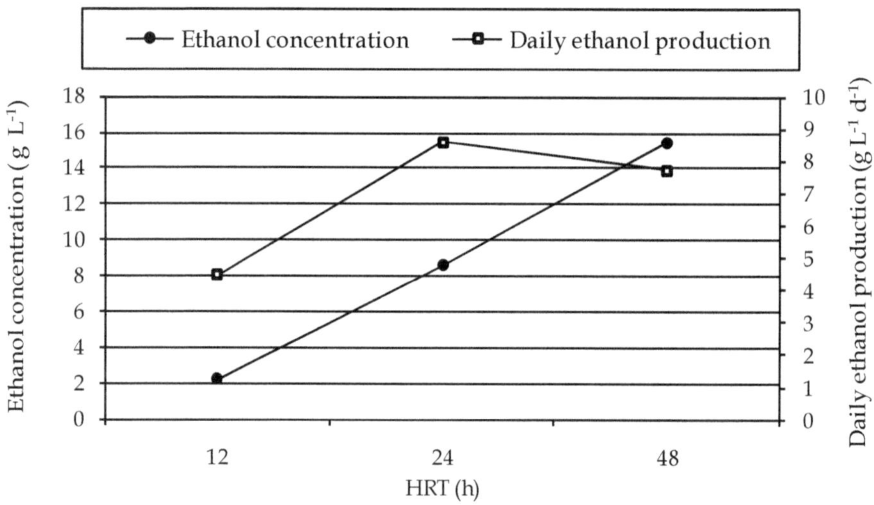

Fig. 6. Effects of HRT on daily ethanol production and ethanol concentration in the effluent.

The negative effect of longer HRT on the daily ethanol productivity could be associated with a negative effect of increasing concentration of ethanol. According to Golubev & Golubev (2004) ethanol concentration of 2 - 4% produces a negative effect on the growth of *Kluyveromyces*. Silveira et al. (2005) observed the growth inhibition of *K. marxianus* when the ethanol concentration increased from 10 g L^{-1} to 20 g L^{-1}. de Gultz (2009) studied alcohol tolerance of direct whey fermenting yeasts (four strains of *K. marxianus*) and indirect whey fermenting yeasts (three strains of *S. cerevisiae*). From the results it can be seen that some strains of *K. marxianus* showed considerable alcohol tolerance of 71 - 81 g L^{-1} with fermentation times ranging from 11 to 32 h , while alcohol tolerance for *S. cerevisiae* reached 85 - 148 g L^{-1} with fermentation time ranging from 29 to 64 h.

Increasing in the HRT, increased the ethanol yield (g ethanol g^{-1} consumed lactose), (Fig. 7). The ethanol yield obtained in this study was 0.089, 0.203, 0.325 g g^{-1} at the HRT of 12, 24, 48 h, respectively. Silveira et al. (2005) obtained a higher ethanol yield of 0.52 g g^{-1} with the initial lactose concentration of 50 g L^{-1} with the yeast strain of *K. marxianus*.

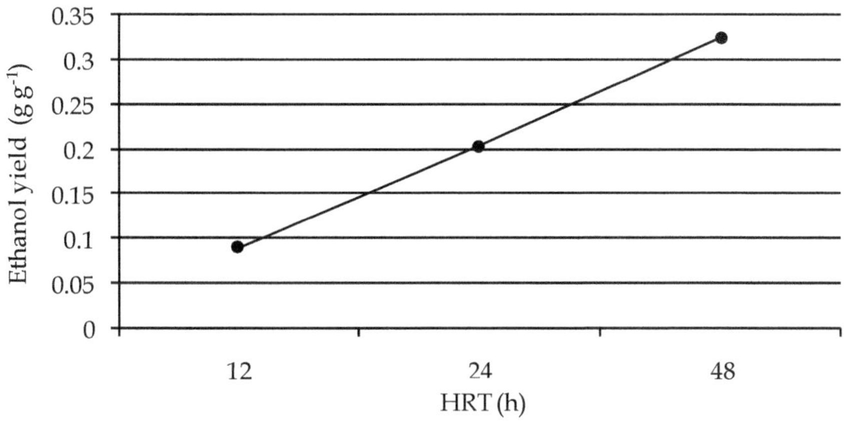

Fig. 7. Effect of HRT on the ethanol yield.

Ozmihci & Kargi (2007) stated, that biomass concentration is an important parameter affecting the ethanol formation efficiency. The volumetric rate of sugar utilization can increase with biomass concentration. They found that when the biomass concentration was 510 mg L^{-1}, the rate of sugar utilization was about 1580 mg L^{-1} h^{-1} at HRT of 30 h. The maximum sugar utilization rate of 2200 mg L^{-1} h^{-1} they obtained at the biomass concentration of 1020 mg L^{-1} and HRT of 24 h. In this study the concentration of yeast in the UASB reactors ranged from 705 to 869 mg L^{-1} by the duration of the experiment. The volumetric rate of sugar utilization was as high as 927, 1576, 1724 mg L^{-1} h^{-1} at HRT of 12, 24 and 48 h, respectively.

3.1.3 Conclusions

The utilization of whey UF permeate to ethanol in continuous fermentation is possible. *Kluyveromyces marxianus* was able to metabolize lactose and the total fermentation efficiency was as high as 95%. The lactose utilization and ethanol production were connected with the HRT. The maximum daily ethanol production of 8.61 g $L^{-1}d^{-1}$ was achieved when the HRT was 24 h and the ethanol yield was 0.203 g g^{-1}. The results indicated that the HRT should be 24 h to obtain high rates of ethanol formation and to avoid product inhibition. Doubling the HRT to 48 h did not contribute to a noticeable increase of ethanol production and the daily ethanol production decreased to 7.73 g $L^{-1}d^{-1}$ while the ethanol yield was 0.325 g g^{-1}.

3.2 Bioethanol production by *Saccharomyces cerevisiae*

3.2.1 Materials and methods

3.2.1.1 Microorganisms

The yeast *Saccharomyces cerevisiae* B-4 obtained from Institute of Agricultural and Food Biotechnology Warsaw, Poland, was used for assessment ultrasound exposition to ethanol production. The yeast cultures were cultivated on YPG slants (2% glucose, 2% peptone, 1% yeast extract) supplemented with 2% agar, at pH 5.0 and 30 °C for 24 h. The active cultures for inoculation were prepared by growing the yeast in a 1 L baffled shake-flask containing sterile water and 100 mL YPG medium at 30 °C for 24 h on orbital shaker table at 200 rpm to a concentration of approximately 10^8 cells mL^{-1}. The cultures in baffled shaken flasks of 100 mL were used to prepare the inocula. After 24 h of incubation at 30 °C, the precultures were centrifuged at 3800 rpm for 10 min and the cells were resuspended in sterile water to obtain 10^6 cells mL^{-1}. Enzyme β-D-galactosidase (optimum temperature 30 °C, optimum acidity pH 4.5-5.0, activity 8.7 AU mg^{-1} d.m. of the preparation), from *Aspergillus oryzae* manufactured by the SIGMA company (USA), was used for co-immobilization process. The amount of yeast and enzyme was 3% free cell inoculum and 4 cm^3 enzyme solution. The yeast culture was co-immobilized in the 2% (w/v) sodium alginate by dropping yeast and enzyme into 150 cm^3 0.09 mol L^{-1} solution of $CaCl_2$ with 10% glucose. The cell beads were washed with sterile water and were stored as a fermentation medium in physiological solution at 8°C.

3.2.1.2 Fermentation medium and experimental system

UF whey permeate (non-deproteinized, non diluted and non-sterilized) with the average lactose concentration of 50 g L^{-1} from the Dairy Plant in Nowy Dwór Gdański, Poland, was used as a fermentation substrate (Table 2).

Continuous fermentation was carried out in the laboratory-scale plant consisted of the two UASB reactors with a working volume of 5 L each (Fig. 8). These two reactors were used to enable parallel test series with and without ultrasound irradiation. The fermentation medium was pumped continuously to the bottom part of the reaction tank by means of the peristaltic pumps. The necessary mixing was achieved through the upward wastewater flow. The reactors were water-jacketed and operated at a constant temperature of 30±1 °C. The pH of mixed liquid in the reactors was controlled automatically at pH 5.1 ± 0.2 with 2 M NaOH.

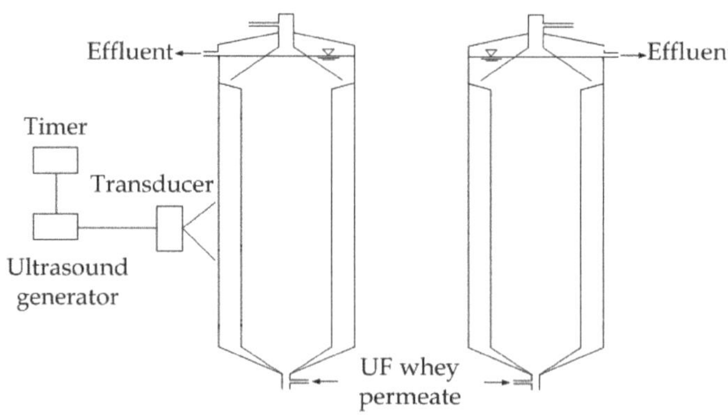

Fig. 8. A scheme of the research station.

The reactors were inoculated with 40% (v/v) solid beads containing the immobilized cells which corresponded to 39.4 g cells dry weight - DW L^{-1} of working bioreactor volume. After adding the cell beads inoculum to the bioreactors, before starting continuous feeding, a batch fermentation was conducted for 24 h under additional gentle agitation (100 rpm). Next the reactors worked at different HRTs of 12, 24 and 36 h. At each HRT the reactor was operated till it has reached the steady-state (the steady-state conditions were evidenced when the standard deviations of the ethanol and lactose concentrations in the effluent distillate were within 3%), thus 30 days of each fermentation step (step 1 - HRT of 12 h, step 2 - HRT of 24 h, step 3 - HRT of 36 h). The fresh inoculum was added to the reactors before each fermentation step and the aged one was removed.

The ultrasound irradiation of the reactor with yeasts was made by a special ring with a transducer (Intersonic S.C. Poland) that was attached at the bottom of the reactor. The range of the frequency generator was adjustable between 20–25 kHz and the maximum power of 50 W. The experiments were carried out with the stable sonication power of 1 W L^{-1} and the frequency of 20 kHz.

3.2.1.3 Analytical methods

Lactose and ethanol concentrations in the effluent distillate were determined according to Standard Methods (PN-67/A-86430; PN-A-79528-3:2007). The samples were analyzed in triplicates and results were reproducible within 3% deviation.

All fermentation steps connected with different HRTs were carried out in triplicate. Significant differences between the effects obtained in the two reactors with and without ultrasound exposure were analysed using an ANOVA *F*-test (Statistica 7.1 software, Statsoft Inc.) A 5% probability level was applied for all the tests. If $p<0.05$ from an ANOVA *F*-test, the differences between the effects were considered to be significantly different from one another.

3.2.2 Results and discussion

3.2.2.1 Sonification parameters

In the experiment, the frequency of applied ultrasounds was 20 kHz and the power input was 1.0 W L^{-1}. The initial experiments were done to find the best irradiation period. The experiments revealed, that continuous low energy ultrasound irradiation during 12, 24 and 36 h did not enhance ethanol productivity by co-immobilized *S. cerevisiae*, moreover the ethanol yield coefficients were lower than those obtained in experiments without ultrasound irradiation. The subsequent experiments were carried out with time intervals with and without ultrasonic irradiation in order to obtain the positive influence of ultrasound on biological activity of *S. cerevisiae*. The results showed that the culture should have been sonicated for 1 min every 6 h. It was similarly to results obtained by Marques et al. (2006). They investigated the effect of ultrasound pulses on enzymatic activity of *S. cerevisiae*. Their results showed that the ultrasound pulse at low frequency (20–25 kHz) for a short sonification period of 1 and 2 min increased cell permeability, and the viability rate of yeasts was over 95%. However, in the 4 min sonification, the rate decreased to 46%.

The use of ultrasounds to stimulate biological activity and ethanol production by *S. cerevisiae* are reported by Schläfer et al. (2000). After testing several different frequencies and power levels, they carried out the experiments at 25 kHz, 0.3 and 12 W L^{-1}. At an ultrasound intensity of 12 W L^{-1} there was no recognizable difference in the biological activity of yeasts with and without ultrasound. The authors stated that some pauses are needed between ultrasound exposure to obtain positive effects on biological activity of yeast *S. cerevisiae*. Moreover, an increase in biological activity appeared after irradiation and high activity of ultrasound activated cultures stopped for some hours after irradiation. The authors stated, that discontinuous ultrasonic irradiation of *S. cerevisiae* was more beneficial for activating fermentation than the continuous exposure, because only a few steps in intracellular metabolisms are supported by ultrasound and others are not or even inhibited.

Liu et al. (2007) investigated the changes of biological activity of aerobic activated sludge after ultrasonic irradiation. The activity of microorganisms rose sharply after ultrasonic exposure of 0.3 W cm^2, 35 kHz for 10 min, and reached a peak level in 8 h after exposure (100% higher than that of the initial level immediately after exposure). Then it dropped rapidly in the next 8 h. In 24 h after ultrasonic irradiation, the enhancement effect induced by ultrasound almost disappeared, and the cells activity returned to the normal state as control cells without ultrasound stress. The authors stated that the enhancement might be due to defense response of microorganisms evoked by the mechanical stress. That reactions are usually observed when cells are challenged by biotic or abiotic stresses.

Pitt & Ross (2003) used ultrasonic irradiation to increase the growth rate of bacterial cells attached to a polyethylene surface. It was found that low frequency ultrasound (70 kHz) of low intensity (<2 W cm^{-2}) increased the growth rate of the cells compared to growth without ultrasonic waves. They stated that ultrasounds can increase the rate of transport of oxygen and nutrients to the cells and the rate of transport of waste products away from the cells, thus enhancing their growth.

Xie et al. (2009) studied the enhancement effect of low-intensity ultrasound (35 kHz) on anaerobic sludge activity. The experiments showed, that the optimal ultrasonic intensity and irradiation period were 0.2 W cm^2 and 10 min, respectively.

To sum up, the optimal ultrasonic intensity and irradiation period are varied in each biological process enhanced by ultrasound and should be find experimentally.

3.2.2.2 Effect of HRT on ethanol fermentation

In order to estimate an optimal fermentation time under ultrasonic exposure in this study, parameters such as ethanol concentration, ethanol volumetric productivity, ethanol yield and lactose consumption were investigated.

The maximum values of ethanol concentration and lactose consumption were achieved when the HRT was 36 h. Under the HRT of 36 h in the ultrasound-assisted fermentation, the average ethanol concentration of 26.30 g L^{-1}, ethanol yield of 0.532 g g^{-1} lactose and lactose consumption of 98,9% were obtained (Fig. 9–11). Using *S. cerevisiae* without ultrasound exposure gave the results as 23,60 g L^{-1}, 0.511 g g^{-1}, 92,4%, respectively and the differences were statistically significant ($p<0.05$). Shortening the HRT to 24 h allowed remaining high ethanol yield of 0.520 g g^{-1} with sonicated *S. cerevisiae*, but in the control fermentation unit it was as low as 0.487 g g^{-1} ($p<0.05$). When the HRT was 12 h the ethanol yields were 0.318 and 0.365 g g^{-1} depending on using ultrasounds device (Fig. 11). From the economic viewpoint, shortening the fermentation time (HRT) could reduce costs of industrial ethanol production. The study showed that there is no need to extend the HRT over 36 h or more, because most of the lactose was converted into ethanol during 24 h (95.6% in the ultrasound-assisted fermentation. Nikolić et al. (2010) stated that optimal fermentation time for free and immobilized *S. cerevisiae* was 38 h. Ozmihci & Kargi (2008) studied ethanol production from cheese whey powder (CWP) solution containing 50 g sugar L^{-1} at six different HRTs varying between 17.6 and 64.4 h by *Kluyveromyces marxianus* strains. Percent sugar utilization,

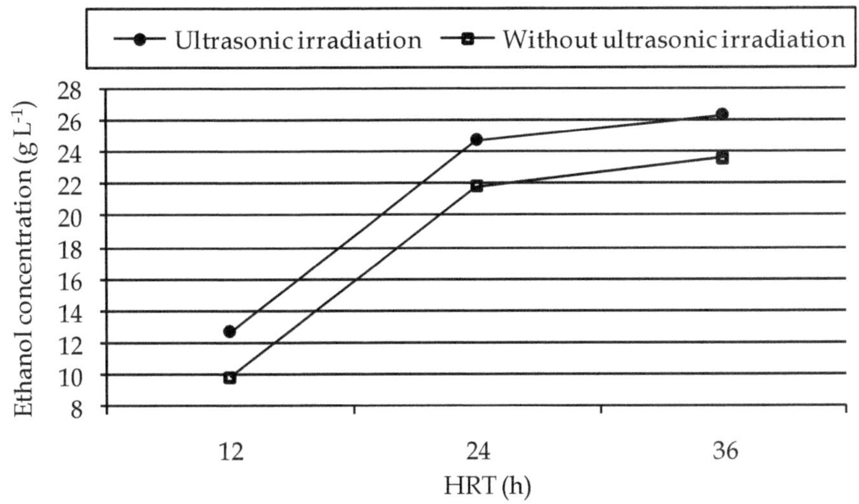

Fig. 9. Effects of HRT and ultrasound irradiation on the ethanol concentration

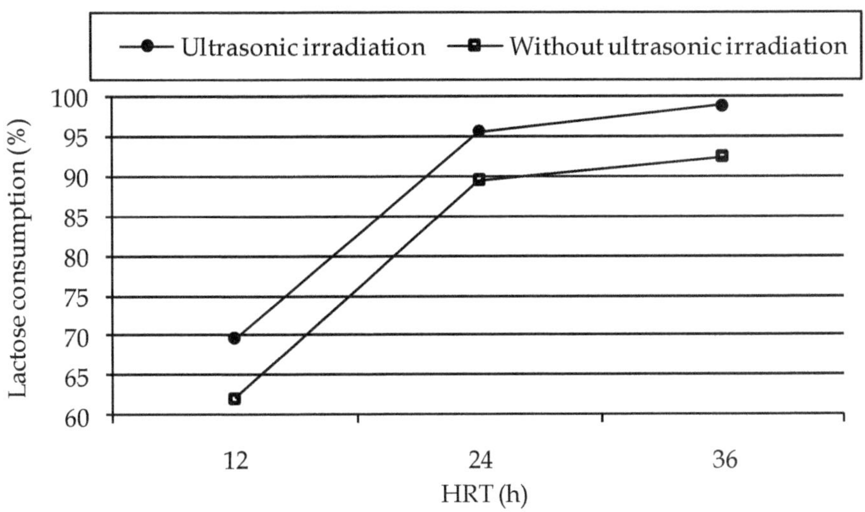

Fig. 10. Effects of HRT and ultrasound irradiation on the lactose consumption by co-immobilized *S. cerevisiae*

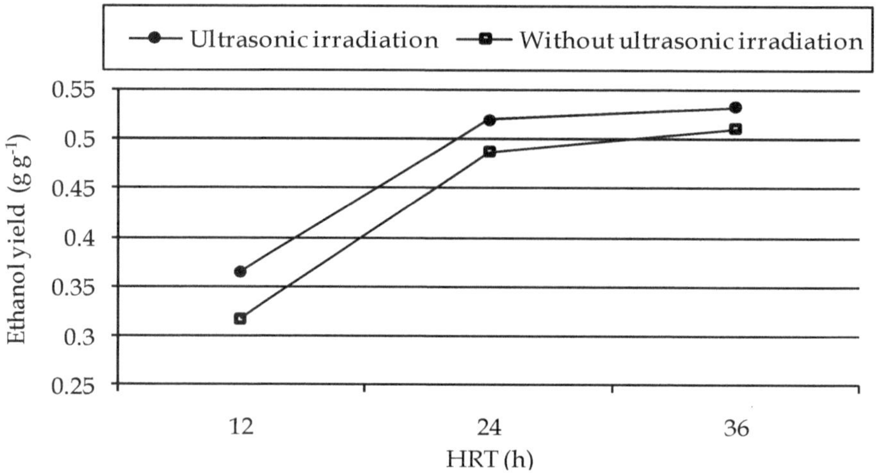

Fig. 11. Effect of HRT and ultrasound irradiation on the ethanol yield

effluent ethanol concentration and ethanol yield increased with increasing HRT from 17.6 to 50 h. Further increasing in HRT to 64.4 h resulted in decrease of the analyzed parameters. Moreover, the time for fermentation decreased at higher initial substrate concentration (Guimarães et al., 2008a; Nikolić et al., 2010; Ozmihci & Kargi, 2008). According to Guimarães et al. (2008a) the fermentations with 50-150 g lactose L^{-1} reached completion in about the same time of 27 h but the maximum ethanol concentration increased linearly with increasing initial lactose concentration from 6.5 g ethanol L^{-1} with 20 g lactose L^{-1} to 57 g L^{-1} with 200 g L^{-1}. They also stated that increasing lactose concentration led to incomplete fermentation and impair the fermentation due to nutrient limitation.

Interestingly, the volumetric productivities of ethanol decreased at longer HRT (Table 3). Maximum productivity of ethanol of 1.060 g L^{-1} h^{-1} was observed under the HRT of 12 h when the culture has been sonicated and 0.908 g L^{-1} h^{-1} under the HRT of 24 h in the fermentation process without ultrasound irradiation ($p<0.05$). The volumetric ethanol productivity in the ultrasound-assisted fermentation obtained in this work was higher than that reported for batch or fed-batch fermentations with *S. cerevisiae* strains: 0.3 g L^{-1} h^{-1} (Rubio-Texeira et al., 1998), 0.46 g L^{-1} h^{-1} (Guimarães et al., 2008b), 0.14 - 0.6 g L^{-1} h^{-1} (Ramakrishnan & Hartley, 1993), 1 g L^{-1} h^{-1} (Compagno et al., 1995). Ozmihci & Kargi (2007) using *Kluyveromyces marxianus* to ferment concentrated cheese whey powder solution obtained higher volumetric ethanol productivity over 2 g L^{-1} h^{-1}, but after 120 h fermentation.

HRT	Ethanol volumetric productivity in the ultrasound-assisted fermentation system (g L^{-1} h^{-1})	Ethanol volumetric productivity in the control fermentation system (g L^{-1} h^{-1})
12 h	1.060	0.822
24 h	1.035	0.908
36 h	0.730	0.655

Table 3. Effects of HRT on the ethanol volumetric productivity

3.2.2.3 Effect of ultrasounds on ethanol fermentation

In all HRTs, significant higher ethanol productions in the ultrasound-assisted fermentation process than in the control fermentation process were recorded ($p<0.05$). When the HRT was 12 h, the ethanol concentration without ultrasonic treatment was 9.87 g L^{-1} and it was significant lower by 2.85 g L^{-1} than the production in the process stimulated with low intensity ultrasounds ($p<0.05$) (Fig. 9). Lactose consumption was only 62.1%, but application of ultrasound increased it to 69.7% ($p<0.05$) (Fig. 10). The best results were obtained with the longest HRT of 36 h. Ethanol concentration increased to the value of 26.30 g L^{-1} when the culture has been sonicated, while in the fermentation process without ultrasound irradiation was only 23.60 g L^{-1} ($p<0.05$), (Fig. 9). Lactose consumption was as high as 98.9% in ultrasound-assisted fermentation unit and was significant higher by 6.5% than the consumption in the reactor without ultrasonic irradiation ($p<0.05$) (Fig. 10). High ethanol production and lactose consumption were observed with shortening HRT to 24 h. *S. cerevisiae* stimulated with low intensity ultrasound produced 24.85 g ethanol L^{-1}, while the lactose consumption was 95.6% (Fig. 9–10). In the control fermentation unit there was 21.79 g L^{-1} and 89.5%, respectively. The differences were statistically significant ($p<0.05$). Under the HRT of 36 h, in the fermentation process with ultrasound irradiation the maximum ethanol yield of 0.532 g g^{-1} lactose was observed, whereas using biocatalyst *S. cerevisiae* without ultrasound exposure gave the result as 0.511 g g^{-1} (Fig. 11) ($p<0.05$). Shortening the HRT to 24 h allowed remaining high ethanol yield of 0.520 g g^{-1} with sonicated *S. cerevisiae,* but in the control fermentation process it was as low as 0.487 g g^{-1} ($p<0.05$). When the HRT was 12 h the ethanol yield was only 0.365 and 0.318 g g^{-1}, respectively ($p<0.05$).

There were only few experiments investigating the enhancing ethanol production by ultrasonic stimulation of *S. cerevisiae*. Schläfer et al. (2000) improved biological activity of *S. cerevisiae* by low energy ultrasound assisted bioreactors operated at a frequency of 25 kHz and a power input of 0.3 W L^{-1}. The ethanol production without ultrasonic treatment varied between 3–12 g L^{-1}, while ultrasonic stimulation increased it to 30 g L^{-1}. The highest ethanol concentrations were obtained with a cycle regime of ultrasound exposure and a pause, because during continuous ultrasound irradiation no stimulation in the ethanol fermentation process was recorded.

Lanchun et al. (2003) investigated the influence of low intensity ultrasound on physiological characteristic of *S. cerevisiae*. The results of their study showed, that ultrasounds in the frequency of 24 kHz and the power efficiency of 2 W with 1 s irradiation time every 15 s and 30 min duration cycle, stimulated the material transport and improved the cell's metabolism by changing the osmotic pressure of membrane. Consequently, transfer of substance was speeded up, enzyme synthesis was driven up and enzyme activity was enhanced.

The positive results of the ultrasound treatment on the ethanol production by co-immobilized *S. cerevisiae* seemed to be a combination of different processes, including activating the yeast by improving the mass transfer rate of nutrients in the liquid, enhancing the uptake of foreign substances and the release of intracellular products in cells, improving the cell growth and degassing of CO_2 (Lanchun et al. 2003; Liu et al., 2007; Liu et al. 2003b). Stimulating enzyme activity is done by increasing in the mass transfer rate of the reagents to the active site (Liu et al., 2007). Ultrasounds irradiation can cause thermal and mechanical stress to biological materials (Liu et al., 2003b). High energy ultrasonic waves break the cells and denaturate enzymes (Liu et al., 2007; Pitt & Ross, 2003). Low energy ultrasounds can produce a variety of effects on biological materials, including the inhibition or stimulation cellular metabolisms, enzyme activity, alteration of cell membranes and other cellular structures (Liu et al., 2007; Liu et al. 2003a). According to Xie et al. (2008), cavitation is the primary basis of biological effects of low intensity ultrasound. Cavitation bubbles produced by low intensity ultrasound can cause acoustic microstreaming (Xie et al., 2008). The microstreaming surrounding the cells can cause shear stress and enhance the mass transfer, which may stimulate metabolic activities inside the cells (Liu et al., 2003b; Pitt & Ross, 2003; Xie et al., 2008). When ultrasonic intensity is sufficiently low, a stable cavitation occurs and leads to the enhancement of mass transfer and fluid mixing, which produces positive effects on the rate of biological reactions in the exposure systems (Liu et al., 2007).

The growth activity of yeast cells is hardly changed within the early period of sonication regardless of either damage to cell wall, or complete inactivation of the yeast located in the cavitation zone (Tsukamoto et al., 2004). Short sonication time up to 5 min of irradiation indicated bactericidal effects, but the cells were able to repair the damages. According to Guerrero et al. (2005) yeasts, inclusive with *S. cerevisiae*, are highly resistant to ultrasound damage. Moreover, at relatively low intensity of ultrasounds, microorganisms can adapt to the irradiation exposure and their biological activity increases (Liu et al., 2007). With relatively short irradiation period, cell damage and membrane permeability induced by ultrasounds appear to be temporary and reversible. Lanchun et al. (2003) also stated that sonication cannot influence on fermentation strength of *S. cerevisiae* descendants.

3.2.3 Conclusions

The utilization of milk permeate to ethanol in continuous fermentation by co-immobilized *S. cerevisiae* is possible. The optimal ultrasonic intensity and irradiation period are varied in each biological process enhanced by ultrasound and should be find experimentally. According to this experiment, stimulation of yeasts activity could be achieved in the presence of low intensity ultrasound (1 W L^{-1}, 20 kHz), and 1 min every 6 h irradiation period is favorable to increase ethanol production efficiency. Moreover, the short exposure of yeast to ultrasound could reduce the operation costs comparing with continuous irradiation.

For the continuously operating bioreactors, the maximum rates of sugar utilization were 98.9 and 92.4% for the yeast with ultrasound exposure and without ultrasound exposure ($p<0.05$), respectively. The maximum ethanol yield was 0.532 g g^{-1} lactose, while using *S. cerevisiae* without ultrasound exposure 0.511 g g^{-1}. The study showed that there is no need to extend the HRT over 36 h or more, because most of the lactose was converted into ethanol during 24 h (95.6% in the ultrasound-assisted fermentation).

All results obtained here raises the new perspectives for disposal UF whey permeate.

4. Biohydrogen and methane production in two-stage fermentation process

Hydrogen is an eco-friendly, clean energy alternative because its combustion by-product is only water, so does not contribute to the greenhouse effect. Hydrogen has a high energy yield (122 kJ g^{-1}), therefore in recent times a great deal of attention is being paid to the usage of hydrogen as a fuel. However, a major doubt on hydrogen as a clean energy alternative is that most of the hydrogen gas is currently generated from fossil fuels by thermochemical processes, such as hydrocarbon reforming, coal gasification and partial oxidation of heavier hydrocarbons (Castellò et al., 2009; Mohan et al., 2007). These methods are considered to be energy intensive and not environmental friendly. It is well known that only biological hydrogen production processes from the fermentation of renewable substrates, such as organic wastewater or other wastes are the promising alternative for hydrogen generation. Several strategies for the production of biohydrogen by fermentation in lab-scale have been found in the literature: photo-fermentation (Gadhamshetty et al., 2008), dark-fermentation (Krupp & Widmann, 2009) and combined-fermentation, which refers to the two fermentations combined (Nath & Das, 2009). However, no strategies for industrial scale productions have been found. In order to define the industrial scale biohydrogen production, more information from laboratory scale experiments are needed, especially related to design and optimization process, and operating parameters. Moreover, generation of biohydrogen by acidogenic phase of anaerobic process (dark-fermentation) is connected with incomplete degradation of organic material into organic acids, so there is a need to utilize by-products of the fermentation process.

As a result, the fermentative hydrogen production could be coupled with subsequent anaerobic digestion step with the conversion of remaining organic content to biogas. A two-stage fermentation process, in which acidogenesis and methanogenesis occur in the separate reactors may offer several advantages, such as improved total wastewater degradation and enhancing biohydrogen and methane production (Venetsaneas et al., 2009).

The dairy industry produces highly concentrated, carbohydrate-rich wastewaters, but their potential for biohydrogen generation has not been extensively studied. There were some experiences working with cheese whey as the substrate for biohydrogen production (Azbar et al. 2009; Castellò et al., 2009; Venetsaneas et al., 2009). The objectives of this work were: (1) to check the ability to produce biohydrogen using raw, unsterilized UF whey permeate, (2) to combine biohydrogen dark-fermentation process with methane fermentation of biohydrogen production by-products (mainly organic acids) in two-stage continuous fermentation process.

4.1. Materials and methods

4.1.1 Inoculum and wastewater

Anaerobic granular sludge from a full-scale UASB reactor treated fruit juice processing wastewater in fruit juice industry, Olsztynek, Poland, was used as an inoculum for biohydrogen and methane production. Prior to inoculation of the hydrogenogenic reactor, the granular sludge was washed with three volumes of tap water and then boiled for 2 h to inactivate hydrogen consuming microorganisms. A final concentration of inoculum was 151 g TSS L^{-1}. No pre-treatment of the granular sludge used for methane production was carried out prior to its inoculation in methanogenic reactor. Initial concentration of inoculum for methane production was 94 g TSS L^{-1}. Both reactors were initially inoculated at a ratio of 20% (by volume).

UF whey permeate was obtained from a cheese production factory in Nowy Dwór Gdański, Poland. It was received from the factory once a week, was stored at -20°C and was thawed before used. The average composition of the feedstock was shown in Table 2. Prior to being fed into the reactor, the substrate was diluted with tap water to a COD concentration of 10,000 mg L^{-1} in accordance with the required OLR (10 or 20 g COD L^{-1} d^{-1}) in the hydrogenogenic reactor. Diluting UF whey permeate was maintained at a temperature 4°C until used.

4.1.2 Hydrogen production process setup and operation

A 5 L UASB reactor with 4.5 L working volume (R1) was made of stainless steel and cylindrical in shape (Fig. 12). The reactor was constantly stirred at 50 rpm. The pH of mixed liquid in R1 was controlled automatically with 6 M NaOH. The temperature in R1 was maintained at 37°C by inserting the reactor in the thermostatic chamber. For start-up, the reactor was filled with undiluted UF whey permeate and was operated anaerobically at a batch mode. When hydrogen production reached its peak value, the bioreactor feeding mode was turned to a continuous one at a HRT of 24 h (OLR of 10 g COD L^{-1} d^{-1}) or at a HRT of 12 h (OLR of 20 g COD L^{-1} d^{-1}). The R1 performance (biogas production and composition in H_2 and CH_4, COD, Total Volatile Fatty Acids - TVFA concentration, pH) was monitored twice a week throughout the experimental period (84 days).

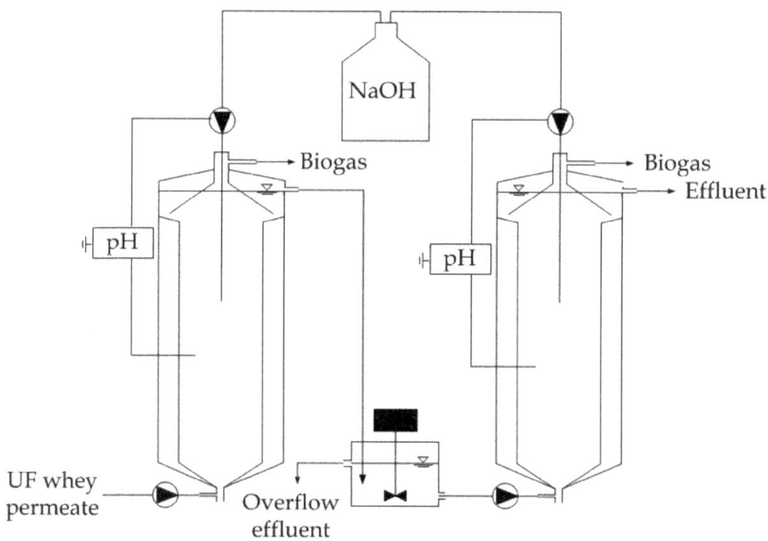

Fig. 12. Schematic of the two-stage fermentation system

4.1.3 Methane production process setup and operation

A 5 L UASB reactor with 4.5 L working volume (R2) was made of stainless steel and cylindrical in shape (Fig. 12). The reactor was constantly stirred at 50 rpm. The pH of mixed liquid in R2 was controlled automatically at pH 7.2 (±0.05) with 6 M NaOH. The temperature in R2 was maintained at 37°C by inserting the reactor in the thermostatic chamber. The R2 was fed with the effluent from R1, which was collected in a 3 L container used as a storage tank which was constantly stirred at 50 rpm (Fig. 12). The temperature in the tank was maintained at 37°C by placing it in the thermostatic chamber. Overflow effluent flowed out in the top part of the storage tank and was collected in a separate container. The R2 was operated at an HRT of 3 d. The R2 performance (biogas production and composition in CH_4, COD, TVFA concentration, pH) was monitored twice a week throughout the experimental period, from day 51 to 84. Before R2 was fed with R1 effluent, the diluted UF whey permeate had been used as a feedstock to reach the OLR of 2 g COD L^{-1} d^{-1} and HRT of 3 d.

4.1.4 Analytical methods

Determinations of COD, TSS, lactose, TVFA concentrations were carried out according to the Standard Methods for the Examination of Water and Wastewater (PN-74/C-04578.03; PN-78/C-04541; PN-67/A-86430; PN-75/C-04616.04). The measurement of the pH was continuously measured by the membrane electrodes, model ESAgP-301W, Eurosensor, placed in the liquid phase of reactor. Biogas composition (CH_4, H_2 and CO_2) was analyzed by using an electronic analyzer (Gas Data GFM 430, Gas Data Ltd.). Biogas production was measured with a water displacement meter.

4.2 Results and discussion

The hydrogenogenic reactor (R1) operation started with a HRT of 24 h, OLR of 10 g COD L^{-1} d^{-1} and pH 5.8. Under these conditions hydrogen production was as low as 0.12 L H_2 d^{-1} (0.027 L H_2 L^{-1} d^{-1}) (Fig. 13). It was noticed methane presence in biogas up to 19% v/v, while hydrogen concentration was still very low (up to 10% v/v). In order to inhibit methane production, the HRT was reduced to 12 h and OLR increased to 20 g COD L^{-1} d^{-1} on day 11. After 5 days, the HRT was increased to 24 h. Hydrogen content in biogas increased to the average value 15.7% v/v and methane was still present (<8% v/v). According to Yang et al. (2007), HRT shorter than 24 h does not favor the biohydrogen generation from cheese whey wastewater, but other researchers stated that short HRT could help to control the methanogenic reaction in hydrogenogenic phase (Castellò et al., 2009). Then, the HRT was again set at 12 h, OLR increased to 20 g COD L^{-1} d^{-1} and pH was decreased to 5.2 (day 24). It was seen that although the methanogenic bacteria was assumed to be washed out from the bioreactor under short HRT, the inhibition of methanogenic bacteria activity should be coupled with pH decrease. Liu et al. (2006) found that pH is the most critical factor for inhibition of methanogenesis and the optimum pH should be around 5.0 - 5.5. According to the literature, the optimum pH range for lactose (or whey) acidogenesis is between 6 and 6.5 (Venetsaneas et al. 2009). Antonopoulou et al. (2008) reported, that high concentration of hydrogen (over 25% v/v) in the gas phase was when the pH in the reactor was maintained at 5.2 ± 0.1. Wang et al. (2006) stated, that pH of 5.5 should be avoided in the biohydrogen fermentation process because at that level of pH, the propionic-acid type fermentation commonly occurred. The accumulation of propionic acid can lead to lower efficiency of methanogenic phase followed the hydrogenogenic phase. Mohan et al. (2007) found that pH 6 was the optimum for effective H_2 yield. However, maintaining pH at 6 or above is difficult because of large amounts of fatty acids generation. At this study, the pH in the hydrogenogenic reactor was initially maintained on the level of 5.8. As the pH was reduced down to 5.2 (coupled with HRT shortening and OLR increasing), methane content in biogas faded out.

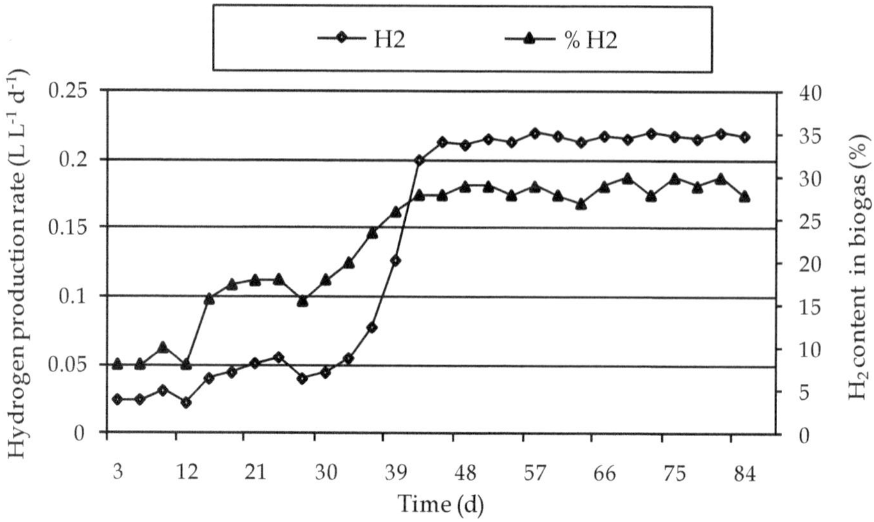

Fig. 13. Hydrogen production in R1 throughout the experimental period

The experimental results demonstrated by Azbar et al. (2009) showed that OLR did not result in any statistically significant change in hydrogen production rate from cheese whey when anaerobic reactors were operated using a range of OLR of 21, 35, 45 g COD L^{-1} d^{-1} and a constant HRT of 24 h. Moreover, lower HRT values (e.g. 1 d) increased hydrogen production rate. Mohan et al. (2007) found, that increasing OLR lead to periodic decreasing in biohydrogen production rate from dairy wastewater. This was attributed to the adaptation of microbial inoculum to higher substrate loading rate which is parallel to this results (Fig. 13). High substrate loading rate shows higher availability of substrate resulting in active substrate metabolism leading to higher H_2 yield (Mohan et al., 2007).

During the experiment, after reaching steady-state conditions (the standard deviations of monitored parameters were within 5%) on day 51, the average rate of hydrogen production was 0.97 L H_2 d^{-1} (0.21 L H_2 L^{-1} d^{-1}) (Fig. 13), hydrogen content in biogas increased to 29,8% v/v and methane was present in a low concentration (<1% v/v). Yang et al. (2007) reported hydrogen production ranging from 0.264 to 0.312 L H_2 L^{-1} d^{-1} (0.396 - 0.468 L H_2 d^{-1}) for OLR of 10 g COD L^{-1} d^{-1} and a HRT of 24 h using wastewater made from dry whey permeate powder. Biohydrogen contents in biogas fluctuated between 22 and 26% v/v. Castellò et al. (2009) obtained 0.55 L H_2 d^{-1} (0.122 L H_2 L^{-1} d^{-1}) for HRT of 12 h and OLR of 20 g COD L^{-1} d^{-1} from cheese whey. Biohydrogen content in biogas ranged from 20 to 30%. Hydrogen production between 0.3 and 7.9 L H_2 L^{-1} d^{-1} (2.5 L H_2 L^{-1} d^{-1} on average) from dairy wastewater was reported by Azbar et al. (2009). Venetsaneas et al. (2009) operated two-stage anaerobic reactors using cheese whey as a fermentation substrate. During the hydrogenogenic stage they achieved biohydrogen production rate of 0.96 L H_2 d^{-1} (1.92 L H_2 L^{-1} d^{-1}) and the percentage of hydrogen in the gas phase was 32.0±1.9% v/v (OLR 15 g COD L^{-1} d^{-1}, HRT 24 h). Liu et al. (2006) achieved hydrogen production rate of 0.64 L H_2 d^{-1} (1.6 L H_2 L^{-1} d^{-1}) and hydrogen content in the biogas of 42% v/v from household solid waste in the two-stage fermentation process.

The effluent from R1 was collected in the storage tank from where it was pumped to the R2. The average pH at the outlet of R1 was 5.0 with a standard deviation of 0.3. According to the literature, optimum pH for methanogenic bacteria is between 6.0 and 7.5 (Mohan et al. 2007). In R2 the pH was maintained on the level of 7.2 (±0.05). The R2 was fed with the R1 effluent from 51 day of experimental period, after the R1 had reached the steady-state. The average COD concentration of the R1 effluent was 5770 mg L^{-1} thus the R2 reactor was operated at OLR of about 2 g COD L^{-1} d^{-1}. The majority of COD was composed of VFA (total of all acids was 5147 mg L^{-1} on average) generated during acidogenic fermentation step in R1. Fig. 14 shows the generation of biogas and methane produced throughout the experimental period. After the first 12 days, the reactor performance was stable and the average biogas and methane production rates were 3.0 L d^{-1} (0.67 L L^{-1} d-1) and 2.2 L CH_4 d^{-1} (0.47 L CH_4 L^{-1} d^{-1}), respectively. The methane content in biogas approached 71% v/v. Venetsaneas et al. (2009) found that about 1 L CH_4 d^{-1} was achieved in two-stage process for cheese whey fermentation resulting in 68% v/v methane production which is comparable to this study. The methanogenic digester was operated at an HRT of 20 d and OLR of 2.5 g COD L^{-1} d^{-1}. Liu et al. (2006) achieved better methane production from household solid waste in the two-stage fermentation process of 11.5 L CH_4 d^{-1} with a 65% v/v methane concentration in biogas.

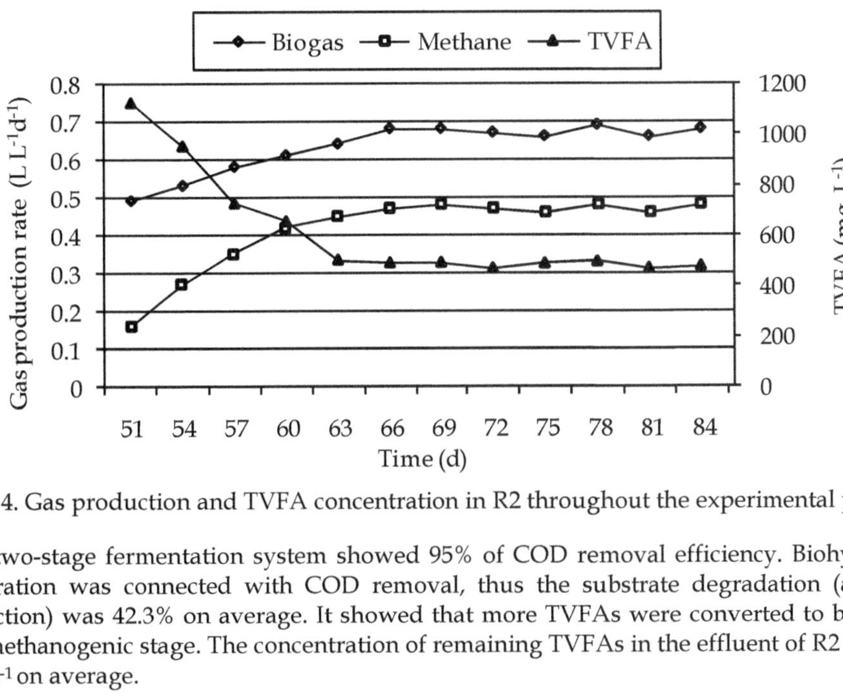

Fig. 14. Gas production and TVFA concentration in R2 throughout the experimental period

The two-stage fermentation system showed 95% of COD removal efficiency. Biohydrogen generation was connected with COD removal, thus the substrate degradation (as COD reduction) was 42.3% on average. It showed that more TVFAs were converted to biogas in the methanogenic stage. The concentration of remaining TVFAs in the effluent of R2 was 485 mg L^{-1} on average.

During the study it was noticed a partial fermentation of raw UF whey permeate, although it was maintained at 4°C before used. VFA concentration increased periodically from 55 mg L^{-1} to 2220 mg L^{-1} after thawing for 20 hours. It seems that the UF whey permeate feedstock should be supplemented with $NaHCO_3$ to increase its alkalinity level before frozen (Castellò et al., 2009). It allows to reduce the lactic acid concentration in the influent, because it was reported, that the conversion efficiency of lactic acid to hydrogen is much lower than that of glucose or sucrose (Guo et al. 2010). The coexistence of LAB (lactic acid bacteria) and the hydrogen producing bacteria was investigated by Noike et al. (2002). They found, that hydrogen fermentation was replaced by lactic acid fermentation caused by LAB present in the raw wastewater. Their inhibitory effect on hydrogen production could be explained by excretion of bacteriocins (Noike et al., 2002).

4.3 Conclusions

Raw, unsterilized UF whey permeate could be a substrate for biohydrogen production. Exploitation of UF whey permeate as a feedstock for H_2 and biogas production is an attractive and effective way of wastewater treatment with simultaneous renewable energy producing. Several factors influenced H_2 production in continuous bioreactors, especially pH of operation, HRT and OLR. The experiment showed, that pH 5.8 in the hydrogenogenic phase was not sufficient to inhibit the activity of methanogenic bacteria. Lower HRT values (<24 h) should be apply to eliminate methanogenic bacteria. Moreover, the regulation of pH in the hydrogenogenic reactor should be coupled with the influent pH adjustment in order to avoid lactic acid production in the raw wastewater. The HRT of 12 - 24 h and OLR of 10 - 20 g COD L^{-1} d^{-1} were found to be sufficient for effective H_2 yield.

This study demonstrated that biohydrogen production from UF whey permeate can be efficiently coupled with methane production in a subsequent step. For hydrogen production stage in stable conditions (HRT of 12 h, OLR of 20 g COD L^{-1} d^{-1}, pH 5.2) hydrogen production rate was 0.97 L H_2 d^{-1} and hydrogen content in the biogas reached 29,8% v/v. In the methanogenic step, the average biogas and methane production rates were 3.0 L d^{-1} and 2.2 L CH_4 d^{-1}, respectively, while the methane content in biogas approached 71% v/v. The two-stage fermentation system reached 95% of COD removal efficiency.

The results of these investigations made the base for further studies to find the optimal operating conditions for higher biohydrogen production in UASB reactor in two-stage fermentation process with respect to determine the optimal OLR, HRT and pH parameters of the reactor giving stable long-term generation of H_2 from raw UF whey permeate.

5. General conclusions

Nowadays, an excessive use of fossil fuels has led to significant emissions of CO_2 in the atmosphere which is responsible for causing extensive climate changes (Soccol et al., 2010). As a result of this with the increase in fossil fuel prices direct the efforts towards utilizing renewable energy sources. Considerable progress in searching for alternative energy sources has been made since the oil crisis of 1973. However, it must be noted that an only the renewable biomass used for energy production contributes to the reduction of negative environmental impacts, e.g. decreased GHG emissions.

Currently, commercial biofuels production, such as ethanol and biogas, relies mostly on the fermentation of cane sugar, molasses or glucose derived from corn, sugar beet, wheat or potatoes. It is not economically accepted because these biomass production for biofuels competes for the limited agricultural land needed for food and feed production. Much of the hydrogen produced in the world is obtained from natural gas, which is not environmental friendly. Therefore, a significant increase in biofuels production would be possible only if technologies are developed to convert the waste biomass.

Dairy industry, like most other food industries, generates strong wastewaters characterized by high COD concentrations representing their high organic content (Demirel et al., 2005). Whey is by-product of milk processing and is abundantly obtained during cheese production. According to Najafpour et al. (2008), worldwide cheese production generates more than 145 million tonnes of liquid whey per year. In the case of deproteination of whey for the production of a valuable human food additive, the residual whey permeate is still a waste with high COD and must be treated before disposal. Due to its lactose major component, whey permeate is a well defined and suitable substrate for anaerobic digestion (Kourkoutas et al., 2002; Najafpour et al., 2008; Venetsaneas et al., 2009; Zafar & Owais, 2006). UF whey permeate fermentation in UASB reactors to produce biofuels (bioethanol, biogas, biohydrogen) has been successfully tested in this study.

6. Acknowledgements

The study of bioethanol production was supported by a grant N523 049 32/1753 from Ministry of Science and Higher Education, Poland in 2007 - 2008.

The study of biohydrogen production was supported by a grant N N 523 555138 from Ministry of Science and Higher Education, Poland in 2010 - 2013.

7. References

Agricultural Market Agency (August 2009). Dairy Market in Poland (in Polish), ISBN 978-83-60012-74-1, Warsaw, Poland

Aktaş, N.; Boyacı, İ.; Mutlu, M. & Tanyolaç A. (2006). Optimization of lactose utilization in deproteineted whey by Kluyveromyces marxianus using response surface methodology (RSM). *Bioresource Technology*, Vol.97, No.18, (December 2006), pp. 2252-2259, ISSN 0960-8524

Angelidaki, I.; Ellegaard, L. & Ahring, B.K. (2003). Applications of the Anaerobic Digestion Process. Biomethanation II. *Advances in Biochemical Engineering/Biotechnology*, Vol.82, pp. 1-32, ISSN 0724-6145

Antonopoulou, G.; Gavala, H.N.; Skiadas, I.; Angelopoulos, K. & Lyberatos, G. (2008). Biofuels generation from sweet sorghum: Fermentative hydrogen production and anaerobic digestion of the remaining biomass. *Bioresource Technology*, Vol.99, No.1, (January 2008), pp. 110-119, ISSN 0960-8524

Azbar, N.; Dokgöz, F.; Keskin, T.; Korkmaz, K.S. & Syed, H.M. (2009). Continuous fermentative hydrogen production from cheese whey wastewater under thermophilic anaerobic conditions. *International Journal of Hydrogen Energy*, Vol.34, No.17, (September 2009), pp. 7441-7447, ISSN 0360-3199

Balat M. & Balat H. (2009). Recent trends in global production and utilization of bio-ethanol fuel. *Applied Energy*, Vol.86, No.11, (November 2009), pp. 2273-2282, ISSN 0306-2619

Castellò, E.; Garcĭa y Santos, G.; Iglesias, T.; Paolino, G.; Wenzel, J.; Borzacconi, L. & Etchebehere, C. (2009). Feasibility of biohydrogen production from cheese whey using a UASB reactor: Links between microbial community and reactor performance. *International Journal of Hydrogen Energy*, Vol.34, No.14, (July 2009), pp. 5674-5682, ISSN 0360-3199

Chisti, Y. (2003). Sonobioreactors: using ultrasound for enhanced microbial productivity. *Trends in Biotechnology*, Vol.21, No.2, (February 2003), pp. 89-93, ISSN 0167-7799

Choung, Y.K. & Jeon, S.J. (2000). Phosphorus removal in domestic wastewater using anaerobic fixed beds packed with iron contactors. *Water Science and Technology*, Vol.41, No.1, pp. 241-244, ISSN 0273-1223

Compagno, C.; Porro, D.; Smeraldi, C. & Ranzi, B.M. (1995). Fermentation of whey and starch by transformed *Saccharomyces cerevisiae* cells. *Applied Microbiology and Biotechnology*, Vol.43, No.5, (October 1995), pp. 822-825, ISSN 1432-0614

Coté, A.; Brown, W.A.; Cameron, D. & van Walsum, G.P. (2004). Hydrolysis of lactose in whey permeate for subsequent fermentation to ethanol. *Journal of Dairy Science*, Vol. 87, No.6, (June 2004), pp. 1608-1620, ISSN 0022-0302

de Glutz, F.N. (2009). Fuel bioethanol production from whey permeate. These EPFL No.4372, 24.04.2009, Available from: <http://biblion.epfl.ch/EPFL/theses/2009/4372/EPFL_TH4372.pdf>

Demirel, B. ; Yenigun, O. & Onay T.T. (2005) Anaerobic digestion of dairy wastewaters : a review. *Process Biochemistry*, Vol.40, No.8., (July 2005), pp.2583-2595, ISSN 1359-5113

Gadhamshetty, V.; Sukumaran, A.; Nirmalakhandan, N. & Myint, M.T. (2008). Photofermentation of malate for biohydrogen production - A modeling approach. *International Journal of Hydrogen Energy*, Vol.33, No.9, (May 2008), pp. 2138-2146, ISSN 0360-3199

Ghaly, A.E. & El-Taweel, A.A. (1997). Kinetic modelling of continuous production of ethanol from cheese whey. *Biomass and Bioenergy*, Vol.12, No.6, pp. 461-472, ISSN 0961-9534

Gnansounou, E. (2010). Production and use of lignocellulosic bioethanol in Europe: Current situation and perspectives. *Bioresource Technology*, Vol.101, No.13, (July 2010), pp. 4842-4850, ISSN 0960-8524

Golubev, W.I. & Golubev, N.W. (2004). Selection and study of potent lactose-fermenting yeast. *Applied Biochemistry and Microbiology*, Vol.40, No.3, (May 2004), pp. 280-284, ISSN 0003-6838

González Siso M.I. (1996). The biotechnology utilization of cheese whey: a review. *Bioresource Technology*, Vol.57, No.1, (July 1996), pp. 1-11, ISSN 0960-8524

Guerrero, S.; Tognon, M. & Alzamora, S.M. (2005). Response of *Saccharomyces cerevisiae* to the combined action of ultrasound and low weight chitosan. *Food Control*, Vol.16, No.2, (February 2005), pp. 131-139, ISSN 0956-7135

Guimarães, P.M.R.; François, J.; Parrou, J.L.; Teixeira, J.A. & Domingues L. (2008a). Adaptive evolution of a lactose-consuming *Saccharomyces cerevisiae* recombinant. *Applied and Environmental Microbiology*, Vol.74, No.6, (March 2008), pp. 1748-1756, ISSN 1098-5336

Guimarães, P.M.R.; Teixeira, J.A. & Domingues L. (2008b). Fermentation of high concentrations of lactose to ethanol by engineered flocculent *Saccharomyces cerevisiae*. *Biotechnology Letters*, Vol.30, No.11, (November 2008), pp. 1953-1958, ISSN 1573-6776

Guo, X.M.; Trably, E.; Latrille, E.; Carrère, H. & Steyer, J.P. (2010). Hydrogen production from agricultural waste by dark fermentation: A review. *International Journal of Hydrogen Energy*, Vol.35, No.19, (October 2010), pp. 10660-10673, ISSN 0360-3199

Jeon, S.J.; Kim, H.S. & Lee, Y.W. (2003). Effect of iron media on the treatment of domestic wastewater to enhance nutrient removal efficiency. *Process Biochemistry*, Vol.38, No.12, (July 2003), pp. 1767-1773, ISSN 1359-5113.

Kargi, F. & Ozmıhcı, S. (2006). Utilization of cheese whey powder (CWP) for ethanol fermentations: Effects of operating parameters. *Enzyme and Microbial Technology*, Vol.38, No.5, (March 2006), pp. 711-718, ISSN 0141-0229

Karri, S.; Sierra-Alvarez, R. & Field J.A. (2005). Zero valent iron as an electron-donor for methanogenesis and sulfate reduction in anaerobic sludge. *Biotechnology and Bioengineering*, Vol.92, No.7., (December 2005), pp. 810-819, ISSN 1097-0290

Kourkoutas, Y.; Dimitropoulou, S.; Kanellaki, M.; Marchant, R.; Nigam, P.; Banat, I.M. & Koutinas, A.A. (2002). High-temperature alcoholic fermentation of whey using *Kluyveromyces marxianus* IMB3 yeast immobilized on delignified cellulosic material. *Bioresource Technology*, Vol. 82, No.2, (April 2002), pp. 177-181, ISSN 0960-8524

Kourkoutas, Y.; Bekatorou, A.; Banat, I.M.; Marchant, R. & Koutinas, A.A. (2004). Immobilization technologies and support materials suitable in alcohol beverages production: a review. *Food Microbiology*, Vol.21, No.4, (August 2004), pp. 377-397, ISSN 0740-0020

Krupp, M. & Widmann, R. (2009). Biohydrogen production by dark fermentation: Experiences of continuous operations in large lab scale. *International Journal of Hydrogen Energy*, Vol.34, No.10, (May 2009), pp. 4509-4516, ISSN 0360-3199

Lai, P.; Zhao, H.Z.; Wang, C. & Ni, J.R. (2007). Advanced treatment of coking wastewater by coagulation and zero-valent iron processes. *Journal of Hazardous Materials*, Vol.147, No.1-2, (August 2007), pp. 232-239, ISSN 0304-3894

Lanchun, S.; Bochu, W.; Liancai, Z.; Jie, L.; Yanhong, Y. & Chuanren, D. (2003). The influence of low-intensity ultrasonic on some physiological characteristics of *Saccharomyces cerevisiae*. *Colloids and Surfaces B: Biointerfaces*, Vol.30, No.1, (July 2003), pp. 61-66, ISSN 0927-7765

Lewandowska, M. & Kujawski, W. (2007). Ethanol production from lactose in a fermentation/pervaporation system. *Journal of Food Engineering*, Vol.79, No.2, (March 2007), pp. 430-437, ISSN 0260-8774

Ling, K.Ch. (2008). Whey to Ethanol: A Biofuel Role for Dairy Cooperatives?, *Research Report 214, USDA Rural Development, US Department of Agriculture (USDA)*, Washington 2008, Available from: < http://www.rurdev.usdu.gov/rbs/pub/RR214.pdf>

Liu, D.; Liu, D.; Zeng, R.J. & Angelidaki, I. (2006). Hydrogen and methane production from household solid waste in the two-stage fermentation process. *Water Research*, Vol.34, No.11, (June 2006), pp. 2230-2236, ISSN 0043-1354

Liu, H.; Yan, Y.; Wang, W. & Yu, Y. (2007). Low intensity ultrasound stimulates biological activity of aerobic activated sludge. *Frontiers of Environmental Science & Engineering in China*, Vol.1, No.1, (February 2007), pp. 67-72, ISSN 1673-7520

Liu, Y.; Takatsuki, H.; Yoshikoshi, A.; Wang, B.C. & Sakanishi A. (2003a). Effects of ultrasound on the growth and vacuolar H^+ -ATPase activity of aloe *arborescens* callus cells. *Colloids and Surfaces B: Biointerfaces*, Vol.32, No.2, (October 2003), pp. 105-116, ISSN0927-7765

Liu, Y.; Yoshikoshi, A.; Wang, B.C. & Sakanishi A. (2003b). Influence of ultrasonic stimulation on the growth and proliferation of *Oryza sativa* Nipponbare callus cells. *Colloids and Surfaces B: Biointerfaces*, Vol.27, No.4, (March 2003), pp. 287-293, ISSN 0927-7765

Marques, L. L. M.; Buzato, J.B. & Celligoi, M.A.P.C. (2006). Effect of raffinose and ultrasound pulses on invertase release by free and immobilized *Saccharomyces cerevisiae* in loofa (*Luffa cylindrica*) sponge. *Brazilian Archives of Biology and Technology*, Vol.49, No. 6, (November 2006), pp. 873-880, ISSN 1516-8913

Meyer-Aurich, A.; Schattauer, A.; Hellebrand, H.J.; Klauss, H.; Plöchl, M. & Berg, W. (2012). Impact of uncertainties on greenhouse gas mitigation potential of biogas production from agricultural resources. *Renewable Energy*, Vol.37, No.1, (January 2012), pp. 277-284, ISSN 0960-1481

Mohan, S.V.; Babu, V.L. & Sarma, P.N. (2007). Anaerobic biohydrogen production from dairy wastewater treatment in sequencing batch reactor (AnSBR): Effect of organic loading rate. *Enzyme and Microbial Technology*, Vol.41, No.4, (September 2007), pp. 506-515, ISSN 0141-0229

Muthaiyan, A. & Ricke S.C. (2010). Current perspectives on detection of microbial contamination in bioethanol fermentors. *Bioresource Technology*, Vol.101, No.13, (July 2010), pp. 5033-5042, ISSN 0960-8524

Najafpour, G.D., Hashemiyeh, B.A., Asadi, M. & Ghasemi, M.B. (2008). Biological treatment of dairy wastewater in an Upflow Anaerobic Sludge-Fixed Film Bioreactor. *American-Eurasian Journal of Agricultural and Environmental Sciences*, Vol.4, No.2, pp. 251-257, ISSN 1818-6769

Nath, K. & Das, D. (2009). Effect of light intensity and initial pH during hydrogen production by an integrated dark and photofermentation process. *International Journal of Hydrogen Energy*, Vol.34, No.17, (September 2009), pp. 7497-7501, ISSN 0360-3199

Nikolić, S.; Mojović; L.; Pejin, D.; Rakin, M. & Vukašinović, M. (2010). Production of bioethanol from corn meal hydrolyzates by free and immobilized cells of *Saccharomyces cerevisiae* var. *ellipsoideus*. *Biomass and Bioenergy*, Vol.34, No.10, (October 2010), pp. 1449-1456, ISSN 0961-9534

Noike, T.; Takabatake, H.; Mizuno, O. & Ohba, M. (2002). Inhibition of hydrogen fermentation of ogranic wastes by lactic acid bacteria. *International Journal of Hydrogen Energy*, Vol.27, No.11-12, (November-December 2002), pp. 1367-1371, ISSN 0360-3199

Ozmihci, S. & Kargi, F. (2007). Kinetics of batch ethanol fermentation of cheese-whey powder (CWP) solution as function of substrate and yeast concentrations. *Bioresource Technology*, Vol.98, No.16, (November 2007), pp. 2978-2984, ISSN 0960-8524

Ozmıhcı, S. & Kargi, F. (2008) Ethanol production from cheese whey powder solution in a packed column bioreactor at different hydraulic residence times. *Biochemical Engineering Journal*, Vol.42, No.1, (November 2008), pp. 180-185, ISSN 1369-703X

Pitt, W.G. & Ross S.A. (2003) Ultrasound increases the rate of bacterial cell growth. *Biotechnology Progress*, Vol.19, No.3, (January 2003), pp. 1038-1044, ISSN 1520-1633

Pöschl, M.; Ward, S. & Owende, P. (2010). Evaluation of energy efficiency of various biogas production and utilization pathways. *Applied Energy*, Vol.87, No.11, (November 2010), pp. 3305-3321, ISSN 0306-2619

Prochnow, A.; Heiermann, M.; Plöchi, M.; Linke, B.; Idler, C.; Amon, T. & Hobbs, P.J. (2009). Bioenergy from permanent grassland - A review: 1. Biogas. *Bioresource Technology*, Vol.100, No.21, (November 2009), pp. 4931-4944, ISSN 0960-8524

Ramakrishnan, S. & Hartley, B.S. (1993). Fermentation of lactose by yeast cells secreting recombinant fungal lactase. *Applied and Environmental Microbiology*, Vol.59, No.12, (December 1993), pp. 4230-4235, ISSN 1098-5336

Rubio-Texeira, M.; Castrillo, J.I.; Aadam, A.C.; Ugalde, U.O. & Polaina, J. (1998). Highly efficient assimilation of lactose by a metabolically engineered strain of *Saccharomyces cerevisiae*. *Yeast*, Vol. 14, No.9, (June 1998), pp. 827-837, ISSN 1097-0061

Sarkar, N.; Ghosh, S.K.; Bannerjee, S. & Aikat, K. (2012). Bioethanol production from agricultural wastes: An overview. *Renewable Energy*, Vol.37, No.1, (January 2012), pp. 19-27, ISSN 0960-1481

Schläfer, O.; Sievers, M.; Klotzbűcher, H. & Onyeche, T.I. (2000). Improvement of biological activity by low energy ultrasound assisted bioreactors. *Ultrasonics*, Vol.38, No.1-8, (March 2000), pp. 711-716, ISSN 0041-624X

Silveira, W.B.; Passos, F.J.W.; Mantovani, H.C.; Passos F.M.L. (2005). Ethanol production from cheese whey permeate by *Kluyveromyces marxianus* UFV-3: A flux analysis of oxido-reductive metabolism as a function of lactose concentration and oxygen levels. *Enzyme and Microbial Technology*, Vol.36, No.7, (May 2005), pp. 930-936, ISSN 0141-0229

Soccol, C.; Vandenberghe, L. & Medeiros A. (2010). Bioethanol from lignocelluloses: status and perspectives in Brazil. *Bioresource Technology*, Vol.101, No.13, (July 2010), pp. 4820-4825, ISSN 0960-8524

Stephenson, A.L.; Dupree, P.; Scott, S.A.; Dennis, J.S. (2010). The environmental and economic sustainability of potential bioethanol from willow in the UK. Bioresource Technology, Vol.101, No.24, (December 2010), pp. 9612-9623, ISSN 0960-8524

Tsukamoto, I.; Constanyinoiu, E.; Furuta, M.; Nishimura, R. & Maeda, Y. (2004). Inactivation effect of sonication and chlorination on *Saccharomyces cerevisiae*. Calorimetric Analysis. *Ultrasonics Sonochemistry*, Vol.11, No.3-4, (May 2004), pp. 167-172, ISSN 1350-4177

Van Ginkel, S.; Oh, S. & Logan B. (2005). Biohydrogen gas production from food processing and domestic wastewater. *International Journal of Hydrogen Energy*, Vol.30, No.15, (December 2005), pp. 1535-1542, ISSN 0360-3199

Venetsaneas, N.; Antonopoulou, G.; Stamatelatou, K.; Kornaros, M. & Lyberatos, G. (2009). Using cheese whey for hydrogen and methane generation in a two-stage continuous process with alternative pH controlling approaches. *Bioresource Technology*, Vol.100, No.15, (August 2009), pp. 3713-3717, ISSN 0960-8524

Vlyssides, A.; Barampouti, E.M. & Mai S. (2009). Influence of ferrous iron on the granularity of a UASB reactor. *Chemical Engineering Journal*, Vol.146, No.1, (January 2009), pp. 49-56, ISSN 1385-8947

Volkland, H.P.; Harms, H., Wanner, O. & Zehnder, A.J.B. (2001). Corrosion protection by anaerobiosis. *Water Science and Technology*, Vol.44, No.8, pp. 103-106, ISSN 0273-1223

Wang, L.; Zhou, Q. & Li, F.T. (2006). Avoiding propionic acid accumulation in the anaerobic process for biohydrogen production. *Biomass and Bioenergy*, Vol.30, No.2, (February 2006), pp. 177-182, ISSN 0961-9534

Xie, B.; Liu, H. & Yan Y. (2009). Improvement of the activity of anaerobic sludge by low-intensity ultrasound. *Journal of Environmental Management*, Vol.90, No.1, (January 2009), pp. 206-264, ISSN 0301-4797

Xie, B.; Wang, L. & Liu, H. (2008). Using low intensity ultrasound to improve the efficiency of biological phosphorus removal. *Ultrasonics Sonochemistry*, Vol.15, No. 5, (July 2008), pp. 775-781, ISSN 1350-4177

Yadvika, S.; Sreekrishnan, T.R.; Kohli, S. & Rana, V. (2004). Enhancement of biogas production from solid substrates using different techniques - a review. *Bioresource Technology*, Vol.95, No.1, (October 2004), pp. 1-10, ISSN 0960-8524

Yang, P.; Zhang, R.; McGarvey, J.A. & Benemann, J.R. (2007). Biohydrogen production from cheese processing wastewater by anaerobic fermentation using mixed microbial communities. *International Journal of Hydrogen Energy*, Vol.32, No.18, (December 2007), pp. 4761-4771, ISSN 0360-3199

Yu, H.Q.; Fang, H.H.P. & Tay J.H. (2000). Effects of Fe^{2+} on sludge granulation in upflow anaerobic sludge blanket reactors. *Water Science and Technology*, Vol.41, No.12, pp. 199–205, ISSN 0273-1223

Zafar, S. & Owais, M. (2006). Ethanol production from crude whey by *Kluyveromyces marxianus*. *Biochemical Engineering Journal*, Vol.27, No.3, (January 2006), pp. 295-298, ISSN 1369-703X

11

Study on Manufacturing Technology and Performance of Biogas Residue Film

Chen Haitao, Li Lixia, Wang Hanyang and Liu Lixue
Northeast Agricultural University, China

1. Introduction

Plastic mulching cultivation technology originated from Japan in 1955. This technology was introduced to China in 1978, and it was comprehensively spread after a series of tests. Plastic film mulching technology was widely used due to increasing temperature, preserving soil moisture, increasing yield, preventing soil erosion etc. To 1986, the using area of plastic film in China had leap to the first in the world. The application of mulching cultivation technology is an effective way to improve crop yield, and has promoted the development of agriculture. The application of plastic film was known as the third revolution following fertilizer, seeds in agriculture, also called "white revolution"(Zhang Ying,2005).

However, plastic is a polymer material, which is non-perishable, difficult degradation, experiments show that the degradation of plastics needs 200 years in the soil (Bian Yousheng, 2005). With the unceasing expanding of mulching area of plastic film and the increasing of using years, the residual plastic film that was not degraded and continued to be accumulated in farmland, large numbers of residual film formed barrier layer gradually, which could hinder the development of crop root and the absorption of moisture and nutrients, hinder machinery tillage, damage crops growth environment, lead to soil compaction , poor permeability, reduction of crop yield and severe environmental pollution, the phenomenon that a large number of plastic film left in farmland make "white revolution " which has brought Gospel to agricultural production be transformed into "white pollution" (Liu Junke, 2000). In order to protect farmland ecological environment, domestic and international measures were actively developed and a variety of environmentally friendly biodegradable plastic film which can produce oxidized photochemical and biochemical effects were promoted, the use of biodegradable plastic film would be an effective way to completely solve the "white pollution" (Yang Huidi, 2000).

In recent years, biodegradable plastic film material and degradation process has become a research focus (Li Xianfa, 2004; Sun Jianping, et al, 2000; Zhang Chunhong et al, 2007). Currently, photolysis film, biodegradable film, photolysis/biodegradable film were researched more, the degradation process of film depended on biodegradation, photo-degradation and chemical degradation, and the effect of efficiency, synergy and coherence between the three main degradation process. Although a variety of biodegradable plastic film were developed in China, due to the high cost, the poor economy and the difficulty of controlling the degradation of biodegradable film, the promotion and development were hindered. So, the development of degradable film which has good economy and low cost has become the main research direction.

In China, the construction of biogas started in the 1970s, and so far, with over 30 years of history. Biogas technology mainly used manure, straw and other organic substances as raw material to produce biogas by anaerobic fermentation (Zhang Yongmei, 2008). In recent years, the promotion and application of biogas technology was rapid in China, at the end of 2005, 17 million household biogas digesters and 140000 sewage purification digesters were built in rural areas, with an annual output of about 80 billions m^3 gas. By the end of 2006, the total number of the rural household biogas has reached 22 millions, biogas construction had entered a new stage of rapid development. Currently, biogas residue were mainly used for planting, breeding, sideline and processing industry (Lu Mei et al, 2007; Guo Qiang, et al, 2005; RK Gupta, 2002), which was difficult to get high value resources utilization, so if its solid waste after anaerobic digestion could achieve high value resource utilization, the economic efficiency of industrial production of biogas could be improved and good ecological and social benefits would be produced (Tian Xin, 2008; Ye Xujun, et al, 2000).

In addition, because ruminant animals mainly fed on coarse fodder, such as rice straw, wheat straw, corn stalk, etc. The conversion rate of straw fibre was not high, the main reason for this phenomenon was that 20% to 70% of the fibre can not be degraded by rumen bacteria of ruminants, the result was that the biogas residue which was from ruminant manure fermentation contained large amounts of wood cellulose (An Juan, 2005).

Therefore, the full biodegradable biogas residue fibre film made of ruminant animal manure or fibre residue of straw fermentation for biogas and a certain percentage of plant fibre, using cleaner production processes, had moisture conservation and weeding property. The film could be completely degraded by microorganisms in the period of crop growth, restore in the soil, improve soil organic matter content, and meanwhile not pollute the environment, all of these would lay the foundation for applications and promotion of biodegradable film, to sustain agricultural sustainable development of our country had important realistic and historical significance.

1.1 Research status

1.1.1 Biogas residue utilization status

The research results showed that biogas residue was one of the residue after organic matter anaerobic fermentation for biogas, which was mainly composed of undecomposed raw materials solid and new generation microbial biomass (Lin Jianfeng, 2003).As cellulose , hemicellulose , lignin and other substance of fermentation materials remained in the biogas residue in the fermentation process, so biogas residue basically retains all the components of anaerobic fermentation production besides the gas. Biogas residue generally contains organic matter 36% to 49.9%, humic acid 10.1% to 24.6%, crude protein 5% to 9%, total nitrogen 0.8% to 1.5%, total phosphorus 0.4% to 0.6%, and total potassium 0.6% to 1.2%. The requirement of N,P,K of anaerobic fermentation process is very low, therefore, the majority of N,P,K and other elements of fecaluria were not being used, and eventually left in the biogas residue and biogas slurry(Bian Yousheng,2005;Xie Tao, 2007). As microbial groups and undecomposed raw materials, so the biogas residue had its unique property (Zhang Quanguo, 2005).

Organic matter of biogas residue is not only a good fertilizer, but also conducive to microbial activity and the formation of granular structure, the organic matter surface can absorb amounts of soluble effective nutrients, under the soil microorganism's action, it can continuously provide compatible nutrients for growing (Zhang Wudi, 2003). Currently, the utilization of biogas residue is mainly as following (Guo Qiang et al., 2005):

1. Biogas residue fodder

Biogas residue contain 24 kinds of amino acid, many kinds of trace elements, B vitamins and other nutrients, biogas residue can be used to feed pigs, 50kg fodder can be saved and one month of fattening period can be shortened when fattening one pig. Feeding fish with biogas residue can not only improve fish yield and quality, but also reduce the occurrence of fish diseases, breeding earthworms with biogas residue can provide fodder of high-quality protein for livestock, while improve the utilization value of biogas residue.

2. Biogas residue fertilizer

Biogas residue is a high quality fertilizer, it can effectively improve soil physical and chemical property, increase soil organic matter and nutrient content, improve soil porosity, bulk density and water retention (Xu Shiwen, 1987).

3. Biogas residue adsorbent

The research which was made by C.Namasivayam (1995) showed that biogas residue can absorb heavy metal Cr^6 in wastewater better at pH was 1.5; biogas residue can absorb the "Direct red 12 B" stain in industry wastewater better at pH 2.7; biogas residue can absorb P_b ^{2} in wastewater better, and adsorption capacity can reach 28mg/g.

4. Biogas residue brewing

Using artificial culture and old kiln mud as bacteria, together with anaerobic biogas residue, the yield rate of wine increased 10.5% than without biogas residue (Lu Baoqing, 1997).

5. Biogas residue compost

Mixing biogas residue and straw with a 1 : 1 ratio was used for mushroom matrix. The biggest biogas industry was built in Nyirbator of Hungary (S.Ranik, 2004).

1.1.2 Biodegradable film research status

1.1.2.1 Research status of biodegradable film materials

Biodegradable materials research began in the 1960s. The initial study was mainly to add natural polymers with biodegradable properties (such as starch, etc.) to generic plastic, then get the so-called biodegradable materials. St.Lawrence starch-company developed a starch-polyethylene or polypropylene blends in Canada (Qiu Weiyang, 2002). With human understanding of biodegradability of macromolecule, the research focus began to turn to biodegradable materials (Qiao Haijun, 2007), which can be classified as microorganism synthetic polymer, chemical synthetic polymer and natural polymer.

Currently, the research of biodegradable film mainly focuses on the following aspects: (1) photo-degradation film, it is made of resin mixed with photo-sensitizer and accelerator. (2)

biodegradable film, it includes structural degradation film, biodegradable film containing inorganic salts and adding starch. (3) photo- biodegradable film; (4) plant fibre film.

In China, the major research was additive photo-degradation film and synthetic photo-degradation film. The research focused on using light stabilizers to control degradation period. Since 1997, 944-polymeric efficient light stabilizer, BW-6911 new light stabilizer were developed, which replaced the severe irritation and sensitization GW-504/2002 anti-aging system. American Dupont CO. , Ltd produced copolymers of ethylene and CO, American OCC and DOW CO. , Ltd had used this technology to produce film and develop industrial production (Xiong Hanguo, 2004). The disadvantages of photo-degradation film were susceptible to external environment, which was difficult to control the degradation period , and covering field, the part into soil can not be degraded, so its application was limited (Xu Xiangchun, 2006).

The degradation of biodegradable film was caused by microbes in natural environmental condition. It was divided into additive biodegradable film and completely biodegradable film according to degradation mechanism and damage style.

At present, additive biodegradable film was composed of plastic, starch, compatibility agents, self-oxidants, processing additives. Typical varieties were polyethylene starch biodegradable film (Liu Ming et al., 2008). There were institutes of physics and chemistry, Beijing University of Technology, Guangdong biodegradable plastic CO., Ltd more than 20 research institutes. The research focused adding starch or modified starch into PE.

The main varieties of completely biodegradable film were PLA, PCL, and PHB and so on. United States used PCL to produce synthetic polymer biodegradable film (KAM Abd I J-kader, 2002). Warner-Lambert developed a new type of resin, which was made of 70% amylopectin starch and 30% amylose starch (He Aijun, 2002). It had good biodegradability, was considered a significant development in material science.

Photo-degradation film was made of additive photo-sensitizer, auto-oxidants, and anti-oxidants as microbial culture medium in general polymer.

Plant fibre film has good ventilation, wet and dry strength and good biodegradability. Chinese academy of agricultural sciences successfully developed the environmentally-friendly hemp film (Fu Dengqiang, 2008). In addition, paper films composed of different materials were produced. South China Technology University used sugar cane and starch as materials to manufacture a kind of fully degradable film (Tan Chengrong, 2002). Japan manufactured biodegradable film with 1%-10% chitosan cloth softwood mechanical pulp original paper in 1990. The demand of environmental film increased in Washington State University, France, Germany, Italy, Canada, Netherlands and South Korea and other countries, leading to the environmentally-friendly film industry rapid development (Han Yongjun, 2008).

1.1.2.2 Research status of degradability of biodegradable film

In 1996, biodegradability of plant fibre paper was studied, which lower mechanical properties under certain environmental conditions, and eventually fragmented or completely degraded. Weight reduction and observation methods were employed by Gao Yujie (1996), Zhang Wenqun (1994), Wang Weigang (2003), Li Zhiming (2004) and Wang

Wei to study the biodegradability (2009), observation method only can describe the film degradation process and the weight reduction method can quantitatively explain the degradation process of biodegradable film.

In this chapter, the physical properties and chemical composition of the biogas residue produced by anaerobic fermentation using ruminant feces were determined and analyzed; manufacturing technology and performance of biogas residue film was studied by the methods of central composite quadratic rotatable orthogonal experiment; biogas residue fibre mulching for planting eggplant was studied with the method of random plot experiment.

2. The physical components of biogas residue and the chemical components of biogas residue fibre

2.1 The physical components of biogas residue

The determination of physical components of biogas residue was shown in table2-1.

Composition	Quality/ g	Percentages/ %
Fibre	477	64
Non-mineral impurity	265	35
Mineral	8	1

Table 2-1. Biogas residue physical composition and their mass percentages

Table 2-1 showed that biogas residue was composed of three parts, fibre proportion was maximum, mineral proportion was minimum. There were mainly plastic, hair and grass seeds in the non-mineral impurities

2.1.1 Each component fibre and mineral content of biogas residue

The determination of each component fibre and mineral content was shown in table 2-2.

Group /mm	0.25~0.5	0.5~1	1~2	2~5	>5	total
Fibre quality /g	158	124	59	126	10	477
Minerals quality /g	0.8	5	1.3	0.7	0.2	8
Fibre quality percentages /%	21.10	16.50	7.90	16.80	1.40	64
Minerals quality percentages /%	0.11	0.67	0.17	0.09	0.03	1

Table 2-2. The fibre and minerals content and their mass percentages of each group

Table 2-2 showed that fibre quality percentage of 0.25 mm to 0.5 mm was maximum; fibre quality percentage of more than 0.5 mm was min; minerals quality percentage of 0.5 to 1 mm was maximum, minerals quality percentage of more than 5 mm was min.

2.1.2 Fibre morphology

Fig. 2-1 showed the fibre morphology of each group.

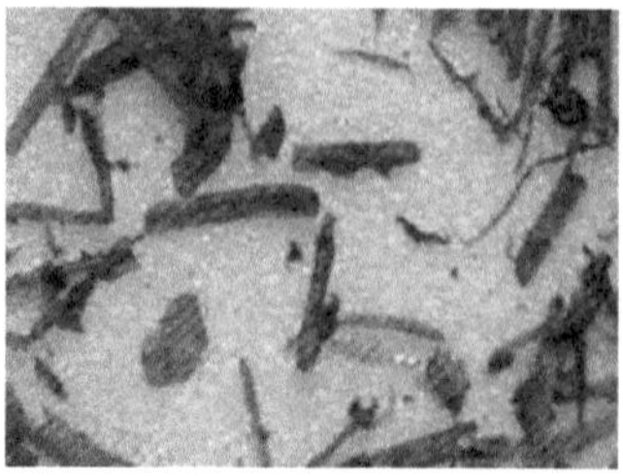

0.25 mm~0.5 mm(Eyepiece10X,Objective 2X)

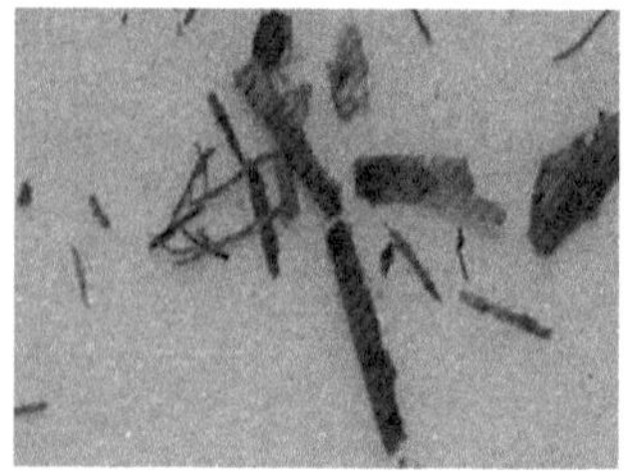

0.5 mm~1 mm (Eyepiece 10X,Objective 1X)

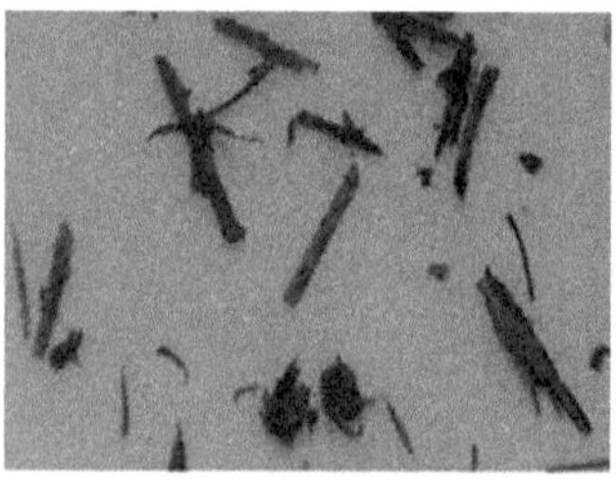

1 mm~2mm
(Eyepiece10X,Objective 1X)

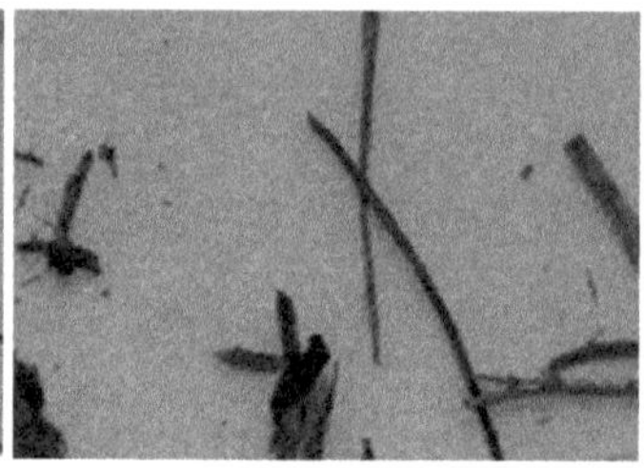

2 mm~5 mm
(Eyepiece 10X,Objective 1X)

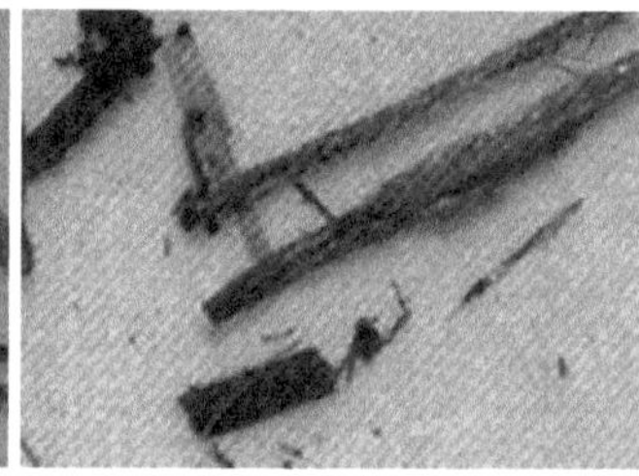

>5 mm
(Eyepiece 10X, Objective 1X)

Fig. 2-1. Each component fibre morphology of the different amplification under microscope

2.1.3 The fibre length to width ratio

Determination results of the fibre length to width ratio of each group were measured and analysed by Motic Images Plus and shown in table 2-3.

Group /mm	Avg. L mm	L SD	L D range mm	Avg. W mm	W SD	W D range mm	Avg. L to W ratio	L to W ratio SD	L to W ratio D range
0.25~0.5	0.718	0.0990	0.52~0.92	0.060	0.0162	0.03~0.09	11.875	3.9790	3.92~19.83
0.5~1	1.940	0.1254	1.69~2.19	0.164	0.0432	0.08~0.25	11.845	1.6714	8.50~15.19
1~2	1.853	0.0873	1.68~2.03	0.142	0.0571	0.03~0.26	13.086	2.1457	8.79~17.38
2~5	2.493	0.3461	1.80~3.19	0.140	0.0345	0.07~0.21	17.726	4.2136	9.30~26.15
>5	3.491	0.6534	2.18~4.80	0.157	0.0318	0.09~0.22	22.174	4.9820	12.21~32.14

Notes: distribution estimates according to $\bar{x} \pm 2\sigma$, Avg. =average, L=length, W=width, D=distribution.

Table 2-3. Determination results of the fibre length to width ratio

Table 2-3 showed that average length distribution range was 0.52 mm to 4.80 mm, which of 0.25 mm to 0.5 mm was minimum, which of more than 5 mm was maximum, average width distribution range was 0.03 mm to 0.26 mm, which of 0.25 mm to 0.5 mm was maximum, which of 0.5 mm to 1 mm was maximum; average length to width ratio distribution range was 3.92 to 32.14, which of 0.5 mm to 1 mm was minimum, which of more than 5 mm was maximum. These showed that biogas residue length to width ratio had a greater dispersion. Although the average length to width ratio distribution range was smaller than straw fibre, which showed that it had a certain available value.

2.2 The chemical components of biogas residue fibre

Fig.2-2 showed the quality percentage of cellulose, hemi-cellulose, and lignin in biogas residue.

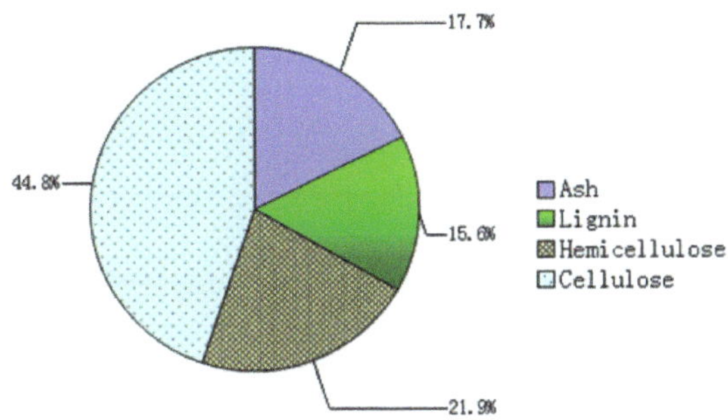

Fig. 2-2. Chemical compositions of biogas residue fibre

According to research results (Gao Zhenghua, 2008; Chen Hongzhang, 2008), the cellulose, hemi-cellulose, lignin of straw, wheat and corn stalks compared with these of biogas residue fibre were shown in table 2-4.

Species	Cellulose/%	Hemi-cellulose/%	Lignin/%	Ash/%
Straw	36.5	27.7	12.3	13.3
Wheat stalk	38.6	32.6	14.1	5.9
Corn stalk	38.5	28	15	4.2
Biogas residue fibre	44.8	21.9	15.6	17.7

Table 2-4. Chemical composition of biomass

The comparative analysis results showed that, cellulose quality percentage of biogas residue fibre after anaerobic fermentation was 5% higher than straw, wheat stalk, corn stalk; while hemi-cellulose quality percentage was 5%lower than straw, wheat stalk, corn stalk; lignin quality percentage did not change. The result showed that anaerobic fermentation to lignin content was not influence, hemi-cellulose relatively reduced, cellulose relatively increased, it is positive to the resources utilization of biogas residue.

3. The study on manufacturing technology of biogas residue film

Biogas residue fibre film samples were prepared with the method of clean pulping and paper-making process. The optimization of the technological parameters were studied by the method of the central composite quadratic orthogonal rotational experiment, beating degree, grammage, rosin, bauxite and wet strength agent were selected as input variables, and dry tensile strength, wet tensile strength, degradation period were chosen as response functions.

Factors and its levels of experiment were shown in table 3-1. Experimental plan and results were shown in table 3-2.

Factor level	Beating degree /SR°	Grammage /g/m²	Rosin /%	Bauxite /%	Wet strength agent /%
z_j	x_1	x_2	x_3	x_4	x_5
γ(+2)	50	110	1.2	6	3.0
(+1)	45	95	1	5	2.4
(0)	40	80	0.8	4	1.8
(-1)	35	65	0.6	3	1.2
-γ(-2)	30	50	0.4	2	0.6

Table 3-1. Factors and its levels of experiment

Run	Factors					Response functions		
	Beating degree SR°	Grammage g/m²	Rosin %	Bauxite %	Wet strength agent %	Dry tensile strength N	Wet tensile strength N	Degradation period day
	x_1	x_2	x_3	x_4	x_5	y_{1i}	y_{2i}	y_{3i}
1	35	65	0.6	3	2.4	29.8	12.0	26.7
2	45	65	0.6	3	1.2	24.2	9.9	25.9
3	35	95	0.6	3	1.2	33.0	15.1	30.3
4	45	95	0.6	3	2.4	35.8	17.4	34.1
5	35	65	1	3	1.2	18.3	9.5	26.5
6	45	65	1	3	2.4	20.3	10.7	27.8
7	35	95	1	3	2.4	35.4	15.7	30.1
8	45	95	1	3	1.2	30.1	16.1	30.1
9	35	65	0.6	5	1.2	23.1	9.6	20.6
10	45	65	0.6	5	2.4	25.5	11.0	27.9
11	35	95	0.6	5	2.4	30.9	18.0	38.3
12	45	95	0.6	5	1.2	35.2	9.8	33.3
13	35	65	1	5	2.4	20.1	11.1	29.5
14	45	65	1	5	1.2	23.0	9.0	20.1
15	35	95	1	5	1.2	32.9	13.0	32.1
16	45	95	1	5	2.4	35.5	17.0	38.9
17	30	80	0.8	4	1.8	26.3	14.7	30.5
18	50	80	0.8	4	1.8	23.4	12.0	31.1
19	40	50	0.8	4	1.8	18.0	9.7	25.0

Run	Factors					Response functions		
	Beating degree SR°	Grammage g/m²	Rosin %	Bauxite %	Wet strength agent %	Dry tensile strength N	Wet tensile strength N	Degradation period day
	x_1	x_2	x_3	x_4	x_5	y_{1i}	y_{2i}	y_{3i}
20	40	110	0.8	4	1.8	40.5	19.5	37.0
21	40	80	0.4	4	1.8	33.8	16.4	29.4
22	40	80	1.2	4	1.8	26.4	12.6	30.6
23	40	80	0.8	2	1.8	25.9	11.4	31.1
24	40	80	0.8	6	1.8	26.9	11.1	32.5
25	40	80	0.8	4	0.6	22.1	8.7	26.4
26	40	80	0.8	4	3	26.6	15.5	37.2
27	40	80	0.8	4	1.8	28.0	13.4	32.8
28	40	80	0.8	4	1.8	32.3	13.3	33.4
29	40	80	0.8	4	1.8	29.6	12.7	33.2
30	40	80	0.8	4	1.8	30.6	13.8	31.6
31	40	80	0.8	4	1.8	28.8	11.8	33.6
32	40	80	0.8	4	1.8	27.6	12.3	36.8
33	40	80	0.8	4	1.8	31.1	14.5	36.7
34	40	80	0.8	4	1.8	33.2	14.9	35.1
35	40	80	0.8	4	1.8	31.8	14.4	34.6
36	40	80	0.8	4	1.8	30.7	14.7	41.1

Table 3-2. Experimental plan and results

3.1 Degradation period test

The arrangement of the degradation period test was shown in Fig.3-1. Degradation state of film samples during degradation period was shown in Fig.3-2

Fig. 3-1. The arrangement of the degradation period test

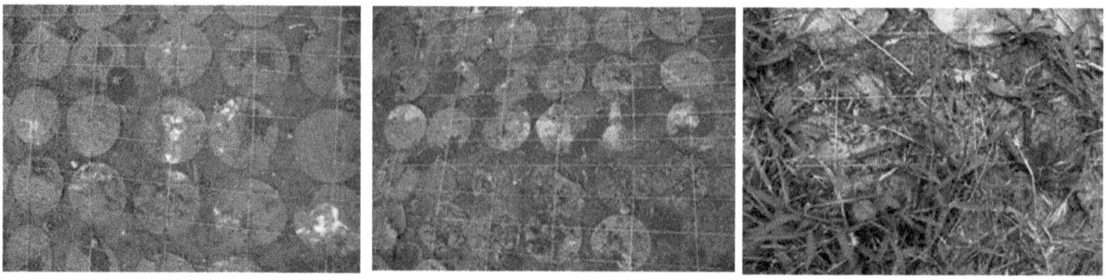

(a)part of film samples after 20 days (b) part of film samples after 35 days (c) part of film samples after 50 days

Fig. 3-2. Mulching degradable process in the different time during the period of degradation test

The fibre film of biogas residue for degradation discovered that the appearance and performance of the samples changed a lot because of light, air temperature, air humidity, wind, rain and other weather factors, coupled with the soil temperature, humidity, combined effect of microorganisms. According to the observation, the degradation was divided into several stages, initially, the sample surface appeared holes or small cracks, called induction period of the film degradation; over time, holes and cracks gradually expanded, the edge glued to the soil surface, the role of soil microorganisms on the film samples increased, resulted in an increasing number of small holes, broken into fragmentation period of the film samples, the film samples effected by various types of micro-organisms would become increasingly thin, the mechanical strength decreased gradually, until the film entered into the fast degradation period of the samples. Especially, after rain, the increasing of air humidity and soil humidity would make mechanical strength of the samples decrease rapidly, so soil moisture is an important impact factor of the film degradation.

During degradation of the film, the dry tensile strengths were regularly measured, according to scatter, the trends of dry tensile strength (N) and date (d) were available, according to trend line, the time of dry tensile strength at zero of each group was estimated, which was defined degradation period. The result was shown in table 3-2.

3.2 Response model

3.2.1 Response model

Response models of dry tensile strength, wet tensile strength and degradation period at α=0.05, were significant and the models were shown as equation 3-1, 3-2 and 3-3.

$$y_1=30.38+9.135\times10^{-3}x_1+5.4x_2-1.52x_3+0.067x_4+0.94x_5-1.02x_1^2-0.64x_4^2-1.15x_5^2+1.14x_1x_4+1.24x_2x_3+0.96x_3x_4-1.11x_4x_5 \quad (3\text{-}1)$$

$$y_2=13.81-0.36x_1+2.47x_2-0.35x_3-0.36x_4+1.43x_5-0.62x_4^2-0.42x_5^2+0.64x_1x_3+0.46x_2x_5-0.44x_3x_5+0.65x_4x_5 \quad (3\text{-}2)$$

$$y_3=34.95+0.22x_1+3.59x_2+0.017x_3+0.5x_4+2.33x_5-1.12x_1^2-1.07x_2^2-1.32x_3^2-0.87x_4^2-0.87x_5^2+1.67x_2x_4+1.41x_4x_5 \quad (3\text{-}3)$$

3.2.2 Analysis of importance of various factors on response functions

Importance of various factors on response functions was shown in table 3-3.

Source	Importance		
	Dry tensile strength (N)	Wet tensile strength (N)	Degradation period (D)
Beating degree	1.645	1.046	0.905
Grammage	1.838	1.323	2.341
Rosin	1.094	1.333	0.931
Bauxite	1.344	1.951	2.099
Wet strength agent	1.526	2.839	2.253

Table 3-3. Importance of each factor

3.3 Effect of interaction factors on dry tensile strength

3.3.1 Effect of beating degree and bauxite on dry tensile strength

Fig.3-3 showed the effect of beating degree and bauxite on dry tensile strength when other factors were held at 0 level. With the increase of beating degree and bauxite, dry tensile strength firstly increased and then slowly decreased, the maximum occurred when the two factors were held at 0 level. The reason was that with the beating degree increased, the fibre sub-wire broom degree was high, the exposure of hydrogen bonding of the fibre surface increased, the bonding forces between the fibres enhanced, so that the film strength increased; when beating degree was more than a certain value, the single fibre strength was destroyed, the bonding force decreased, led to the decrease of strength, adding bauxite excessively to strength had side effect, thus resulting in strength decreased.

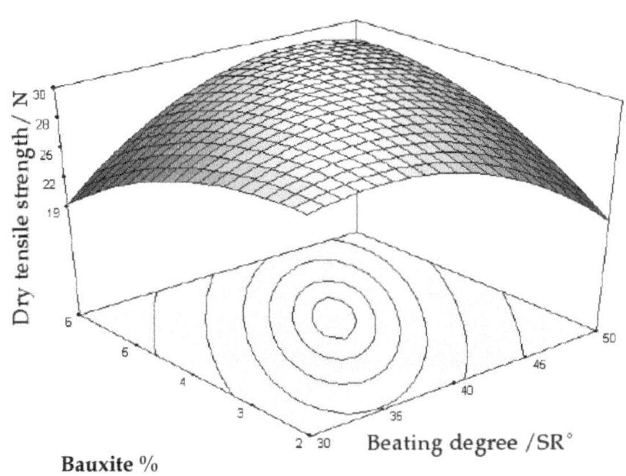

Fig. 3-3. Response surface and contour plots for the effects of beating degree and bauxite on dry tensile strength: grammage was held at 80 g/m^2 and rosin was held at 0.8%, wet strength agent was held at 1.8%

3.3.2 Effect of grammage and rosin on dry tensile strength

Fig.3-4 showed the effect of grammage and rosin on dry tensile strength when other factors were held at 0 level. Dry tensile strength significantly increased with the increase of grammage; when grammage was small, dry tensile strength slowly decreased with the amount of rosin increases, this was because bonding effect of the additive rosin to fibre became larger, when the grammage was large, dry tensile strength slowly increased with the amount of rosin increase, the maximum occurred when rosin was held at 1.2%, and grammage was held at 110 g/m^2. With the grammage increased, the number of fibre in per area increased, bonding between the fibres enhanced, the strength increased, at this moment, the positive effect of grammage on strength was much greater than the negative impact of rosin.

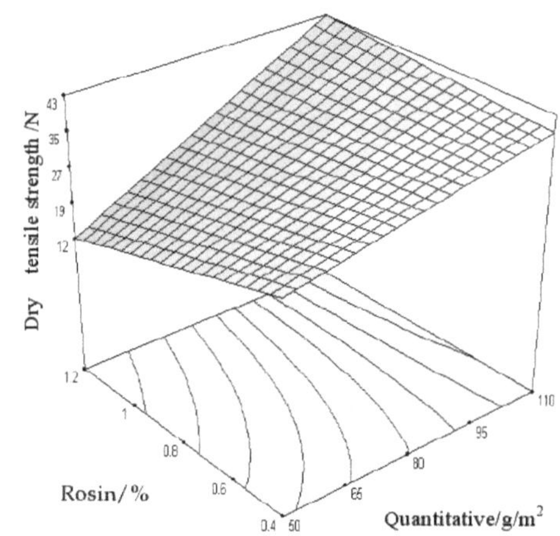

Fig. 3-4. Response surface and contour plots for the effects of grammage and rosin on dry tensile strength: beating degree was held at 40SR°, bauxite was held at 4%, wet strength agent was held at 1.8%

3.3.3 Effect of rosin and bauxite on dry tensile strength

Fig.3-5 showed the effect of grammage and rosin on dry tensile strength when other factors were held at 0 level. Adding the amount of bauxite at a low level, the dry tensile strength decreased with the increase of added rosin amount; adding the amount of bauxite at a high level, the added rosin amount almost had no effect on the dry tensile strength, maximum of the dry tensile strength occurred when bauxite was held at 4%, and rosin was held at 0.4%, because rosin adsorption has been saturated, there was no effect on the strength.

3.3.4 Effect of bauxite and wet strength agent on dry tensile strength

Fig.3-6 showed the effect of bauxite and wet strength agent on dry tensile strength when other factors held at 0 level. When bauxite was near 0 level, the dry tensile strength increased with the increase of wet strength agent, when bauxite was higher than the zero level, with the wet strength agent increased, the dry tensile strength first increased and then decreased, the maximum occurred when wet strength agent was held at 2%, and bauxite was held at 3.5%. This is because with the adding of the wet strength agent, the adsorption of the fibre system to wet strength agent had already been saturated, and it no longer played a role in increasing strength, anionic trash in absorption system impacted the combination between the fibre, leading to strength decreased.

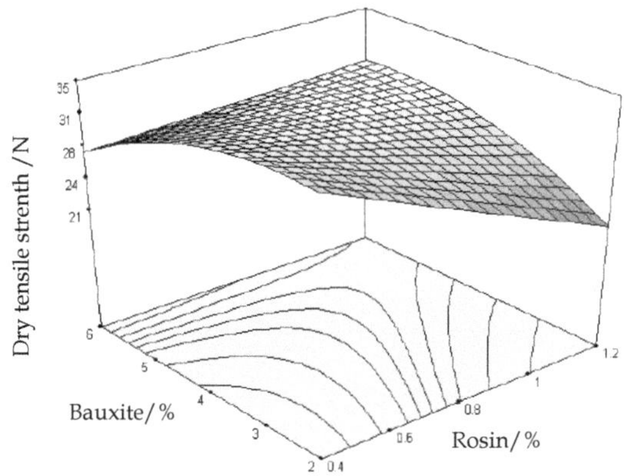

Fig. 3-5. Response surface and contour plots for the effects of rosin and bauxite on dry tensile strength: beating degree was held at 40SR°, grammage was held at 80 g/m^2, wet strength agent was held at 1.8%

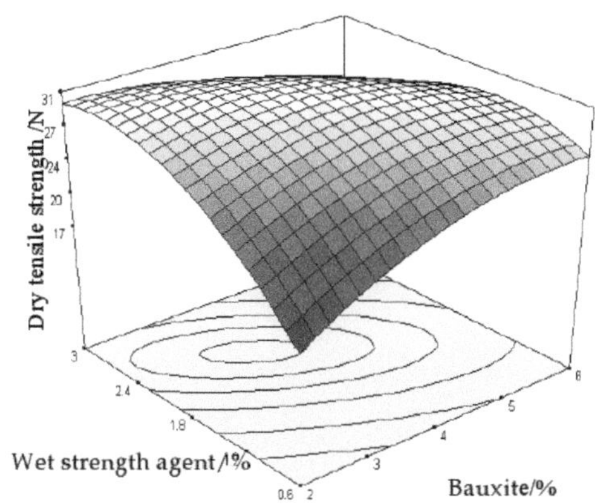

Fig. 3-6. Response surface and contour plots for the effects of bauxite and wet strength agent on dry tensile strength: beating degree was held at 40SR°, grammage was held at 80 g/m^2, rosin was held at 0.8%

3.4 Effect of interaction factors on wet tensile strength

3.4.1 Effect of beating degree and rosin on wet tensile strength

Fig.3-7 showed the effect of beating degree and rosin on wet tensile strength when other factors were held at 0 level. When the beating degree was lover than 0 level, wet tensile strength decreased with the increase of rosin; when the beating degree was higher than 0 level, wet tensile strength increased with the increase of beating degree and rosin, the maximum value occurred when beating degree was held at 30SR°, and rosin was held at 0.4%. This is because added rosin impacted adsorption effect of fibre to wet strength agents, wet tensile strength decreased, but with the continuing increase of beating degree, the fibre sub-wire broom degree further enhanced, the adsorption effect of fibre on wet strength agent was over than the rosin.

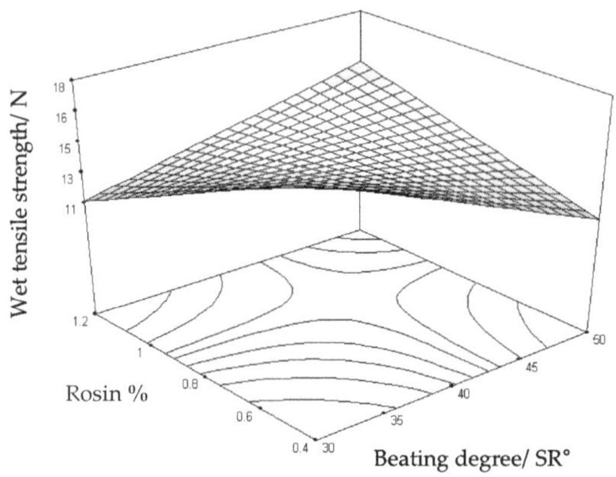

Fig. 3-7. Response surface and contour plots for the effects of beating degree and rosin on wet tensile strength: grammage was held at 80 g/m^2, bauxite was held at 4%, wet strength agent was held at 1.8%

3.4.2 Effect of grammage and wet strength agent on wet tensile strength

Fig.3-8 showed the effect of grammage and wet strength agent on wet tensile strength when other factors were held at 0 level. Wet tensile strength gradually increased with the increase

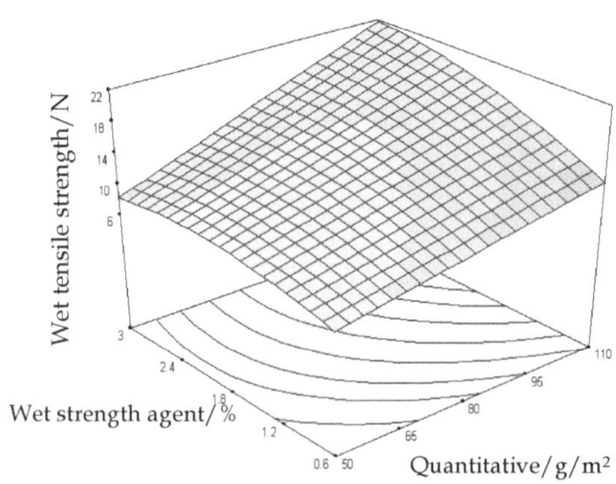

Fig. 3-8. Response surface and contour plots for the effects of grammage and wet strength agent on wet tensile strength: beating degree was held at 40SR°,rosin was held at 0.8%, bauxite was held at 4%

of the grammage and wet strength agent; the maximum occurred when wet strength agent was held at 3%, and grammage was held at 110 g/m2, this is because the number of fibre increased and bonding effect of fibre enhanced, when wet tensile strength increased with the increase of grammage, at the same time, wet strength agent provided cationic charge, fibre strongly adsorbed wet strength agent added because of pulp fibre with anionic charge, so that the wet tensile strength of film increased.

3.4.3 Effect of rosin and wet strength agent on wet tensile strength

Fig.3-9 showed the effect of rosin and wet strength agent on wet tensile strength when other factors were held at 0 level. When rosin was lover than 0 level ,wet tensile strength increased with the increase of wet strength agent; When rosin was higher than 0 level, the increase of wet tensile strength became flat with wet strength agent increased, the maximum occurred when wet strength agent was held at 3%, and rosin was held at 0.4%. The reason was that the increase of rosin, affected the adsorption of the fibre to wet strength agent, to a certain extent, reduced the effect of wet strength agent, wet tensile strength would not increase or decrease.

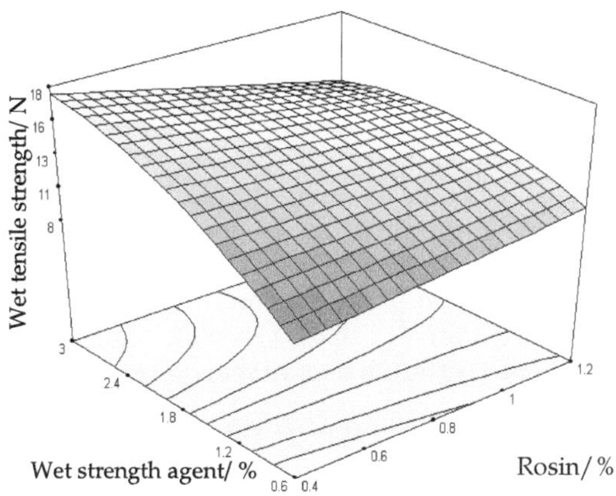

Fig. 3-9. Response surface and contour plots for the effects of rosin and wet strength agent on wet tensile strength: beating degree was held at 40SR°, grammage was held at 80 g/m2, bauxite was held at 4%

3.4.4 Effect of bauxite and wet strength agent on wet tensile strength

Fig.3-10 showed the effect of bauxite and wet strength agent on wet tensile strength when other factors were held at 0 level. When bauxite was at any level, wet tensile strength gradually increased with the increase of wet strength agent ;When bauxite was lower than 0 level, wet tensile strength increased, When bauxite was higher than 0 level, wet tensile strength decreased, the maximum occurred when wet strength agent was held at 3%, and bauxite was held at 4.5%.The reason was that the increased amount of added bauxite in the slurry system, leaded to adsorption of anionic trash in fibre system, affected the adsorption to the wet strength agent , resulted in decrease of wet tensile strength.

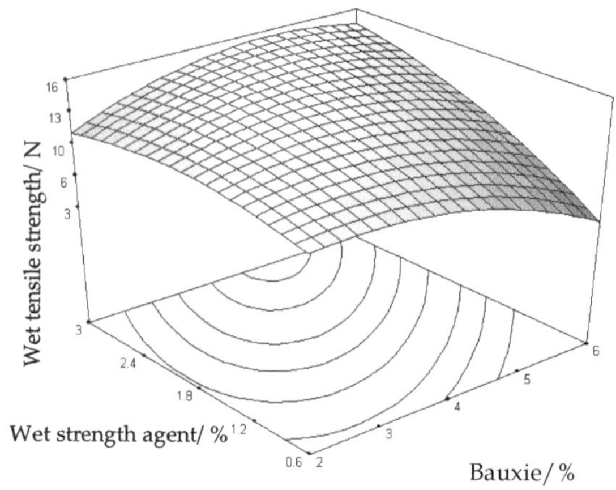

Fig. 3-10. Response surface and contour plots for the effects of bauxite and wet strength agent on wet tensile strength: beating degree was held at 40SR°, grammage was held at 80 g/m2,rosin was held at 0.8%

3.5 Effect of interaction factors on degradation period

3.5.1 Effect of grammage and bauxite on degradation period

Fig.3-11 showed the effects of grammage and bauxite on degradation period when other factors were held at 0 level. Degradation period gradually increased with the increase of bauxite and grammage, the maximum occurred when bauxite was held at 6%, and grammage was held at 110 g/m2, because of the increase of grammage, the number of fibres grew, the bonding capacity between fibres enhanced, the amount of bauxite increased, the ability of fibre that absorbing additives enhanced, which made dry tensile strength and wet tensile strength of film become larger, the degradation period of film increase.

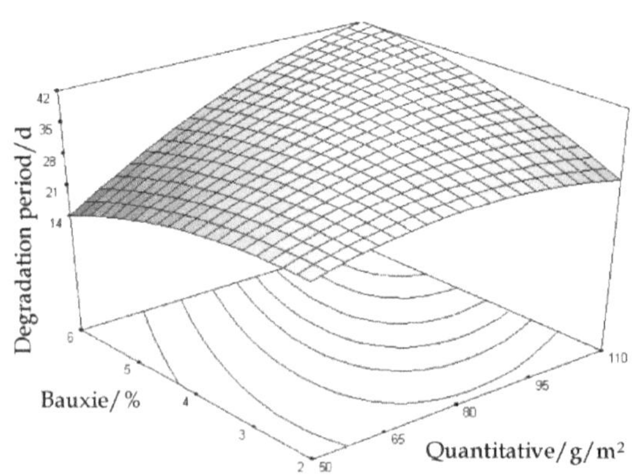

Fig. 3-11. Response surface and contour plots for the effects of grammage and bauxite on the degradation period: beating degree was held at 40SR°, rosin was held at 0.8%,wet strength agent was held at 1.8%

3.5.2 Effect of bauxite and wet strength agent on degradation period

Fig.3-12 showed the effects of bauxite and wet strength agent on degradation period when other factors were held at 0 level. Degradation period gradually increased with the increase of bauxite and wet strength agent, the maximum occurred when bauxite was held at 6%, and wet strength agent was held at 3%, wet strength agent at a suitable amount could improve the wet tensile strength of film, and make degradation period grow.

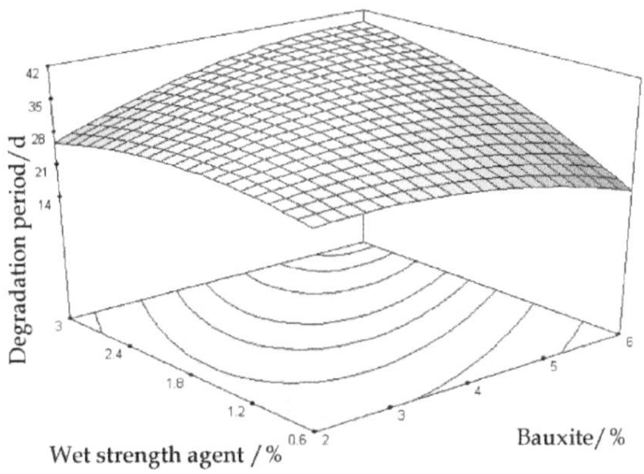

Fig. 3-12. Response surface and contour plots for the effects of bauxite and wet strength agent on the degradation period: beating degree was held at 40SR°, grammage was held at 80 g/m2,rosin was held at 0.8%

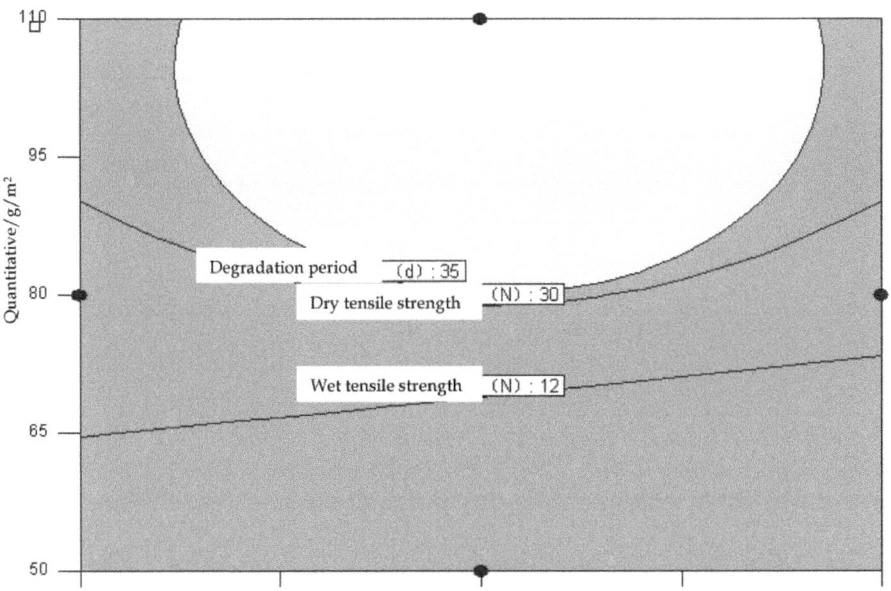

Fig. 3-13. Optimum analysis plot: rosin was held at 0.8%, bauxite was held at 4%, wet strength agent was held at 1.8%.

3.6 Optimization

The rule of optimization based on film performance meeting the agronomic requirement to reduce energy consumption and save raw materials as much as possible was applied to determine the optimum combination of the factors. The result was that when rosin was held at 0.8%, bauxite was held at 4%, wet strength agent was held at 1.8%, beating degree was held at 35 SR°, grammage was held at 80 g/m^2, the performance that was dry tensile strength was greater than 30N, wet tensile strength was greater than 12N, the degradation period was 35 days to 60 days could be obtained, seeing in Fig 3-13.

3.7 Manufacturing technology

Based on the above research results, manufacturing technology of the biogas residue fibre film was obtained, seeing Fig 3-14.

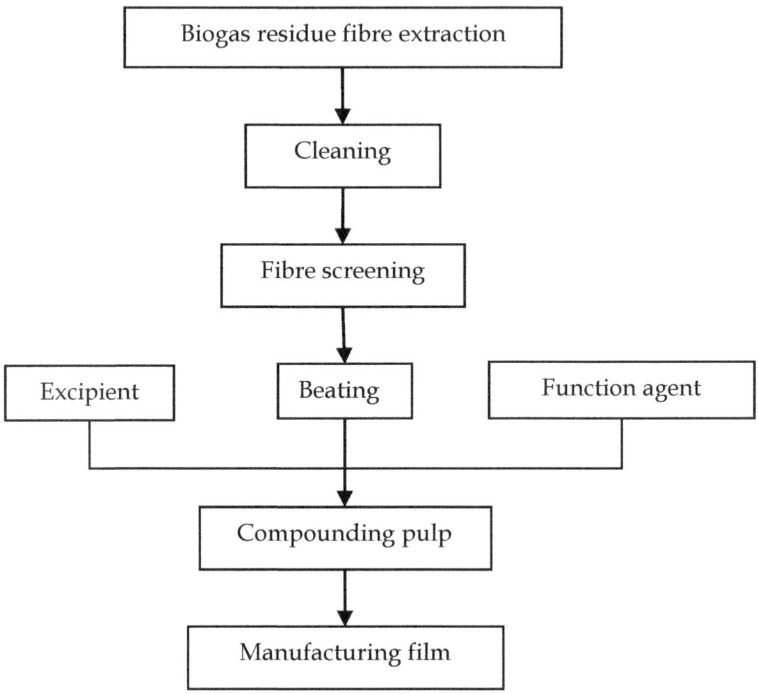

Fig. 3-14. Manufacturing technology of the biogas residue fibre film

4. The experiment of cultivating eggplants with biogas residue fibre film

In order to test the performance differences among the three kinds of biogas residue fibre mulch, black biodegradable mulching added in activated carbon, plastic film and control (without mulching), to cultivate eggplant, a comparative field test through the multiple comparisons for the soil moisture content, soil temperature, weed growth amount and eggplant yield was employed. Performance index of the treatments was shown in table4- 1.

Treatment	Grammage g/m²	Dry tensile strength N	Wet tensile strength N	Thickness mm	Width cm
A	65	33.70	12.46	0.248	70
B	80	36.30	14.40	0.316	70
C	100	30.69	10.40	0.385	70
D	55	39.16	13.46	0.099	70
E	9	1.50	-	0.011	70

Notes: A,B,C were biogas residue fibre mulch of different grammage, respectively; D was black biodegradable mulching added in activated carbon; E was plastic film.

Table 4-1. Performance index of the treatments

4.1 Effect of different treatments on the weed amount

The effect scene of different treatments on the weed amount in the case of cultivating eggplant was shown in Fig.4-1. Weed amount of mulching was markedly less than the control, and weed amount of plastic film was more than biogas residue films. In the period of observation, weeds were flourished under the plastic film or even broke plastic mulch. Weeds mainly grew from transplanted hole and rupture to all the treatments, and making the film rupture expands.

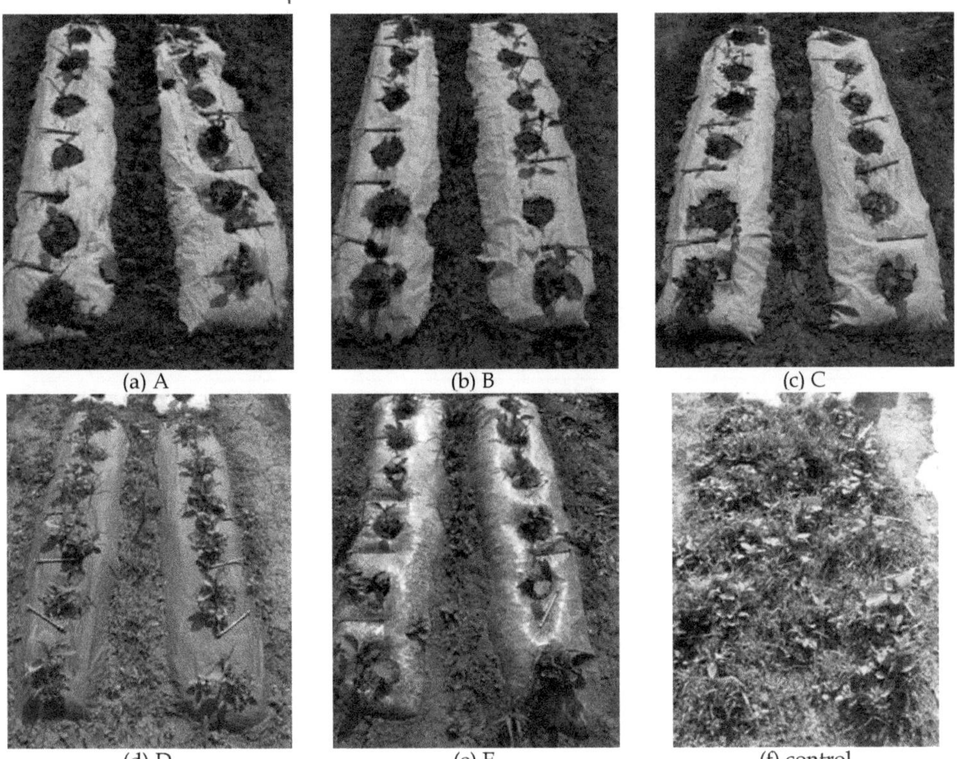

(a) A (b) B (c) C (d) D (e) E (f) control

Fig. 4-1. The effect scene of different treatments on the weed amount in the case of cultivating eggplant

The measured results of total weeds of each treatment were shown in table 4-2.

Interval\Treatment	Weed amount (g)					
	A	B	C	D	E	F
1	63.01	79.02	48.15	69.19	430.7	528
2	90.02	66.88	79.26	72.6	204.99	376.05
3	88.96	79.06	79.62	87.38	328.85	358.63
Mean	80.66	74.99	69.01	76.39	321.51	420.89

Table 4-2. The measured results of total weeds of each treatment

Analysis of variance of the weed amount was shown in table 4- 3.

Source	SS	DF	MS	F	P-value	$F_{0.05}$
Interval	9089.895	2	4544.947	1.290108	0.317365	4.1028
Treatment	365347.3	5	73069.46	20.74117	5.51E-05	3.3258
Error	35229.19	10	3522.919			
Total	409666.4	17				

Table 4-3. Analysis of variance of the weed amount

There were significant differences among the treatments, according to analysis of the multiple comparisons among treatments, seeing table 4-4.

Treatment	Mean difference (g)						LSD 0.05	LSD 0.01
	$\bar{y}_i$	$\lvert\bar{y}_i - \bar{y}_F$	$\lvert\bar{y}_i - \bar{y}_E\rvert$	$\lvert\bar{y}_i - \bar{y}_D\rvert$	$\lvert\bar{y}_i - \bar{y}_C\rvert$	$\lvert\bar{y}_i - \bar{y}_B\rvert$		
A	80.66	340.23**	240.85**	4.27	11.65	5.67	108	154
B	74.99	345.9**	246.52**	1.4	5.98			
C	69.01	351.88**	252.5**	7.38				
D	76.39	344.5**	245.12**					
E	321.51	99.38						
F	420.89							

Notes: $\bar{y}_i$ was mean of the i treatment; $LSD_{0.05}$ and $LSD_{0.01}$was significant at 0.05, not significant at 0.01.

Table 4-4. Multiple comparisons of the weed amount among treatments

The results showed that there were significant differences between three kinds of biogas residue fibre film, black film and plastic film, control; there were significant differences between three kinds of biogas residue fibre films and black film. Weed amount of A biogas residue fibre film decreased 81% as compared with control, and decreased 75% as compared with the plastic film; weed amount of B biogas residue fibre film decreased 82% as compared with control, and decreased 77% as compared with the plastic film; weed growth amount of C biogas residue fibre film decreased 84% as compared with control, and decreased 79%as compared with the plastic film. From this, three kinds of residue fibre films had significant effect of suppressing weeds.

4.2 Effect of different treatments on soil moisture

The soil moisture content of each treatment and analysis of variance of the soil moisture content were respectively shown in table 4-5 and table 4-6.

Interval\Treatments	A	B	C	D	E	F
1	19.49	19.29	19.31	18.06	20.97	17.82
2	20.53	20.33	19.35	18.25	19.43	15.4
3	19.21	18.52	20.15	18.15	21.04	16.47
Mean	19.74	19.38	19.61	18.16	20.48	16.56

Table 4-5. The soil moisture content of each treatment

Source	SS	DF	MS	F	P-value	$F_{0.05}$
Interval	0.2636	2	0.1318	0.1777	0.8398	4.1028
Treatments	29.7137	5	5.9427	8.0105	0.0028	3.3258
Error	7.4187	10	0.7419			
Total	37.3960	17				

Table 4-6. Analysis of variance of the soil moisture content

Table 4-6 showed that there were significant differences among treatments. Multiple comparisons of the soil moisture content among treatments were shown in table 4-7.

Treatment	Mean difference (%)						$LSD_{0.05}$	$LSD_{0.01}$
	$\bar{y}_i$	$\lvert\bar{y}_i - \bar{y}_I$	$\lvert\bar{y}_i - \bar{y}_I$	$\lvert\bar{y}_i - \bar{y}_I$	$\lvert\bar{y}_i - \bar{y}_C$	$\lvert\bar{y}_i -$		
A	19.74	3.18**	0.74	1.58*	0.13	0.36	1.56	2.23
B	19.38	2.82**	1.1	1.22	0.23			
C	19.61	3.05**	0.87	1.45*				
D	18.16	1.6*	2.32**					
E	20.48	3.92**						
F	16.56							

Notes: The significance of symbols was as same as table 4-4.

Table 4-7. Multiple comparisons of the soil moisture content among treatments

The results showed that there were significant differences between three kinds of biogas residue fibre film, plastic film and control; there were significant differences between black film and control; there were significant differences between plastic film and black film. There were significant differences between A, C of biogas residue fibre film and black film;

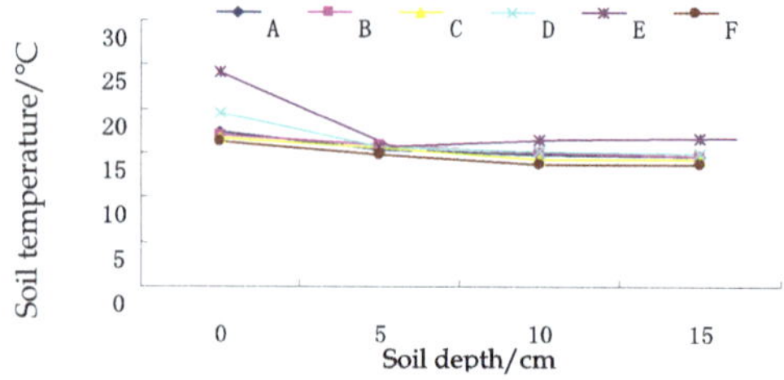

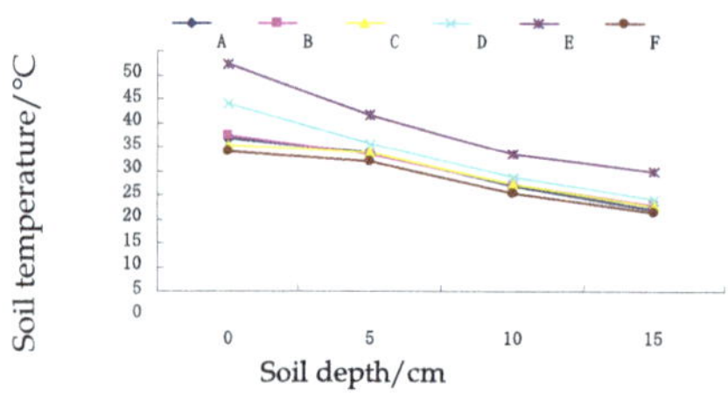

(b) Seedling stage pm14:00

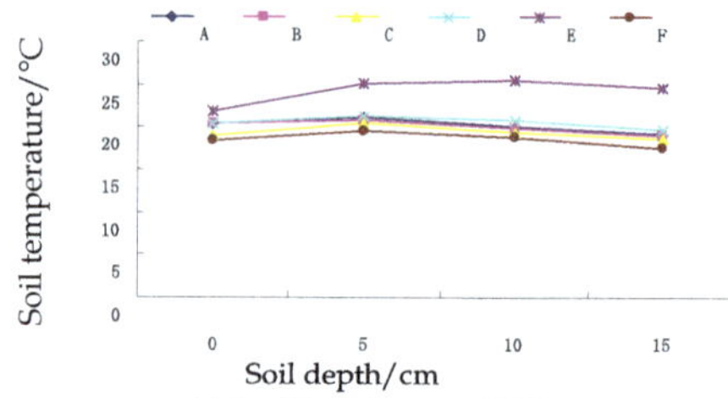

(c) Seedling stage pm 19:00

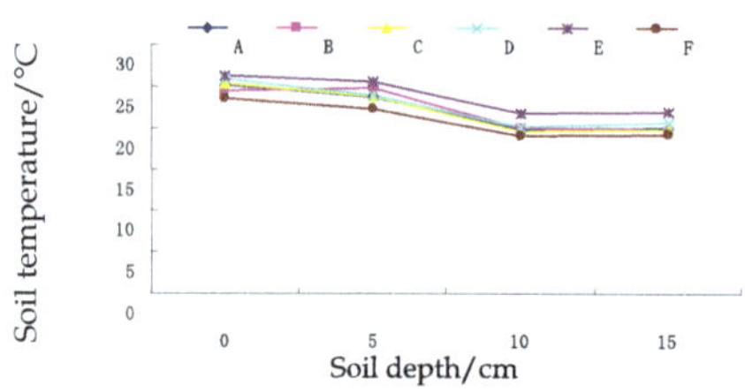

(d) Florescence am 8:00

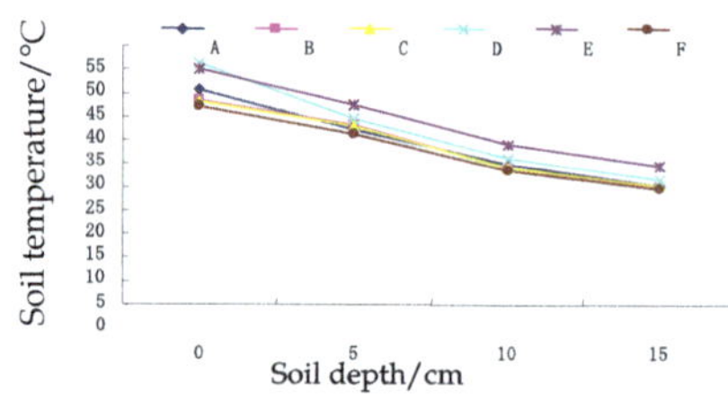

(e) Florescence pm 14:00

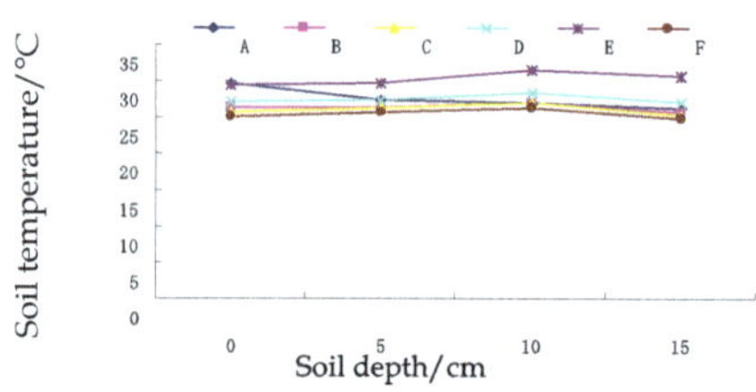

(f) Florescence pm 19:00

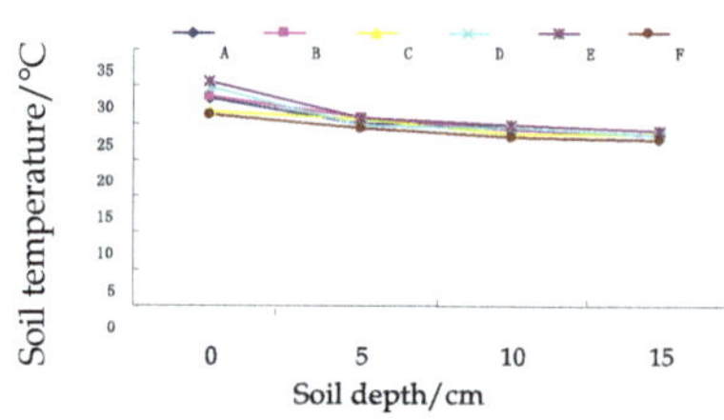

(g) Maturity am 8:00

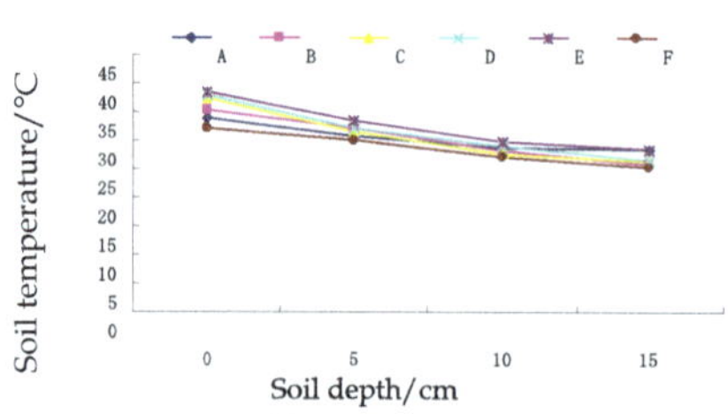

(h) Maturity pm 14:00

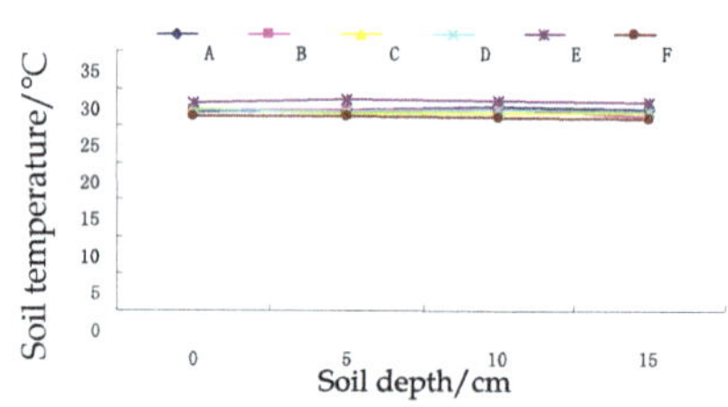

(i) Maturity pm 19:00

Fig. 4-2. The soil temperature that varied with soil depth of the six treatments

there were not significant differences between B of biogas residue fibre film and black film; there were not significant differences between three kinds of biogas residue fibre film and plastic film. It can be seen, soil moisture of biogas residue fibre film was significantly higher than control and black film, and there were not significant difference as compared with plastic film.

4.3 The effect of different treatments on soil temperature

4.3.1 The effect of soil depth on soil temperature among treatments in different growth stages

The soil temperatures that varied with soil depth of the six treatments were shown in Fig.4-2.

Figure 4-2 showed that the mulching temperature was higher than the control during the whole growth period, the temperature of plastic film was the highest, the temperature of black film was higher than the biogas residue fibre film, and the temperature of A, B, C of the biogas residue fibre film was slightly higher than the control. It could be seen from the temperature curve of am7:00 and pm 14:00, soil temperature gradually decreased with the soil deepening, but, the temperature of soil surface decreased at pm 19:00, the soil temperature slightly increased with soil deepening in the stage of revival and flowering, and the soil temperature was constant in maturity; mulching had a certain warming effect in the stage of revival and flowering, and had not warming effect in maturity, the reason was that the film had been degraded.

4.3.2 Effect of different treatments on total accumulated temperature

The measured results of the total accumulated temperature of each treatment were shown in table 4-8.

Soil depth cm	Interval\treatments	A	B	C	D	E	F
0	1	435.6	438.4	427.5	463.2	526.9	396.1
	2	408.7	403.9	415.8	430	479.6	401.8
	3	418.1	410.2	369.6	431.3	511.8	369.1
	Mean	420.8	417.5	404.3	441.5	506.1	389
5	1	395	380.5	395.6	410	435.7	380.8
	2	393.1	397.2	386.8	392.4	425.4	377.6
	3	380.7	397.4	391.8	405.7	446.5	370.5
	Mean	389.6	391.7	391.4	402.7	440.1	374.7
10	1	356.5	356	343.2	366.4	381.3	350.4
	2	365.8	367.1	358.5	378.2	407.2	335.7
	3	362.8	354.8	359.4	367.2	413	344.4
	Mean	361.7	359.3	353.7	370.6	400.5	343.5
15	1	339.2	332.4	330.5	346.7	370.4	337.1
	2	348.1	346	346.6	359.8	385.5	342.3
	3	342.9	341.3	335.4	348.6	398.8	323.8
	Mean	343.4	339.9	337.5	351.7	384.9	334.4

Table 4-8. The total accumulated temperature of each treatment

Analysis of variance of the total accumulated temperature of each treatment was shown in table 4-9.

Soil depth cm	Source	SS	DF	MS	F	P-value	$F_{0.05}$
0	Interval	3016.57	2	1508.29	6.3092	0.0169	4.1028
	Treatment	25517.2	5	5103.44	21.3477	4.84E-05	3.3258
	Error	2390.63	10	239.06			
	Total	30924.4	17				
5	Interval	58.83	2	29.42	0.3999	0.6806	4.1028
	Treatment	6241.93	5	1248.39	16.9726	0.000133	3.3258
	Error	735.53	10	73.55			
	Total	7036.3	17				
10	Interval	324.96	2	162.48	2.1843	0.1633	4.1028
	Treatment	5774.55	5	1154.91	15.526	0.000195	3.3258
	Error	743.86	10	74.39			
	Total	6843.37	17				
15	Interval	432.25	2	216.13	4.113	0.0497	4.1028
	Treatment	5264.62	5	1052.92	20.037	6.42E-05	3.3258
	Error	525.47	10	52.55			
	Total	6222.34	17				

Table 4-9. Analysis of variance of the total accumulated temperature of each treatment

Multiple comparisons of the total accumulated temperature of different soil depths among treatments were shown in table 4-10.

Table 4-10 showed that there were significant differences between plastic film and three kinds of biogas fibre residue film, black film and control while the soil depth was 0cm and 10cm. There were significant differences between the black film and control; there were significant differences between A, B treatment of biogas residue fibre film and control; there were no significant differences between C treatment of biogas residue fibre mulch and control; there were significant differences between black film and C treatment of biogas residue fibre film. While the soil depth was 5 cm, there were significant differences between plastic film and three kinds of biogas fibre residue film, black film and control; there were significant differences between A, B treatment of biogas fibre residue film, black film and control. While soil depth was 15 cm, there were significant differences between plastic film and three kinds of biogas fibre residue film, black film and control; there were significant differences between the black film and C treatment of biogas residue fibre film and control.

While the soil depth was 0 cm, 5 cm, 10 cm, the total accumulated temperature of A, B treatment of biogas residue film obviously increased than control, and C did not significantly increase; while the soil depth was 0cm, the total accumulated temperature of A treatment increased 31.8 °C, B increased 28.5 °C, and C increased 15.3°C than control; while the soil depth was 5cm, the total accumulated temperature of A treatment increased 17°C, B increased 17.3°C, and C increased 15.2°C than control; while the soil depth was 10cm, the total accumulated temperature of A treatment increased 18.2 °C, B increased 15.8 °C, and C increased 10.2 °C than control. While the soil depth was 15 cm, the total accumulated temperature of three kinds of biogas fibre residue film nearly closed to control, and was less than the black film and plastic film. There was no significant difference between the three kinds of biogas residue fibre film.

Soil depth (cm)	Treat ment	Mean difference (℃)						LSD0.05	LSD0.01
		$\bar{y}_i$	$\lvert\bar{y}_i-\bar{y}_F\rvert$	$\lvert\bar{y}_i-\bar{y}_E\rvert$	$\lvert\bar{y}_i-\bar{y}_D\rvert$	$\lvert\bar{y}_i-\bar{y}_C\rvert$	$\lvert\bar{y}_i-\bar{y}_B\rvert$		
0	A	420.8	31.8*	85.3**	20.7	16.5	3.3	28.13	40.01
	B	417.5	28.5*	88.6**	24	13.2			
	C	404.3	15.3	101.8**	37.2*				
	D	441.5	52.5**	64.6**					
	E	506.1	117.1**						
	F	389							
5	A	391.4	17*	48.7**	11.3	1.8	0.3	15.6	22.19
	B	391.7	17.3*	48.4**	11	2.1			
	C	389.6	15.2	50.5**	13.1				
	D	402.7	28.3*	37.4**					
	E	440.1	65.7**						
	F	374.4							
10	A	361.7	18.2*	38.8**	8.9	8	2.4	15.68	22.32
	B	359.3	15.8*	41.2**	11.3	5.6			
	C	353.7	10.2	46.8**	16.9*				
	D	370.6	27.1**	27.1**					
	E	400.5	57**						
	F	343.5							
15	A	343.4	9	41.5**	8.3	5.9	3.5	13.19	18.76
	B	339.9	5.5	45**	11.8	2.4			
	C	337.5	3.1	47.4**	14.2*				
	D	351.7	17.3*	33.2**					
	E	384.9	50.5**						
	F	334.4							

Notes: The significance of symbols was as same as table 4-4.

Table 4-10. Multiple comparisons of the total accumulated temperature of different soil depths among treatments

4.4 The effect of different treatments on eggplant yield

Analysis of variance of the yield was shown in table 4-11.

Source	SS	DF	MS	F	P-value	$F_{0.05}$
Interval	0.5030	2	0.2515	2.4037	0.1405	4.1028
Treatment	3.7059	5	0.7412	7.0833	0.0045	3.3258
Error	1.04637	10	0.1046			
Total	5.2553	17				

Table 4-11. Analysis of variance of the yield date

Table 4-11 showed that there were significant differences between the yields of each treatment. Multiple comparisons of the eggplant yield among treatments were shown in table 4-12.

Treatment	Mean difference (kg)						LSD0.05	LSD0.01
	$\bar{y}_i$	$\lvert\bar{y}_i - \bar{y}_F\rvert$	$\lvert\bar{y}_i - \bar{y}_E\rvert$	$\lvert\bar{y}_i - \bar{y}_D\rvert$	$\lvert\bar{y}_i - \bar{y}_C\rvert$	$\lvert\bar{y}_i - \bar{y}_B\rvert$		
A	1.95	0.42	1.00**	0.50	0.07	0.30	0.58	0.82
B	2.25	0.72*	0.70*	0.20	0.37			
C	1.88	0.35	1.07**	0.57				
D	2.45	0.92**	0.50					
E	2.95	1.42**						
F	1.53							

Notes: The significance of symbols were as same as table4-4.

Table 4-12. Multiple comparisons of the eggplant yield among treatments

The results showed that there were significant differences between B treatment of biogas residue fibre film and control, no significant differences between A, C treatment and control, significant differences between black film and control. There were extremely significant differences between the plastic film, A, C treatment and control, there were significant differences between plastic film and B treatment, no significant differences between plastic film and the black film. It can be seen, the yield of B treatment increased 47% as compared with control.

5. Conclusions

1. Biogas residue are mainly composed of the fiber, non-metallic minerals, minerals, their mass percentage are 64%, 35% and 1% respectively; cellulose, hemicellulose, lignin and ash are 44.8%,21.9%,15.6% and 17.7% respectively in the biogas residue fiber. The mass percent of biogas residue cellulose is over 5% more than that of rice straw, wheat straw and corn stalk; the mass percent of the hemicellulose is 5% less than that of rice straw, wheat straw and corn stalk.
2. An optimum factors combination is rosin 0.4%, bauxite 4% and wet tensile strength 1.8%, beating degree 40SR°, grammage 80 g/m^2, in this case, for the biogas residue fibre film, dry tensile strength can attain more than 30N , wet tensile strength can attain more than12N, the degradation period can attain 35 days to 60 days.
3. The rank of importance of the five factors on the dry tensile strength: grammage, beating degree, wet strength agent, bauxite and rosin; on the wet tensile strength: wet strength agent, bauxite, rosin, grammage and beating degree; on the degradation: grammage, wet strength agent, bauxite, rosin and beating degree.
4. Biogas residue fibre film has significant effect against weeds as compared with plastic film and bare field.
5. There was no significant difference of conserving moisture between biogas residue fiber mulching and the plastic film.
6. In the period of eggplant growth, the rank of the total accumulated temperature: plastic film, black film, biogas residue fibre film and bare field.
7. The eggplant yield of biogas residue fiber mulching whose grammage is 80g/m^2 is 47% more than bare field, and slightly lower than that of the plastic film.
8. It is feasible to make biodegradable mulching from the biogas residue, in accordance with the agronomic requirement

6. References

An Juan, Zhao Xiaogang, J. (2005). Several factors of the effect of ruminant digestion on cellulose Food industry 2:21-23 (in Chinese)

Bian Yousheng , M. (2005) The waste processing and recycling in ecological agriculture Chemical industry press 533-542 (in Chinese)

Chen Hongzhang, M. (2008) Biomass science and engineering Chemical industry press 89-132 (in Chinese)

C. Namasivayam, R.T.Yamuna, J. (1995) Adsorption of Chromium (VI) by a Low-cost Adsorbent :Biogas Residual Slurry Chemosphere, 30 (3) : 561-578

C. Namasivayam, R.T.Yamuna, J. (1995) Adsorption of Direct Red 12 B by Biogas Residual Slurry: Equilibrium and Rate Processes Environmental Pollution, 89:1-7

C. Namasivayam, R.T.Yamuna , J. (1995) Waste Biogas Residual Slurry as an adsorbent for the Removel of Pb (II) from Aqueous solution and Radiator Manufacturing Industry Wastewater Bioresource Technology, 52 (2) : 125-131

Fu dengqiang, D. (2008) Water conservation and heat preservation properties of bast fiber film mulching and its effects on crops Chinese academy of agricultural sciences (in Chinese)

Gao Yujie, J. (1996). Plant fibre paper biodegradable Paper and paper making 45 (in Chinese)

Gao Yujie, J. (1998). Paper-based practical application of agricultural plastic film Tianjin paper making 11-13 (in Chinese)

Gao Zhenhua, Di Mingwei, M. (2008). Biomass material and applications Chemical industry press 273-281 (in Chinese)

Gao Qiang, Niu Dongjie, Cheng Haijing, J. (2005). Comprehensive utilizing of biogas residue China resources comprehensive utilization. (12) : 11-15 (in Chinese)

Han Yongjun, Chen Haitao, J. (2008) .A review on the studies and applications of mulching paper Journal of agricultural mechanization research 12:244-246 (in Chinese)

He Aijun, J. (2002) Development of degradable plastic New chemical materials 30 (3) : 1-6 (in Chinese)

Jia Weigui, Pan Xuefeng, Kang Binglong, Liu Rongzhong, J. (2000) Effect of corn mulching degradation Anhui agricultural science bulletin 6 (4) : 32-33 (in Chinese)

Jin Yi, M (2007) Experimental design and statistical analysis China agriculture press 139-157 (in Chinese)

KAM Abd I J-Kader, S F Abdel Hamied, J. (2002) .Preparation of Poly (vinyl alcohol) Films with Promising Physical Properties in Comparison with Commercial Polyethylene film Joumal of Applied Polymer Scinence, 86 (5) : 1219-1226

Li Xianfa, J. (2004). Study status on degradable mulching film Plastics 33 (1): 76-81 (in Chinese)

Li Chunxi, Jiang Lina, Shao Yun,Wang Wenlin, M . (2005) Biostatistics 186-220 (in Chinese)

Li Zhiming, Wu Minge, Zhou Jinghui, J. (2004) Kenaf Mulch:Heat Preservation and Biodegradation China pulp&paper 23 : 20-22 (in Chinese)

Lin Jianfeng, J. (2003). Apptication technique for biogas fermenting product Renewable energy 4:36-38 (in Chinese)

Liu Junke, M. (2000) .Plastic waste recycling and use technology China petrochemical press 26-27 (in Chinese)

Liu Ming, Huang Zhanbin, Yang Yujiao, J. (2008) A study on status and developmental trend of biodegradable plastic film Chinese agricultural science bulletin 24 (9) : 439-443 (in Chinese)

Lu Mei, Mao Yurong, Yang kanglin, J. (2007) The use of biogas slurry and biogas residue Agricultural services 24(5) : 37-39 (in Chinese)

Qiao Haijun, D. (2007) Degradation of BiodegradableMulch Films and Its Effects on the Growth of Corn Gansu agricultural university master's degree thesis (in Chinese)

Qiu Weiyang, Qiu Xianping, M. (2002) Starch plastics Beijing chemical industry publishing 116-117 (in Chinese)

R. K. Gupta, V. R. Sharma, K. N. Shrma, J. (2002) Increase the Yield of Paddy and Wheat with the Application of Biogas Slurry Progressive Farming, 39 (10) : 22-24

S. Ranik, R.Nandi, J. (2004) Effect of Supplementation Ofrice Straw with Biogas Residual Slurry Manure on the Yield.Protein and Mineral Contents of Oyster Mushroom Industrial Crops and Products, (20) :311-319

Sun Jianping, Chen Xinghua, Hu Youhui, J. (2000) PROGRESS IN THE REASEARCH OF AGRICUL AND TECHNOLOGY 1:25 (in Chinese)

Tan Chengrong, Wang Shuangfei,Song Hainong, J. (2002) Preparation of degradable mulch film frombagasse Modern Chemical Industry 22 (11) : 24-28 (in Chinese)

Tan Xin, D. (2008) Research on overall evaluation system of large and medium-sized bio-gas project Beijing chemical university (in Chinese)

Wang Wei, Yu Chunling, Bian Guiqin, Liu Baojing, Li Wenjun , J. (2009) The degradation experimental study of plant fiber regeneration film Environmental protection and circular economy 29-30 (in Chinese)

Wang Weigang, J. (2003) The degradation process and degradation test report of biodegradable plastic film Xinjiang Agricultural Science and Technology 41-42 (in Chinese)

Xiong hanguo, Zhang Meiling, J. (2004) Biodegradable plastic field characteristics research [J] Hubei agricultural sciences 5 : 37-38 (in Chinese)

Xu Xiangchun, Wang Chaoyun, J. (2006) The status and development trend of cultivation mulch film at home and abroad China flax industry,1:6-11 (in Chinese)

Xu Shiwen, Jin Liangzao, J. (1987) Fertilizer effect of biogas fermentation residue China Biogas,2:1-3.

Yang Huidi, M. (2000) Plastic film and ecological environment protection Chemical industry press 110-113

Zhang Chunhong, Wang Ronghua, Chen Qiuling, Sun kewei, J. (2007). Development of application of plant fiber in full-biodegradable composites..Materials review.21(10): 35-38

Zhang Quanguo, M. (2005) Biogas technology and application Chemical industry press 123-156 (in Chinese)

Zhang Wenqun, Jin Weixu, J. (1994) Degradation fragments and soil water movement SOILS ANDFERTILIZERS 3: 12-15 (in Chinese)

Zhang Wudi, Song Hongchuan, Yin Fang, M. (2003) The comprehensive utilization of biogas fermentation residue Yunan science and technology press 35-45 (in Chinese)

Zhang Ying, Wang Xiaohui, M. (2005) Agricultural solid waste utilization Chemical industry press 120-124 (in Chinese)

Zhang Yongmei, M. (2008) New rural construction biomass energy utilization China electric power press 109-112 (in Chinese)

12

Digestate: A New Nutrient Source - Review

Marianna Makádi, Attila Tomócsik and Viktória Orosz
Research Institute of Nyíregyháza, RISF, CAAES, University of Debrecen, Hungary

1. Introduction

Digestate is the by-product of methane and heat production in a biogas plant, coming from organic wastes. Depending on the biogas technology, the digestate could be a solid or a liquid material.

Digestate contains a high proportion of mineral nitrogen (N) especially in the form of ammonium which is available for plants. Moreover, it contains other macro- and microelements necessary for plant growth. Therefore the digestate can be a useful source of plant nutrients, it seems to be an effective fertilizer for crop plants. On the other hand, the organic fractions of digestate can contribute to soil organic matter (SOM) turnover, influencing the biological, chemical and physical soil characteristics as a soil amendment.

Besides these favourable effects of digestate, there are new researches to use it as solid fuel or in the process of methane production.

2. Origin of digestate

For protection of the environment, the recycling of organic materials has essential role. The anaerobic digestion (AD) is an important method to decrease the quantity of organic wastes by utilization them for energy and heat production. The by-product of this process is the digestate.

In an AD process, different organic materials could be used alone or in mixture of animal slurries and stable wastes, offal from slaughterhouse, energy crops, cover crops and other field residues, organic fraction of municipal solid wastes (OFMSW), sewage sludge. The quality of digestate as a fertilizer or amendment depends not only on the ingestates but also on the retention time. The longer retention time results in less organic material content of the digestate because of the more effective methanogenesis (Szűcs et al., 2006).

Biogas technology is known to destroy pathogens. The thermophilic AD increases the rate of elimination of pathogenic bacteria, therefore the amounts of fecal coliforms and enterococcus fulfilled the requirements of EU for hygienic indicators (Paavola & Rintala, 2008). Mesophilic digestion alone may not be adequate for correct hygienization, it needs a separate treatment (70 °C, 60 min., particle size<12 mm) before or after digestion, especially in the case of animal by-products (Bendixen, 1999; Sahlström, 2003).

Two types of digestate are the liquid and the solid ones which are distinguished on the bases of their dry matter (DM) content. The liquid digestate contains less than 15% DM content, while the solid digestate contains more than 15% DM. Solid digestate can be used similar to the composts or could be composted with other organic residues and can be more economically transported over grater distances than the liquid material (Møller et al., 2000).

3. Composition of digestate

The quality of a digestate is determined by the digestion process used and the composition of ingestates therefore the agricultural use and efficacy of the nascent materials could be different. Nevertheless, some common rules can be found in the course of the digestion process which allow us to evaluate the results of a digestion process.

3.1 pH of digestate

Generally, the pH of digestate is alkaline (Table 1). Increases in pH values in the course of the AD may have been caused by the formation of $(NH_4)_2CO_3$ (Georgacakis et al., 1992).

Type of ingestate	Type of digestion process	pH of ingestate	pH of intermedier stage	pH of digestate	Source of data
Pharmaceutical industry sludge	mesophilic, solid type digester	7.0	7.5	7.8	*Gómez et al., 2007*
Cattle manure	mesophilic, liquid type digester	6.9	7.2	7.6	*Gómez et al., 2007*
Primary sludge from municipal waste water treatment plant and organic fractions of municipal solid wastes	thermophilic (co-digestion), liquid type digester	3.5	5.0	7.5	*Gómez et al., 2007*
Energy crops, cow manure slurry and agro-industrial waste	thermophilic (co-digestion), liquid type digester	4.8	7.5	8.7	*Pognani et al., 2009*
Energy crops, cow manure slurry, agro-industrial waste and OFMSW	thermophilic (co-digestion), liquid type digester	4.0	8.1	8.3	*Pognani et al., 2009*

Table 1. Changes of the pH in different digestion systems

The pH is increased under the digesting process, but its range depends on the quality of ingestate and the digestion process. The end values are irrespective of the starting value.

The alkaline pH of digestate is a useful property because of the worldwide problem of soil acidification.

3.2 Macroelement content of digestate

The other characteristics of digestate also are differed depending on the source materials and the digestion process. In Table 2 some major properties of different liquid digestates can be seen, but these are mean values which could be altered in the course of the digestion process. Therefore regular monitoring of digestate properties is needed in the case of its agricultural use.

Type of ingestate	Type of digestion process	Total-N (N_t)	NH_4-N	Total-P	Total-K	Source of data
Swine manure	mesophilic	2.93 (g L^{-1})	2.23 (g L^{-1})	0.93 (g L^{-1})	1.37 (g L^{-1})	*Loria et al., 2007*
Liquid cattle slurry	mesophilic	4.27 (% DM)	52.9 (‰ N_t)	0.66 (% DM)	4.71 (% DM)	*Möller et al., 2008*
Energy crops, cow manure slurry and agro-industrial waste	thermophilic	105 (g kg^{-1} TS)	2.499 (g L^{-1})	10.92 (g kg^{-1} TS)	-	*Pognani et al., 2009*
Energy crops, cow manure slurry, agro-industrial waste and OFMSW	thermophilic	110 (g kg^{-1} TS)	2.427 (g L^{-1})	11.79 (g kg^{-1} TS)	-	*Pognani et al., 2009*
Cow manure, plant residues and offal	mesophilic and thermophilic	0.2013 (%m/m, fresh matter)	0.157 (%m/m, fresh matter)	274.5 mg kg^{-1} (fresh matter)	736.45 mg kg^{-1} (fresh matter)	*Makádi et al.,2008b*
Clover/grass or pea straw or cereal straw or silage maize and clover/grass silage (mean)	mesophilic	0.253 (%m/m, fresh matter)	0.176 (%m/m, fresh matter)	0.62 (% DM)	18.5 (% DM)	*Stinner et al., 2008*

Table 2. Characteristics of liquid digestates from different origin

Nitrogen (N) is a major plant nutrient and is the most common plant growth limiting factor of agricultural crops. The fertilizing effect of added N is decreased by the inadequate synchrony of crop N demand and soil N supply (Binder et al., 1996; Möller & Stinner, 2009). The advantage of digestate application is the possibility of reallocation of the nutrients within the crop rotation from autumn to spring, when crop nutrient demand arises (Möller et al., 2008). The higher N content of a digestate comparing to the composts is the consequence of the N concentration effect because of carbon degradation to CO_2 and CH_4 and N preservation during AD (Tambone et al., 2009).

The NH_4 content of the digestate is about 60-80% of its total N content, but Furukawa and Hasegawa (2006) reported 99% of NH_4-N of the digestate originated from kitchen food wastes. Generally, the NH_4-N concentration is increased by the protein-reach feedstock (Kryvoruchko et al., 2009) like diary by-products and slaughterhouse waste (Menardo et al., 2011). The conversion of organic N to NH_4-N allows its immediate utilization by crops (Hobson and Wheatley, 1992). The higher amount of NH_4-N and the higher pH predominate over the factors (lower viscosity, lower dry matter content) which could reduce the ammonia volatilization from the digestate (Möller & Stinner, 2009). The emission of ammonia could be decreased by different injection techniques which lower the air velocity above the digestate and because of the bound of gaseous ammonia to soil colloids and soil water (McDowell and Smith, 1958). The application depth has a significant effect on ammonia volatilization. Surface application of a liquid biofertiliser caused the loss of 20-35% of the applied total ammoniacal N while disc coulter injection into 5-7 cm depth reduced the ammoniacal loss to 2-3% (Nyord et al, 2008). This method should be used also in the case of digestate application to reduce ammonia volatilization.

Digestate has higher phosphorus (P) and potassium (K) concentration than that of composts (Tambone et al., 2010) therefore it is more suitable for supplement of these missing macronutrients in soils. Furthermore, Börjesson and Berglund (2007) assumed all phosphorus in the digestate to be in available forms, therefore digestate seems to be a useful material for supplement missing nutrients of soil, especially of the P and K. The average phosphorus-potassium ratio of digestates is about 1:3 which is excellent for grain and rape. Accumulation of P and K in soil could be avoided by the reduction of the applied digestate dose but in this case, for the supplement of nitrogen gap, the artificial fertilizer has to be used.

3.3 Microelement content of digestate

Plants, animals and humans require trace amounts of some heavy metals like copper (Cu), zinc (Zn), while others like cadmium (Cd), chromium (Cr), mercury (Hg), lead (Pb) are toxic for them. Heavy metal content of the feedstock usually originates from anthropogenic source and is not degraded during AD. The main origins of the heavy metals are animal feed additives, food processing industry, flotation sludge, fat residues and domestic sewage.

With a N load of 150 kg ha^{-1}, the heavy metals load into the soil (Cd, Cr, Cu, Ni, Pb, Zn) were lower in the case of digestate addition comparing to the compost and sewage sludge treatments while were higher in some heavy metals (Cu, Ni, Pb, Zn) comparing to the mineral fertilizer (Pfundtner, 2002).

3.4 Organic matter content of digestate

The amounts of organic dry matter and the carbon content of digestate are decreased by the decomposition of easily degradable carbon compounds in the digestors (Stinner et al., 2008). Menardo et al. (2001) found the degree of organic matter (OM) degradation between 11.1%

and 38.4% in the case of different ingestates, highest loading rates and hydraulic retention times while Marcato et al. (2008) found this value of 53%. If the organic loading rate of biogas plant is high and the hydraulic retention time is short, the digestate will contain a considerable amount of undigested OM, which is not economic and not results a good amendment material. However, the OM content of digestate is more recalcitrant and therefore the microbial degradation and soil oxygen consumption can be decreased by its application (Kirchmann & Bernal, 1997).

The adequacy of digestate as soil amendment is based on its modified OM content. Most OM is converted into biogas, while the biological stability of remaining OM was increased during AD with the increase of more recalcitrant molecules like lignin, cutin, humic acids, steroids, complex proteins. These aliphatic and aromatic molecules are possible humus precursors with high biological stability (Tambone et al., 2009). Pognani et al. (2009) found the increase of these macromolecules' quantities in the course of AD as it can be seen in Table 3.

Type of ingestate	Total solid (TS) ($g\ kg^{-1}$ ww)		Lignin ($g\ kg^{-1}$ TS)		Hemicelluloses ($g\ kg^{-1}$ TS)		Celluloses ($g\ kg^{-1}$ TS)	
	Inge-state	Dige-state	Inge-state	Dige-state	Inge-state	Dige-state	Inge-state	Dige-state
Energy crops, cow manure slurry and agro-industrial waste	127	35	49	280	35	42	50	68
Energy crops, cow manure slurry, agro-industrial waste and OFMSW	143	36	72	243	27	54	71	79

Table 3. Changes in macromolecules content on the course of AD *(Data from Pognani et al., 2009)*

Similarly, the rate of lignin-C, cellulose-C and hemicellulose-C are increased in the organic matter content after AD of cattle and pig dung (Kirchmann & Bernal, 1997). The increase of these macromelecules-C were 2.4-26.8 %, 14.2-13.9 % and 7.3 % in the manures, respectively. The hemicellulose-C content in the anaerobically treated pig dung was decreased by 23.8 %. However, the increase of non-decomposable carbon content of digestate is always smaller than that of composts (Gómez et al., 2007). On the other hand, improving the fertilizer effect of a digestate with its higher decomposable carbon content results in an increase in roots and crop residues which may have an important effect on the soil organic matter content.

4. Effects of digestate on soil properties

Digestate is a very complex material therefore its using has effect on the wide range of physical, chemical and biological properties of the soil, depending on the soil types (Makádi et al., 2008). The recycled organic wastes are suitable for contribution to maintain the soil nutrient levels and soil fertility (Tambone et al., 2007). Among the organic amendments the ratio of liquid digestate in the agriculture is known to be around of 10%. It can be applied as a fertilizer, but it could be appropriate as a soil quality amendment (Schleiss and Barth, 2008). Comparing to the other organic materials, the amendment properties rank sequentially as compost ~ digestate > digested sludge >> ingestate, on the bases of OM degradability (Tambone et al., 2010).

4.1 Effect of digestate on soil pH

Odlare et al. (2008) have not found significant change in the pH after 4-year-long biogas residue application rate. The pH of soils were 5.6 and 5.7 in the control and biogas residue treated samples, respectively. Similar results were reported by Fuchs & Schleiss (2008), because they have found an enhance of soil pH for about ½ unit after harvesting of maize. Because of the alkaline pH of digestates, an increase of the soil pH should be supposed. However, digestate might contain various acidic compounds (e.g. gallic acid). The polycondensation, connection to organic and inorganic colloids and transformation of these acids can have an effect also on the soil chemical properties and finally the decrease of soil pH (Tombácz et al., 1998, 1999), more particularly at the soils with high organic and inorganic colloid contents. Therefore the regular monitoring of soil pH is necessary in case of long-term digestate application.

4.2 Effect of digestate on soil macroelement content

One of the main problem of digestate (and other N fertilizer) application is the N leaching. However, Renger & Wessolek (1992) and Knudsen et al. (2006) found that the N leaching was dependent on the use of cover crops. Similar results were reported by Möller & Stinner (2009) who did not find differences in the soil mineral N content among different manuring systems in the case of winter wheat, rye and spelt in autumn, before use of cover crops. That means that the use of cover crops is an appropriate method to avoid N leaching and to compensate for higher N application. From the same experiment, Möller et al. (2008) reported average soil mineral N content in spring. In this case they found significant higher soil mineral N content of the digested slurry treated samples (Table 4).

Treatments	Soil mineral N (kg N ha^{-1}), 0-90 cm soil layer
Farmyard manure	65.7 a
Undigested slurry	71.1 ab
Digested slurry	89.2 c
Digested slurry + field residues	81.3 bc
Digested slurry + field residues + clover/grass and silage maize mixture	83.6 bc

Table 4. Average soil mineral N content in spring in 0-90 cm with the main crops spelt, rye and spring wheat from 2003-2005 *(Data from Makádi et al., 2007)*. a, b, c indexes mean the different values ($P<0.05$).

Digestate contains high proportion of NH_4-N therefore it would be expected to increase NH_4-N content of treated soil. However, digestate applied in the fall could easily be nitrified by early spring (Rochette et al., 2004; Loria et al., 2007). This predisposed N loss with occurrence of wet conditions.

Generally, the digestate application does not cause any significant changes in the total-N and available P content, while the available K content was increased by the application of biogas residue (Olsen et al., 2008). Similar results have found Vágó et al. (2009), who reported the significant increase of 0.01 M dm^{-3} $CaCl_2$ extractable P content even after 5 L m^{-2} digestate treatment, while the K content of soil was significantly increased by 10 L m^{-2} digestate dose only.

4.3 Effect of digestate on soil microelement

After the application of the digestate in 5 and 10 L ha^{-1} dosages, the Cd, Co, Cu, Ni and Sr content of soil solutions did not change. The Zn content decreased significantly, while the amount of manganese (Mn) increased by almost 40% (Vágó et al., 2009) (Table 5).

Element	Control	5 L ha^{-1} digestate	10 L ha^{-1} digestate
Cd	0.063	0.067	0.055
Co	0.064	0.071	0.057
Cu	0.089	0.112	0.118
Mn	25.5	35.1	35.5
Ni	0.50	0.52	0.35
Sr	8.56	8.60	8.62
Zn	1.40	0.98	0.062

Table 5. Microelement content of soil samples (mg kg^{-1}) treated with liquid digestate (extraction with 0.01 M dm^{-3} $CaCO_3$). (*Data from Vágó et al., 2009*).

The increasing soluble P content of digestate treated soil decreased the available Zn content in the soil solution by building slightly soluble zinc-phosphate residue (Vágó et al., 2009).

4.4 Effect of digestate on soil organic matter content

Soil OM decreases in crop soils in Europe and in other continents therefore using amendments for increasing the soil OM content has a particular interest.

Digestate contains high amount of volatile fatty acid (C2-C5) which could be decomposed within few days in the soil (Kirchmann & Lundwall, 1993). The greatest rate of decomposition were observed in the first day after the treatment (Marcato et al., 2009) but the mineralization rate were high during the first 30 days (Plaza et al., 2007). Moreover, the C-mineralization values from the soil incubation assay showed that the results of raw slurry were similar to the effect of compost being in the start of composting process while the digested slurry had similar C-mineralization rate in the soil samples than that of the matured compost (Marcato et al., 2009).

4.5 Effect of digestate on the microbiological activity of soil

Soil microbial community has an important role in the fertility of soil and its alteration after intervention to the soil (e.g. manuring, soil improving, soil pollution) could be indicate more sensitive these changes than changes in the soil physical and chemical properties.

Among the different organic wastes like compost, biogas residue, sewage sludge and different manures with and without mineral N, the biogas residue was more efficient for promoting the soil microbiological activity. The high amount of easy-degradable carbon increased the substrate induced respiration (SIR), which was enhanced by the higher carbon content resulted from the higher litter and root exudates of higher plant growth. In accordance with these results, the largest proportion of active microorganisms was found in the digestate treated samples (Odlare et al., 2008; Kirchmann, 1991). Similarly, the activity of invertase was significantly higher in the digestate treated samples than that in control ones (Makádi et al., 2006).

Besides the macro- and micronutrient content of digestate which are important not for the crops but for soil microorganisms too, it contains growth promoters and hormones, also. Therefore it could be used for stubble remains to facilitate their decomposing. Makádi et al. (2007) compared the effect of digestate and Phylazonit MC bacterial manure on the growth of silage maize (*Zea mays* L. 'Coralba') as a second crop after winter wheat and on the enzyme activities of soil. Digestate was used at the rate of 50% of the total N demand of silage maize while the Phylazonit MC was used at 5 L ha^{-1} dose. Their results of the changes in enzyme activities are summarized in Table 6.

Treatments	Enzyme activity (mean±S.D.)	
	16/08/2006.	27/09/2006
Invertase activity (mg glucose 1 g^{-1}soil 4 h^{-1})		
a) Control	5,618±1,392^{a}	3,767±2,030^{b}
b) Phylazonit MC	7,437±1,945^{a}	4,095±0,901^{b}
c) Phylazonit MC+digestate	6,613±2,230^{a}	1,584±0,748^{a}
d) Digestate	6,024±1,486^{a}	6,206±0,997^{c}
Catalase activity (mg O_2 1 g^{-1} dry soil 1 h^{-1})		
a) Control	1,468±0,118^{b}	1,797±0,289^{b}
b) Phylazonit MC	1,160±0,144^{ab}	1,410±0,050^{a}
c) Phylazonit MC+digestate	0,983±0,275^{a}	1,205±0,117^{a}
d) Digestate	1,961±0,395^{c}	1,288±0,063^{a}

Table 6. Invertase and catalase activities of soil on the 3rd and 9th week after digestate and Phylazonit MC treatment (*Data from Makádi et al., 2007*). a, b, c indexes mean the different statistical groups according to Tukey's test ($p<0.05$).

The maximum of the degradation of disaccharides, indicated by the invertase activity, was found in the 3rd week after Phylazonit MC treatment, while it was found only after the 9th week in the digestate treated soil samples. The Phylazonit MC contains only bacteria and promoting agents of bacterial activity for degrading the soil OM. Contrarily, in the digestate treated samples the degradation of disaccharides takes place at similar rate through 9 weeks because of the OM content of digestate used. Changes in catalase activity indicate the effect of nutrient content of digestate to the increasing microbial metabolism.

5. Effects of digestate on crop yield

On the bases of the plant reaction on the digestate treatment, plants could be classify into the sensitive (alfalfa, sunflower, soybean) and the non-sensitive (winter wheat, triticale, sweet corn, silage maize) groups. The sensitive plants can be treated by digestate only in their certain life stages, for example, young alfalfa is very sensitive after sowing while old alfalfa is very sensitive before cutting. In the case of sensitive plants the burning effect of digestate can be observed but it follows a strong and quick recovering process. For the non-sensitive plants the digestate can be used in any developmental stage. It is favourable, because for example, in rainy period the digestate technically could not be applied (Makádi et al., 2008).

The right application rate of liquid or solid digestate depends on the plant nitrogen demand. It should be applied when plant N demand arises. This time for non-legume scpecies is the late winter and spring (Stinner et al., 2008). Similarly, Wulf et al. (2006) used 70% of the digestate in spring and 30% in autumn, while Makádi et al. (2008) and Nyord et al. (2008) split into two and three the applied rate through the vegetation period.

Because of its high available nutrient content, digestate application resulted in significantly higher aboveground biomass yields in the case of winter wheat and spring wheat than the farmyard manure and undigested slurry treatment. The effectiveness of a digestate depends on the composition of co-digestied material, the treated plant species and the treatment methodology. Co-digestion of different organic materials results in more effective digestate. (Möller et al., 2008; Stinner et al., 2008).

After the burning effect of digestate the soybean plants recovered and grew more, but lower sprouts. These sprouts were very productive, the number of pods was also higher in the treated samples, therefore the yield and thousand seed weight were also higher (Table 7, Makádi et al., 2006)

Digestate ($L\ m^{-2}$)	Height of plants (cm)	Weight of sprout ($g\ m^{-2}$)	Weight of pods ($g\ m^{-2}$)	Weight of grain ($g\ m^{-2}$)	Thousand seed weight (g)
	mean±S.D.				
0	74.3± 1.15a	218.0± 33.08a	351.2± 69.69a	233.2± 40.61a	134.3± 1.71a
5	71.8± 2.68a	214.4± 4.98a	521.0± 20.30b	335.2± 43.46b	172.2± 6.61b
10	70.2± 7.73a	234.4± 7.73a	811.0± 33.09c	566.5± 25.05c	191.0± 8.69c

Table 7. Yield parameters of soybean after digestate treatment (*Data from Makádi et al., 2008).* a, b, c indexes mean the different statistical groups according to Tukey's test ($p<0.05$).

These yield parameters are close correlations with some soil parameters changing after digestate amendment. Increasing in important nutrient contents contribute to the better development of plants (Makádi et al., 2008b, Table 8).

Comparing the effect of liquid digestate and the equal quantity of water to the yield of sweet corn and silage maize, significantly higher yields were found in the digestate treatment. In this case the applied digestate on the bases of plants N demand was split into two parts (Makádi et al., 2006). That means that the favourable effects of digestate are caused by its soluble macro- and micronutrient content.

		NO_3-N	AL-P	AL-K	AL-Mg
Number of pods	Pearson Corr.	0.712*	0.798*	0.622	0.850**
	Sig. (2-tailed)	0.031	0.01	0.074	0.004
Weight of pods	Pearson Corr.	0.755*	0.824**	0.693*	0.839**
	Sig. (2-tailed)	0.019	0.006	0.039	0.005
Weight of grain	Pearson Corr.	0.742*	0.832**	0.739*	0.810**
	Sig. (2-tailed)	0.022	0.005	0.023	0.008
Thousand seed weight	Pearson Corr.	0.695*	0.690*	0.827**	0.595
	Sig. (2-tailed)	0.038	0.040	0.006	0.091

* Correlation is significant at the 0.05 level; ** Correlation is significant at the 0.01 level.

Table 8. Correlations between soil and plant parameters in digestate treatment experiment. (*Data from Makádi et al., 2008b*)

Comparing the effect of digestate and a bacterial manure (Phylazonit MC, the experimental conditions can be found in the section 4.5). The Phylazonit MC treatment increased the green weight of silage maize by 47.18% while the digestate by 142.34%, comparing to the control. The results obtained can be seen in Table 9 (Makádi et al., 2007).

Treatments	Green weight, t ha^{-1} mean±S.D.
Control	6,448±2,580^{a}
Phylazonit MC	9,490±4,081^{ab}
Phylazonit MC + digestate	13,997±0,493^{bc}
Digestate	15,626±2,293^{c}

Table 9. Green weight of silage maize as a second crop after digestate and Phylazonit MC treatment of stubble. (*Data from Makádi et al., 2007*). a, b, c indexes mean the different statistical groups according to Tukey's test ($p<0.05$).

The positive effect of Phylazonit MC treatment was the result of its microbes, plant growth promoters and microelement content, while the favourable effect of digestate treatment was caused by its macro- and microelement and high water content and the increase of soluble macroelement content of soil because of the increased microbial activity.

6. Effects of digestate on the quality of crops

Crop yield is very important economical parameter of plant production but nowadays the quality of foods is becoming more and more important. Digestate treatment seems to be very effective to increase the protein content of plants. Banik and Nandi (2004) investigated biogas residual slurry manures (solid digestate) used as supplement with rice straw for preparation of mushroom beds. The application of biomanure increased the protein content of mushroom 38.3-57.0%, while the carbohydrate concentrations were decreased. Results can be seen in Table 10.

Similar results were reported by Makádi et al., (2008b) who found significant increase of protein content of treated soybean. They have found 30.65±1.42% protein in control plants, while these values were 34.83±1.50% and 35.67±1.81% for 5 and 10 L m^{-2} treatments, respectively. Changes in amino acid composition of test plants were also very favourable, because almost every essential and non-essential amino acid quantity was increased significantly after digestate treatment. In line with these results the oil content of the treated plants decreased significantly.

Treatments	Protein (%)	Carbohydrate (%)	Lipid (%)	Increase of protein over control (%)
Straw (100%)	21.56	28.81	10.43	0
Straw + cowdung biomanure	29.81	20.21	13.73	38.3
Straw + poultry litter biomanure	33.57	21.45	7.96	55.7
Straw + jute caddis biomanure	33.84	21.79	13.93	57.0

Table 10. Effect of supplementation of rice straw with solid digestate on major nutrient contents of mushroom (*Pleurotus sajor caju*). (*Data from Banik and Nandi, 2004*)

Qi et al (2005) examined the effect of fermented waste as organic manure in cucumber and tomato production in North China. Before the vegetables transplantation, the diluted fermented residual dreg was applied 20-30 cm below the soil surface at a rate of 37,500 kg ha^{-1}, while liquid digestate was sprinkled to the soil surface in three vegetables growing stages and on the vegetable leaves once time. They found increasing yield (18.4% and 17.8%) and vitamin C content (16.6% and 21.5%) of treated cucumber and tomato, respectively.

As the results show, the digestate application in solid or liquid form could result significant improvement in the quality of foods without damaging the environment, which is very important for the sustainable environment and healthy life.

7. Legislation of digestate utilization in agriculture

Sustainable recycling of organic wastes demands clear regulations of recycled wastes, the used recycling methods and the controlling of products. These regulation processes for the digestate are different in certain countries, respected the elaboration and the used limits.

In Hungary, the digestate is regarded as other non-hazardous waste if the ingestate does not contain sewage or sewage sludge, while in the presence of these materials the conditions of the digestate utilisation depend on the quality of the given material.

In Scotland the BSI PAS110:2010 digestate quality assurance scheme is applied. If a digestate complies with the standards for the quality, the usage criteria and the certification system stated in the worked scheme, the Scottish Environment Protection Agency (SEPA) does not apply the waste regulatory control for it.

In Swiss the digestate which suits the limits, can be used as soil conditioner and fertilizer in "bio"-agriculture.

In Germany the origin of the input materials determines the quality label of digestate product by biowaste and renewable energy crops. Digestates have to fulfil the minimum quality criteria for liquid and solid types which determine the minimum of nutrients and the maximum of pollutions in the digestate. Pollutions mean toxic elements, physical contaminants and pathogen organisms. The quality of digestate products is regularly controlled by "Bundesgütegemeinschaft Kompost e.V." (BGK) (Siebert et al., 2008).

8. Future prospects

Beside the fertilizer or amendment properties of digestate, nowadays there are some other ways to utilize it. These new methods are very creative and make the possibility of proper utilization of digestates with different quality.

A new promising alternative of the digestate utilization is its use as solid fuel after drying. Kratzeisen et al. (2010) used liquid digestate originated from silage maize co-digestion with different field crops and animal residues. After drying the digestate, the water content of pellets made was 9.2-9.9%. Their mechanical durability fulfilled the requirements of standards for pellets. Moreover, the calorific value of these pellets was similar to the calorific value of wood. Therefore digestate fuel pellet seems to be a good alternative fuel for wood.

Another interesting possibility of digestate utilization is the using of digestate effluent to replace freshwater and nutrients for bioethanol production. Gao & Li (2011) found that ethanol production was enhanced with digestate effluent by as much as 18% comparing to the freshwater utilization.

Digestate can be separated to liquid and solid fraction. Liquid fraction is suitable for irrigation and it has high N and K content. Solid fraction contains a great amount of volatile solid and P (Liedl, et al., 2006) and - by its fertilizer effect - has also high biogas and methane potential, therefore it could be used as a co-ferment for anaerobic digestion (Balsari et al., 2009)

9. Conclusion

The use of anaerobic digestion for treatment of solid and liquid organic wastes has vastly increased world-wide. The by-product of this process is the digestate, a liquid or solid material with high nutrient and organic matter content. These properties of the digestate make possible to use it as plant nutrients and to characterize it as a fertilizer. On the other hand, a biomass, reach in recalcitrant molecules is characterized by a high biological stability degree which is suitable for soil improving. The utilization of digestate as fertilizer provides economic and environmental benefits because of its higher stabile organic matter content, the hygienization effect of anaerobic digestion process and the reduced quantity of the artificial fertilizers needs for plant production. Moreover, the alkaline pH of digestate could contribute to the decrease of soil acidification, which is a serious problem of the world. Using digestate in place of artificial fertilizers could contribute to maintain the fertility of soil.

As the results show, the digestate application in solid or liquid form could result significant improvement of the quantity and quality of foods through the even nutrient supply harmonizing with the necessity of plants and through its microelement content in the available forms for plants. In this way, digestate application in agriculture could contribute to the healthy life of humans.

Microbiological activity of soil could be increased by application of digestate which is also an important condition of soil fertility.

Beyond these "classical" application possibilities of digestate, there are new promising alternatives for its utilization which means more opportunities to use this valuable matter for making better our environment and our life.

10. Acknowledgment

Our thanks to Dr. Judit Dobránszki at the University of Debrecen, Research Institute of Nyíregyháza and Prof. György Füleky at the Szent István University, Department of Soil Sciences and Agrochemistry for their enthusiastic checking of the English and for their valuable comments on improvements of the manuscript.

11. References

Balsari, P., Menardo, S., Gioelli, F. & Dinuccio, E. (2009). Il progetto europeo EU Agrobiogas: finalitá, obviettivi e primi risultati ottenuti. In: Proceedings of IX Convegno Nazionale dell'Associazione Italiana di Ingegneria Agraria - Ricerca e innovazione nell'ingegneria dei biosistemi agro-territoriali, Ischia (NA)

Banik, S. & Nandi, R. (2004). Effect of supplementation of rice straw with biogas residual slurry manure on the yield, protein and mineral contents of oyster mushroom. *Industrial Crops and Products*, Vol. 20, No. 3, (November 2004), pp. 311-319, ISSN 0926-6690

Bendixen, H.J. (1999) Hygienic safety: results of scientific investigations in Denmark (sanitation requirements in Danisc Biogas Plants). In: Proceedings of the IEA Workshop: Hygienic and Environmental Aspects of Anaerobic Digestion: Legislation and Experiences in Europe. Universität Hohenheim, Stuttgart, pp. 27-47.

Binder, D.L., Sander, D.H., Frank, K.D. & Shires, W.L. (1996). Nitrogen fertilizer equivalency of anaerobical digested municipal sludge. In: Proc. North Centr. Axt.-Ind. Soil Fert. Conf. 26th, St. Louis, MO. 20-21 Nov.1996. Vol. 12. pp. 108-114. Potash and phosphate Inst., Brookings, SD.

Börjesson, P. & Berglund, M. (2007). Environmental systems analysis of biogas systems—Part II: The environmental impact of replacing various reference systems. *Biomass and Bioenergy*, Vol. 31, No. 5, (May 2007), pp. 326-344, ISSN 0961-9534

Fuchs, J.G. & Schleiss, K. (2008). Effects of compost and digestate on environment and plant production - Results of two research project. Proceedings of the Internationale Conference ORBIT 2008, Wageningen, 13-16 October, 2008. CD-ROM (*ISBN 3-935974-19-1*)

Furukawa, Y. & Hasegawa, H. (2008). Response of spinach and komatsuna to biogas effluent made from source-separated kitchen gabage. *Journal of Environmental Quaity*, Vol. 35, No. 5, (September-October 2006), pp. 1939-1947, ISSN 0047-2425

Gao, T. & Li, X. (2011). Using thermophilic anaerobic digestate effluent to replace freshwater for bioethanol production. *Bioresource Technology*, Vol. 102, No. 2, (January 2011), pp. 2126-2129, ISSN 0961-9534

Georgacakis, D., Sievers, D.M. & Ianotti, E.L. (1982). Buffer stability in manure digesters. *Agricultural Wastes*, Vol. 4, No. 6 , (November 1982), pp. 427-441, ISSN 09608524

Gómez, X., Cuetos, M.J., García, A.I. & Morán, A. (2007). An evaluation of stability by thermogravimetric analysis of digestate obtained from different biowastes. *Journal of Hazardous Materials,* Vol. 149, No.1, (October 2007) pp. 97-105, ISSN 0304-3894

Hobson, P. & Wheatley, A. (1992). Anaerobic digestion - modern theory and practice. Elsevier Applied Science, 269.

Kirchmann, H. (1991). Carbon and nitrogen mineralization of fresh, aerobic and anaerobic animal manures during incubation with soil. *Swedish Journal of Agricultural Research,* Vol. 21, No. 4 , pp. 165-173, ISSN 0049-2701

Kirchmann, H. & Lundwall, A. (1993). Relationship between N immobilization and volatile fatty acids decomposition in soil after application of cattle and pig slurry. *Biology and Fertility of Soils,* Vol. 15, No. 3 , (March 1993), pp. 161-164, ISSN 0178-2762

Kirchmann, H. & Bernal, M.P. (1997). Organic waste treatment and C stabilization efficiency. *Soil Biology and Biochemistry,* Vol. 29, No. 11/12, (November-December, 1997), pp. 1747-1753, ISSN 0038-0717

Knudsen, M.T., Kristensen, J.B.S., Bernsten, J., Petersen, B.M. & Kristensen, E.S. (2006). Estimated N leaching losses for organic and conventional farming in Denmark. *Journal of Agricultural Science* Vol. 144, No. 2, (Online February 2006), pp. 135-149, ISSN 0021-8596

Kratzeisen, M., Starcevic, N., Martinov, M., Maurer, C. & Müller, J. (2010). Applicability of biogas digestate as solid fluel. *Fuel,* Vol. 89, No. 9, (September 2010), pp. 2544-2588, ISSN 0016-2361

Kryvoruchko, V., Machmüller, A., Bodiroza, V., Amon, B. & Amon, T. (2009). Anaerobic digestion of by-products of sugar bett and starc potato processing. *Biomass and Bioenergy,* Vol. 33, No. 4 , (April 2009), pp. 620-627, ISSN 0961-9534

Liedl, B.E., Bombardiere, J. & Chaffield, J.M. (2006). Fertilizer potential of liquid and solid effluent from thermophilic anaerobic digestion of poultry waste. *Water Science and Technology,* Vol. 53, No. 8 , pp. 6979, ISSN 0273-1223

Loria, E.R., Sawyer, J.E., Backer, D.W., Lundwall, J.P. &Lorimor, J.C. (2007). Use of anaerobically digested swine manure as a nitrogen source in corn production. *Agronomy Journal,* Vol. 99, No. 4, (July-August 2007), pp. 1119-1129, ISSN 0002-1962

Makádi, M., Tomócsik, A., Orosz, V., Lengyel, J., Biró, B. & Márton, Á. (2007). Biogázüzemi fermentlé és Phylazonit MC baktériumtrágya hatása a silókukorica zöldtömegére és a talaj biológiai aktivitására. (Effect of digestate and Phylazonit MC on the yield of silage maize and the biological activity of the soil) *Agrokémia és Talajtan* Vol. 56, No. 2, (December, 2008), pp. 367-378, ISSN 0002-1873

Makádi, M., Tomócsik, A., Lengyel, J. & Márton, Á (2008). Problems and successess of digestate utilization on crops. Proceedings of the Internationale Conference ORBIT 2008, Wageningen, 13-16 October, 2008. CD-ROM (*ISBN 3-935974-19-1*)

Makádi, M., Tomócsik, A., Kátai, J., Eichler-Loebermann, B. & Schiemenz, K. (2008b): Nutrient cycling by using residues of bioenergy production - effects of biogas-digestate on plant and soil parameters. *Cereal Research Communications, Cereal Research Communications,* Vol. 36, Supplement 5, (August 2008), pp. 1807-1810, ISSN 0133-3720

Marcato, C.E., Pinelli, E., Pouech, P.,Winterton, P. & Guiresse, M. (2008). Particle size and metal distribution in anaerobically digested pig slurry. *Bioresource Technology,* Vol. 99, No. 7, (May 2008), pp. 2340-2348, ISSN 09608524

Marcato, C.E., Mohtar, R., Revel, J.C., Pouech, P., Hafidi, M. & Guiresse, M. (2009). Impact of anaerobic digestion on organic matter quality in pig slurry. *International*

Biodeterioration & Biodegradation. Vol. 63, No. 3, (April 2009), pp. 260-266, ISSN 0964-8305

McDowell, L.L. & Smith, G.E, (1958). The retention and reaction of anhydrous ammonia on different soil types. *Soil Science of America Journal*, Vol. 22, No. 1, pp. 38-42, ISSN 0361-5995

Menardo, S., Gioelli, F. & Balsari, P. (2011). The methane yield of digestate: Effect of organic loading rate, hydraulic retention time and plant feeding. *Bioresource Technology*, Vol. 102, No. 3 , (February 2011), pp. 2348-2351, ISSN 09608524

Møller, H., Lund, I. & Sommer, S. (2000). Solid-liquid separation of livestock slurry: efficiency and cost. *Bioresource Technology*, Vol. 74, No. 3, (September 2000), pp. 223-229, ISSN 0960852

Möller, K., Stinner, W., Deuker, A. & Leithold, G. (2008). Effects of different manuring systems with and without biogas digestion on nitrogen cycle and crop yield in mixed organic diary farming systems. *Nutrient Cycling in Agroecosystems* Vol. 82, No. 3, (November 2008), pp. 209-232, ISSN 13851314

Möller, K. & Stinner, W. (2009). Effects of different manuring systems with and without biogas digestion on soil mineral nitrogen content and on gaseous nitrogen losses (ammonia, nitrous oxides). *European Journal of Agronomy*, Vol. 30, No. 1, , (January 2009), pp. 1-16, ISSN 1161-0301

Nyord, T., Søgaard, H.T., Hansen, M.N & Jensen, L.S. (2008). Injection methods to reduce ammonia emission from volatile liquid fertiliser applied to growing crops. *Biosystem Engineering*, Vol. 100, No. 2, (June 2008), pp. 235-244, ISSN 1537-5110

Odlare, M., Pell, M. & Svensson, K. (2008). Changes in soil chemical and microbiological properties during 4 years of application of various organic residues. *Waste Management*, Vol. 28, No. 7, (January 2008), pp. 1246-1253, ISSN 0956-053X

Paavola,T. & Rintala, J. (2008). Effects of storage on characteristics and hygienic quality of digestates from four co-digestion concepts of manure and biowaste. *Bioresource Technology*, Vol. 99, No. 15 , (October 2008), pp. 7041-7050, ISSN 0960852

Pfundtner E. (2002). Limits and merits of sludge utilisation – Land application. Conference Proceedings of Impacts of Waste Management. Legislation on Biogas Technology. Tulln, 2002. pp.1-10.

Plaza, C., Garcia-Gil, J.C. & Polo, A. (2007). Microbial activity in pig slurry-amended soils under aerobic incubation. *Biodegradation*, Vol. 18, No. 2, (April 2007), pp. 159-165, ISSN 0923-9820

Pognani, M., D'Imporzano, G., Scaglia, B. & Adani, F. (2009). Substituting energy crops with organic fraction of municipal solid waste for biogas production at farm level: A full-scale plant study. *Process Biochemistry*, Vol. 44, No. 8, (August 2009), pp. 817-821, ISSN 1359-5113

Qi, X., Zhang, S., Wang, Y., Wang, R. (2005). Advantages of the integrated pig-biogas-vegetable greenhouse system in North China. *Ecological Engineering*, Vol. 24, No. 3, (February 2005), pp. 175-183, ISSN 0925-8574

Renger, M. & Wessolek, G. (1992). Qualitative und Quantitative Aspekte zur Nitratverlagerung. Mitteilungen der Deutschen Bodenkundlichen Gesellschaft 68, 111-114.

Rochette, P., Angers, D.A., Chantigny, M.H., Bertrand, N. & Côtè, D. (2004). Carbon dioxide and nitrous oxide emissions following fall and spring applications of pig slurry to an agricultural soil. Soil Science of Soc. Am. J. Vol. 68, Vol. 68, No. 4, pp. 1410-1420, ISSN 0361-5995

Sahlström, L. (2003): A review of survival of pathogenic bacteria in organic waste used in biogas plants. *Bioresource Technology,* Vol. 87, No.2, (April 2003), pp. 161-166, ISSN 0960852

Schleiss, K. & Barth, J. (2008): Use of compost and digestate: choosing the product depending of utilizastion, strategy and aim. In: Fuchs Jacques G., Thomas Kupper, Lucius Tamm & Kaarina Schenk (Eds.) (2008): Compost and digestate: sustainability, benefits, impacts for the environment and for plant production. Proceedings of the international congress CODIS 2008, February 27-29, 2008, Solothurn, Switzerland. pp. 199-208.

Siebert, S., Thelen-Jüngling, M. & Kehres, B. (2008). Development of quality assurance and quality characteristics of composts and digestates in Germany. Proceedings of the Internationale Conference ORBIT 2008, Wageningen, 13-16 October, 2008. CD-ROM (*ISBN 3-935974-19-1*)

Stinner, W., Möller, K. & Leithold, G. (2008). Effect of biogas digestion of clover/grass-leys, cover crops and crop residues on nitrogen cycle and crop yield in organic stockless farming system. *European Journal of Agronomy,* Vol. 29, No. 2-3, (August 2008), pp. 125-134, ISSN 1161-0301

Szűcs, B., Simon, M. & Füleky, G. (2006). Anaerobic pre-treatment effects on the aerobic degradability of waste water sludge. Proceedings of the Internationale Conference ORBIT 2006, Weimar, 13-15 September, 2006. Part 2, pp. 425-434.

Tambone, F., Genevini, P. & Adani, P. (2007). The effect of short-term compost application on soil chemical properties and on nutritional status of maize plant. *Compost Science & Utilization,* Vol. 15, No. 3, (July 2007), pp. 176-183, ISSN 1065-657X

Tambone, F., Genevini, P., D'Imporzano, G. & Adani, F. (2009). Assessing amendment properties of digestate by studying the organic matter composition and the degree of biological stability during the anaerobic digestion of the organic fraction of MSW. *Bioresource Technology,* Vol. 100, No. 12, (June 2009), pp. 3140–3142, ISSN 09608524

Tambone, F., Scaglia, B., D'Imporzano, G. Schievano, A., Orzi, V., Salati, S. & Adani, F. (2010). Assessing amendment and fertilizing properties of digestates from anaerobic digestion through a comparative study with digestated sludge and compost. *Chemosphere,* Vol. 81, No. 5, (Oktocer 2010), pp. 577-583, ISSN 0045-6535

Tombácz, E., Szekeres, M.,.Baranyi, L. & Micheli, E. (1998). Surface modification of clay minerals by organic polyions. *Colloids and Surfaces* A., Vol. 141, No. 3, (November 1998), pp. 379-384, ISSN 0927-7757

Tombácz, E., Filipcsei, G., Szekeres, M, & Gingl, Z. (1999) Particle aggregation in complex aquatic systems, *Colloids and Surfaces,* A, Vol. 151, No. 1, (June 1999), pp. 233-244, ISSN 0927-7757

Vágó, I., Kátai J., Makádi M. & Balla Kovács A. (2009). Effects of biogas fermentation residues on the easily soluble macro- and microelement content of soil. Trace elements in the food chain. Vol. 3. Deficiency or excess of trace elements in the environment as a risk of health. Pp. 252-256. Publ.: Working Committe on Trace Elements and Institute of Materials and Environmental Chemistry of the Hungarian Academy of Sciences, Budapest. Eds.: Szilágyi M, Szentmihályi K. ISBN 978-963-7067-19-8.

Wulf, S., Jäger, P. & Döhler, H. (2006). Balancing of greenhouse gas emissions and economic efficiency for biogas-production through anaerobic co-fermentation of slurry with organic waste. *Agriculture, Ecosystems and Environment, Agriculture, Ecosystems and Environment,* Vol. 112, No. 2-3, (February 2006), pp. 178-185, ISSN 0167-8809

13

Microbiological Methods of Hydrogen Generation

Krystyna Seifert, Michał Thiel, Ewelina Wicher, Marcin Włodarczak and Marek Łaniecki
Faculty of Chemistry, A. Mickiewicz University, Poznań, Poland

1. Introduction

As long as any country economy is based on fossil fuels, the prosperity of many nations is in danger. Rapidly growing prices of oil and natural gas can lead to the worldwide economic crisis. Therefore the search for new, clean, cheap and renewable sources of energy and energy carriers is urgently required. Although many different methods are suggested to solve this problem the use of hydrogen as the future energy carrier is necessary. Application of biochemistry in generation of energy is a challenge both for academia and industry. Different types of biomass pyrolysis and/or fermentative processes can partially solve the problems of renewable energy generation. Although other solutions are at the moment much more technologically advanced (e.g. hydropower or wind farms) the future of energetic will belong to the biological systems. Generation of biogas or biohydrogen under anaerobic conditions are the very promising processes, especially at local environment. Different types of agriculture and food industry wastes can serve here as an excellent source of organic carbon in microbiological processes.

It is well known that burning of hydrogen either chemically or electrochemically (e.g. fuel cells) generates large quantities of energy and it is environmentally friendly. Application of biohydrogen in local environment (farms, small communities, etc.) certainly will improve local energy distribution and will lower costs of used energy.

This review paper describes basic principles of fermentative and phofermentative hydrogen generation. Biophotolysis of water, anaerobic dark and photofermentative processes in presence of organic substances, as well as the hybrid systems used in microbiological methods of hydrogen generation are described. The description of the applied microorganisms and enzymes is presented.

2. General

Biophotolysis of water, fermentation and photofermentation of organic substrates are considered to be the best biological methods of hydrogen generation. Reversibility, lack of toxic substances generated in these processes, mild conditions for microbiological reactions, as well as operation at low pressure of these processes are the conditions required for modern microbiological systems. Moreover, the possibility of application of different waste waters (containing organic carbon) in these processes is an additional benefit.

Fermentation is the process generating basically two gaseous metabolites: hydrogen and carbon dioxide. The volatile fatty acids (VFA) and alcohols represent liquid metabolites of dark fermentation. The low yield of generated hydrogen and high concentration of CO_2 (almost 50%) in gaseous products are the main disadvantages of microbiological hydrogen generation. In contrary, high reaction rate and possibility of biodegradation of many organic substances can be assigned to the benefits of this process.

In photofermentation, the photosynthetic heterotrophoic bacteria under anaerobic conditions and in the absence of nitrogen generate hydrogen in presence of organic compounds. Nitrogenase is the enzyme catalyzing hydrogen generation reaction. Presence of molecular nitrogen or nitrogen compounds directs the reaction route towards ammonia formation. The possibility of application of wide spectrum of light (400-950 nm), lack of methabolism generating molecular oxygen, as well as possibility of use of organic substances originating from wastes are the main advantages of photobiological method of hydrogen generation.

Both fermentation and photofermentation require presence of anaerobic microorganisms and the light in case of photofermentation. Photosynthesis, and in consequence also photofermentation is the series of complex reactions transforming energy of light into chemical energy.

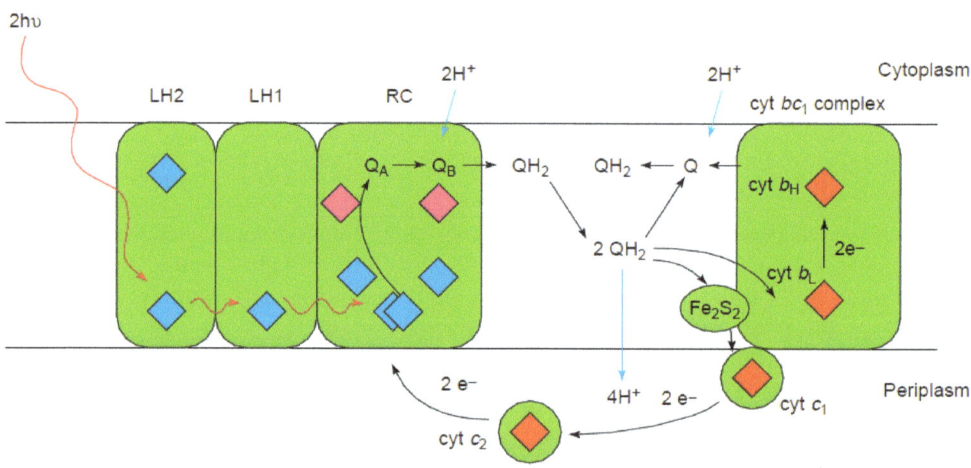

Fig. 1. Scheme of photoinduced cyclic flow of electrons in photosystem of *Rhodobacter sphaeroides* bacteria (Vermeglio, 1999).

The photosynthetic apparatus is localized in invaginations of the cytoplasmic membrane. The photosystem is built of three multimeric (transmembrane) proteins: antennas making the light-harvesting complex (LHC), the reaction centre (RC) and the complex of cytochromes bc_1 (Fig.1) (Vermeglio, 1999). The LHC antennas contain molecules of bacteriochlorophyll and carotenoides. The carotenoides play a double role in LHC systems;

they absorb light from the visible part of the light spectrum in which bacteriochlorophyll is not active and protect the antenna system against damage by singlet oxygen (Isaacs, 1995, Jones, 1997). The majority of the purple bacteria have two different antenna complexes known as LH1 and LH2. The number of LH2 complexes depends on such parameters like light intensity and partial pressure of oxygen, while the number of LH1 complexes is directly correlated with that of the reaction center (RC) to form RC-LH1 center. High ratio of pigment molecules to RC (e.g. 100 molecules of chlorophyll to one RC) increases the area capable of light absorption. Upon absorption of photon by LHC, the reaction centre becomes excited with simultaneous charge separation in a time shorter than 100 picoseconds (ps). The high reaction rate of this process is a consequence of the mutual arrangement of LH1 and RC: one RC is surrounded by a ring of 15-17 LH1 subunits. The closed structure of LH1 complexes in combination with the dense packing of bacteriochlorophyll molecules ensures fast delocalization of the excited state and possibility of energy transfer towards the reaction centre from every point of the ring (Vermeglio, 1999). The reaction centre is an integral part of protein membrane composed of three polypeptides (subunits L, M and H), containing four molecules of bacteriochlorophyll *a* (P_A, P_B, B_A, B_B), two molecules of bacteriofeophityne *a* (H_A, H_B), two molecules of ubichinone(Q_A, Q_B), one molecule of carotenoid (Crt) and one atom of non-heme iron (Fig.2).

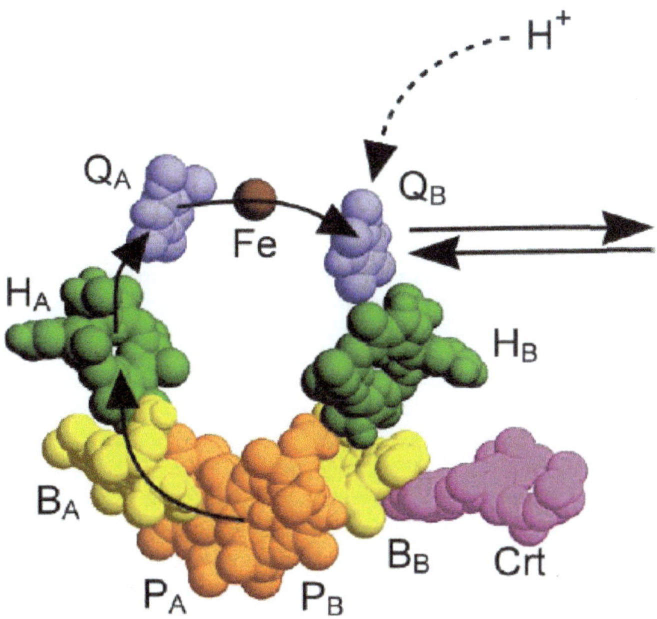

Fig. 2. Reaction center (RC) of photosystem in *Rhodobacter sphaeroides* bacteria (Isaacs, 1995)

All pigments are linked to the heterodimeric protein skeleton of L and M subunits forming five transmembrane protein helixes (Paschenkoa, 2003). The main source of electrons is the "special pair" of the excited bacteriochlorophylls *a* located close to side of the cytoplasmic membrane. The excitation is realized by direct absorption of light by the "special pair" of bacteriochlorophylls absorbing at 870 nm and by energy transfer from other pigment molecules located at RC or LHC. The transfer of electrons from the special pair to bacteriopheophytin, located in the middle of the dielectric cytoplasmic membrane occurs in 3-4 ps. This reaction is probably intermediated by a transient product of monomeric bacteriochlorophyll B_A. In the next 200 ps the electron is transferred to ubiquinone Q_A (connected with RC) and subsequently to ubiquinone Q_B. The transfer of electron to ubiquinone Q_B is accompanied by its protonation. The full reduction of ubiquinone Q_B requires two subsequent cycles in RC after which electrons finally leave RC with electrostatically neutral doubly reduced ubiquinol QH_2 (Jones, 1997). The two protons required for protonation originate from cytoplasmic space. In the next step ubiquinol is oxidized by the bc_1 cytochrome complex. This complex caused reduction of the $[Fe_2S_2]$ unit which is a part of cytochrome (part of Rieske unit) and releases two protons to periplasmic space. Then the cycle of electron transfer is closed by recombination of cytochrome c_2 by reduction of the special pair of bacteriochlorophylls. The cyclic transfer of electrons is accompanied by transfer of protons from cytoplasm to periplasm leading to the proton gradient between the two sides of cytoplasmic membrane, which is the most important effect of photosynthesis because it stimulates ATP synthesis and reduction of NAD^+ (Vermeglio, 1999). Protons accumulated on the periplasmic space of the membrane return to the cytoplasmic space through the ATP synthase channel, which closes the transfer of protons (Paschenkoa, 2003).

3. Microorganisms

Hydrogen generation in microbiological processes can be realized both by eucariota (green algae), procariota (cyanobacteria) in direct or indirect splitting of water under illumination, as well as in the fermentation and photofermentation reactions in presence of organic substances and numerous strains of bacteria. Due to very low yields of hydrogen obtained in presence of algae and cyanobacteria this paper will concentrate only on fermentative processes.

Dark fermentation process towards hydrogen is performed in presence of organothrophic bacteria. Large variety of microorganisms is involved in these reactions, therefore this paper will focus only on the description of three groups of microorganisms.

The first group belongs to anaerobic, gram-negative, mesophilic bacteria of *Clostridium* and *Bacillus* type. *C. acetobutylicum* (Chin, 2003), *C. butyricum* (Masset, 2010, Cai, 2010) *C. pasterianum, C. bifermentants* (Wang, 2003), *C. beijerinckii* (Skonieczny, 2009), *C.tyrobutyricum* (Jo, 2008), *C. saccharoperbutylacetonicum* (Alalayaha, 2008), *B. lichemiformis* (Kalia, 1994), *B.coagulans* (Kotay, 2007) are the most popular representatives of this group. This strains of bacteria can form spores capable to survive in extreme conditions such as low and high temperature, different pH, irradiation, extreme dry conditions or presence of deadly chemical compounds (eg. NaCl). Bacteria cells under these conditions goes to anabiosis: complete reduction of metabolic processes. The separation of already divided DNA occurs at this stage with simultaneous surrounding by two cytoplasmic membranes. The endospore formed under unfavourable conditions can return to normal activity under appropriate conditions. In this process the external protection of outer coating is destroyed. Appropriate temperature, pH and presence of feed compounds facilitate formation of vegetative cells and their growth (Setlow, 2007). In some cases presence and increase of concentration of specific compounds is the biochemical signal to stop the endospore phase. Presence of alanine, serine, cysteine together with lactic acid accelerate germination *C. botulinum* bacteria (Plowman, 2002).

The second group of fermentative bacteria active in hydrogen generation belongs to anaerobic gram-negative bacteria. The best activity in biohydrogen generation *via* dark fermentation was found for the following strains: *Enterobacter asburiae* (Jong-Hwan, 2007), *Enterobacter cloacae* (Mandal, 2006), *Enterobacter aerogenes* (Jo, 2008), *Escherichia coli* (Turcot, 2008), *Klebsiella oxytoca* (Wu, 2010) or *Citrobacter Y19* (Oha, 2003). These strains of bacteria can tolerate oxygen in environment. Here, in aerobic condition the oxygen respiration can occurs. The change of metabolic pathway provides method for survival under variable conditions of environment. These bacteria show better biological activity in comparison with those active only in completely anaerobic conditions. However, in aerobic conditions no hydrogen formation is observed. This effect is caused by inhibition of hydrogenase, enzyme catalyzing hydrogen generation.

Thermophilic bacteria operating at 60-85 ^{o}C belongs to the third group of bacteria generating hydrogen in fermentative processes (Zhang, 2003). The following strains of thermophilic bacteria of *Thermoanaerobacterium thermosaccharolyticum* (Thonga, 2008) and hyperthermophilic of *Thermatoga neapolitana* (Mars, 2010, Eriksen, 2008), *Thermococcus kodakaraensis* (Kanai, 2005), or *Clostridium thermocellum* (Lewin, 2006) can generate hydrogen in presence of organic substrates at relatively high temperatures. It was established that thermophilic bacteria are the most effective from all those already described.

Application of *C. saccharolyticus* and *Thermatoga elfii* thermophilic bacteria results in 80 % yield of the theoretical one (theoretically 4 moles of glucose can be transformed into acetic acid with 100% yield) while applying saccharose or glucose (Vardar-Schara,2008), respectively. High yield in hydrogen generation is explained by Guo *et al.* (Guo, 2010) who assumes that high temperature can accelerate hydrolysis of substrates engaged in this process. At the same time Valdes-Vazquez et al. (Valdes-Vazquez, 2005) demonstrates that such results are not surprising, because optimal activity of hydrogenase is 50-70 °C. Unfortunately, the high yield of hydrogen generation with thermophilic bacteria is not equivalent to total amount of generated gas (Hallenbeck, 2009). In this situation the construction of bigger reactors is required what in consequence increase total costs. Moreover, reaction performed at higher temperatures require additional thermal energy supplied to the bioreactor.

Photofermentation in hydrogen generation is the process which requires appropriate strain of bacteria, organic substances (mainly VFA) and light with appropriate intensity. The following strains of bacteria indicate activity in photoproduction of hydrogen: *Rhodobacter sphaeroides* (Koku, 2002), *Rhodobacter capsulatus* (Obeid, 2009), *Rhodovulum sulfidophilum* (Maeda, 2003), *or Rhodopseudomonas palustris* (Chen, 2008). The research of new strains active in photogeneration of hydrogen is performed in numerous laboratories all over the world. These efforts were recently awarded by discovery of activity in *Rheudopseudomanas faecalis* (Ren, 2009).

Rhodobacter sphaeroides belong to the group of bacteria the best recognized in hydrogen generation. These gram-negative bacteria belongs to the purple non-sulfur (PNS) *Proteobacteria* subgroup (Porter, 2008). The morphology is different because the shape of these bacteria as well as their dimensions strongly depends on the medium (see Fig. 3). In medium containing sugars the dimensions are limited to 2.0-2.5 x 2.5-3.0 μm, whereas under other conditions they can vary from 0.7 to 4.0 μm (Garrity, 2005).

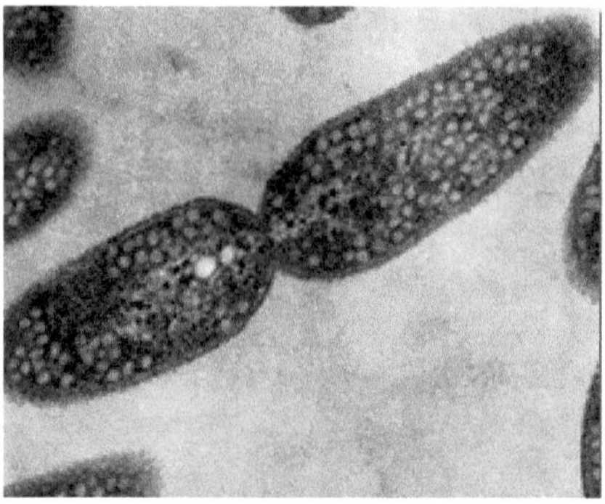

Fig. 3. *Rhodobacter sphaeroides* ATCC 17039 (Garrity, 2005).

Rhodobacter spheroides indicate strong chemotaxis with certain sugars, aminoacids and several organic acids (Packer, 2000). They are also capable to accept molecular nitrogen. Their metabolism is very elastic because they can germinate both in aerobic conditions (with or without light) as well as in anaerobic environment, in presence of light.

Under aerobic conditions this strain is used in purification of animal wastes (Huang, 2001) and biotransformation of toxins present in plant extracts (Yang, 2008). In the absence of oxygen *Rhodobacter spheroids* can be used in synthesis of carotenoides (Chen, 2006) and the most of all in hydrogen generation (Kars, 2010).

4. Enzymes

All biological processes require presence of specific enzymes. Processes of reduction of protons as well as oxidation of hydrogen (see reaction below)require at least presence of three enzymes: iron hydrogenase, nickel-iron hydrogenase and nitrogenase.

$$H_2 \leftrightarrow 2H^+ + 2e^- \quad (1)$$

Hydrogenase exists in a number (~ 40) of prokariota, both aerobic and anaerobic, as well as in certain eukariota, e.g. in photosynthetic algae (Nickolet, 2000). Hydrogenases show different but significant sensitivity towards oxygen and light. Among more than 100 discovered hydrogenases essentially only those containing Fe and Ni atoms in active center are considered as the most attractive:

- [Fe] hydrogenase - containing only Fe atoms is the most sensitive towards oxygen inhibition but almost 100 times more active than [Ni-Fe] hydrogenases
- [Ni-Fe] or [Ni-Fe-Se] - these two types of hydrogenases indicate much higher affinity towards hydrogen than [Fe] hydrogenase (Darensbourg , 2000).

Active centers of both hydrogenases are composed from iron-sulfur clusters coordinated by carbonyl (CO) or cyanide (CN^-) ligands.

Iron hydrogenases are the two directions enzymes because they catalyze both reduction of protons to the molecular hydrogen and the reverse reaction. There are three forms of these enzymes: monomeric – build only from the subunit controlling catalysis, dimeric, trimeric and tetrameric. Active centers located in these enzymes are not uniform either, however, all of them contain H-cluster (see Fig. 4) (Nicolet, 2000). Applying FTIR, EPR and XRD spectroscopy for analysis of monomeric hydrogenase, isolated from *Clostridium pasterianum* , it was found that H-cluster is composed from two basic units: [4Fe-4S] single group, responsible for electron transport, and the unique arrangement of [2Fe] capable to perform the reverse oxidation reaction of hydrogen. The regular cluster [4Fe-4S] is linked with four cysteine and sulfur atom of one of these forms the bridge bond between [4Fe-4S] and [2Fe]. In this dimeric system, the octahedral iron atoms are linked through two sulfur atoms (see Fig. 5) (Darensbourg, 2000). Moreover, it was found that these atoms are coordinated with five non-protein ligands (CO and CN-1) and water molecule. The bridge sulfur atoms forms additionally the 1,3- propanodithiol structure. The presence of covalent bond between sulfur atoms influence the charge of H-cluster and electric properties (Nicolet, 2000).

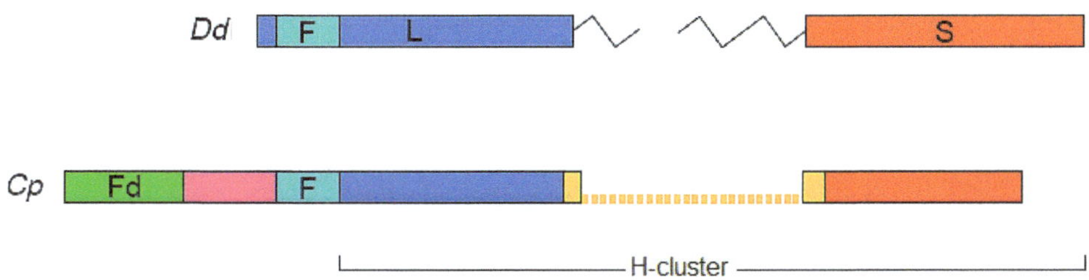

Fig. 4. Scheme of iron hydrogenase in *Desulfovibrio desulfuricans (Dd)* and *Clostridium pasterianum (Cp)*. F - double cluster of [4Fe-4S], L-large subunit of H-cluster, S - small subunit of H cluster, Fd- [2Fe-2S] cluster related to ferredoxin. Pink color represents the unique structure of [4Fe-4S]. In *Dd* hydrogenase large and small subunits are connected via cysteine, whereas in Cp hydrogenase these units are linked with protein chain.

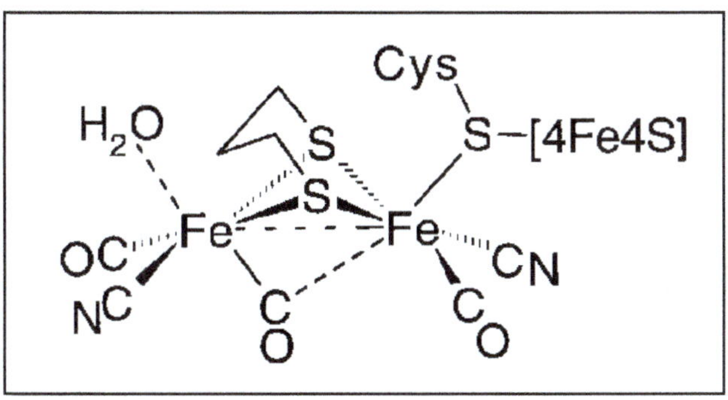

Fig. 5. Scheme of active center of iron hydrogenase (Darensbourg, 2000)

In active center of hydrogenase it is possible to identify such aminoacids as methionine and histidyne (Das, 2006). These two amino acids become attached to active center during formation of channels (for H_2 and H^+) connecting enzyme surface with reaction slit. The comparison of H-clusters in two strains of bacteria *Clostridium pasteurianum (Cp)* and *Desulfovibrio desulfuricans (Dd)* shows that in both cases the [2Fe] group is involved in hydrogen bond formation with lysine. However, when the second iron atom in *Cp* is engaged with serine, in the case of *Dd,* alanine is involved instead. In the case of fermentative bacteria of the *Clostridium* family in the large unit of monomeric iron hydrogenase it was confirmed a presence of three excessive systems: the [2Fe-2S] structure, rarely existing [4Fe-4S] structure with slit and space constructed from two [4Fe-4S] systems (Vignais, 2006).

Nickel–iron hydrogenase isolated from *Desulfovibrio gigas* and *Desulfofibrio vulgaris* is composed from large subunit α (60 kDa) containing Ni-Fe active center and small subunits β (30 kDa) equipped with three iron-sulfur clusters. These clusters are involved in electron transfer between active centers, donors and acceptors. All these clusters are located in the strait lines in which [3Fe-4S] appears between two [4Fe-4S] structures (Vignais, 2006).

The active center f [Ni-Fe] hydrogenase exhibits the unique location of ligands (see Fig. 6)

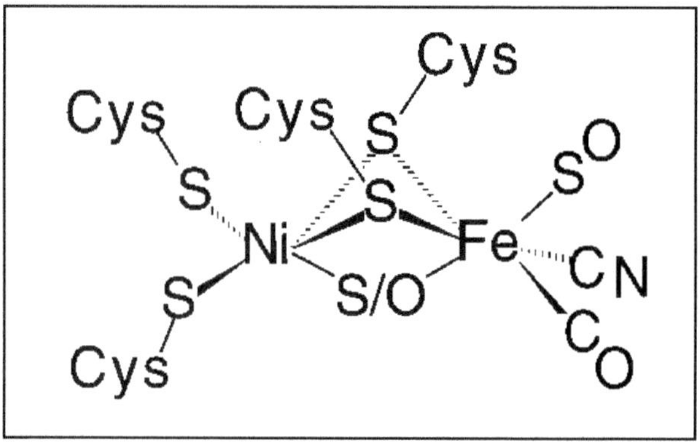

Fig. 6. Scheme of active center of nickel-iron hydrogenase (Darensbourg, 2000).

Here, four molecules of cysteine coordinate one three valent nickel atom. Two of them coordinate simultaneously iron, also located in active center. This kind of arrangement induce formation of sulfur bridges between nickel and iron atoms. Moreover , non-protein ligands such as SO, CO, CN and CO, CN are located in active centers of *D. vulgaris* and *D. gigas*, respectively. Nickel and iron atoms are bonded with monoatomic sulfur (*D. vulgaris)* or oxygen (*D.gigas*) bridges. Generated space is an ideal place for hydrogen reduction with electrons transported by iron-sulfur clusters from the surface of enzyme. The change of nickel valance form III to 0 and the return to basic state together with reconstruction of sulfur (or oxygen) bridge is observed in this catalytic cycle.

Nitrogenase is considered as the essential part of nitrogen circulation system in the living world. Nitrogen present in the air, needs to be transformed into compounds acceptable by living organisms. The diazotrophic microorganisms, including the PNS bacteria, are able to transform atmospheric nitrogen into NH_3. There three types of nitrogenases built of two separate protein units: dinitrogenase (either Mo-Fe protein, or V-Fe protein, or Fe-Fe protein) and reductase (Fe protein). The main task of reductase is the delivery of electrons of high reductive potential to nitrogenase which uses them to different reduce N_2 to NH_3. Six electrons are involved in this process to reduce the oxidation degree of nitrogen from 0 to 3. The enzyme also transfers two other extra electrons to protons with final formation of one molecule of H_2. Reduction of nitrogen to ammonia is highly energy consuming process because of the necessity of breaking the stable triple bond in nitrogen molecule and needs 16 ATP molecules per one molecule of nitrogen:

$$N_2 + 8H^+ + 8e + 16ATP \rightarrow 2NH_3 + H_2 + 16ADP + 16P_i \quad (2)$$

Both components, nitrogenase and reductase are iron-sulphur proteins, in which iron is bonded with sulphur both in cysteine and the inorganic sulphide.

Reductase (Figure 7) is a dimer with mass of 30 kDa composed of four iron atoms and four inorganic sulphides (4Fe-4S). The site for ATP/ADP bounding is located on the surface of this subunit. Reductase transfers electrons from the reduced ferredoxin towards dinitrogenase. This process occurs during hydrolysis of ATP with simultaneous dissociation of the complex.

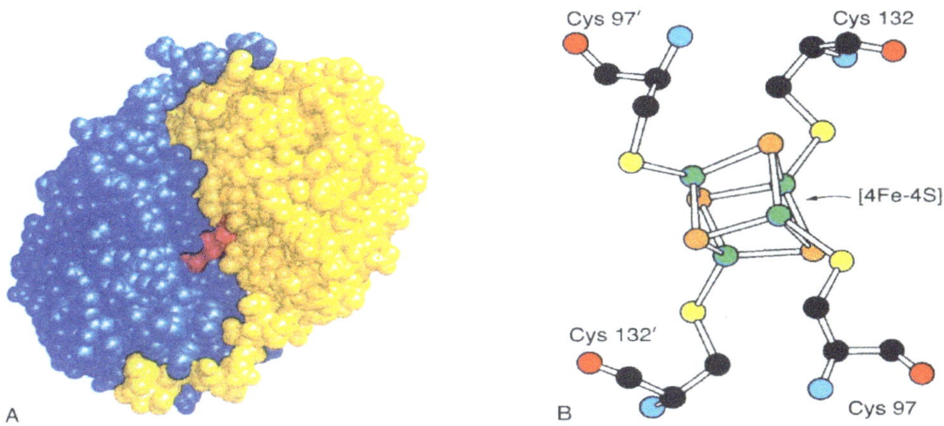

A. Red: ADP molecule obtained during ATP hydrolysis (location at the boundary of two dimers - blue and yellow),

B. [4Fe-4S] cluster located on the boundary of dimers. Green - iron, orange - inorganic sulfur, black - carbon, yellow organic sulfur, blue - nitrogen, red - oxygen,

Fig. 7. Reductase structure- nitrogenase component (Berg, 2002).

Dinitrogenase is a tetramer of the structure $\alpha_2\beta_2$ and molecular weight of 240kDa (Figure 8). At the interface between the α and β subunits there is the P unit through which electrons are able to penetrate. Two cubo-octahedrons of 4Fe-4S are linked *via* sulphur atoms from cysteine residues. The flow electrons is realized from P unit to coenzyme Fe-Mo. This coenzyme is built of two units of M-3Fe-3S linked via sulphur atoms. In one unit M stands for Mo, while in the other one for Fe. Atmospheric nitrogen is transformed in the central part of coenzyme Fe-Mo. Multiple interactions of Fe-N type weaken the triple bond in molecular nitrogen which lowers the activation limit for nitrogen reduction (Berg, 2002). The synthesis of nitrogenase strongly depends on the light access to the medium and its intensity. Catalytic stability of nitrogenase is ensured by alternating light and dark 12-hour periods (day and night sequence) (Meyer, 1978). In the absence of molecular nitrogen and with large quantities of energy provided by ATP (Koku, 2002) nitrogenase catalyses hydrogen generation (see eq.3). Nitrogenase acts as a safety valve regulating cell reduction potential (Kars, 2010).

$$2H^+ + 2e + 4ATP \rightarrow H_2 + 4ADP + 4P_i \quad (3)$$

There are two main inhibitors of nitrogenase during hydrogen photobiogeneration: molecular oxygen and nitrogen. In the presence of molecular nitrogen occurs competitive nitrogen fixation reaction and this stops almost completely hydrogen evolution. Ammonium ions at concentrations higher than 20 μmol are successful but reversible inhibitors of hydrogen generation (Waligórska, 2009) as well. The nitrogen necessary for the cell functioning is usually provided by ethanolamine and glutamate.

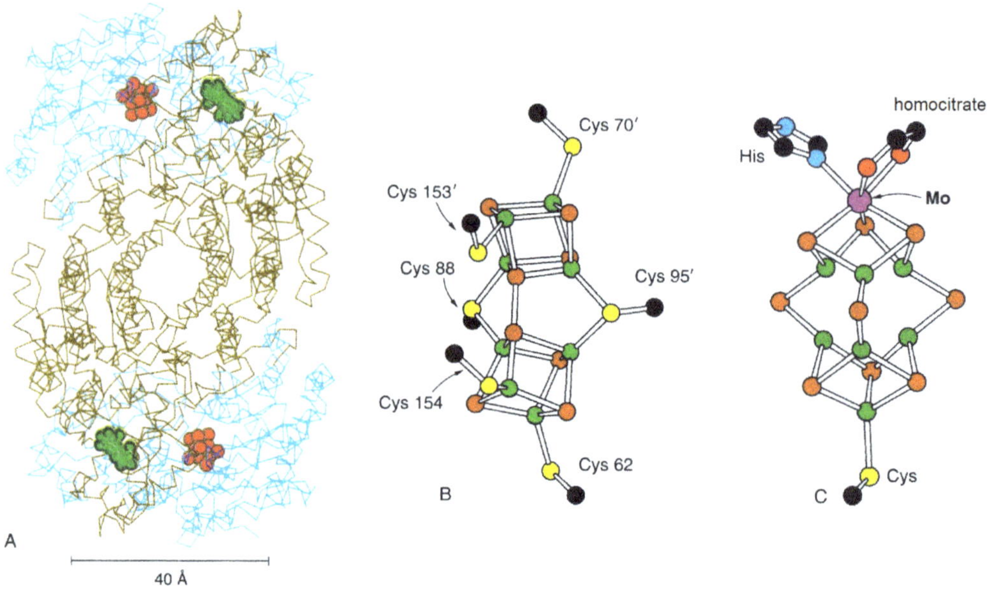

A. Blue lines - two chains of α tetramer, yellow lines – β subunits, green – P group, red – Fe-Mo coenzyme.
B. P group composed from two subunits of [4Fe-4S]. Colors description the same as in Figure 7.
C. Mo-Fe coenzyme. Mo bonded with homocitrate of histidine, Fe bonded with cysteine.

Fig. 8. Dinitrogenase construction (Berg, 2002).

However, glutamate can be the source of nitrogen inhibiting hydrogen evolution similarly as non-ammonium compounds. It can when glutamate becomes the source of carbon after the other sources are exhausted (Koku, 2002). In order to avoid such situation a medium with a relatively high ratio of organic carbon to nitrogen should be applied (e.g. malate to glutamate =15/2 (Eroglu, 1999).

5. Substrates and metabolism

The metabolism of carbon during fermentation process towards hydrogen is based on transformation of pyruvate in presence of majority of microorganisms active in this reaction. The first step of dark fermentation is based on glycolysis occurying in cytosol of cell, also known as **Embden-Mayerhof-Parnas** (EMP) pathway (Stryer, 1999). This pathway is initiated by one molecule of glucose, catalyzed by different enzymes and further transformed into 2 molecules of pyruvate. The energy liberated during oxidation of 3-phosphoglycerol aldehyde is sufficient for phosphorylation of generated acid towards 1,3-bisphosphoglycerol and reduction of NAD^+ to NADH. This reaction is catalyzed by 3-phosphoglycerol dehydrogenase. Transformation of glucose to pyruvate is during glycolysis is accompanied by formation two molecules of ATP and two molecules of NADH.

Glucose is not the only substrate in glycolysis. Simple sugars such as fructose or galactose as well as complex sucrose - saccharose, lactose, maltose, cellobiose or cellulose can be used as the initial substrate for glycolysis. However, the incorporation of these complex sugars into glycolysis pathway require initial hydrolysis to the simple carbohydrates.

Glycerol can be considred as a good substarate for glycolysis. A part of glycerol is oxidized into dihydroxyacetone by glycerol dehydrogenase. Next, dihydroxyacetone is phosphorylated into phosphodihydroxyacetone in the presence of dihydroxyacetone kinase. Thanks to triozophosphate isomerase phosphodihydroxoacetone is transformed into 3-phosphoglycerol aldehyde and further participate in EMP pathway.

There are known also other anaerobic pathways transforming glucose into pyruvate as e.g. Entner-Daudoroff or phosphate pentose pathway (Schlegel, 2003, Dabrock, 1992, Vardar-Schara, 2008, Chin, 2003).

Entner-Doudoroff pathway goes from glucose to pyruvate and is known also as 2-keto-3-detoxy-6-phosphogluconate. Here, glucose-6-phosphate is transformed with phosphogluconate dehydrogense into 6-phosphogluconate. In the next step, the removal of water from 6-phosphogluconate leads to formation of 2-keto-3-deoxy-6-phosphogluconate. This process is followed by formation of pyruvate and 3-phosphoglycol phosphate. These transformations are analogous to glycolitic pathway already described. One molecule of glucose is transformed into molecules of pyruvate with simultaneous formation of one NADPH (reduced dinucleotide nicotinoamine adenine phosphate and one molecule of ATP (Schlegel, 2003).

Pentophosphate pathway is based on initial phosphorylation of glucose to glucose-6-phosphate with help of hexokinase. Further steps are more complicated. The glucose-6-phosphate dehydrogenaze transfer hydrogen to NAD simultaneously forming of gluconolactone. The phosphate gluconolactone dehydrogenase helps to generate 6-phosphogluconate acid. The last phase is based on decarboxylation of the acid into ribuloso-6-phosphate. The transfer of this compound into riboso-5-phosphate and xylulose-5-phosphate starts a non-oxidative phase. At this stage of reaction the reversible reaction between these compounds occurs with formation of sedoheptulose-7-phosphate and 3-phosphoglycerate aldehyde. Subsequent reactions can generate fructose-6-phosphate an erythrose-4-phosphate. In further reactions erythrose-4-phosphate is transformed into 3-phosphoglycerate aldehyde and fructose-6-phosphate. Thus, one cycle of pentophosphate pathway generates 2 molecules of fructose-6-phosphate, one molecule of 3-phosphateglycerol aldehyde, 3 molecules of CO_2 and 6 molecules of NADPH. The pentophosphate pathway with glycolysis leads finally to the pyruvate formation (Schlegel, 2003).

In the next steps in anaerobic conditions, the oxidative decarboxylation of pyruvate occurs with acetylo-CoA and CO_2 formation. This reaction is catalyzed by pyruvate oxyreductase and the reduced form of ferredoxin appears as a step in final oxidation catalyzed by hydrogenase. Here, electrons reduce protons to molecular hydrogen. The reduced ferredoxin is also formed in glycolysis as the result of NADH oxidation to NAD (Dabrock, 1992). Carbon dioxide, acetic acid, lactic acid ethanol, butanol and acetone accompany hydrogen formation:

$$C_6H_{12}O_6 + 2H_2O \rightarrow 2CH_3COOH + 2CO_2 + 4H_2 \quad (4)$$

$$C_6H_{12}O_6 + 2H_2O \rightarrow CH_3COCH_3 + 3CO_2 + 4H_2 \quad (5)$$

$$C_6H_{12}O_6 \rightarrow CH_3CH_2CH_2CH_2COOH + 2CO_2 + 2H_2 \quad (6)$$

$$C_6H_{12}O_6 \rightarrow CH_3CH_2CH_2\ CH_2OH + 2CO_2 + H_2O \quad (7)$$

$$C_6H_{12}O_6 \rightarrow 2CH_3CH_2OH + 2CO_2 \quad (8)$$

$$C_6H_{12}O_6 \rightarrow 2CH_3CHOHCOOH \quad (9)$$

These reactions indicate that theoretical yield of hydrogen should 4 moles of hydrogen per one of glucose when acetone or acetic acid are among the products (Vardar-Schara, 2008).

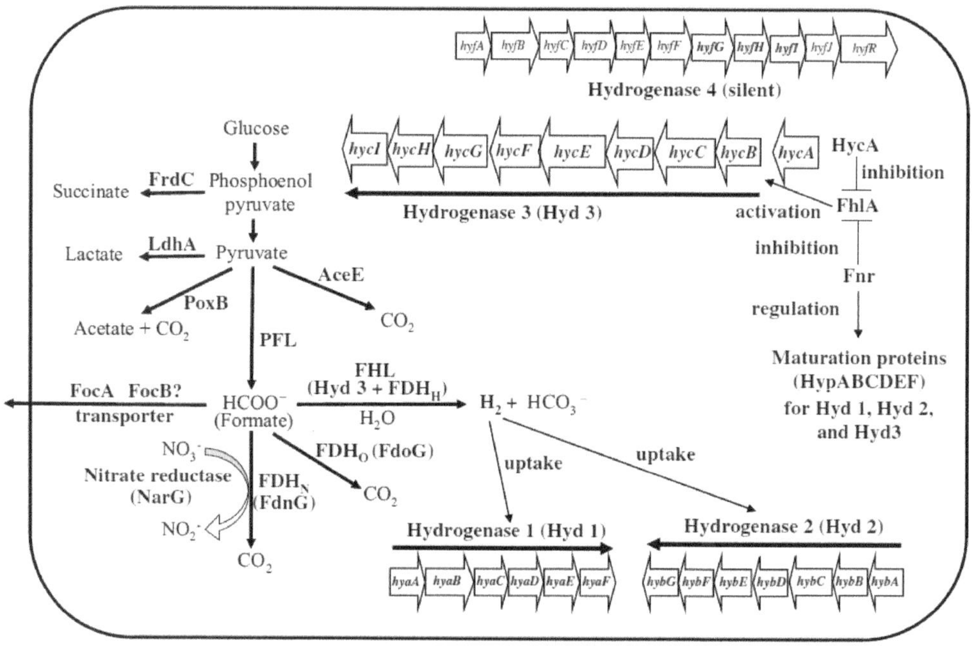

Fig. 9. Scheme of fermentative hydrogen production in *E. coli* (Maeda, 2008).

Cells metabolize glucose into phosphoenolpyruvate, pyruvate, and formate. Phosphoenolpyruvate is converted to succinate by fumarate reductase (FrdC), and pyruvate is converted to either lactate by lactate dehydrogenase (LdhA), to carbon dioxide (CO_2) and acetate by pyruvate oxidase (PoxB), to carbon dioxide by pyruvate dehydrogenase (AceE), or to formate by pyruvate formate lyase (PFL). Hydrogen is produced from formate by the formate hydrogen lyase (FHL) system consisting of hydrogenase 3 (Hyd 3) and formate dehydrogenase-H (FDHH); the FHL is activated by FhlA that is regulated by Fnr and repressed by HycA. Evolved hydrogen is consumed through the hydrogen uptake activity of hydrogenase 1 (Hyd 1) and hydrogenase 2 (Hyd 2). Formate is exported by FocA and/or FocB and is metabolized by formate dehydrogenase-N (FDHN; FdnG), which is linked with nitrate reductase A (NarG) and formate dehydrogenase-O (FDHO; FdoG). HypABCDEF are maturation proteins for hydrogenases 1, 2, and 3 (Maeda, 2008)

Transformation of pyruvate to acetylo-CoA and formic acid occurs in the presence of puruvate-formate liase with relatively anaerobic microorganisms. Formic acid is then transformed into hydrogen and CO_2 in the presence of formic-hydrogen lyase. Here, 2 molecules of hydrogen from one molecule of glucose can be generated. Similarly as in the case of completely anaerobic bacteria, pyruvate can form lactic acid (reaction 9), whereas acetylo-CoA into ethanol and acetic acid (reactions 8 and 4). These processes can lower the theoretical amounts of generated hydrogen. Additional negative effect comes from the formation of succinic acid. Namely, formate-hydrogen lyase. become active only at low values of pH what in consequence is caused by formation of acids. Thanks to the decomposition of formic acid further fermentation towards other acids can proceed (Dabrock, 1992, Hallenbeck, 2009).

The absence of photosystem II in purple non-sulphur bacteria eliminates the problem of oxygen inhibition in hydrogen generation. However, in order to decompose water molecule and generate an electron in the photobiological process, the PNS bacteria need simple organic and inorganic compounds for photosynthesis. Organic compounds are a source of carbon and electrons. The PNS bacteria can use also CO_2 as a source of carbon after transformation of metabolism into photoautotrophic one. However, if the light intensity is too low to reduce CO_2 then the cell can use H_2 and even H_2S (at low concentrations) as a source of electrons (Kars, 2010). However, CO_2 absorption is the basic metabolic process in the cell developing either in autotrophic or heterotrophic systems. The removal of RuBisCO enzyme *via* genetic modification of PNS bacteria results in the decline of photoheterotrophic development (Akkerman, 2002). Hydrogen generation with PNS bacteria can be realized in the presence of such simple organic molecules as acetate, lactate, malate or glucose. The maximum theoretical yields of conversion of these compounds to photogenerated hydrogen are described by the following equations:

(acetic acid) $$C_2H_4O_2 + 2H_2O \rightarrow 2CO_2 + 4H_2 \tag{10}$$

(lactic acid) $$C_3H_6O_3 + 3H_2O \rightarrow 3CO_2 + 6H_2 \tag{11}$$

(malic acid) $$C_4H_6O_5 + 3H_2O \rightarrow 4CO_2 + 6H_2 \tag{12}$$

(glucose) $$C_6H_{12}O_6 + 6H_2O \rightarrow 6CO_2 + 12H_2 \tag{13}$$

The theoretical amounts are usually much higher than those observed in experiments. The conversion of lactate and malate occurs easily with relatively high yields, but that of acetate and glucose is much more difficult and gives low yields of hydrogen (Kars, 2010). The discrepancies between theoretical values in hydrogen productivity and those obtained in experiments can be explained by different metabolic pathways of carbon in PNS bacteria (Figure 10, Koku, 2002).

The amount of electrons generated on absorption of organic compounds depends on the source of organic carbon. Even a slight difference in the molecular structure can lead towards completely different metabolic pathway. For example, D- and L-isomers of malate (after conversion into pyruvate) can easily join the TCA cycle. In this way the energy demand for hydrogen generation is met, whereas such a substrate as acetate is used in the other metabolic pathways: e.g. glyoxylate cycle, citramalate cycle, and ethylmalonyl-CoA pathway (Kars, 2010). The excess of electrons generated during assimilation of such substrates as glycerol or butyrate must be accepted during CO_2 photoreduction. Therefore, when the only source of carbon is glycerol, it is not assimilated in significant amounts, which is changed after supplementation of glycerol with malate. Initially, malate is assimilated from the medium and evolution of CO_2 occurs. In the second step of reaction, the evolved CO_2 permits the use of glycerol as a substrate (Pike, 1975).

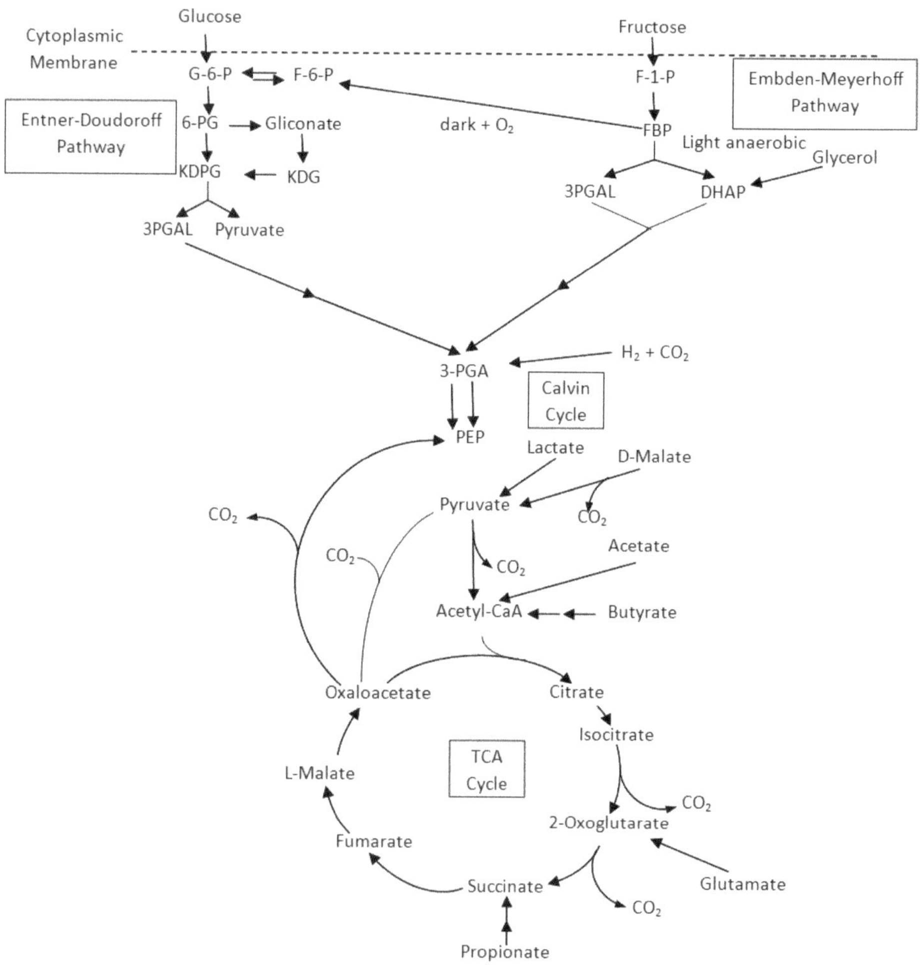

Fig. 10. Simplified scheme of carbon metabolism in *Rhodobacter sphaeroides* bacteria (Kotay, 2008).

Although the large variety of substrates can be used by photosynthetic bacteria, only a few fulfill the requirements for fast reaction rate and high yield of photogenerated H_2. In general, the preferred substrates as anions of organic acids, whereas carbohydrates do not meet the above criteria (Koku, 2002).

6. Microbiological methods of hydrogen generation

Biological methods of hydrogen generation from water in presence of microalgae and cyanobacteria are known since seventies of XX century. This process can performed in direct and undirect biophotolysis.

Direct biophotolysis

Cells of certain algae (eg. *Chlamydomonas reinhardtii, Chlorealla fusca*) or cyanobacteria are capable to split water into molecular hydrogen and oxygen under illumination.

$$2H_2O + \text{light energy} \rightarrow 2H_2 + O_2 \quad (14)$$

This process require absolutely anaerobic conditions. Light energy with wavelength lower than 680 nm is absorbed by photosystem II (PSII) and generate stream of electrons and protons originating from water. Other photosystem (PSI) is induced with light wavelength lower than 700 nm. This allows for transportation of electrons from PSII to PSI *via* chain of reductors called cytochrome *bf*. Electrons from PSI system are transferred *via* ferrodoxine to hydrogenase (algae or cyanobacteria) or nitrogenase (cyanobacteria) and these enzymes reduce protons to molecular hydrogen. In direct biophotolysis neither CO_2 nor liqid metabolites are observed. Hydrogenase is very sensitive to oxygen and irreversibly inhibits its activity, therefore constant removal of oxygen is required (Das, 2008). Recent studies concentrate on elimination of sensitivity of algae towards oxygen (Benemann, 1997).

Indirect biophotolysis

This process can be performed with certain cyanobacteria (e.g. *Anabeana variabilis*) in two steps:

$$12H_2O + 6CO_2 + \text{light energy} \rightarrow C_6H_{12}O_6 + 6O_2 \quad (15)$$

$$C_6H_{12}O_6 + 12H_2O + \text{light energy} \rightarrow 12H_2 + 6CO_2 \quad (16)$$

Different periods of oxygen and hydrogen generation allows to eliminate the inhibiting effect of oxygen on enzymes. Similarly, as in direct biophotolysis, photons activate PSI and PSII. In the presence of RuBisCO enzyme the CO_2 adsorption occurs, what in consequence of photosynthetic reactions generate glucose and oxygen . In the second step in presence of hydrogense and nitrogenase (Kars, 2009) the decomposition of organic compound occurs.

This type of metabolism in industry is difficult to perform because of periodicity of process. The hydrogen yields generated either by direct or indirect photolysis are unfortunately very low in comparison with other fermentative methods (Das, 2008).

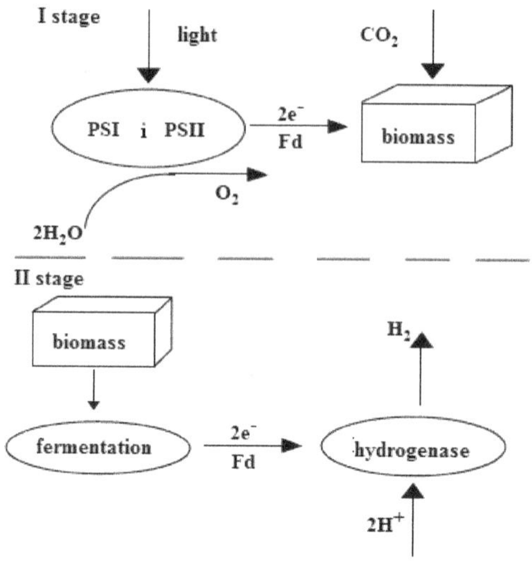

Fig. 11. Scheme of indirect photolysis of water (Maness, 2001)

Photofermentation

This process is based on decomposition of organic compounds to hydrogen in the absence of both oxygen and nitrogen but in presence of photosynthetic bacteria under illumination. Scheme on Fig. 12 shows the process of photofermentation catalyzed by nitrogenase.

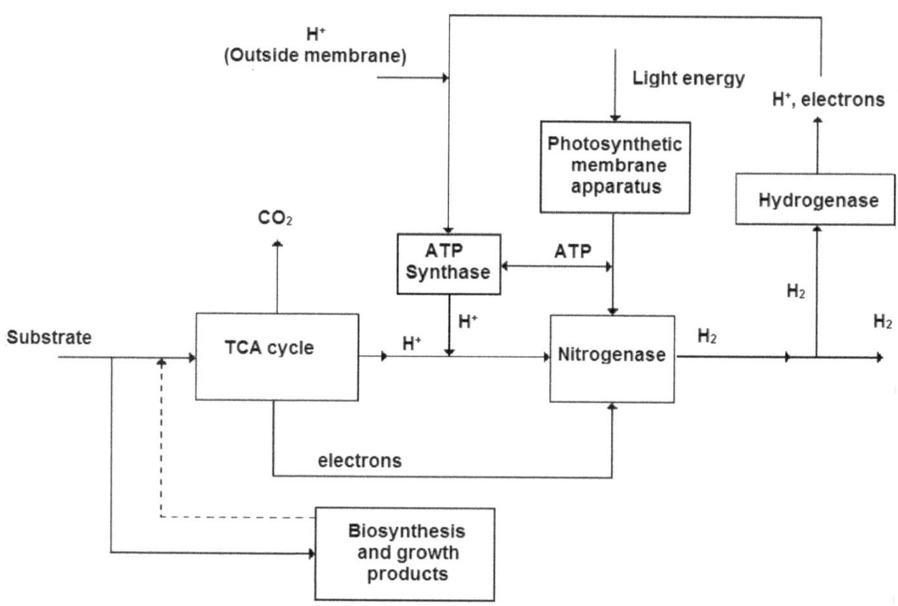

Fig. 12. Scheme of hydrogen generation in photofermentation process (Koku, 2002)

Organic substrate is oxidized to CO_2 in the cycle of tricarboxylic acids (TCA). Generated in this process electrons are transferred to nitrogenase *via* many carriers (e.g. NAD and ferredoxin). Nitrogenase reduce protons to molecular hydrogen. The photosynthetic apparatus acts simultaneously with TCA cycle transforming light into chemical energy. Here, the ATP with protons and electrons are directed towards nitrogenase. Photosynthetic bacteria contain the reverse hydrogenase enzyme which oxidize hydrogen back to protons. The final amount of photogenerated hydrogen is the difference between hydrogen formed in presence of nitrogenase and hydrogen consumed by reversed hydrogenase. The main advantage of this process rely on the high yield of hydrogen while transforming organic compounds to H_2 and CO_2.

Dark fermentation

Dark fermentation can occur in the absence of light. Anaerobic microorganisms are generating hydrogen while transforming biodegradable substances under oxygen free conditions. Unfortunately hydrogen is not the only gaseous product of this process. Carbon dioxide, methane, hydrogen sulfide can be found among generated gases, as well as liquid metabolites such as simple volatile fatty acids (VFA) and simple alcohols. In the presence of hydrogenase an organic compound is transformed in glycolysis (17) process into pyruvate. Next, it is oxidized to acetylo-Co-A with the reduction of ferredoxin (18). In the third step ferrodoxine is oxidized and evolved electrons are directed to protons and formation of molecular hydrogen (19).

$$\text{Glucose} \rightarrow \text{pyruvate} \tag{17}$$

$$\text{Pyruvate} + \text{CoA} + 2\text{Fd(ox)} \rightarrow \text{acetylo-C-A} + 2\,\text{Fd(red)} + CO_2 \tag{18}$$

$$2\text{Fd (red)} \rightarrow 2\text{Fd(ox)} + H_2 \tag{19}$$

Theoretically, one mole of glucose should generate 4 moles of hydrogen and acetic acid in dark fermentation process. In practice, this yield is lower (2.5 - 2.7 moles of H_2). Final amount of generated hydrogen depends on many factors including type and concentration of the substrate, pH value, hydraulic time of retention, substrate to inoculum ratio, Fe ions concentration etc. The relatively high rate of hydrogen production is the important factor influencing possible industrial applications.

Hybrid systems

One-step hybrid system

Application of hybrid systems allows the use of apparently useless and difficult to operate substrates in the photofermentation process. These compounds (e.g. saccharides) are decomposed in dark fermentation process into simple organic acids (e.g. acetic or butyric) which further undergo photofermentation by PNS bacteria. In one-step hybrid systems both types of bacteria grow in one pot. The amount of generated hydrogen comes from two processes occurring almost simultaneously.

Synchronization of activity of both types of bacteria cultures is the main parameter in hybrid system. The rate of reaction is usually much higher in dark fermentation than in photofermentation. As a result of this discrepancy, the excessive accumulation of VFA and alcohols is observed. High concentration of VFA in the medium leads to the substrate inhibition (Kargi, 2010) as well as to the lowering of pH value (Liu, 2010) which consequently decreases the hydrogen yield or completely stops hydrogen generation. In hybrid systems photofermentation is the rate limiting step and slows down the overall reaction rate (Ozmihci, 2010). The unfavorable effect caused by the difference in the reactions rates can be counteracted by appropriate choice of concentrations of different strains of bacteria. The optimum concentration ratios can vary from 1:3.9 (Argun, 2010) even to 1:600 (Liu, 2010) depending on the strains of bacteria and types of substrates. The use of hardly soluble substrates such as e.g. starch leads towards formation of suspensions and flocculation of bacteria cells and further to limited accessibility of organic carbon to the bacteria and decrease in the yield of microbiologically generated hydrogen (Argun, 2009).

The main advantage of one-step hybrid systems is the high rate and much higher yield of hydrogen produced in comparison to those obtained in the process of dark fermentation performed by one culture only. Further increase in the yield of hydrogen generated by hybrid systems can be achieved by application of two-step systems (Argun, Kapdan 2009).

Two-steps hybrid systems

The yield of hydrogen generation in the photofermentation process can be lowered by low access of light, inappropriate concentration of the medium, substrate inhibition, presence of ammonium ions or other contaminants (Ozmihci, 2010). Because a much greater number of parameters influence the yield of hydrogen in photofermentation than in dark fermentation, the former process should be performed in an independent photobioreactor. Application of two-step hybrid systems allows the use of wastes containing inhibitors of photofermentation process (e. g. ammonium ions) (Azbar, 2010). These inhibitors are neutral for bacteria engaged in the dark fermentation.

Moreover, separation of these processes into two-step hybrid systems extends the list of organic substrates as it permits the use of highly thermophilic bacteria operating in temperatures higher than 70 °C (Ozgur, 2010). The natural organic substrates and wastes that can be used in two-step hybrid system. One mole of glucose theoretically generates four moles of hydrogen in dark fermentation, whereas acetic acid is the only side product (Antgenent, 2004). In practice, dark fermentation of liquid wastes generates much lower amounts of hydrogen (2.5-2.7 mole H_2 per mole of glucose in waste) (Ueno, 1998, Yokoi, 2001, Yokoi, 2002). The hybrid systems are much more efficient. These results suggest that further development of two-step hybrid system can lead towards effective, economically feasible commercial applications.

7. Modifications

Genetic engineering is one of the methods for improvement of activity in hydrogen generation by microorganisms. Although the yields of generated hydrogen can be performed by optimization of the reaction conditions, genetic modifications seems to be the appropriate solution at the moment. The main idea of modification rely on implantation of other genes into the bacteria strains containing hydrogenase.

The *E. coli* are very frequently use in genetic modifications due to the well recognized metabolism of these bacteria. The *E. coli* are producing hydrogen as the result of decomposition of formic acid in presence of formate-hydrogen liaze (FHL) representing the set of enzymes localized in the inner cell membrane. Hydrogenase 3 coded as *hycA* and formate dehydrogenase known as *fdhF* are the main components of the FHL. The presence of *hycA* gene limits the synthesis of *fhlA*, responsible for better activity of FHL towards hydrogen. Therefore the removal of hycA increases the *fhlA* gene expression and in consequence hydrogen production by 5-10%. (Hallenback, 2009). The research of the FHL genes expression were performed by Bisaillon *et al.* and other authors (Bisaillon, 2006, Turcot, 2008, Penfold, 2003) and they found almost two times higher rate of hydrogen generation for modified strain of E. coli HD701. Genes responsible for nickel-iron hydrogenases (hydrogenase I and II) coded by *hya* and *hyb* operons were found in the *E. coli* genom as well. It was found that elimination of these enzymes by genetic modification can result with almost 35% higher production of hydrogen (Hallenback, 2009, Bisaillon, 2006, Turcot, 2008). Other profits originating from genetic engineering are related to deactivation of enzymes responsible for transformations of glucose into lactic, succinic and fumaric acids. The removal of *ldhA* (lactic acid) and *frdBC* (succinic and fumaric acids) genes results in increase of hydrogen formation. The 1.4 fold higher amount of hydrogen were found by Yoshida *et al.* (Yoshida, 2006) in this situation. The new mutant strain of SR 15 can produce 1.82 mol H_2/mol glucose what is close to the theoretical value (2 mol H_2/mol glucose). Studies performed by *Maeda et al.* (Maeda, 2007) showed that bacteria BW2513 with seven modified genes (*hyaB, hybC, hycA,fdoG, frdC, ldhA and aceE*) generate 4.6 fold more hydrogen than wild-type strain.

The nitrogenase and uptake hydrogenase play an important role in the photofermentation process of hydrogen generation by PNS bacteria. The engineering of the mutants free of uptake hydrogenase is the basic task of gene modifications. Genes coding hydrogenase (*hup*) can be modified by resistance gene insertion into the *hup* genes or by deletion of *hup* genes (Kars, 2009, Kars, 2008, Kim, 2006). Appropriately modified *Rhodobacter spheroids* can generate hydrogen also in the absence of light (Kim, 2008).

Production of polyhydroxybutyrate (PHB) accompany hydrogen generation by PNS bacteria what applies the excess of reducing equivalents in other metabolic pathway. The PHB is the storage material stored in cytoplasm. This compound is formed in the environment rich in carbon compounds but lean in nitrogen (Kemavongse, 2007). The PHB is unwanted competition product accompanying hydrogen generation. The removal of genes responsible for formation of PHB syntase effectively stops generation of the polymer (Kim, 2006). Low activity in PHB formation not always results in an increase of hydrogen yield. Whereas in presence of lactate, malate or malate the amount of photogenerated hydrogen is not influenced by PHB (Hustede, 1993) the presence of acetate can increase photofermentation towards hydrogen. However, the importance of PHB as biodegradable polymer significantly increased in recent years. Therefore, simultaneous photogeneration of hydrogen and PHB gained economic dimension (Yigit, 1999).

There are genetic modifications influencing changes in the amount of LHC (light harvesting complexes). The reduction of pigment present in LHC diminish the self-shadow effect and therefore better access of light into deeper located cells. The decrease of amount of LH1 (Vasilyeva, 1999) complexes with maximum of absorption at 875 nm or those with absorption maximum at 800 and 850 nm (LH2) (Kim, 2006) can increase the amount of photo generated hydrogen. Genetic manipulations cannot lead to total elimination of the pigments (Kim, 2006).

The negative influence of ammonium ions on nitrogenase is well recognized. Therefore, genetic modifications of nonsensitive to NH_4^+ ions should be the subject for considerations. Among many methods reducing the role of ammonium ions in photofermentation is blockage of Calvin cycle via mutation of genes coding the RuBisCO enzyme. This way the excess electron stream is directed to nitrogenase even in the presence of NH_4^+ ions. Another modification can be achieved by disruption of proteins transporting NH_4^+ ions through cytoplasmic membrane. Strains of this type (e.g. *Rhodobacter capsulatus*) loose their ability to regulate nitrogenase in presence of ammonium ions. (Qian, 1996). Such modifications allow to perform photofermentation even in the presence of molecular nitrogen. Although the amount of generated hydrogen is lower than in nitrogen free atmosphere but economically much more favorable (Yakunin, 2002).

Genetic modifications can be very effective but also troublesome and very expensive. Therefore other methods of process improvement are under investigations. Optimum value of pH equals 7. Photofermentation with *Rhodobacter sphaeroides* starting at pH=6.8 and ending at pH=7.5 results in significant drop of activity (7 times) but PHB concentration is tripled (Jamil, 2009).

Photofermentative bacteria belongs to mesophilic microorganisms and operate between 30 and 35 °C. Therefore, any critical temperatures act against high yield of hydrogen. For example *Rhodobacter capsulatus* operating at temperatures varying from 15-40 °C produce 50% less hydrogen than the same bacteria kept at constant temperature of 30 °C (Özgüra, 2010).

The access of photobacteria to the light with appropriate length and intensity play a crucial role m hydrogen photogeneration. Better access of light induce better phosphorylation and in consequence more effective synthesis of ATP and better yield of photofermentation (Kars, 2010).

Although the PNS bacteria absorb light in wide spectrum 400-950 nm the range of 750-950 nm is the most important (Eroglu, 2009, Ko, 2002). The light intensity is as well important as their wavelength. For *Rhodobacter sphaeroides* the amount of generated hydrogen grows linearly from 270 W/m^2 (4klx) to 600 W/m^2 (~ 10 klx). Below 270 W/m^2 no activity of bacteria is observed (Miyake, 1999, Uyar, 2007).

Application of illumination with wavelength longer than 900 nm results in overheating of the system. This require additional cooling systems because of decrease the amount of generated hydrogen. An application of appropriate filters cutting the unwanted range of spectrum seems to be the only solution in this situation (Ko, 2002). Considering natural irradiation one should remember about day-night periodicity. It was found, however, that amount of generated hydrogen is even higher under periodic irradiation than under the continuous one (Eroglu, 2010, Koku, 2003). The day-night illumination induces better activity of nitrogenase what results from better adjustment of PNS bacteria to live in natural conditions (Meyer, 1978).

The presence of organic compounds, also those containing nitrogen (except NH_4^+ ions) is the key issue for the photofermentation. However, presence of macro and microelements at appropriate concentration can influence the hydrogen productivity. Iron belongs to the most important ones. This element exists mainly as the cofactor of proteins engaged in metabolism. Process of photofermentation, strictly related to the transport of electrons.

There are many electrons carriers such as cytochromes (proteins containing Fe) or ferredoxin. Moreover, the main enzyme in photofermentation - nitrogenase contains 24 atoms of iron in each molecule. The presence of iron ions in medium containing PNS bacteria is one of the very important factors influencing hydrogen productivity. At concentrations of Fe^{2+} ions lower than 2.4 mg/l there is no hydrogen in products. At concentrations higher than 3.2 mg/l the gradual decrease of evolved hydrogen is observed. It was assumed that non physiological coagulation of the cells can occurs (Zhu, 2007). Molybdenum is the second microelement playing an important role in photofermentaive hydrogeneration. The optimal concentration of molybdenum is 16.5 μmol/l (Kars, 2006).

The substrate yield in hydrogen production can be significantly improved by adding other strains of bacteria into the liquid medium. Improvement in photofermentation was achieved by adding halofilic archeons of *Halobacterium salinarum* type. The integral membrane protein - bacteriorhodopsin as the pump for the light excited electrons. The H^+ ions are pumped out from cytoplasm outside the cell . The proton gradient is then engaged in ATP synthesis by *Rhodobacter sphaeroides* and this way increasing hydrogen generation. In this case, it is advised to use strains of PNS bacteria tolerating high concentrations of salts (Zabut, 2006) because of the high activity of bacteriorodopsyne in aqueous solution with high ionic strength.

8. Acknowledgement

This work was supported by Polish Ministry of Science and Higher Education (grant no: N N204 185440)

9. References

Akkerman, I.; Janssen, M.; Rocha, J.; Wijffels, RH. (2002). Photobiological hydrogen production: photochemical efficiency and bioreactor design, *Int J Hydrogen Energy*, vol.27, pp.1195-1208.

Alalayaha, WM.; Kalila, MS.; Kadhuma, AAH.; Jahima, JM.; Alaujb, NM. (2008). Hydrogen production using *Clostridium saccharoperbutylacetonicum* N1-4 (ATCC 13564), *Int J Hydrogen Energy*, vol.33, pp.7392-96.

Antgenent, LT.; Karmi, K.; Al-Dahhan, MH.; Wernn, BA.; Domiguez-Espinoza, R. (2004). Production of bioenergy and biochemicals from industrial and agricultural wastewater, *Trends In Biotechnol*, vol.22, pp.477-485.

Argun, H.; Kargi, F. (2010). Bio-hydrogen production from Grodnu wheat starch by continuous combined fermentation using annular-hybrid bioreactor, *Int J Hydrogen Energy*, vol.35, pp.6170-6178.

Argun, H.; Kargi, F. (2009). Effects of the substrate and cell concentration on bio-hydrogen production from ground wheat by combined dark and photo-fermentation, *Int J Hydrogen Energy*, vol.34, pp.6181-6188.

Argun, H.; Kargi, F.; Kapdan, IK. (2009). Hydrogen production by combined dark and light fermentation of ground wheat solution, *Int J Hydrogen Energy*, vol.34, pp.4305-4311.

Azbar, N.; Tuba, F.; Dokoz, C. (2010). The effect of dilution and L-malic acid addition on bio-hydrogen production with *Rhodopseudomonas palustris* from effluent of an acidogenic anaerobic reactor, *Int J Hydrogen Energy*, vol.35, pp.5028-5033.

Benemann, JR. (1997). Feasibility analysis of photobio;ogical hydrogen production, , *Int J Hydrogen Energy*, vol.22, pp.979-987.

Berg, JM.; Tymoczko, JL.; Stryer, L. (2002). Biochemystry, New York: WH Freeman.

Bisaillon, A.; Turcot, J.; Hallenbech, PC. (2006). The effect of nutrient limitation on hydrogen production by batch cultures of *Escherichia coli, Int J Hydrogen Energy*, vol.31, pp.104-1508.

Cai, G.; Jin, B.; Saint, C.; Monis, P. (2010). Metabolic flux analysis of hydrogen production network by *Clostridium butyricum* W5: Effect of pH and glucose concentration, *Int J Hydrogen Energy*, vol.35, pp.6681-90.

Chen, ChY.; Lu, WB.; Liu, CH.; Chang, JS. (2008). Improved phototrophic H_2 production with *Rhodopseudomonas palustris* WP3-5 using acetate and butyrate as dual carbon substrates, *Bioresource Technology*, vol.99, pp.3609-3616.

Chen, D.; Han, Y.; Gu, Z. (2006). Application of statistical methodology to the optimization of fermentative medium for carotenoids production by *Rhodobacter sphaeroides*, *Process Biochemistry*, vol.41, pp.1773-1778.

Chin, HL.; Chen, CP.; Chou, CP. (2003) Fed batch operation using *Clostridium acetobutylicum* suspension culture as biocatalyst for enhancing hydrogen production, *Biotechnology Progress*, vol.19, pp.383-388.

Dabrock, B.; Bahl, H.; Gottschalk, G. (1992). Parameters affecting solvent production by *Clostridium pasteurianum, Applied and Environmental Microbiology*, vol.58, pp.1233-1239.

Darensbourg, MY.; Lyon, EJ.; Smee, JJ. (2000). The bio-organometallic chemistry of active site iron in hydrogenases, *Coordination Chemistry Review*, vol.206-207, pp.533-561.

Das, D.; Dutta, T.; Nath, K.; Kotay, Das, AS.; Veziroglu, TN. (2006). Role of Fe-hydrogenase in biological hydrogen production. *Current Science*, vol.90, pp.1627-1637.

Das, D.; Veziroglu, TN. (2008). Advances in biological hydrogen production processes, *Int J Hydrogen Energy*, vol.33, pp.6046-6057.

Eriksen, NT.; Nielsen, TM.; Iversen, N. (2008). Hydrogen production in anaerobic and microaerobic *Thermotoga neapolitana, Biotechnology Letters*, vol.30, pp.103-109.

Eroglu, I.; Aslan, K.; Gunduz, U.; Yucel, M.; Turker, L. (1999). Substrate consumption rates for hydrogen production by *Rhodobacter sphaeroides* in a column photobioreactor, *Journal of Biotechnology*, vol.70, pp.103-113.

Eroglu, E.; Eroglu, I.; Gunduz, U.; Yucel, M. (2009). Treatment of olive mill wastewater by different physicochemical methods and utilization of their liquid effluents for biological hydrogen production, *Biomass and Bioenergy*, vol.33, pp.701-705.

Eroglu, E.; Gunduz, U.; Yucel, M.; Eroglu, I. (2010). Photosynthetic bacterial growth and productivity under continuous illumination or diurnal cycles with olive Mill wastewater as feedstock, *Int J Hydrogen Energy*, vol.35, pp.5293-5300.

Garrity, GM.; Brenner, DJ.; Noel, NR.; Staley, JR. (2005) Bergey's Manual of Systematic Bacteriology, Volume Two:The Proteobacteria, Part A 166, Springer-Verlag

Guo, XM.; Trably, E.; Latrille, E.; Carrere, H.; Steyer, JP. (2010). Hydrogen production from agricultural waste by dark fermentation:A review, *Int J Hydrogen Energy*, vol.35, pp.10660-10673.

Hallenbeck, PC. (2009). Fermentative hydrogen production: Principles, progress, and prognosis, *Int J Hydrogen Energy*, vol.34, pp.7379-7389.

Hallenbeck, PC.; Ggosh, D. Advances in fermentative biohydrogen production: the way forward?, *Trends in Biotechnology*, vol.27, pp.287-297.

Huang, JS.; Wu, CS.; Jih, CG.; Chen, CT. (2001). Effect of addition of *Rhodobacter sp.* To activated-sludge reactors treating piggery wastewater, *Water Research*, vol.35, pp.3867-3875.

Hustede, E.; Steinbuchel, A.; Schlegel, HG. (1993). Relationship between photoproduction of hydrogen and the accumulation of PHB in non-sulphur purple bacteria. *Appl Microbiol Biotechnol*, vol.39, pp.87-93.

Isaacs, NW.; Cogdell, RJ.; Freer, AA.; Prince, SM. (1995) Light-harvesting mechanisms in purple photosynthetic bacteria, *Current Opinion in Structural Biology*, vol.5, pp. 794-797.

Jamil, Z.; Suffian, M.; Annuar, M.; Ibrahim, S.; Vikineswary, S. (2009). Optimization of phototrophic hydrogen production by *Rhodopseudomonas palustris* PBUM001 via statistical experimental design, *Int J Hydrogen Energy*, vol.34, pp.7502-7512.

Jo, HJ.; Lee, DS.; Park, D.; Choe, WS.; Park, JM. (2008). Optimization of key process variables for enhanced hydrogen production by *Enterobacter aerogenes* using statistical methods, *Bioresoure Technology*, vol.99, pp.2061-66.

Jones, MR. (1997) Photosynthesis:A new twist to biological solar power, *Current Biology*, vol.7, pp.541-543.

Jong-Hwan, S.; Yoon, H.; Eun, J.; Ahn, K.; Mi-Sun, K.; Sang, JS.; Hyun, PT. (2007) Fermentative hydrogen production by the newly isolated *Enterobacter asburiae* SNU-1, , *Int J Hydrogen Energy*, vol.32, pp.192-199.

Kalia, VC.; Jain, SR.; Kumar, A.; Joshi, AP. (1994). Fermentation of biowaste to H_2 by *Bacillus licheniformis*, *World J Microbiol Biotechnol*, vol.10(2), pp.224-7.

Kanai, T.; Imanaka, H.; Nakajima, A.; Uwamori, K.; Omori, Y.; Fukui, T.; Atomi, H.; Imanaka, T. (2005). Continuous hydrogen production by the hyperthermophilic archaeon, *Thermococcus kodakaraensis* KOD1, *Journal of Biotechnology*, vol.116, pp. 271-282.

Kargi, F.; Ozmihci, S. (2010). Effects of dark/light bacteria ratio on bio-hydrogen production by combined fed-batch fermentation of ground wheat starch, *Biomass and Bioenergy*, vol.34, pp.869-874.

Kars, G.; Gunduz, U. (2010). Towards a super H_2 producer: Improvements in photofermentative biohydrogen production by genetic manipulations, *Int J Hydrogen Energy*, vol.35, pp.6646-6656.

Kars, G.; Gunduz, U.; Yucel, M.; Rakhely, G.; Kovacs, KL.; Eroglu, I. (2009). Evaluation of hydrogen production by *Rhodobacter sphaeroides* O.U.001 and its hupSL deficient mutant Rusing acetate and malate as carbon sources, *Int J Hydrogen Energy*, vol.34, pp.2184-2190.

Kars, G.; Gunduz, U.; Rakhely, G.; Yucel, M.; Eroglu, I.; Kovacs, KL. (2008). Improved hydrogen production by uptake hydrogenase deficient mutant strain of *Rhodobacter sphaeroides* O.U.001, *Int J Hydrogen Energy*, vol.35, pp.3056-3060.

Kars, G.; Gunduz, U.; Yucel, M.; Turker, L.; Eroglu, I. (2006). Hydrogen production and transcriptional analysis of *nif*D, *nif*K and *hup*S genes in *Rhodobacter sphaeroides* O.U.001 grown in media with different concentrations of molybdenum and iron, *Int J Hydrogen Energy*, vol.31, pp.1536-1544.

Kemavongse, K.; Prasertsan, P.; Upaichit, A.; Methacanon, P. (2007). Effect of co substrate on production of poly-β-hydroxybutyrate (PHB) and copolymer PHBV from newly identified mutant *Rhodobacter sphaeroides* U7 cultivated under aerobic-dark condition, *Songklanakarin Journal of Science and Technology*, vol.29(4), pp.1101-13.

Kim, MS.; Baek, JS.; Lee, JK. (2006). Comparison of H2 accumulation by *Rhodobacter sphaeroides* KD131 and its uptake hydrogenase and PHB synthase deficient mutant, *Int J Hydrogen Energy*, vol.31, pp.121-127.

Kim, EJ.; Kim, MS.; Lee, JK. (2008). Hydrogen evolution under photoheterotrophic and dark fermentative conditions by recombinant *Rhodobacter sphaeroides* containing the genes for fermentative pyruvate metabolism of *Rhodospirillum rubrum, Int J Hydrogen Energy*, vol.32, pp.5131-5136.

Kim, EJ.; Kim, JS.; Kim, MS.; Lee JK. (2006). Effect of changes in the level of light harvesting complexes of *Rhodobacter sphaeroides* on the photoheterotrophic production of hydrogen, *Int J Hydrogen Energy*, vol.31, pp.531-538.

Ko, IB.; Noike, T. (2002). Use of blue optical filters for suppression of growth of algae in hydrogen producing non-axenic cultures of *Rhodobacter sphaeroides* RV, *Int J Hydrogen Energy*, vol.27, pp.1297-1302.

Koku, H.; Eroglu, I.; Gunduz, U.; Yucel, M.; Turker, L. (2002). Aspects of the metabolism of hydrogen production by *Rhodobacter sphaeroides, Int J Hydrogen Energy*, vol.27, pp.1315-1329.

Koku, H.; Eroglu, I.; Gunduz, U.; Yucel, M.; Turker, L. (2003). Kinetics of biological hydrogen production by the photosynthetic bacterium *Rhodobacter sphaeroides* O.U. 001, *Int J Hydrogen Energy*, vol.28, pp.381-388.

Kotay, SM.; Das, D. (2007). Microbial hydrogen production with *Bacillus coagulans* IIT-BT S1 isolated from anaerobic sewage sludge, *Bioresoource Technology*, vol.8, pp. 1183-90.

Kotay, SM.; Das, D. (2008). Biohydrogen as a renewable energy resource - Prospects and potentials, *Int J Hydrogen Energy*, vol.33, pp.258-263.

Levin, DB.; Islam, R.; Cicek, N.; Sparling, R. (2006). Hydrogen production by *Clostridium thermocellum* 27405 from cellulosic biomass substrate, *Int J Hydrogen Energy*, vol.31, pp.1496-1503.

Liu, BF.; Ren, NQ.; Tang, J.; Ding, J.; Liu, WZ.; Xu, JF.; Cao, GL.; Guo, WQ.; Xie, GJ. (2010). Bio-hydrogen production by mixed culture of photo-and dark-fermentation bacteria, *Int J Hydrogen Energy*, vol.35, pp.2858-2862

Maeda, I.; Miyasaka, H.; Umeda, F.; Kawase, M.; Yagi, K. (2003). Maximization of hydrogen production ability In high-density suspension of *Rhodovulum sulfidopilum* Celle

Rising intracellular poly(3-hydroxybutyrate) as sole substrate, *Biotechnology and Bioengineering*, vol.81, pp.474-481.

Maeda, T.; Sanchez-Torrez, V.; Wood, K. (2008). Metabolic engineering to enhance bacterial hydrogen production, *Microbiol Biotechnology*, vol.1, pp.30-39.

Maeda, T.; Sanchez-Torrez, V.; Wood, K. (2007). Enhanced hydrogen production from glucose by metabolically engineered *Escherichia coli, Applied Microbiological Biotechnology*, vol.77, pp.879-890.

Maness, PC.; Weaver, PF. (2001). Evidence for three distinct hydrogenase activities in *Rhodospirillum rubrum. Applied Microbiology and Biotechnology*, vol.57, pp.751-756.

Mandal, B.; Nath, K.; Das, D. (2006). Improvement of hydrogen production under decreased partial pressure of H_2 by *Enterobacter cloacae, Biotechnology Letters*, vol.28, pp. 831-835.

Masset, J.; Hiligsmann, S.; Hamilton, C.; Beckers, L.; Franck, F.; Thonart, P. (2010). Effect of pH on glucose and starch fermentation in batch and sequenced-batch mode with a recently isolated strain of hydrogen-producing *Clostridium butyricum* CWBI1009, *Int J Hydrogen Energy*, vol.35, pp.3371-3378.

Meyer, J.; Kelley, BC.; Vignais, PM. (1978). Effect of light on nitrogenase function and synthesis in *Rhodopseudomonas capsulate*, *Journal of Bacterology*, vol.136, pp.201-208.

Miyake, J.; Wakayama, T. (1999). Simulation of the daily sunlight illumination pattern for bacterial photo-hydrogen production. *Journal of Bioscience and Bioenginering*, vol.88(6), pp.659-663.

Nicolet, Y.; Lemon, BJ.; Fontecilla-Camos, JC.; Peters, JW. (2000). A novel FeS cluster in Fe-only hydrogenases, *Trends in Biochem Sciences*, vol.25, pp.138

Obeid, J.; Magnin, JP.; Flaus, JM.; Adrot, O.; Willison, JC.; Zlatev, R. (2009). Modelling of hydrogen production in batch cultures of the photosynthetic bacterium *Rhodobacter capsulatus, Int J Hydrogen Energy*, vol.34, pp.180-185.

Oha, YK.; Seola, EH.; Kimb, JR.; Parka, S. (2003). Fermentative biohydrogen production by a new chemoheterotrophic bacterium *Citobacter sp.* Y19, *Int J Hydrogen Energy*, vol.28, pp.1353-1359.

Ozgur, E.; Mars, AE.; Peksel, B.; Louwerse, A.; Yucel, M.; Gunduz, U.; Pieternel, AM. (2010). Biohydrogen production from beet molasses by sequential dark and photofermentation, *Int J Hydrogen Energy*, vol.35, pp.511-517.

Ozmihci, S.; Kargi, F. (2010). Effects of starch loading rate on performance of combined fed-batch fermentation of ground wheat for bio-hydrogen production, *Int J Hydrogen Energy*, vol.35, pp.1106-1111.

Ozmihci, S.; Kargi, F. (2010) Bio-hydrogen production by photo-fermentation of dark fermentation effluent with intermittent feeding and effluent removal, *Int J Hydrogen Energy*, vol.35, pp.6674-6680.

Özgüra, E.; Uyard, B.; Ozturkb, Y.; Yucel, M.; Gunduz, U.; Eroglu, I. (2010). Biohydrogen production by *Rhodobacter capsulatus* on acetate at fluctuating temperatures, Resources, *Conservation and Recycling*, vol.54, pp.310-314.

Packer, HL.; Armitage, JP. (2000). Inverted behavioral responses in wild-type *Rhodobacter sphaeroides* to temporal stimuli, *FEMS Microbiology Letters*, vol.185, pp.299-304.

Paschenkoa, VZ.; Gorokhova, VV.; Knoxa, PP.; Krasilnikova, PM.; Redlinb, H.; Rengerc, G.; Rubina, Ab. (2003). Energetic and mechanisms of light efficiency of charge separation and electron transfer processes in *Rhodobacter sphaeroides* reaction center, *Bioelectochemistry*, vol.61, pp.73-84.

Penfold, WW.; Forster, CF.; Macaskie, LE. (2003). Increased hydrogen production by Escherichia coli strain HD701 in comparison with the wild-type parent strain MC4100, *Enzyme and Microbial Technology*, vol.33, pp.185-189.

Pike, L.; Sojka, GA. (1975). Glycerol dissimilation in *Rhodopseudomonas sphaeroides, Journal of Bacteriology*, vol.124, pp.1101-1105.

Plowman, J.; Peck, MW. (2002). Use of a newel methods characterize the response of spores of non-proteolytic *Clostridium botulinum* types B,E and F to a wide range of germinates and conditions, *Journal of AppliedMicrobiology*, vol.92, pp.681-694

Porter, SL.; Wadhams, GH.; Armitage, JP. (2008). *Rhodobacter sphaeroides*: complexity in chemo tactic signaling, *Trends in Microbiology*, vol.16, pp.251-260.

Qian, Y.; Tabita, FR. (1996). A Global Signal Transduction System Regulates Aerobic and Anaerobic CO2 Fixation in *Rhodobacter sphaeroides, Journal of Bacteriology*, vol.178(1), pp.12-18.

Ren, NQ.; Liu, BF.; Ding, J.; Xie, GJ. (2009). Hydrogen production with *R. Faecalis* RLD-53 isolated from freshwater pond sludge, *Bioresource Technology*, vol.100, pp.484-487.

Schlegel, G. (2003) , PWN, Warszawa

Setlow, P. (2007). I will survive: DNA protection in bacterial spores, *Trends in Microbiology*, vol.15, pp.172-180.

Skonieczny, MT.; Yargeau, V. (2009) Biohydrogen production from wastewater by *Clostridium Beijerinckii*: Effect of pH and substrate concentration, *Int J Hydrogen Energy*, vol.34, pp.3288-3294.

Stryer, L. (1999). Biochemistry PWN, Warszawa

O-Thonga, S.; Prasertsanb, P.; Karakasheva, D.; Angelidakia. (2008). Thermophilic fermentative hydrogen production by the newly isolated *Termoanaerobacterium thermosaccharolyticum* PSU-, *Int J Hydrogen Energy*, vol.33, pp.1204-1214.

Turcot, J.; Bisaillon, A.; Hallenbeck, PC. (2008). Hydrogen production by continuous cultures of Escherichia coli under different nutrient regimes, *Int J Hydrogen Energy*, vol.33, pp.1465-1470.

Ueno. Y.; Otauka, S.; Morimoto, M. (1998). Hydrogen production from industrial wastewater by anaerobic microflora in hemostat culture, *J Ferment Bioeng*, vol.82, pp.194-197.

Uyar, B.; Eroglu, I.; Yucel, M.; Gunduz, U.; Turker, L. (2007). Effect of light intensity, wavelength and illumination protocol on hydrogen production in photobioreactors, *Int J Hydrogen Energy*, vol.32, pp.4670-4677.

Valdez-Vazquez, Rios-Leal, E.; Esparza-Garcia, F.; Cecchi, F.; Poggi-Varaldo, HN. (2005). Semi-continous solid substrate anaerobic reactors for H_2 production from organic waste: Mesophilic versus thermophilic regime, *Int J Hydrogen Energy*, vol.30, pp.1383-1391.

Vardar-Schara, G.; Maeda, T.; Wood, TK. (2008). Metabolically engineered bacteria for producing hydrogen via fermentation, Microbial Biotechnology, vol.1, pp. 107-125.

Vasilyeva, l.; Miyake, M.; Khatipov, E.; Wakayama, M.; Sekine, M.; Hara, M.; Nakada, E.; Asada, Y.; Miyake, J. (1999). Enhanced Hydrogen Production by a Mutant of *Rhodobacter sphaeroides* Having an Altered Light-Harvesting System, *Journal of Bioscience and Bioengineering,* vol.87(5), pp.619-624.

Vermeglio, A.; Joliot, P. (1999). The photosynthetic apparatus *of Rhodobacter shaeroides, Trends in Microbiology,* vol.7(11), pp.435-440

Vignais, PM.; Billoud, B.; Meyer, J. (2001). Classification and phylogeny of hydrogenases, *FEMS Microbiology Reviews,* vol.25, pp.455-501.

Waligórska, M.; Seifert, K.; Górecki, K.; Moritz, M.; Łaniecki, M. (2009). Kinetic model of hydrogen generation by *Rhodobacter sphaeroides* in the presence of NH_4^+ ions, *Journal of Applied Microbiology,* vol.107, pp.1308-1318.

Wang, CC.; Chang, CW.; Chu, CP.; Lee, DJ.; Chang, BV.; Liao, CS. (2003). Producing hydrogen from wastewater sludge by *Clostridium bifermentans, Journal of Biotechnology,* vol.102, pp.83-92.

Wu, X.; Li, Q.; Dieudonne, M.; Cong, Y.; Zhou, J.; Long, M. (2010). Enhanced H_2 gas production from bagasse using adhE inactivated *Klebsiella oxytoca* HP1 by sequebtial dark-photo fermentation, *Bioresource Technology,* vol.101, pp.9605-9611.

Yakunin, AF.; Hallenbeck, PC. (2002). AmtB Is Necessary for NH4-Induced Nitrogenase Switch-Off and ADP-Ribosylation in *Rhodobacter capsulatus, Journal of Bacteriology,* vol.184, pp.4081-4088.

Yang, G.; Zhang, Z.; Bai, H.; Gong, J.; Wang, Y.; Li, B.; Li, J. (2008). Biotransformation of β-Amyrin acetate by *Rhodobacter sphaeroides, Journal of Bioscience and Bioengineering,* vol.105, pp.558-561.

Yokoi, H.; Saitsu, A.; Uchida, H.; Hirose, J.; Hayashi, S.; Takasaki, Y. (2001). Microbial Hydrogen Production from Sweet Potato Starch Residue, *J Bioscien Bioeng,* vol.91, pp.58-63.

Yokoi, H.; Maki, R.; Hirose, J.; Hayashi, S. (2002). Microbial production of hydrogen from starch-manufacturing wastes, *Biomass Bioenerg,* vol.22, pp.389-395.

Yoshida, A.; Nishimura, T.; Kawaguchi, H.; Inui, M.; Yukawa, H. (2006). Enhanced hydrogen production from glucose using ldh- and *frd*-inactivated Escherichia coli strains, *Applied Microbiological Biotechnology,* vol.73, pp.67-72.

Yigit, D.; Gunduz, U.; Turker, L.; Yucel, M.; Eroglu, I. (1999). Identification of by-products in hydrogen producing bacteria; *Rhodobacter sphaeroides* O.U. 001 grown in the waste water of a sugar refinery, *Journal of Biotechnology,* vol.70, pp.125-131.

Zabut, B.; El-Kahlout, K.; Yucel, M.; Gunduz, U.; Turker, L.; Eroglu, I. (2006). Hydrogen gas production by combined systems of *Rhodobacter sphaeroides* O.U.001 and *Halobacterium salinarum* in a photobioreactor, *Int J Hydrogen Energy,* vol.31, pp. 1553-1562.

Zhang, T.; Liu, H.; Fang H. (2003). Biohydrogen production from starch in wastewater under thermophilic condition, *Journal of Environmental Mangament* , vol.69, pp.149

Zhu, H.;. Fang, HP.; Zhang, T.;. Beaudette, LA. (2007). Effect of ferrous ion on photo heterotrophic hydrogen production by *Rhodobacter sphaeroides, Int J Hydrogen Energy*, vol.32, pp.4112-4118.

14

Photofermentative Hydrogen Generation in Presence of Waste Water from Food Industry

Krystyna Seifert, Roman Zagrodnik,
Mikołaj Stodolny and Marek Łaniecki
Faculty of Chemistry. A. Mickiewicz University, Poznań, Poland

1. Introduction

Constantly increasing demand for energy has created extensive consumption of fossil fuels and the thread of their exhaustion has became a serious concern. At the same time it has been an inspiration for search for new, environmental friendly energy sources, out of which hydrogen seems to be one of the most promising. It is easily accessible, harmless, renewable and effective (high heat of combustion) energy carrier (Ball, 2009). Within the numerous methods of hydrogen production, biological methods (so called "green technology") have gained substantial importance. These methods consist of fermentative decomposition of organic substances, biophotolysis of water by algae and cyanobacteria, decomposition of organic compounds by photosynthetic bacteria and two-stage hybrid systems with fermentative and photosynthetic bacteria (Waligórska, 2006, Koku, 2002, Su, 2009).

Photofermentation represents the process where heterotrophic bacteria in the presence of light decompose organic substances and produce hydrogen and CO_2. It has been already shown that purple non-sulphur bacteria *Rhodobacter sphaeroides* act as efficient biocatalyst in the process of hydrogen production from the wastes coming from breweries and dairy industry. Brewery wastes carry high concentration of organic compounds (COD 0.8-2.5kg/hl of beer) and represent high volumes (1.3-1.8 hl/hl of beer). The amount of waste during beer production is enormous and equals the amount of water applied for production diminished with water present in beer (usually 3-4 hl of waste per 1 hl of beer). A chemical composition of waste strongly depends on the kind of beer produced and fermentation degree. Such waste can contain aminoacids, proteins, organic acids, sugers, alcohols, as well as vitamins of the B group. (Wojnowska-Baryła, 2002, Srikanth, 2009, Cui, 2009) As far as dairy wastes are concerned, they contain an average of 5-50 g O_2 /l. These wastes are mainly composed of remaining of milk, fats and whey. Typical Polish dairy produces 450-600 m^3 of wastes per day, half of which goes directly to rivers, lakes and to the ground. These wastes easily undergo fermentation, which causes acidification, intense oxygen consumption, bottom sedimentation and growth of fungi. The organics in both dairy and brewery wastes represent the efficient substrate for *Rhodobacter sphaeroides* and seem to be a promising source for energy production. The efficient use of food wastes in hydrogen generation with simultaneous degradation of these laborious wastes seems to be a very environmentally friendly solution. The US Department of Energy Hydrogen Program in United States estimates that contribution of hydrogen to total energy market will be 8-10% by 2025 (National Hydrogen Energy Roapmap, 2002). It is predicted that hydrogen will become the main carrier of energy in the near future due to environmental and universal applications reasons. It is clean, highly energetic energy carrier (142.35 kJ/g), with almost tripled gravimetric energy density compared to ordinary hydrocarbons. Although the described method is relatively simple and cheap it still requires optimization due to the obtained unsatisfied yields.

2. Materials and methods

2.1 Inoculum, medium and procedures

Photoheterotrophic bacteria *Rhodobacter sphaeroides* O.U. 001 ATTC 4919 (Fig.1) were cultivated on Van Niel's medium containing: K_2HPO_4 (1.0 g/l), $MgSO_4$ (0.5 g/l), yeast extract (10g/l) and tap water filled up to 1 l and then activated according to the procedure already described (Waligórska, 2006). For hydrogen generation a modified Biebl and Pfennig medium (Biebl, 1981) was applied. This standard medium contained: KH_2PO_4 (0.5 g/l); $MgSO_4$*$7H_2O$ (0.2 g/l); NaCl (0.4 g/l); $CaCl_2$*$2H_2O$ (0.05 g/l), L-malic (2.0 g/l); sodium glutamate (0.36 g/l), ferric tartrate (0.005 g/l); yeast extract (0.17 g/l) and microelements: $ZnCl_2$ (0.07 g/l); $MnCl_2$*$4H_2O$ (0.1 g/l); H_3BO_3 (0.06 g/l); $CoCl_2$*$6H_2O$ (0.2 g/l); $CuCl_2$*$2H_2O$ (0.02 g/l); $NiCl_2$*$6H_2O$ (0.02 g/l); Na_2MoO_4*$2H_2O$ (0.04 g/l); HCl 25% (1ml/1).

The untreated food waste were initially filtered through cotton wool, next sterilized at 120°C by autoclaving for 20 min and re-filtered applying paper filter.

Wastes with different COD values (46 g O_2/l for dairy wastes, 220 and 27 g O_2/l for brewery wastes) after pretreatment were introduced to the medium, which did not contain L-malic acid. The medium was inoculated with bacteria 30% v/v (0.36 g dry wt/l). The process was performed in small vials (25 ml) made from sodium glass and filled with 12.5 ml of inoculated medium. Tightly closed vials were carefully deaerated with argon before starting the illumination. All experiments were carried out at 28 ± 2°C and pH after sterilization and inoculation varied between 7.0 and 7.2. The mercury-tungsten lamp (Ultra-Vitalux –300W from Osram) was applied in all experiments. The intensity of light during hydrogen generation was 9 klx (116 W/m^2). The vials with Biebl and Pfenniga standard medium was used as reference (Biebl, 1981).

2.2 Analytical methods

The content of H_2, CO_2 was measured with gas chromatography (Varian GC-3800 equipped with Carboplot P7 capillary column and TCD). The loss of organic substances was monitored with COD measurement (dichromate method) after centrifugation of biomass (Standard methods, 1995). The biomass content was established spectrophotometrically measuring optical density at 660 nm (DU640 UV-VIS spectrophotometer from Beckmann). Cell dry weight was determined using gravimetric method. Six samples from the same kinetic measurement points at respective time intervals were mixed together, 10 ml of cell suspension was centrifuged at 12000 g for 12 min, the pellet was washed twice with deionized water and dried at 80°C for 4 h. Elemental analysis of the foot wastes (C,H,N,O) was performed in triplicate using an elemental analyser (Vario EL III Elementary). Concentration of Fe, Ca, Mg in purified wastes was measured by ICP OES spectroscopy. The value of pH was measured with glass electrode ERH-11. The intensity of luminance was measured at the external wall of the bottle with a luxometer Lx204 made by Slandi, Poland and a pyranometer CMP3 by Kipp & Zonen (Waligórska, 2006). The light conversion efficiency (η) was calculated based on the following formula (Koku, 2002):

$$\eta(\%) = \frac{33.61 \cdot \rho \cdot V}{I \cdot A \cdot t} \tag{1}$$

where "V" is the volume of produced H_2 in liters, "ρ" is the density of the produced hydrogen gas in g/l, "I" is the light intensity in W/m^2, "A" is the irradiated area in m^2 and "t" is the duration of hydrogen production in hours.

Substrate efficiency Y_{sub} (l /l waste) was calculated as final hydrogen concentration per l of waste:

$$Y_{sub} = \frac{H_{max}}{V_{waste}} \tag{2}$$

where H_{max} is a final hydrogen concentration in l, V_{waste} is waste concentration in l.

Specific efficiency Y_{sp} (l H_2 / g COD) was calculated based on following equation:

$$Y_{sp} = \frac{H_{max}}{COD_{loss}} \tag{3}$$

The modified Gompertz (Eq. 4) was applied for calculations of cumulative amounts of hydrogen and carbon dioxide (Mu, 2007, Nath, 2008, Chen, 2006):

$$H = H_{max} \exp\left\{-\exp\left[\frac{R_{max,H_2} e}{H_{max}}(\lambda - t) + 1\right]\right\} \tag{4}$$

where: H - cumulative hydrogen (l/l_{medium}), H_{max} - maximum cumulative hydrogen (l/l_{medium}), $R_{max,\,H2}$ –maximum rate of hydrogen production (l/l/h), t - fermentation time (h), λ - lag time (h), e - exp = 2.718.

3. Results and discussion

3.1 Pretreatment of wastes

The wastes applied in this series of experiments required high temperature pretreatment (120°C for 20 min), which had significantly increased the efficiency of hydrogen production by removing from the crude waste microorganisms realizing competitive fermentation. The crude wastes were acidic (dairy waste pH 4.2, brewery waste pH 4.7) and contained high concentration of NH_4^+ (40 mg/l dairy waste and 96 mg/l brewery waste), which can significantly reduce hydrogen production (Waligórska, 2009). High concentration of

N-NH_4^+ as well as N_2 inhibits hydrogen production by nitrogenase. In the absence of nitrogen in the system the nitrogenase catalyses the reduction of protons to molecular hydrogen (Melis, 2006, Yakunin, 1988, Pawlowski, 2003, Dubbs, 2004).

The characteristics of applied wastes is given in Table 1. In order to establish the influence of the wastes pretreatment conditions on the final production of hydrogen a series of experiments with non-treated and sterilized waste were performed. These measurement were performed with solution containing brewery waste at concentration of 10% v/v inoculated with 10% and 30% v/v of inoculums. The results of these experiments are shown on Fig.2. The experiments with "raw", undiluted dairy waste failed.

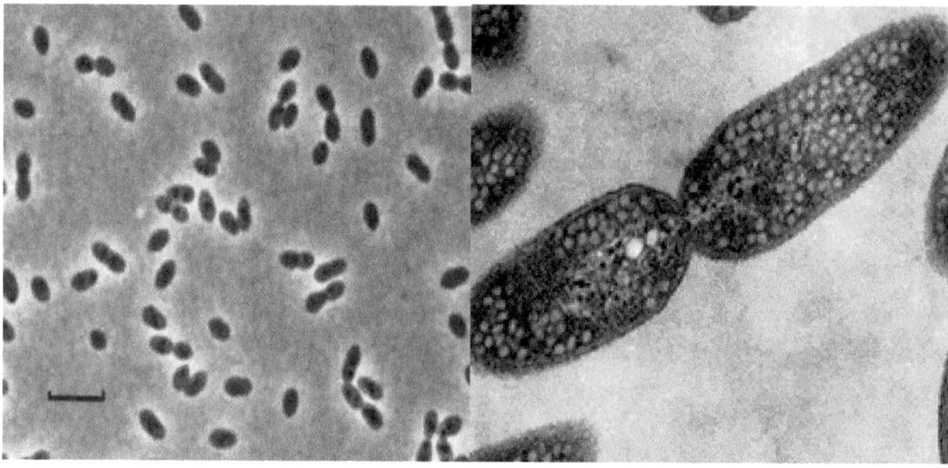

Fig. 1. *Rhodobacter sphaeroides* ATCC 17032. Micrographs performed with electron microscope with phase contrast (PCM). Tab on left micrograph equals 5 μm (Garrity, 2005).

Parameters	pH	COD [g/dm³]	NH_4^+ [g/dm³]	N [%]	C [%]	H [%]	S [%]	Ca [mg/l]	Fe [mg/l]	Mg [mg/l]
Brewery waste I	4.7	220	96	0.7	36.7	7	0.05	37.2	1.04	96
Brewery waste II	8.5	27	12	0.5	13.3	3	0.01	88.4	0.8	58.6
Dairy waste	4.2	46	40	1.05	35.5	6.3	0.08	1043	0.54	80

Table 1. Characteristics of the food wastes.

Application of the sterilized brewery waste with concentration of inoculum 10% v/v resulted in double amount of produced hydrogen. Triplication was observed at higher concentration of inoculums (30% v/v). Many laboratories apply similar pretreatment conditions. Thermal treatment at 95°C for 45 min (Yetis, 2000), filtration or sedimentation (Salih,1989) as well as dilution leads towards removal of fermentation bacteria and solid sediments from medium. Moreover, were applied: illumination with UV radiation, termal treatment at 50°C in presence of 1vol. % of hydrogen peroxide. It was found that only thermal sterilization was successful method.

Many food waste, for example dairy waste, contained significant amounts of whey particles. It was interesting to check whether microorganisms utilize only organic compounds from solution or may be originate from consumption of solid particles as well? The results indicated that at higher concentration of waste the amount of generated hydrogen increased about 40 - 60% when we non-filtered waste (Fig 3). The only exception can be observed In waste with lower concentration 5 % v/v. Here, large differences in hydrogen production are not observed.

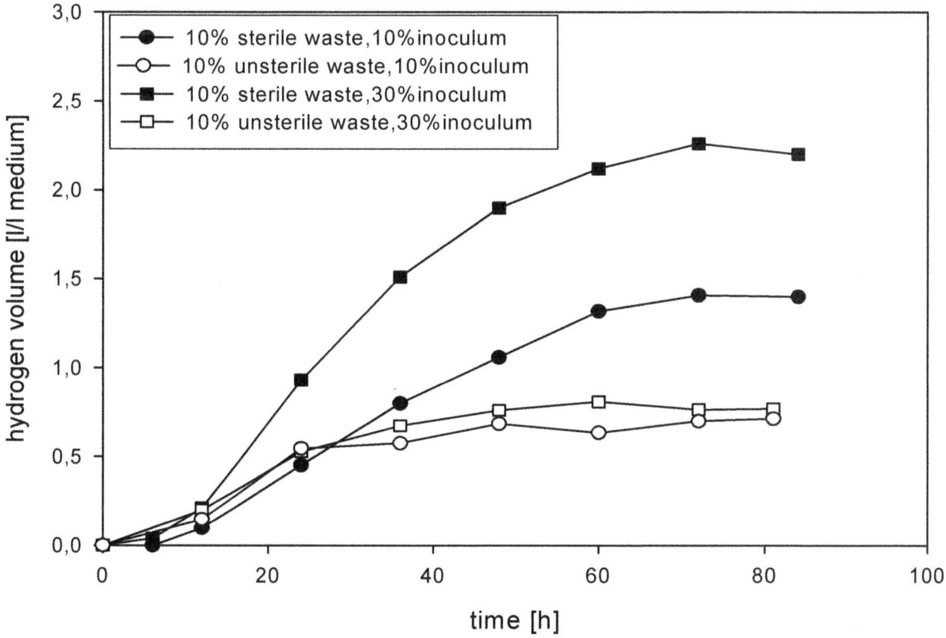

Fig. 2. Influence of sterilization of brewery wastewaters on kinetics of hydrogen generation. (Seifert, 2010)

The whey suspension contains 5 wt.% of lactose, proteins, fats and lactic acid. However, all these components can be an excellent source of organic carbon for *R. sphaeroides* during hydrogen generation (Koku, 2002) due to relatively good solubility. Obeid et al.(2009) used lactic acid as a source of organic carbon in hydrogen generation applying *Rhodobacter capsulatus* and obtained relatively high yield of H_2 (5.5 l H_2/l) but the acclimatization time was long and lasted 24 h. Sugars, proteins as well as fatty acids were already applied as substrates in hydrogen photogeneration (Eroglu, 2004; Yokoi, 2002; Zhu, 1999)

3.2 Light intensity effect

For these series of experiments the medium containing 40% v/v of dairy waste and 10% v/v of brewery waste were applied. The media were inoculated with *Rhodobacter sphaeroides* O.U.001 in concentration 30% v/v 0,36 g dry wt/l. The effect of the light intensity was checked out for 5, 9 and 13 klx. (Fig.4). The highest volumes of hydrogen (3.2 l H_2/l medium for dairy wastes, and 2.3 l H_2/l medium for brewery wastes) were observed when 9 and 13 klx were applied.

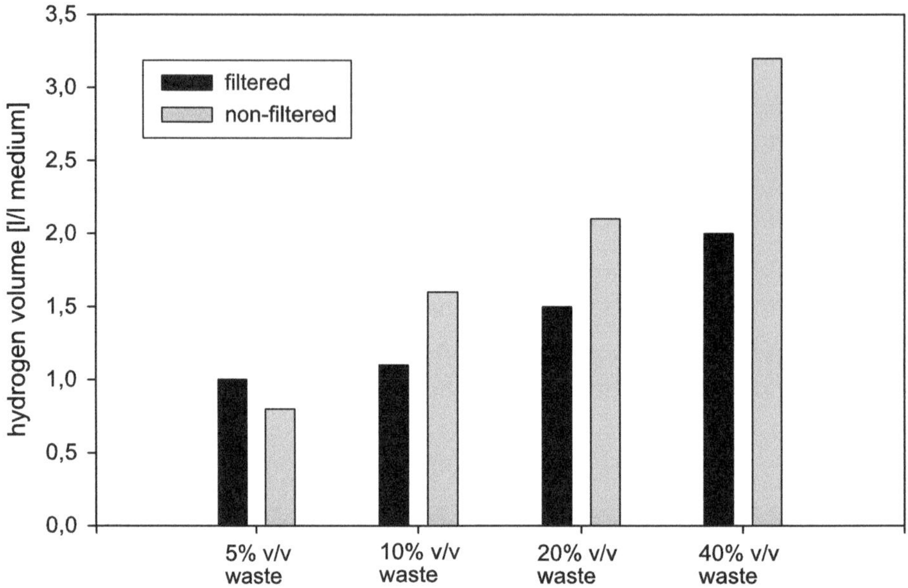

Fig. 3. Influence of filtration of dairy wastewaters on hydrogen generation.

Similar light intensity (8klx) was use by Zhu et al. for tofu wastewaters treated with *Rhodobacter sphaeroides*. Volume of hydrogen obtained for these conditions was 1.5 l H_2/l medium (2.8 l H_2/l medium when glucose was used) (Zhu, 1999). Nath et al. showed the best results of hydrogen generation when 10 klx was applied (Nath, 2009). Li et al., however, studying the photofermantation of glycerol with *Rhodobacter sphaeroides ZX-5* proved the highest hydrogen production with light intensity not exceeding 5 klx (Li, 2009). Surprisingly, high light intensity was tested by Obeid et al. for photofermentation of lactate medium and *Rhodobacter capsulatus IR3* (up to 50 klx). Highest effectiveness and rate of hydrogen production were obtained when 30-50 klx were used. These tests are essential taking into account that light intensity on sunny day can be higher than 100 klx.

Light intensity seems to be an important factor in hydrogen photogenerating process. On one hand increase at light intensity stimulates hydrogen production and biomass growth, on the other hand too high intensity may cause the reduction of nitrogenase activity or even damage of the cells (Asada, 1999, Uvar, 2005). An important parameter which shows the relationship between light intensity, irradiation area, duration of H_2 production and total H_2 amount is the light conversion efficiency (η, equation 1). It is the ratio of the total energy of the obtained hydrogen to the total energy input of the photobioreactor by solar radiation (Eroğlu, 2007). In our tests η reached the highest value when 9 klx was applied (2.4 % for dairy waste, 1.7 for brewery waste, Table 2) Results in Table 2 show that illumination with

5 klx leads also to high light conversion efficiency. However, in this case the duration of the process significantly increases. For this reason 9 klx seemed to be optimal and has been used for further experiments.

	standard			40% dairy waste			10% brewery waste		
Light intensity [klx]	final time [h]	hydrogen [l/l medium]	η [%]	final time [h]	hydrogen [l/l medium]	η [%]	final time [h]	hydrogen [l/l medium]	η [%]
5	106	2.26	1.76	72	1.7	1.9	96	2.0	1.7
9	76	2.3	1.73	60	3.2	2.4	60	2.26	1.7
13	48	2.0	1.33	60	3.15	1.7	60	2.3	1.2

Table 2. The effect of light intensity on duration of the process, hydrogen production and light conversion efficiency (η) (30% v/v inoculum).

3.3 The effect of inoculum concentration

In these series of experiments we tested several concentrations of inoculum introduced to the medium: 5-40 % v/v (0.086 g dry wt/l - 0.48 g dry wt/l) for standard medium and 10% and 30% (0.086 g dry wt/l and 0.36 dry wt/l) in case of medium containing wastes (Fig. 5). The optimum inoculum concentration in all cases turned out to be 30% v/v. Data in Table 3 indicate that the second higher concentration produces more hydrogen, shorter lag phase

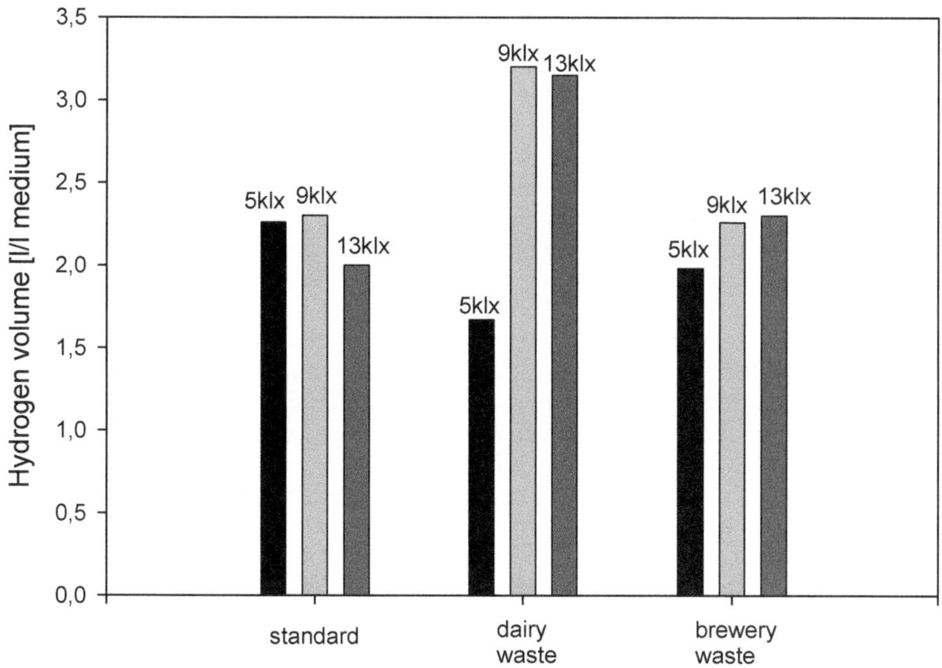

Fig. 4. The effect of light intensity on hydrogen production in photofermentation process (30% v/v inoculum, 40% v/v dairy waste, 10% v/v brewery waste)

and bigger COD loss. Increase to 40% lead to smaller amount of produced H_2. This effect seems to be caused by the fact that with the inoculum, apart from biomass, we also introduced metabolites which in high concentrations negatively influence the efficiency of photofermentation (Waligórska, 2006, Koku, 2002).

Inoculum concentration (g dry wt/l)	H_{max} (l/l)	$R_{max,H2}$ (l/l/h)	λ_{H2} (h)	Y (l H_2/l waste)	pH final	COD loss (g O_2/l)	Bio-mass (g/l)
Dairy waste							
0.086	2.52±0.17	0.057±0.018	18.0±7.6	5.8	7.3	3.5	2.0
0.36	3.23±0.21	0.049±0.007	14.5±4.3	7.6	6.9	4.2	2.2
Brewery waste I							
0.086	1.41±0.04	0.034±0.004	11.6±2.9	13.6	6.1	3.1	2.7
0.36	2.24±0.09	0.061±0.009	9.4±2.6	19	6.2	3.8	2.6

Table 3. Kinetic parameters of cumulative hydrogen production at different concentration of inoculum

3.4 The effect of waste concentration on hydrogen production

The effect of waste concentration was studied with inoculum concentration of 30% (0.36 dry wt/l) and light intensity of 9 klx. The following waste concentration were used: 5, 10, 20, 40, 60% v/v in case of dairy waste, 1, 3, 5, 10, 20% v/v in case of brewery waste I and 5, 10, 20, 40, 80% v/v in case of brewery waste II. The results in Tabl.4 show the maximum hydrogen production of 3.2 l/l medium occurring when 40% of dairy waste was used. When brewery waste with high COD (220 g O_2/l) was applied, 2.2 l of H_2 per l medium was produced (waste concentration10% v/v). In case of brewery waste with low COD (27 g O_2/l) only 0.67 l of H_2 per l medium was produced (waste concentration 80 % v/v). If higher concentrations of wastes were applied, the efficiency of hydrogen production was lower, which was caused by and inhibiting concentration of N-NH_4^+ (40 mg/l for dairy waste and 96 mg/l for brewery waste) (Waligórska, 2009, Melis, 2006). Such concentration of ammonium ions can diminish significantly the overall generation of hydrogen . The presence of ammonium ions as well as N_2 causes reduction of nitrogen *via* nitrogenase into gaseous NH_3 instead of required hydrogen. The amount of evolved CO_2 never exceeded 10 vol. %. Additionally, higher concentrations of wastes caused acidification of medium during the process and darkens the medium, which makes the access of the light into the medium more difficult and negatively impact on hydrogen production. The final pH values presented on fig. 6 show the drop from 7.1 to 5.2 in case of brewery waste and 7.5 to 5.7 in case of dairy waste. This effect is caused mainly by formation of organic acids (lactic and acetic) (Koku, 2002). The higher was the concentration of the waste the higher was the amount of detected acids and lower value of pH. This can be explained by higher ability of transfer of undissociated form of acids towards the cell, followed by dissociation inside the cell, proton release and final inhibition of the process (Van Ginkel, 2005).

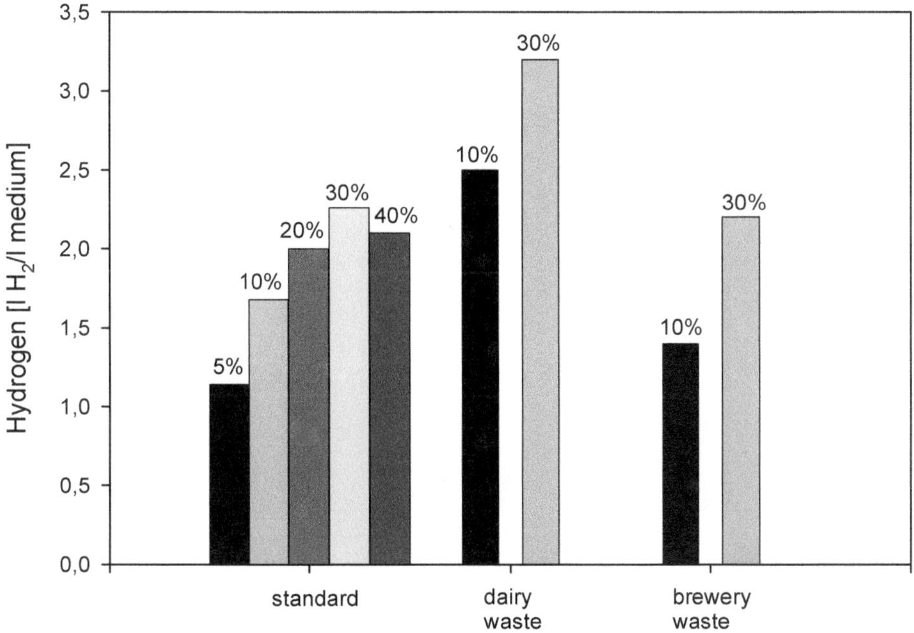

Fig. 5. The effect of inoculum concentration on hydrogen production

Concentration of dairy waste (% v/v)	H_{max} (l/l medium)	COD loss (gO_2/l medium)	Y_{sub} (l H_2/l waste)	Y_{sp} (l H_2/ COD_{loss})
		Dairy waste (COD = 46 g O_2/l)		
5	0.77	1.3	11.3	0.6
10	1.58	1.8	13.7	0.78
20	2.1	2.8	9.4	0.75
40	3.2	4.2	7.6	0.76
60	0	-	-	-
		Brewery waste I (COD = 220 g O_2/l)		
1	0.86	1.9	56	0.45
3	1.17	2.4	29	0.49
5	1.4	2.8	22	0.51
10	2.24	3.8	19	0.59
20	0.52	2.3	1.1	0.23
		Brewery waste II (COD = 27 g O_2/l)		
5	0.38	1.3	3.6	0.29
10	0.4	1.5	2.0	0.27
20	0.4	1.6	1.0	0.25
40	0.56	2.4	0.9	0.23
80 (concentrated)	0.67	2.8	0.59	0.24
		Standard (L-malic acid)		
0.2	2.3	1.9	-	1.2

Table 4. The correlation between waste concentration, amount of hydrogen produced, COD loss and efficiencies.

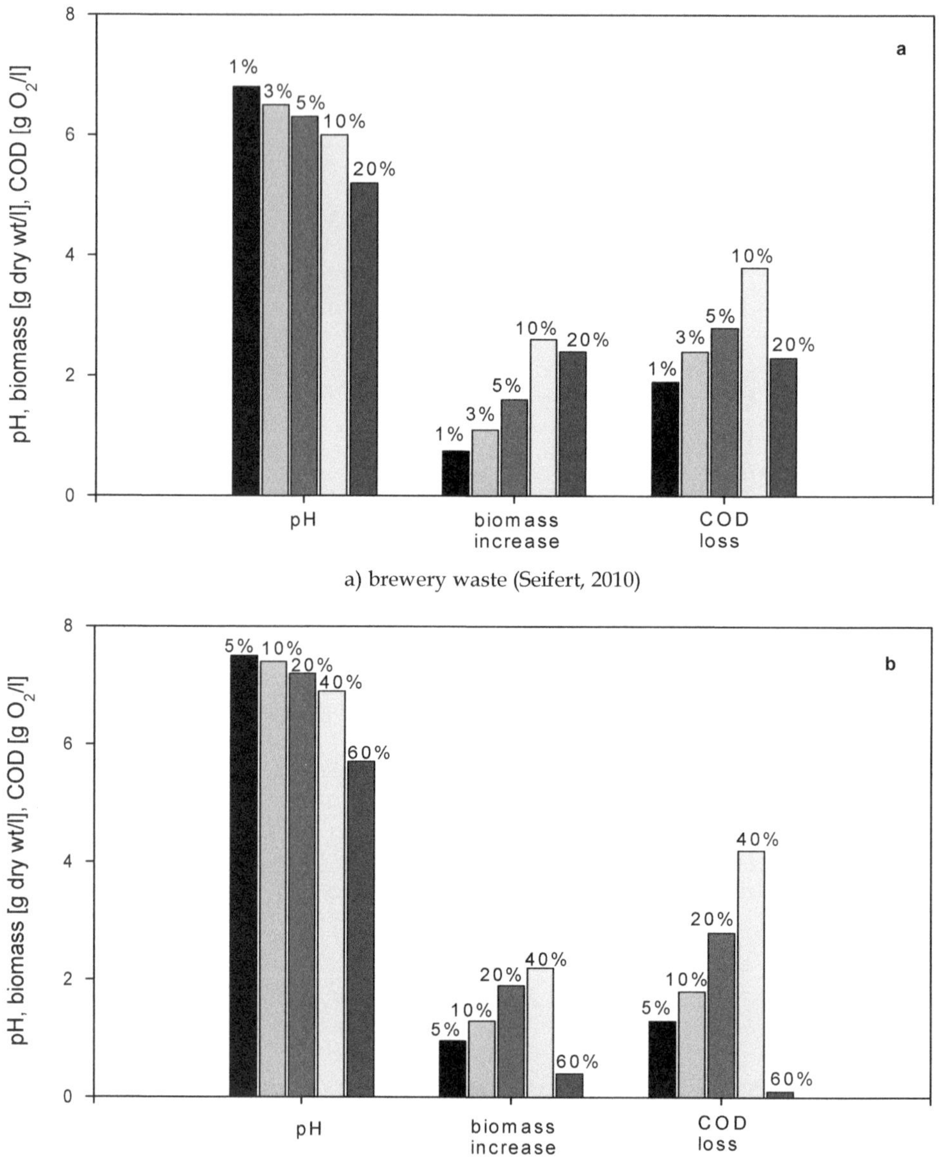

a) brewery waste (Seifert, 2010)

b) dairy waste

Fig. 6. Influence of food wastewater concentration on pH, biomass increase and COD loss.

With the rising concentration of wastes we observed higher COD loss, biomass increase and increase of specific efficiency (Table 4, fig.6). With further increase of waste concentration COD loss and specific efficiency were lower. However substrate efficiency decreases with higher waste concentration. Similar results showed Eroglu et al. obtaining the best substrate efficiency of hydrogen generation (0.1g/l waste) for low waste concentration (olive mill wastewater 2%) however maximum volume of hydrogen production (0.45 l/l) and highest COD loss (40%) were observed when higher waste concentrations were used (Eroğlu, 2004). Also Mohan et al. studding hydrogen production from vegetable based market waste, obtained good specific efficiency when low waste concentrations were used, however highest COD loss (almost 60%) occurred when higher waste concentrations were introduced to the media (Mohan , 2009). Comparing the above results with the ones obtained for hydrogen generation on standard medium with L-malic acid, it can be seen that total amount of produced hydrogen is by 30% higher when dairy waste in concentration of 40%v/v was used and comparable when brewery waste with high COD was used (Table 4, Fig 7). Different papers published so far have proved that organic substrates such as glucose, sucrose, malic acid have been more effective than the waste containing media (Yetis2000, Zhu, 1999, Basak, 2009). However based on our results we can state that wastes studied in this paper represent an effective nutrient for photobiological hydrogen production.

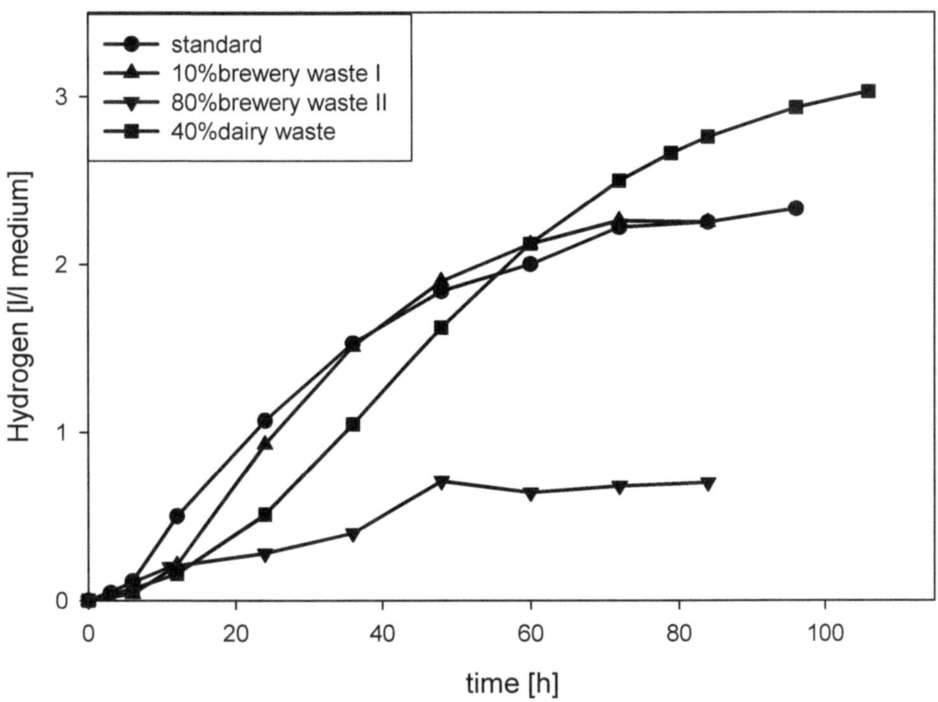

Fig. 7. The effect of optimum waste concentration on hydrogen production (30% v/v inoculum, 9 klx)

3.5 The influence of pH correction on hydrogen production

The untreated "raw" dairy wastewater with low value of pH (4.27) was completely non-active in hydrogen generation by microbiological method. However, we assumed that the same wastewater under controlled pH can generate hydrogen similarly as a sterilized one. Therefore, in order to achieve similar conditions like in bioreactor operating under controlled pH we performed our batch tests in small photoreactors (capacity of 60 ml with working capacity of 30 ml) correcting pH with 0.5M solution of NaOH every 12 h. Medium containing non-sterilized dairy wastewater with concentration of 40 v/v % was inoculated with bacteria at two different concentrations: 0.086g dry wt/l (10 vol.%) or 0.36 g dry wt/l (30 vol.%). Data presented in table 5 indicate that stabilization of the system at pH close to 7 allows for hydrogen generation even from the untreated dairy wastewater. Application of inoculums with concentration at the 0.36 g dry wt/l level generates 3.6 l H_2/l. The four-fold dilution of microorganisms reduces the volume of hydrogen to 2.6 l H_2/l. Although the starting time was relatively long (about 20 h) savings which could arise from the application of untreated waste can be significant. Performing the same experiment with brewery waste II (concentration 40 v/v %) the yield of the generated hydrogen has not been improved. In this case the value of pH rapidly grew to 7.5- 7.9 in the first two days. However , it can not be excluded that in the system with controlled pH this yield could be much higher. Preliminary experiments performed under such conditions confirm this assumption.

Inoculum conc.	H_{max} (l/l)	$R_{max,H2}$ (l/l/h)	λ_{H2} (h)	Y (l H_2/l waste)	pH final	COD loss (g O_2/l)	COD loss (%)	Biomass (g/l)
0.086	2.58±0.16	0.038±0.005	20.7±3.6	6.0	6.8	3.8	20	2.2
0.36	3.62±0.24	0.056±0.009	17.0±4.5	8.6	6.7	4.6	23	2.8

* expressed in g dry wt/l
** biomass increase

Table 5. Kinetic parameters of cumulative hydrogen production for non-treated 40 % dairy wastewater, with correction of pH for different concentration of inoculums (Seifert, 2010).

The results presented in this section suggest that hydrogen generation can be effectively performed under solar radiation in photobioreactor operating under continuous conditions.

3.6 Kinetic of hydrogen generation

The results of kinetic considerations based on modified Gompertz equation (Eq. 4) are shown in table 6. Independently from the kind of food waste (in the active of concentration) it was observed that the increase of the volume of generated hydrogen, small drops in reaction rate and prolongation of the lag phase.

These results showed that higher substrate yield increases the reaction rate. Moreover, these values are well correlated with the lag phase in systems with higher concentration of wastes are caused probably by longer adaptation of microorganisms to the bed.

Concentration of waste (% v/v)	H_{max} (l/l)	R_{max} (l/l/h)	λ_{H2} (h)
Dairy waste			
5	0.77±0.03	0.08±0.05	6.5±3.1
10	1.58±0.11	0.058±0.019	7.3±6.2
20	2.10±0.06	0.055±0.021	10.0±4.8
40	3.23±0.21	0.049±0.007	14.5±4.3
Brewery waste			
1	0.86±0.02	0.046±0.007	8.0±1.4
3	1.17±0.05	0.045±0.009	6.1±2.7
5	1.40±0.05	0.042±0.008	6.1±2.1
10	2.24±0.09	0.061±0.009	9.4±2.6
20	0.52±0.02	0.040±0.015	18.7±2.2
standard	2.3±0.2	0.047±0.004	2.7±1.8

Table 6. Kinetic parameters of cumulative hydrogen production for different initial concentration of food waste

4. Conclusions

The presented results shows that the waste studied in this paper represent a vary good substrate in photophermentation by *Rhodobacter sphaeroides.* Light intensity of 9 klx and inoculum concentration of 0.36 g dry wt/l (30% v/v) were used as the most effective (high light conversion efficiency and short duration of the process). The studied wastes has to be treated with high temperature (20 min in 120°C). This pretreatment significantly increases H_2 production. The optimum concentrations of wastes were estimated: 40% v/v for dairy waste and 10% v/v for brewery waste with high COD. These wastes represent the effective (comparable with L-malic acid) nutrient for hydrogen production. Higher wastes concentrations inhibit the process as it initiate fermentation which starts to compete with hydrogen production and additionally increases NH_4^+ concentration, which also negatively affect the process. Brewery waste with low COD shows low efficiencies and needs to be concentrated to supply sufficient concentration of organic compounds. An application of untreated dairy wastewater containing suspensions in efficient hydrogen generation process can be performed only at controlled acidity (pH = 7.0). Kinetic measurements proved that the rate of hydrogen generation drops with concentration of the waste and prolongs the lag phase.

5. Acknowledgements

These studies were supported by Polish Ministry of Science and Higher Education (grant no: N N204 185440).

6. References

Apha, AWWA, WEF. Standard methods for the examination of water and wastewater. 10th ed. Washington, DC: American Public Health Association; 1995.

Asada, Y.; Miyake, J. (1999). Photobiological hydrogen production. *J Biosci Bioeng*, vol.88, pp.1-6.

Ball, M.; Wietschel, M.(2009). The future of hydrogen - opportunities and challenges, *Inter. J. Hydrogen Energy*, vol.34, pp. 615-627.

Basak, N.; Das, D. (2009). Photofermentative hydrogen production using purple non-sulfur bacteria *Rhodobacter sphaeroides* O.U.001 in an annular photobioreactor: A case study, *Biomass and Bioenergy*, vol.33, pp.911-919.

Biebl, H.; Pfennig, N. (1981). Isolation of members of the family Rhodospirillaceae. In: *The prokaryotes* ed. Starr M.P.,Stolp H., Trüper H.G.,Balows A., Schegel H.G., New York: Springer,1, pp.267-273.

Chen, WU.; Chen, SY.; Khanol, SK.; Sung, S. (2006). Kinetic study of biological hydrogen production by anaerobic fermentation, *Int J Hydrogen Energy*, vol.31, pp.2170-8.

Cui, M.; Yuan, Z.; Zhi, X.; Shen, J. (2009). Optimization of biohydrogen production from beer lees using anaerobic mixed bacteria, *Int J Hydrogen Energy*, vol.34, pp. 7971-9.

Dubbs, JM.; Tabita, FR. (2004). Regulators of nonsulfur purple phototropic bacteria and the interactive control of CO_2 assimilation, nitrogen fixation, hydrogen metabolism and energy generation, *FEMS microbial Rev*, vol.28, pp.353-376.

Eroğlu, E.; Gündüz, U.; Yücel, M.; Türker, L.; Eroğlu, I. (2004). Photobiological hydrogen production by using olive mill wastewater as a sole substrate source, *Int J Hydrogen Energy*, vol.29, pp.163-171.

Garrity, GM,; Brenner, DJ.; Krieg, NR.; Staley, JR. (2005). Gergey's Manual of Systematic Bacteriology, Volume two: The Proteobacteria, Part A, Springer-Verlag pp.166

Koku, H.; Eroğlu, I.; Gündüz, U.; Yücel, M.; Türker, L. (2002). Aspects of the metabolism of hydrogen production by *Rhodobacter sphaeroides, Int. J. Hydrogen Energy*, vol.27, pp.1315-1329.

Li, X.; Wang, Y-H.; Zhang, S-L.; Ming, J.C.; Hung, M-Z. (2009). Enhancement of phototrophic hydrogen production by *Rhodobacter sphaeroides ZX-5* using a novel strategy-shaking and extra-light supplementation approach, *Int J Hydrogen Energy*, vol.34, pp.9677-9685.

Melis, M.; Melnicki, M.R. (2006). Integrated biological hydrogen production, *Int J Hydrogen Energy*, vol.31, pp.1563-1573.

Mohan, S.V.; Mohanakrishna, G.; Goud, R.K.; Sarma, P.N. (2009). Acidogenic fermentation of vegetable based market waste to harness biohydrogen with simultaneous stabilization, *Bioresource Technology*, vol.100, pp.3061-3068.

Mu, Y.; Yu, HQ.; Wang, G. (2007). A kinetic approach to anaerobic hydrogen producing process, *Water Res*, vol.41, pp.1152-60.

Nath, K.; Das, D.(2009). Effect of light intensity and initial pH during hydrogen production by an integrated dark and photofermentation, *Int J Hydrogen Energy*, vol.34, pp.7497-7501.

Nath, K.; Muthukumar, M.; Kumar, A.; Das, D. (2008). Kinetics of two-stage fermentation process for the production of hydrogen. *Int J Hydrogen Energy*, vol.33, pp.1195-203.

National Hydrogen Energy Roadmap. Toward a more secure and cleaner energy future for America, Washington; April 2-3, 2002.

Obeid, J.; Magnin, JP.; Flaus, JM.; Adrot, O.; Willson, JC.; Zlatev, R. (2009). Modelling of hydrogen production in batch cultures of the photosynthetic bacterium *Rhodobacter capsulatus*. *Int J Hydrogen Energy*, vol.34, pp.180-185.

Pawlowski, A.; Riedel, KU.; Klipp, W.; Dreiskemper, P.; Grob, S.; Bierhoff, H.; Drepper, T.; Masepohl, B. (2003). Yeast two-hybrid studies on interaction of proteins involved in regulation of nitrogen fixation in the phototrophic bacterium *Rhodobacter capsulatus*, *J Bacteriol*, vol.185, pp.5240-5247.

Salih, FM. (1989). Improvement of hydrogen photoproduction from *E.coli* pre-treated cheese whey. *Int J Hydrogen Energy*, vol.14, pp.661-663

Seifert, K.; Waligórska, W.; Łaniecki, M. (2010). Brewery waste waters In photobiological hydrogen generation in presence of *Rhodobacter sphaeroides* O.U.001, *Int J Hydrogen Energy*, vol.35, pp.4085-4091.

Srikanth, S.; Venkata Mohan, S.; Prathima Devi, M.; Peri, D.; Sarma, PN. (2009). Acetate and butyrate as substrates for hydrogen production through photo-fermentation: process optimization and combined performance evaluation, *Int J Hydrogen Energy*, vol.34, pp.7513-22.

Su, H.; Cheng, J.; Zhou, J.; Song, W.; Cen, K. (2009). Improving hydrogen production from cassava starch by combination of dark and photo fermentation, *Inter. J. Hydrogen Energy*, vol.34, pp.1780-1789.

Uvar, B.; Eroglu, L.; Yucel, M.; Gunduz, U.; Turker, L. (2005). Effect of light intensity and illumination protocol on biological hydrogen production by Rhodobacter sphaeroides O.U.001, *Proceedings IHEC 2005*, Istanbul, Turkey.

Waligórska, M.; Seifert, K.; Szymańska, K.; Łaniecki, M. (2006). Optimization of activation conditions of *Rhodobacter sphaeroides* in hydrogen generation process. *J Appl Microbiol*, vol.101, pp.775-784.

Waligórska, M.; Seifert, K.; Górecki, K.; Moritz, M.; Łaniecki, M. (2009). Kinetic model of hydrogen generation by *Rhodobacter sphaeroides* in the presence of NH_4^+ ions, *J Appl Microbiol*, vol.107, pp.1308-1318.

Wijnowska-Baryła I,; Zielińska, M.; Babuchowska, A.; Deboung, A. (2002). The biodegradation of brewery waste In a two-stage immobilized system, *Pol J Environ Stud*. Vol.11, pp.571-5.

Van Ginkel, S; Logan, BE. (2005). Inhibition of biohydrogen production by undissociated acetic and butyric acids. *Environ Sci Technol*, vol.39, pp.9351-9356.

Yakunin, AF.; Hallenbeck, PC. (1998). Short-term regulation of nitrogenase activity by NH_4^+ in *Rhodobacter capsulatus*: multiple in vivo nitrogenase responses to NH_4^+ addition, *J Bacteriol*, vol.180, pp.6392-6395.

Yetis, M.; Gündüz, U.; Eroğlu, I.; Yücel, M.; Türker, L. (2000). Photoproduction of hydrogen from suger refinery wastewater by *Rhodobacter sphaeroides* O.U.001, *Int J Hydrogen Energy*, vol.25, pp.1035-1041.

Yokoi, H.; Maki, R.; Hirose, J.; Hayashi, S. (2002). Microbial production of hydrogen from starch-manufacturing wastes, *Biomass and Bioenergy*, vol.22, pp.389-395.

Zhu, H.; Suzuki, T.; Tsygankov, A.A.; Asada, Y. (1999). Hydrogen production from tofu wastewater by *Rhodobacter sphaeroides* immobilized in agar gels, *Int J Hydrogen Energy*, vol.24, pp.305-310.

15

Dairy Farming and the Stagnated Biogas Use in Rungwe District, Tanzania: An Investigation of the Constraining Factors

Agnes Godfrey Mwakaje
Institute of Resource Assessment, University of Dar es Salaam, Tanzania

1. Introduction

Dairy farming plays a key role in the lives of poor, rural people in developing countries, providing a major proportion of their cash income, capital assets, draught power, fuel and fertilizer. Small-scale dairying produces valuable food products and provides a regular income and work. Dairying also provides much of the cash needed to perform other socio-economic activities. Milk production generates reliable incomes to meet household livelihoods (Somda et al., 2005). Possession of dairy animals means also financial security, status, self-confidence and an opportunity to have some control over their live (Ramkumar, 2004). It is also more labour intensive and supports substantial employment in production, processing and marketing. This is partly because dairy production often require the introduction of specialised dairy breeds and increased levels of inputs (nutrition and health care) and good linkages to markets, both for milk sales and input acquisition. In Kenya dairy farming has become a very significant source of income and food for an estimated 625,000 smallholder producer households and for those involved in the marketing of milk, in total some 25% of all households in Kenya benefit from dairy farming (Muriuki et al., 2001). In Tanzania about 700 000 dairy cattle are available under smallholder farmers, with an average of 4 cows per household, there might be 175 households keeping indoor fed dairy cattle in Tanzania. Dairy farming in Tanzania is estimated to grow at a rate of 6% per year and there are about 190,000 registered farmers currently (Swai and Kurimuribo, 2011). Most of these cattle are kept in the highland and relatively cold regions of Arusha, Mbeya, Kagera, Iringa and Morogoro. Smallholder dairy farming in Tanzania has had a significant impact on poverty alleviation in terms of income, education, food security and stabilizing farm incomes (Kisusu et al., 2000).

On the other hand, dairy manure is potential for biogas generation. Dairy manure biogas digester technology has proven to be technically and economically feasible and successful in many applications (Schwengels, 2009). Technology pathways involving biogas, natural gas or electricity are advantageous (Hedegaard *et al* 2008) for rural development. Empirical evidence suggests that each household can realise up to US$ 724 by replacing wood use with biogas, apart from other positive impacts to the environment (Langeni *et al.*, 2010). A study by the Institute of Resource Assessment (IRA), University of Dar es Salaam, in 2005, shows a reduction of firewood consumption from 700 to 145m^3 for Lomwe Secondary

School following the adoption of biogas technology which meant a reduction Energy saved annually is approximately 6.7 Terra Joules (T.J) (a reduction of 78.9%) of CO^2 annually (IRA, 2005).

2. Biogas development trend in Tanzania

Biogas technology utilizing animal waste is not new in Tanzania; it was introduced in the country as early as the 1950s by private stakeholders. In 1975, the government through the Small Industries Development Organisation (SIDO) introduced the Indian design (floating gasholder digester) in primary and secondary schools, rural health centres and a number of other institutions. In 1982, the Parastatal Organization Centre for Agricultural Mechanization and Rural Technology (CAMARTEC) increased the dissemination of this technology in the northern regions. About 1 year later, that is around 1983, technical cooperation between Tanzania and the Federal Republic of Germany led to the introduction of the Biogas Extension Services (BES). CAMARTEC and the Deutsche Gesellschaft fur Technische Zusammenarbeit (GTZ) were in-charge of implementing this project and the latter seconded an interdisciplinary team of social scientists, mechanical engineers and agriculturists to Tanzania (Sasse et al., 1991). Between 1984–1985 more strategies were developed to boost biogas adoption. Household plants were offered with a digester volume of 8, 12 and $16m^3$, and in 1990 the programme comprised standardized plants of sizes 12, 16, 30 and $50m^3$ for households and institutions (Mwakaje, 2008). The development work towards sustainable reliability and user friendliness resulted in extensive integration of biogas plants into the work routines of farmers. Over the period, CAMARTEC were involved in building capacity by training technicians in biogas plant construction. A "biogas unit" scheme was introduced and this integrated biogas plants, livestock housing with a concrete floor (Mwakaje, 2008). CAMARTEC was also providing advice on the utilization of slurry, gas pipeline systems, burners and lamps; and women were specifically instructed on how to use and manage the plants. The Ministry of Energy and Minerals in collaboration with donors was also promoting biogas use in the Dar es Salaam region. Its main activity was to support the dissemination of biogas technology in the region through facilitating training for private craftsmen, built demonstration plants and undertaking monitoring and evaluation. Up to 1989, only 200 units of biogas had been installed all over the country (Sasse et al., 1991) but in 1992 this had increased to 600 plants national-wide. Nevertheless, as Mwakaje (2008) noted despite all the efforts, the biogas technology did not diffuse much to the rural poor communities in many parts of the country where indoor fed dairy cattle are kept. Reasons for this poor diffusion of the biogas technology included high installation and maintenance costs and inadequate awareness about the technology. The conventional units being built in the country were large and expensive, costing approximately US$ 1400 for one unit (Rutamu, 1991) to USD 2200 depending on the size of digester (IRA, 2005). Furthermore, repair and maintenance required highly skilled labour and most component parts, constructed mainly from concrete and steel, were far out of the financial reach of smallholder farmers (Mwakaje, 2008). This slow pace of biogas technology development by CAMARTEC raised a number of criticisms among stakeholders. For example, the Evangelical Lutheran Church of Tanzania (ELCT) blamed CAMARTEC its commercially oriented and strictly standardized dissemination programme. The ELCT claimed that the programme had not been adapted to Tanzanian conditions as it only served the rich farmers (Sasse et al., 1991). But also most of the CAMATERC activities were concentrated mainly in the two regions of Kilimanjaro and Arusha in a country with more than 20 regions. On the other hand, the Ministry of energy and minerals' activities were concentrated in the Dar es Salaam region where unfortunately indoor fed dairy cattle are limited to a few households.

Reacting to some of these criticisms, the government of Tanzania changed the biogas technology dissemination strategy in the country. In the years starting 2000 polythene tubular digesters were promoted to reduce production cost through using local materials and simplified installation and operation costs (Mwakaje, 2008). The type of plastic needed for polythene was locally manufactured in Tanzania, maintenance and repair were simple, cheap, and did not require skilled labour and the cost of construction was low. A model promoted by the Sustainable Rural Development (SURUDE) was a low-cost design suitable for poor farmers (CEBITEC, 2003) in rural areas. The material cost was about US$ 100. However, this type of biodigester had one major disadvantage in that it could be easily sabotaged (torn out). This is because the plastic materials of the biodigestor are normally placed on the surface outside the house and therefore could easily be destroyed (Mwakaje 2008).

2.1 Dairy sector and biogas use in Rungwe district

Rungwe district lies between latitudes $8^0 30$ E and $9^0 30$ E and longitudes 33^0S and 34^0 S. It is one of the six districts of Mbeya Region, located in the Southern Highlands of Tanzania. The other districts are Kyela, Chunya, Ileje, Mbeya Rural and Mbozi. Rungwe district has a total area of 2211 sq. km of which 75% is arable land (URT, 1997). Of the remaining area, 44.5 sq. km is covered by forest while 498.3 sq. km is either mountainous or residential areas.

The district is one of the densely populated districts in Tanzania (URT, 2002) with a population of 307,270, which is equivalent to 139 persons per square kilometre with an annual growth rate of 0.9% (URT 2010). The district has limited natural vegetation which varies from upper montane forest at higher elevations to the wet woodland (Miombo) at lower elevations. Forestry reserve accounts for 43,749.9 ha and other forests about 65,813 ha (URT, 2008). In recent years, much of this natural vegetation has been cleared/transformed for agriculture, for habitation, and firewood. Most of the remaining natural vegetation is found in government forest reserves and in locally protected areas, though even these areas have been subjected to varying degrees of people driven disturbances.

Rungwe district put great importance to livestock development particularly dairy cattle as one of the major economic activities. In 2005 the district had 26,137 indoor fed dairy cattle with milk production estimated to be 41,000,000 litres per year. The district has 74,450 households and almost half of the households keep some cattle or pigs in their homestead with an average of between 2-6 cattle (Mwakaje, 2008). Smallholder dairy production is an important undertaking and, if adequately supported by appropriate policies and adaptive research technologies, it may contribute significantly towards the household economy, self-sufficiency in milk and national gross domestic product (Swai and Kimambo, 2011). Walshe et al (1991) comments that where there is access to a market, dairying is preferred to meat production since it makes more efficient use of feed resources and provides a regular income to the producer.

Promotion of smallholder dairy farming can solve the problem of rural poor accessing to clean energy like biogas.

The district is also famous for keeping pigs. Rungwe district has about 44,334 pigs which also contribute significantly to the household's economy and nutrition.

Studies in several African countries, provides a rough sense of the likely economics of introducing biodigesters (Schwengels, 2009) where 2 cows or 1 cow and other livestock like pigs can be appropriate for a family to meet the need of cooking biogas while other research findings suggest that farming households, having 2 (zero-grazed) to 10 cattle or 8 to 40 pigs (or a combination) are enough to produce gas for a household. This means that available number of indoor fed dairy cattle of more than 26,000 and over 44000 pigs, the district can have the capacity of having more than 20000 biodigester, this is about 27% of the district's households.

However, despite the high level of indoor fed dairy cattle in Rungwe District and the potential to generate biogas as well as the efforts to promote biogas use in the country since 1970s by the government and donors, biogas technology has not well developed in the district to date. The trend of biogas technology in the district shows that the technology started in 1993 when one person adopted installed a biogas plant (Mwakaje, 2008). In 1996, 12 households got the service by contributing half of the cost. This was a pilot project by the Danish Volunteers that intended to raise awareness of the technology. With the exception of the year 1996, adoption of the biogas technology has remained low and more or less declining (URT, 2005). Up to 2007 there were about 100 biogas plants, an equivalent to only 0.13% of the total households in the district. This is even more surprising as the district has limited fuelwood sources as well as other clean energy sources. Available information shows that the district has a demand of cooking energy of 600,000 m^3 per annum, while the capability to supply is about 400,000 m^3 (URT, 2005), a 33% deficit (Mwakaje, 2008). The scarcity of fuelwood has increased its cost in terms of purchasing price and time used for fetching (Mwakaje, 2008). The use of other clean energy like electricity and solar power is limited due to both cost and reliability (Mwakaje, 2008).

Why the pace of biogas adoption and use in the district has remained stagnant is the main interest of this study. Although, a study by Mwakaje (2008) highlighted some of the constraining factors, it was not exhaustive. The study focused more on the environmental benefits of adopting biogas technology while other equally important issues related to biogas use and adoption such as socio-economic, institutions; awareness as well as policies were not adequately explained. The main objective of the chapter was to come up with an understanding of the reasons for the stagnated biogas use in Rungwe district despite the availability of large number of dairy cattle and other livestock and in an area with highly inadequate fuelwood supply. Specifically, the chapter investigated issues relates to investment costs, expertise availability, role of institutions and policies in influencing biogas use and level of awareness of biogas use among the Rungwe dwellers. Findings from this study will add to the body of knowledge, inform policy makers, donors, service providers, environmentalists and researchers.

3. Methods

Data were collected from both primary and secondary sources. Secondary data were collected through literature review using published documents and internet material. There was also a review of policies related to energy in Tanzania. Secondary data helped to establish what has been done in the subject and to read what were the remained gaps for field work were. Institutions supporting biogas development were consulted for understanding their performance and constraining factors they are facing. Primary data were collected in areas related to investment cost, awareness, household energy demand, technology service providers, and expertise. In addition, there were consultations with service providers to get information on cost, demand as well as factors constraining the spread of the biogas technology in the District. Furthermore, there was a consultation with local and district institutions and authorities for detailed information on biogas use in the district and whether there has been any efforts to facilitate the adoption of biogas.

The sample frame for this study involved respondents with dairy cattle/biogas use and those with dairy cattle but have not installed biogas plants. Also respondents with access to electricity and other clean energy sources such as LPG were included in the sample. A total of 3 villages were selected for the household sample. These were *Isagilo*, *Kyimo* and *Mpandapanda*. The selection of the villages based on the availability of dairy cows, adoption of biogas technology, availability of other energy sources, socio-economic status and accessibility. The households were selected purposively for those with biogas as well as those with access to electricity as they are few but random for the rest of the dairy keepers.

The total number of households (n) to be surveyed was estimated using the formula below:

$$n = \frac{N}{1 + Ne^2} \tag{1}$$

Where: n = sample size between 5 and 10%
N = total number of households in the village; and
e = desired margin of error.

A sample size of about 10% was selected making a total sample of 120 households. Out of this, 35 had biogas facilities and the remaining 85 had dairy cattle without biogas facility (Table 1). Village roster were used to select the sample households. Data were collected using structured and semi-structured questionnaires and analysed using Statistical Package for Social Sciences (SPSS) as well as livelihoods models. Results have been presented in tables and figures.

Village Characteristics	Characteristics	HH With biogas	Biogas selected for interview	HH without biogas selected for interview	Total Sample selected
Isagilo	Biogas project started free of charge in 1996 and 12 HH installed biogas plants	22	19	35	54
Kyimo	Large population of dairy cows and have electricity services	13	11	28	39
Mpandapanda	Large population of dairy cattle but limited number of biogas users.	7	5	22	27
Total		42	35	85	120

Table 1. Village Characteristics and Sample Size (households)

4. Results and discussion

4.1 Wealth ranking of the respondents

To establish the socio-economic profile of the respondents a wealth ranking approach was used. Wealth ranking was important in this study to determine whether there is any relationship between biogas technology adoption and wealth of the household into wealth ranks using a set of pre-established criteria (Afonja, 1992). Since its introduction in the 1980's, rapid rural appraisal (RRA) wealth ranking has become an increasingly accepted means of assessing relative socio-economic status in the context of applied research projects and development programs (Chambers, 1994). In this study the members of village governments were involved in wealth ranking for their respective villagers. Criteria for wealth ranking was adopted as perceived locally it included aspects of, food security, livestock, dairy cattle and other assets ownership, land and annual incomes (Table 2). Results from wealth ranking (based on communities perception) show that there were no people who were really *well off* in the sample of households but only the so called *slightly well-off*. Out of the 120 respondents 18.3% were *slightly well-off* while *less poor* and *poor* were 61.3% and 20% respectively (Table 2).

	Slightly Well-Off	*Less Poor*	*The Poor*
Food security	Generally food secure all the time	Only rarely may experience seasonal food insecurity	Experience food insecurity for about 2 months a year
Assets	Many own various household assets, (cars, motorcycle, bicycles, TV/radio).	Generally have necessities and relatively few other household assets. such as TV, radio, motorcycles	Limited assets
Livestock	Own dairy, indigenous cows, pigs, goats, chicken.	Own some dairy, indigenous cows, pigs, goats, chicken.	Keep some livestock, especially dairy cows/indigenous cows, pigs, goats, chickens
Land	Relatively large land owners (>2 hectares)	Little land (betwee1-2 hectares)	Little land (<1 hectares) or landless
Work for food	No HH member works for food	Occasionally may sell less than 30 days labour/year	Sell more than 30 days of labour per year. May participate in "food-for-work. Household workforce is mainly comprised of children, women and the elderly who command a low daily wage.
Business/ employment	Have a good business or employed	None	None
Annual cash income	>USD 700	Between USD 400-700	Less than USD 400
Sample respondents	22	74	24
%	*slightly well off* 18.3%	*less poor* 61.7%	*poor* 20%

Table 2. Characteristics of each Wealth rank group

4.2 Characteristics of the respondents by wealth ranks

The empirical evidence suggest that the probability of a household adopting biogas technology increases with decreasing age of the head of household, increasing household income, increasing number of cattle owned, increasing household size, male head of household and increasing cost of traditional fuels (Walekhwa et al., 2009). Also economics, material shortage, operation, and the people's acceptance are considered to be the main factors preventing the diffusion of biogas technology (Taşdemiroglu 1988).

Findings on education show the *slightly well-off* respondents to had relatively good education than other categories although the post secondary education was generally low across the three categories. Post secondary education such as vocational and other training is important as it creates professionals and experts including biogas experts in rural areas. The extremely poor spend very little in education hovering around 2% of household budgets (Banerjee (2007). The reason for low spending in education is that children in poor households typically attend public schools or other schools that do not charge a fee even if the education quality is poor. Poor parents are not reacting to the low quality of these schools, either by sending their children to better and more expensive schools or by putting pressure on the government to do something about quality in government schools. This partly occurs because quite often they are illiterate themselves and therefore may have a hard time recognizing that their children are not learning much (Banerjee, 2007).

Regarding family size respondents from *slightly well off* had small family size (3.3 persons) compared to the *less poor* (4.6 persons) and the *poor* (5.9 persons) (Table 3). This could be explained partly by the low levels of education of the poor. The less educated are more likely to start family life early than educated ones and therefore have high chances of having several children in their reproductive life time. These findings are consistent with Banerjee (2007) observation that family size is large for the extremely poor respondents.

Wealth Category	*Slightly Well-off*	*Less Poor*	*The Poor*
Family size (persons)	3.3	4.6	5.9
Married respondents (%)	78.9	82.6	87.9
Female respondents (%)	22.4	40.7	35.7
Respondent's age (years)	48.4	53.5	44.0
Education			
No formal education (%)	0	1.3	5.1
Completed primary education (%)	50	59.8	66.7
Completed secondary education (%)	33.3	25.5	20.5

Table 3. Characteristics of the respondents

4.3 Sources of energy for cooking

The source of energy varied from one category to another across the three wealth ranks. Nevertheless, fuelwood dominated energy sources in all the three categories, where over 77% of the respondents were using fuelwood for cooking (Table 4), followed by biogas, very few of the respondents were using charcoal. No-one was using electricity for cooking.

The respondents were asked whether they would like to have a biogas facility in their homes or not, and almost all (96%) said yes, they are willing to install biogas facilities.

	The Poor	Less Poor	Slightly Well-off	Average
Fuelwood	89.5	83.3	60.6	77.8
Biogas	15.9	18.6	30.3	21.6
Biogas and charcoal	2.7	6.1	16.7	8.5
Electricity/LPG	0	0	0	0

Source: Survey data 2006

Table 4. Wealth Categories and Sources of Energy for Cooking (%)

4.4 Sources of fuelwood

The sources of fuelwood in the district are communal forests, private forests, farms and timber residues. The distance to the fuelwood sources ranged from 1.1km for the less poor and slightly well-off households and 2.25 km for the 'poor' categories. The average distance for all respondents was 1.40km. The short distance for accessing fuelwood by the slightly well off and less poor is partly because a high proportion of them have private forests near their homes. Most people in the district use fuelwood from their own planted trees. Communal land is very limited in the district.

A high proportion of the communal forests have been severely degraded which makes fuelwood not easily available. Women spend 3-4 hours looking for fuelwood. This means that households with biogas facility were saving 3-4 hours wasted in collecting fuelwood. The saved time is used for other economic activities (e.g. farming and marketing) as well as leisure (e.g. resting and listening to the news and other entertainment). On the other hand, if the use of biogas for cooking will increase the demand for fuelwood in the district may decrease which is likely to benefit the poor because most of them do not have dairy cattle for biogas plant installation.

4.5 Awareness and cost of installing biogas facility

Findings show that one household plant could cost USD 550- 675 with wide standard deviation suggesting a high variation for the cost of installation depending on the expertise availability and the size of the biogas facility. The size for biogas plants ranged from 6-12m^3.

A comparison across wealth ranks shows a significant difference (Table 5). The *slightly well-off* respondents had significantly less installation cost compared to *less poor* ($p<5\%$) and the *poor* (p<2%) categories. However, there was no significant difference of cost of installation between the *less poor* and the *poor* respondents. A major explanation to this is that a high proportion of the *slightly well off* respondents benefited from the pilot project in 1996 when the biogas facilities were installed at half cost by the Danish volunteers. This was a strategy used to sensitise and raise awareness and demand for the biogas facilities. Unfortunately, many people from the *less poor* and the *poor* categories could not take up this opportunity because of many reasons, one of them being risk averse. They wanted to learn from others how it worked and what the advantages were to be. However, by the time they were convinced by the technology and started adopting it, the price had gone back to the market price levels. Another reason for not adopting it during the promotion period was that they had other more pressing issues than biogas, such as a need for cash to carter farming activities and paying for education and health services. Various studies have shown that poor people are always risk averse and therefore it takes time for them to adopt a new technology. Many of the studies about technology adoption conclude that the pace of adopting a new technology in developing countries has been slow among the poor.[1] Feder et al., (1985) have identified factors such as aversion to risk and limited access to information as reasons that could partly explain why adoption is slow. Individual characteristics such as education, access to credit, the capacity to bear risk, availability of other inputs and access to information may play a big role in the adoption of the technology.

Wealth Category	*Slightly Well-off*	*Less Poor*	
Comparison 1	Slightly Well-off	Less Poor	Level of significance
	550 (215)	635 (125)	**
Comparison 2	*Less Poor*	*The Poor*	
	670 (192)	675 (250)	NS
Comparison 3	*Slightly Well-off*	*The Poor*	
	550 (250)	675 (176)	***

NS =not significant, ** Significant at p<5%, *** Significant at p<2%

Table 5. A comparison of cost (USD) of installation across wealth ranks

4.6 Factors constraining biogas use

In a multiple response question, respondents were asked to mention the constraining factors towards biogas use adoption. The main factors mentioned were that the installation cost was too high (95.8%) and lack of credit facility (95%). Other reasons were lack of expertise (91.7%) and inadequate water (60%) to run the plants. Only a small proportion of 3.3% out of the 120 respondents said they do not need the facility. This may suggest that if the access to biogas is facilitated either through subsidy or access to credits many households in the district could adopt the technology. A comparison across the categories suggest that

[1] Giné and Klonner , 2005

respondents from *the poor* and *less poor* were more in demand for credits and also mentioned the facility to be too costly (Table 6), suggesting a different kind of approach to induce them adopt the technology.

Wealth Category	*Slightly Well-off*	*Less Poor*	*The Poor*	Total
Too costly	81.8	98.7	100.0	95.8
Inadequate expertise	95.5	91.7	95.8	91.7
Inadequate water	45.5	54.2	95.8	60
Lack of credit facilities	86.4	98.6	100.0	95
Not aware	9.1	6.9	20.8	10
Do not need	9.1	1.4	4.2	3.3

Table 6. Factors constraining biogas adoption in Rungwe district (%)

4.7 Energy policy and biogas technology promotion

Tanzania's energy demand is characterised by a low per capita consumption of commercial energy (petroleum and electricity) and a high dependence on non-commercial energies, including biomass fuels in the form of firewood, charcoal and bio-waste. Renewable energy technologies currently in use in the country include improved wood-fuel stoves and charcoal production practices, biogas, windmills, and solar thermal and photovoltaics (PV). The applications of these technologies are at various stages of development in terms of demonstration and commercialization.

Tanzania has no renewable energy policy at the moment but only the general energy policy framework for all kinds of energy.

The National Energy Policy (2003) objectives are to ensure availability of reliable and affordable energy supplies and their use in a rational and sustainable manner in order to support national development goals. The National Energy Policy, therefore, aims to establish an efficient energy production, procurement, transportation, distribution and end-use systems in an environmentally sound and sustainable manner (URT, 2003). It also supports research and development of renewable energy and promotes the use of efficient biomass and end-use technologies. The main elements of the policy are:

- development of domestic energy sources and economic energy pricing,
- encouragement of private sector participation in the energy market,
- enhancement of energy efficiency and energy reliability

The New Electricity Act of 2008, provide room for more private sector in energy production and that increases a chance to utilize renewable energy especially on small scale targets. Markets for rural household lighting with solar home systems, biogas, and small hydro-power have expanded through rural entrepreneurship, government programmes, and donor assistance, serving a number of households (Martino *et al* 2002).

The Act also provides roles and relations of the different actors, the ministry; regulators and operators of the sector are determined by legislation. The Ministry of Energy and Minerals (MEM) is responsible for energy development. It supervises the implementation of the energy policy, which is the main guidance for change, backed by legislation and regulation. The ministry also facilitates mobilisation of resources into areas where market forces fail to ensure adequate energy services. The policy put guidance for licensing operators, monitoring markets and performance; and applying any other necessary regulatory measures.

Within the Ministry of Energy and Minerals there is a Rural Energy Agency (REA) for rural electrification. The policy acknowledges that around 80% of the population has very low purchasing power and depends mainly on wood-fuel for cooking and kerosene for lighting, which have negative consequences to the environment and the quality of life, especially to the rural poor. Rural electrification is a case of long-term national interest and a prerequisite for a balanced socio-economic growth for all in Tanzania through enabling rural poor accessing sustainable clean energies.

However, energy policy has attracted criticism in different ways. Stakeholders feel that consideration of improving clean energy by rural poor needs to be on the application of appropriate technologies that are affordable, environmentally sound and well adapted to local needs as explained in the Policy. Also, while gender issues have received attention at micro level in terms of technological interventions such as cookstoves, biogas, solar cookers, and wood plantations, they have yet to be addressed in macro level policies. Women's needs for energy vary depending on whether they are in urban or rural areas, their stage of economic development and whether they are economically active. Parikh (1995) makes a plea to include gender issues in macro level energy policies such as energy investment, imports and pricing. Also there is inadequate information and data on how the ongoing and planned power sector reform can be modified to address the existing challenges, particularly with regard to electrification of the poor (Karekezi and Kimani, 2002). A study by Barnes and Floor (1996) highlights constraints towards improving clean energy in rural development and these include the widespread inefficient production and use of traditional energy sources fuelwood and charcoal which pose economic, environmental, and health threats. Also the highly uneven distribution and use of modern energy sources such as electricity, petroleum products and liquefied or compressed natural gas, pose important issues of economics, equity, and quality of life. The policy does not provide adequate strategies on overcoming these. Many developing countries including Tanzania has general energy policies pertaining to the development of electricity, oil and renewable energy sub-sectors for the benefit of the public and the economy. However, the absence of sharply focused, pro-rural energy policy and/or their policy instruments has been the major challenges towards the observed stagnation of some initiatives like the biogas (Habtetsion and Tsighe, 2002). The Energy Policy formulation in Tanzania takes place in the context of great uncertainty, due to mainly pressures exerted by conflicting interests (Mwandosya and Luhanga 1993).

Within the Energy Policy, biogas has received a low profile or recognition. There is no specific policy statement to explain and strategies for the promotion of biogas technology in rural Tanzania; rather everything is dumped in the category of renewable energy. Omer and Fadalla (2003) recommends that biogas technology must be encouraged, promoted, invested, implemented, and demonstrated, but especially for remote rural areas.

The main challenges facing biogas technology is inappropriate institutional structure and/or gaps in the structure, in addition to lack of corporate culture; poor incentives; and, poor linkages among the various stakeholders concerned in energy for rural development (Habtetsion and Tsighe, 2002). Progressive government intervention is needed to shift reform process towards a more responsible development path of renewable energy (Wamukonya, 2004). Generally speaking, the database for the context of renewable energy in Tanzania is not well documented and the renewable energy technology including biogas is still at an infant stages. So many efforts have been done by individuals of which, most of them have not been documented. The financial capital coupled with poor technology (Mwerangi, 2008) and lack of sustainable institutional framework for renewable energy developments hinders the development of biogas. This trend tallies with Uddin (1999) comment that lack of policy mechanisms, institutional development and financing exist as major barriers for Thailand

Another policy issues is lack of credits. A high proportion of the respondents in this study area indicated high cost and that there were no credit facilities in the area of study. There is also an issue of awareness and culture. A study by Mwakaje (2005) show that a large number of people who have not accessed biogas technology especially from the Muslim community have a perception that biogas is a dirty thing. However, being close to Lomwe Secondary School in Kilimanjaro Region, Tanzania and observing physically the functioning of biolatrine, many neighbour households including the Muslims were motivated to adopt the technology. The challenge was the amount of waste to feed the biodigestor and of course the cost to incur. Improving credit accessibility may have significant impact on biogas adoption in Rungwe district and Tanzania at large. Factors influencing socio-political and community acceptance are increasingly recognized as being important for understanding the apparent contradictions between general public support for renewable energy innovation and the difficult realization of specific projects (Wüstenhagen et al., 2007).

5. Conclusion and recommendations

Despite the over 60 years of biogas promotion in Tanzania the technology has not well developed in Rungwe district to date. This study revealed a number of issues that led to the stagnation of the technology. One, energy policy framework has put low profile of biogas in the rural energy development strategies. The technology has been dumped in the cluster of renewable energy which basically concentrates on major types of energy such as biomass (liquid biofuel and fuelwood). Today, there is a lack of adequate indigenous capacity to design, manufacture, market and distribute as well as install and maintain biogas technologies. Two, the cost of installing biogas facility of USD 550-675 is high for many of the rural poor to afford. Three, there is a tendency of risk averse among the poor to adopt new technologies including biogas. Demonstrating the technology and its related benefits might change the pace of adoption. Four, there is also an issue related to water availability. Where water is far from home creates another burden especially for women who at the end of the day they have to choose between running the biogas facilities or producing food for the family, definitely the latter will prevail. Five, the poor performance of milk marketing is linked with poor government policies, low level of management, inadequate milk markets and difficulties arising from the predominance of direct marketing (Kisusu et al., 2000. Other constraints facing dairy producers include lack of improved technology at farm level and weak institutional support (Somda et al., 2004) small size of farms and their distance from markets, animal health and reproductive problems and lack of good-quality animal feeds in sufficient quantities (Swai and Kurimuribo, 2011). Smallholder dairy producers often face problems of high transaction costs when it comes to the question of marketing their small quantities of milk to distant markets.

The recommendations are that the government should accommodate and institutionalize the planning of biogas technology dissemination energy in rural areas. Sensitisation should be enhanced, and support services should be provided towards optimisation of the biogas production process so that potential benefits are realized (Langeni, 2010). In this regard, addressing technical as well as non-technical factors is essential for the sustainability of biogas development and for decision making processes in the energy sector. The government should facilitate access to credit through providing information and also guarantee farmers to get credits. The government should help the farmers access milk markets through providing marketing information and selling of processed products. Modalities of the arrangements should be to link farmers to markets need to take into account socio-economic and agro-climatic diversities (Chakrabarti and Mukhopadhyay, 2009). There should be educational and awareness campaigns on biogas benefits and successes, the provision of financial and non-financial incentives to households could bolster wider biogas energy acceptance in developing countries (Walekhwa, 2009). Lastly, the government in collaboration with stakeholders should provide water near homes as strategy to facilitate biogas adoption.

6. Acknowledgement

The author would like to thank the Research on Poverty Alleviation (REPOA) for funding this study, without their support this work could have not been accomplished. The University of Dar es Salaam is appreciated for granting permission to undertake this study. Furthermore, thanks should go to the district authorities for facilitating this study. Last and not least, the author is grateful to the farmers for their patience and participating in this research interviews.

7. References

Afonja S. A (1992): "Rapid assessment methodologies: application to health and nutrition programmes in Africa", in: Scrimshaw, N. Gleason, G. (Eds), *Rapid Assessment Procedures: Qualitative Methodologies for Planning and Evaluation of Health Related Programmes*. (Boston: International Nutrition Foundation for Developing Countries, 1992).

Barnes D.F & Floor W.M (1996): Rural energy in developing countries: A Challenge for Economic Development. *Annual Review of Energy and the Environment*. Volume. 21, 497-530

CEBITEC (2003): Centre for Excellence on Biogas Technology, Biogas Bulletin Issues N0. 001

Banerjee A. V (2007): The Economic Lives of the Poor. *Journal of Economic Perspective*, Volume 21 No 1, 141–16.

Chakrabarti S & Mukhopadhyay S (2009): Innovations for Linking Farmers to Markets: Indian Experience. *Working Paper Series. Social science research network*

Chambers R. (1994): "The origins & practice of participatory rural appraisal", *World Development,* Volume. 22, No. 7 pp. 953-969.

Feder, G. R, Just, E., & Zilberman, D. (1985). Adoption of agricultural innovations in developing countries: a survey, *Economic Development and Cultural Change* 33, 255-297.

Habtetsion S & Tsighe Z (2002): The energy sector in Eritrea—institutional and policy options for improving rural energy services. *Energy Policy.* Volume 30, Issues 11-12, 1107-1118

Hedegaard K , Thyø K.A & Wenze H (2008): Life Cycle Assessment of an Advanced Bioethanol Technology in the Perspective of Constrained Biomass Availability. *Environ. Sci. Technol.,* 2008, 42 (21), pp 7992–7999.

IRA (2005): UNDP small grant programme ex-post project study for biogas projects at Lomwe and Kaole Secondary Schools. *A report prepared by the Institute of Resource Assessment (IRA). University of Dar esSalaam, Tanzania,* 2005.

Karekezi S & Kimani J (2002): Status of power sector reform in Africa: impact on the poor. *Energy Policy,* Volume 30, Issues 11-12,

Kisusu R. W.; Mdoe, N. S. Y.; Turuka, F. M & Mlambiti, M. E (2000): Contribution of smallholder dairy production to food security, household income and poverty alleviation: the case of Mvumi dairy development project, Dodoma, Conference paper. *Tanzanian Veterinary Journal* 2000 Volume 20,. 89-99 .

Langeni L. N., Mkiramweni B & Mshoro I.B (2010): Estimating the Potential for Biogas Production and Application in Morogoro Region, Tanzania Energy and 26 environment, pp 1357-1368.

Martinot E., Chaurey A., Lew D, Moreira J.R & Wamukonya N (2002): Renewable energy markets in developing countries. *Annual Review of Energy and the Environment.* Volume. 27, 309-348

Muriuki H.G.; Mwangi, D.M. & Thorpe, W (2001): How smallholder dairy systems in Kenya contribute to food security and poverty alleviation: results of recent collaborative studies. URI: http://hdl.handle.net/10568/1725

Mwakaje A.G. (2008): Dairy farming and Biogas Use in Rungwe District, Southwest Tanzania: A study of opportunities and constraints. Renewable and sustainable *Energy Review.* Volume 12, No 8, 2240-2252.

Mwakaje A.G. (2005): Ex-post study for UNDP-GEF Small Grant Funded Projects. The Case of Lomwe Secondary School Biogas Project. A Report for United Nation Development Programme (UNDP)-GEF.

Mwandosya M.J & LuhangaM.L (1993): Energy and development in Tanzania: Issues and perspectives. *Energy Policy,* Volume 21, Issue 5, 441-453

Mwerangi I (2008): Status of renewables in Tanzania. *Renewable Energy Technologies.* http://renewablestechnologies.blogspot.com/2008/10/status-of-renewables-in-tanzania.html

Omer A.M & Fadalla Y (2003): Biogas energy technology in Sudan. *Renewable Energy,* Volume 28, No 3

Parikh J., (1995): Gender issues in energy policy. *Energy Policy*. Volume 23, No 9, September 1995, 745-754

Ramkumar S., Rao S.V.N & Waldie K (2004): Dairy Cattle Rearing by Landless Rural Women in Pondicherry: A Path to Empowerment. *Indian Journal of Gender Studies June 2004 volume 11 No. 2 205-222*

Rutamu I (1999): Low cost biodigesters for zero grazing smallholder dairy farmers in Tanzania. *Livestock Rural Dev* 1999; Volume 11 No 2.

Sasse L., Kellner, C &Kimaro, A. (1991). Improved Biogas unit for Developing countries. GATE Publication, Eschborn.

Schwengels P., (2009): Dairy Manure Biogas Opportunitiesin Manica Province, Mozambique. A Preliminary Assessment. Prepared for Land O' Lakes International Development Smallholder Dairy Development Programin Manica Province, Mozambique

Somda J., KamuangaM & Tollens E (2005): Characteristics and economic viability of milk production in the smallholder farming systems in. *The Gambia. gricultural Systems* Volume 85, Issue 1, July 2005, pp 42-58

Swai, E.S & Karimuribo, E.D (2011): Smallholder dairy farming in Tanzania. Current profiles and prospects for development. **Source:** Outlook on Agriculture, Volume 40, No 1, March 2011 , pp. 21-27(7)

Taşdemiroğlu E (1988): Review of the biogas technology in Turkey. *Biomass*. Volume 17, No 2, 1988, pp 137-148

Uddin S.N., Taplin R & Yu X (1999): Towards a sustainable energy future—exploring current barriers and potential solutions in Thailand. *Environment, Development and Sustainability*. Volume 1 / 1999 - Volume 13 / 2011

URT (2010): United Republic of Tanzania. Rungwe Socio-economic profile.

URT (2008). District development plan for 2005/2006 and projections for 2006/2007–2007/2008; 2005.

URT (2005): Rungwe District Development Plan for 2005/2006 and projections for 2006/07-2007/08, 2005.

URT (2003): Tanzania Energy Policy of 2003

URT (2002). Population and housing census general report. Central Census Office, National Bureau of Statistics, President's Office Planning and Privatization. Dar es Salaam, Tanzania: Government Printer; 2002.

URT (1997): United Republic of Tanzania, Rungwe socio-economic profile.

Wamukonya N (2004): Corrigendum to "Power sector reform in developing countries: mismatched agendas". *Energy Policy*, 31, 1273–1289]

Walekhwa P.N., Mugisha J & Drake L (2009): Biogas energy from family-sized digesters in Uganda: Critical factors and policy implications. *Energy Policy*. Volume 37, No 7, July 2009, 2754-2762.

Walshe, M.J. Grindle, J. Nell A. and M. Bachmann (1991): Dairy development in Sub-Sahara Africa: A study of issues and options. Working papers from World Bank-Technical Papers

Wüstenhagen R, Wolsink M & Bürer M.J (2007): Social acceptance of renewable energy innovation: An introduction to the concept. *Energy Policy*. Volume 35, No 5, 2683-269.

16

Biogas Plant Constructions

M. Samer
Cairo University, Faculty of Agriculture,
Department of Agricultural Engineering,
Egypt

1. Introduction

The chapter concerns with the constructions of the commercial biogas plants as well as the small and household units. Furthermore, the chapter aims at providing a clear description of the structures and constructions of the anaerobic digesters and the used building materials. Ultimately, the chapter answers an important question: how to build a commercial biogas plant and a household unit, and what are the construction steps?

2. Chapter description and contents overview

The chapter describes the construction steps and operation of biogas plant, which include:

a. Planning the biogas plant layout and designing the digesters, where the rules of thumb for planning the layout of a commercial biogas plant are elucidated and a methodology for specifying the dimensions of both digester(s) and residue storage tank(s) is illustrated, and they are: internal and external diameters of the tanks, wall thickness of the tank, height …etc.
b. Undertaking the project, i.e. carrying out the excavation (digging) works, preparation of the bottom plate of the digester, integrating the heating tubes, building the fermenter, installing the insulation, and technology installation.
c. Running the biogas plant including the mechanization of the biogas plant such as: solids feeder, gas processing unit, mixing technology …etc.
d. System control, i.e. how the individual facility components are monitored by computer technology even from afar as well as on-site using a computer system.

3. Overview

3.1 Components of the biogas unit

The components of a biogas unit are:

1. Reception tank
2. Digester or fermenter
3. Gas holder
4. Overflow tank

3.2 Size of biogas unit

The size of a biogas unit depends on several factors, which are:

1. The amount and type of organic waste to be disposed in the digester
2. The objective of treating the organic waste (the production of energy and/or organic fertilizer)
3. Demand of natural gas and consumption pattern
4. On-site nature of the soil and the level of ground water
5. Air temperature in the region and wind direction throughout the different seasons
6. The training level of the staff on farm and home regarding operation of biogas units

The amount of manure fed into a digester each day has an important effect on its operation. This is measured by volume added in relation to the volume of the digester, but the actual quantity fed to the digester also depends on the temperature at which the digester is maintained. In order to determine the unit size of a biogas unit, the following mathematical equation must be achieved:

$$\text{Digester size (m}^3\text{)} = \text{Daily feed-in (m}^3\text{ day}^{-1}\text{)} \times \text{Retention time (day)} \tag{1}$$

The digester size can be defined as the total size of the biogas unit, which includes the effective size of any volume occupied by the fermented material and the volume of gas storage. Size of the daily feed-in is the size of a mixture of dung with water added to the digester once daily or several times and the average concentration of total solids of 10%, where mixing the organic wastes with water depends on its water content. In the case of wet animal wastes, such as manure the proportion of mixing is 1:1. Generally, Storage capacity has to be calculated by average live weight of animals kept in husbandry systems, amount of added water, periods of no fertilization of crops, and the animal species.

In order to plan a biogas plant and to design a digester, several design parameters must be determined which are: ratio of gathered waste from manure canals to total waste, number of cows in farm, amount of manure produced by a cow which is usually 1.8 m^3 cow^{-1} $month^{-1}$, quantity of daily liquid organic matter deposition into the digester, hydraulic retention time, density and quantity of daily dry organic matter deposition into the digester, and digester load which is usually 2-4 kg m^{-3} day^{-1}. The aforementioned design parameters are used to determine the total volume of the materials that are intended to be stored in the tank and are equal to the internal volume of the tank. Additionally, the designer should take into consideration that a part of the tank (about 10%) is empty and the substrates should not fill it, because it is the place where the gas will accumulate. Even in case of designing other storage tanks (e.g. liquid organic matter tank) it is required to leave 10% of the tank volume empty.

3.3 Types of digester

During the last century a number of different types of flows in simple digester have been developed and they can be of the following kinds: (1) batch flow, (2) continuous flow, (3) continuously expanding, (4) plug flow, and (5) contact flow. The conventional digesters are those utilized to process liquid raw materials with a high content in solids, also called rural digesters, the fermentation chamber having a volume below 100 m^3. Conventional digesters are installed without any type of mechanism to reduce the retention time during which the biomass remains inside are predominant; these systems are fed discontinuously and known as discontinuous-flow i.e. batch digesters, or fed periodically and known as continuous-flow digesters.

Batch digesters are loaded at once, maintained closed for a convenient period, and the organic matter is fermented and then unloaded at a later time. It is quite a simple system with small operational requirements. Installation can be made in an anaerobic tank or in a series of tanks, depending on the biogas demand, availability and amount of raw materials to be utilized. Batch flow is most suitable for dry organic matters (solid materials), e.g. solid vegetable waste. This type of biowastes is fed into the digester as a single batch. The digester is opened, digestate is removed to be used as biofertilizer and the new batch replaces the digestate. The tank is then resealed and ready for operation. Depending on the waste material and the operating temperature, a batch digester will slowly start producing biogas and increase the production with time and then drop-off after 4 to 8 weeks. Batch digesters are therefore best operated in groups, so that at least one digester is always producing biogas.

Continuous digesters are usually requiring daily loading and residue management. The process is referred to as continuous since to every daily load corresponds a similar volume load of fermented material. The biomass inside the digester moves through by the difference in hydraulic heat, between the substrate entering the digester and the digestate coming out when unloading. Each load requires a retention time, usually between 14 to 40 days. Continuous digesters can have their retention period reduced by the introduction of agitation and heating. The disadvantage of these models is that the raw material needs to be diluted. The great advantage of these digesters over the batch type is that a single unit allows a continuous supply of biogas and biofertilizer and the continuous treatment of small amounts of waste (Florentino, 2003). Biogas production can be accelerated by continuously feeding the digester with small amounts of waste daily. If such a continuous feeding system is used, then it is essential to ensure that the digester is large enough to hold all the material that will be fed into the digester in the whole digestion cycle. One key issue is to implement two digesters, i.e. accomplishing the biodegradation of the organic waste through two stages, with the main part of the biogas is being produced in the first stage and the second stage serves as finishing stage of the digestion at a slower rate.

Regarding the continuously expanding flow, the digester starts one third full and then filled in stages and later emptied. Concerning the plug flow, the wastes are added regularly at one end and over-flows the other. In the contact flow, a support medium is provided.

Two simple biogas digester designs have been developed, the Chinese fixed dome digester and the Indian floating cover biogas digester. The digestion process is the same in both digesters but the gas collection method is different in each. In the Indian-type digester, the water sealed cover of the digester rises as gas is produced and acts as a storage chamber, whereas the Chinese-type digester has a lower gas storage capacity and requires efficient sealing in order to prevent gas leakage. Both have been designed for use with animal waste or dung. Additionally, there are also Philippine and Sri Lankan digesters.

3.3.1 Indian digester

The Indian-type digester (Fig. 1) basically is comprised of a cylindrical body, gasometer, feed pit and outlet pit (Florentino, 2003). The digester is made using burnt-clay bricks and cement. The cylindrical dome is made of metal sheets and moves up and down as it stores and releases the biogas. The digester is operated in continuing method and often vertically, almost cylindrical built. The putridity space filled the ground and it has a dividing wall. This dividing wall improves and holds back the fresh slime gush again through short way. The gas is gathered in floating gas lock. The steel gas lock is provided with stir elements. The periodic destruction of swimming layer is performed using the manual stirring of gas lock. The requested gas pressure arises from the heaviness of the swimming gas lock. The gas pressure can basically be changed in the practice by putting things on the gas lock.

This type is suitable for the homogeneous materials, as for the animals' excrements that do not tend to build sinking layers. The green waste must be split. If it is mixed with huge allotments, then it will threat the digester with blockage. Generally, there are several designs of Indian digesters, thereof: floating gas holder type biogas plant (KVIC model), Deenbandhu model, and Pragati model. The KVIC model is composite unit of a masonry digester and a metallic dome, where the maintenance of constant pressure by upward and downward movement of the gas holder. The Deenbandhu model consists of segments of two spheres of different diameters joined at their base, where this model requires lower costs in comparison to KVIC model. The Pragati model is a combination of Deenbandhu and KVIC designs, where the lower part of the digester is semi spherical with conical bottom and the floating drum acts as gas storage.

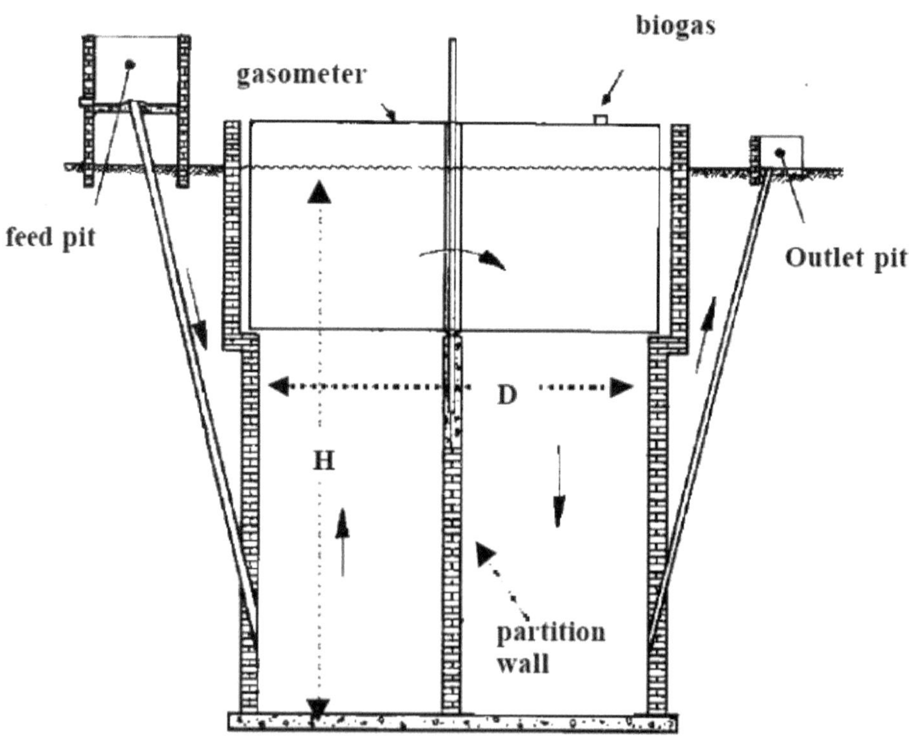

Fig. 1. Indian-type digester (Florentino, 2003)

3.3.2 Chinese digester

The Chinese-type model digester (Fig. 2) is comprised of a cylindrical body, two spherical domes, inlet pit, outlet pit and an inspection opening (Florentino, 2003). The digester is made using cement and bricks and it is a permanent structure. Just as in the Indian digester this has two drains to feed waste and to collect the composted waste.

The biogas is collected in the upper chamber and the waste decomposes in the lower chamber. If the gas pressure exceeds the atmospheric pressure (1 bar) and there is no gas extracted from the dome, then the rot substrate squeezed from the reactor into the filled pipe, but often in the pool of counterpoise. If the produced gas is more than the up used gas, then the slime level will increase. If the up used gas is more than the produced gas during the gas extraction, then the slime level will sink and the rot slime will flow back. The volume of the counterpoise pool must be huge so that the repressed rot substrate can be digested at the highest gas volume. The gas pressure is not constant in the practice. It increases with the quantity of the stored gas. The gas must be regularly produced; therefore the gas pressure organizer or the swimming gas repository room is important.

Owing to the fact that the biogas dome digesters are completely buried underground, the fermentation temperature should be under a day/night temperature change, only in a tolerance range from about ± 2 °C. The difference between summer and winter is large and is subject to the climate zone. The biogas dome digester can be provided with stir. In small family household units, a mix concoction for the biogas dome digester is installed. Different building and construction forms of biogas dome digesters were proved for the Chinese digesters; so that there is a big number of building methods are used.

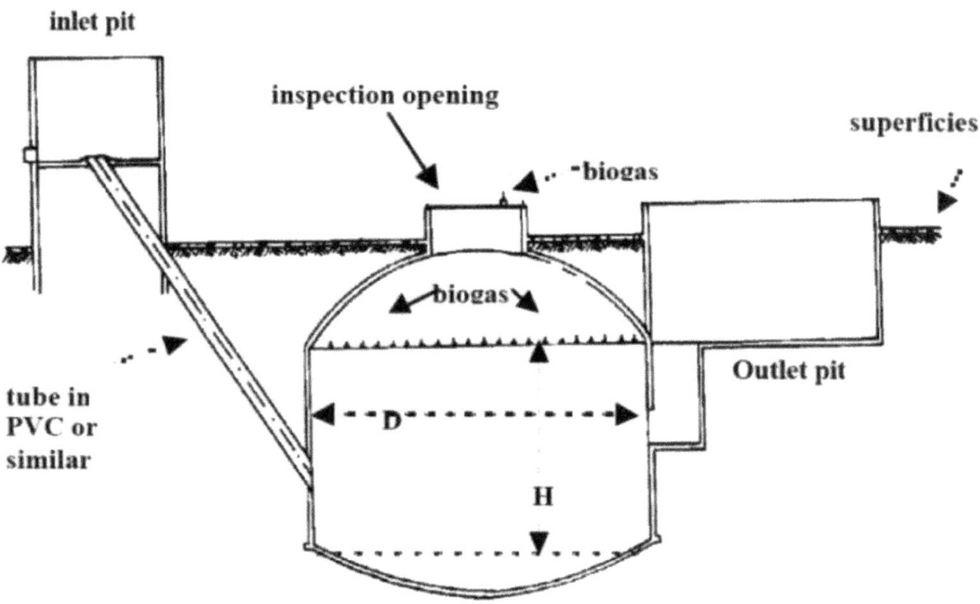

Fig. 2. Chinese-type digester (Florentino, 2003)

3.4 Designs of digester

The most common digester design is cylindrical. Digesters can be classified in horizontal and vertical designs (Fig. 3). Currently, vertical concrete or steel digesters with rotating propellers or immersion pumps for homogenization are widespread. Vertical tanks simply take feedstock in a pipe on one side, whilst digestate overflows through a pipe on the other side. In horizontal plug-flow systems, a more solid feedstock is used as a plug that flows through a horizontal digester at the rate it is fed-in. Vertical tanks are simpler and cheaper to operate, but the feedstock may not reside in the digester for the optimum period of time. Horizontal tanks are more expensive to build and operate, but the feedstock will neither leave the digester too early nor stay inside the digester for an uneconomically long period.

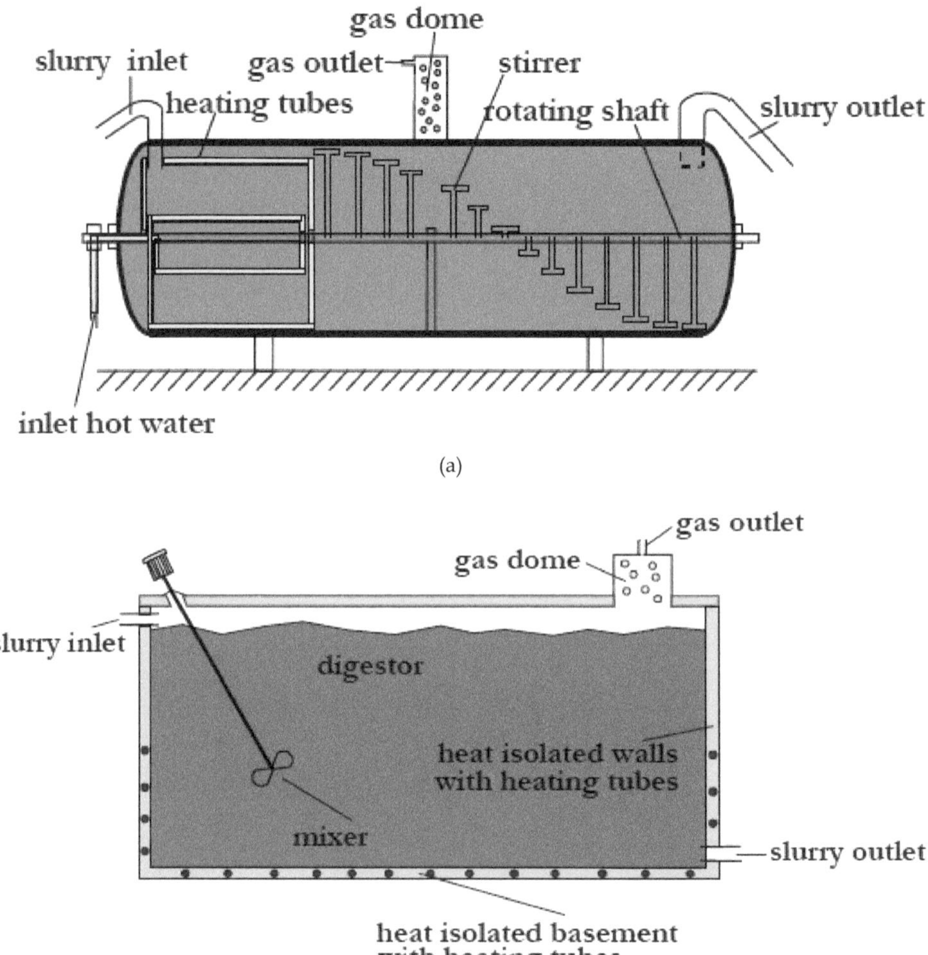

Fig. 3. Horizontal (a) and vertical (b) digester (Gronauer and Neser, 2003)

Anaerobic digesters can be built either above or under the ground. An alternative is that a part of the digester can be buried. Anaerobic digesters constructed above ground are steel structures to withstand the pressure; therefore, it is simpler and cheaper to build the digester underground. Maintenance is, however, much simpler for digesters built above ground and a black coating will help provide some solar heating.

4. Building materials and dimensions

Reinforced concrete is obtained by adequately mixing specific proportions of aggregates (gravels and sand), cement, and water (Bartali, 1999). The water:cement ratio is 0.53 L kg^{-1} and the cement:sand:gravel mass ratio is 1:2.2:3.7 for floors, driveways, structural beams, and columns (Lindley & Whitaker, 1996). Cylindrical cast-in-place concrete tanks are commonly used in biogas plants for storing liquid manure during long periods. A serviceable tank should be watertight to prevent groundwater pollution and corrosion of the reinforcing rods. Therefore, these tanks should be designed to withstand different design loads, i.e. the loads of the soil outside the digester which is buried underground level and loads of the liquid stored inside the digester. Liquid manure is often stored in large cylindrical concrete tanks, which are partially underground. The dimensions of these tanks vary from 18 to 33 m in diameter with heights from 2.4 to 4.9 m and a uniform wall thickness varying from 150 to 200 mm (Ghafoori & Flynn, 2007; Godbout et al., 2003).

The internal volume of the tank can be calculated by multiplying the volume of substrates that should be stored in the tank by 1.10 in order to consider 10% as headspace. The cement mass (kg), gravels volume (m^3), and sand volume (m^3) required to build the tank can be calculated by multiplying the concrete volume of the tank by the constants C, G, and S, respectively, where C represents the mass of cement required to make 1 m^3 of concrete (325 kg m^{-3}), G is the volume of gravel required for 1 m^3 concrete (0.8 m^3 of gravel per m^{-3} of concrete), and S is the volume of sand required for 1 m^3 concrete (0.4 m^3 of sand per m^{-3} of concrete). The type of iron rods should be selected. The different types (NØD m^{-1}, where N is the number of iron rods per meter length, and D is the diameter of the iron rod) are 6Ø6 m^{-1} (0.666 kg m^{-1}) and 6Ø8 m^{-1} (0.888 kg m^{-1}). In the case of constructing a tank without a concrete top, both types can be used. On the other side, in the case of building a tank with a concrete top, the type 6Ø8 m^{-1} must be used with two iron grids (Samer, 2008, 2010, 2011; Samer et al., 2008). The thickness of digester wall should be 35 cm and is built using reinforced concrete to bear the loads of the materials stored in the digester. Tables 1 through 3 show the typical digester specifications for a commercial biogas plant, the required quantities of construction materials to build the digester, and the quantities of the substrates.

Specification	Value	Unit
Internal diameter of the digester	23	m
External diameter of the digester	23.7	m
Internal height of the digester	6	m
Buried part of the digester	2	m
Wall thickness of the digester	0.30	m
Capacity	11820	m^3

Table 1. Typical digester specifications for a commercial biogas plant

Material	Quantity	Unit
Rebar	36	Ton
Cement	320	Ton
Sand	400	m^3
Gravels	800	m^3

Table 2. Required quantities of construction materials to build the digester

Material	Quantity	Unit
Raw slurry storage[1]	18	m^3
Liquid organic matter[2]	21	m^3
Liquid substrate[3]	80	m^3
Dry organic matter[4]	267	m^3 day^{-1}
Total substrates[5]	10750	m^3

[1]Consider a duration of 3 days for mixing and pumping, daily manure deposition of 6 m^3 day^{-1}, 1.8 m^3 cow^{-1} $month^{-1}$, and 100 cows
[2]Consider a storage duration of 7 days and liquid organic matter deposition of 3 m^3 day^{-1}
[3]Consider 40 days of storage duration and liquid substrate deposition of 2 m^3 day^{-1}
[4]Consider digester load of 4 kg m^3 day^{-1} and density of 1.2 kg m^{-3}
[5]Total quantity of substrates (10750 m^3) that should be stored in a digester having a capacity of 11820 m^3

Table 3. Quantities of the substrates

5. Construction steps

5.1 Facility layout

The anaerobic digestion can be accomplished in one digester, and thus the facility is called 'single-stage biogas facility'. In other facility layouts, the anaerobic digestion can be carried out in two stages, i.e. in two different tanks in order to optimize the operating conditions, and thus the facility is called 'two-stage biogas facility'. The single-stage facility is a simple design with a longer track record, and has lower capital costs and technical problems. The two-stage facility has shorter retention time as each stage design is optimized. There is a potentially higher biogas production from two-stage facilities, but they require higher capital costs.

Subsequent to the site investigations such as the soil specifications ground water level, the facility layout should be planned. The commercial biogas plants consists of a fermenter and a secondary fermenter or so called "follow-up fermenter", where both have identical dimensions, usually as follows: height of 6 m, internal diameter of 23 m, and external diameter of 23.70 m. This implies that the thickness of the fermenter wall is usually 35 cm. A residue storage tank is annexed to the fermenters, where the tank has an internal diameter of 25 m, external diameter of 25.70 m and a height of 6 m (Fig. 4). The solids feeder is located adjacent to the fermenter, and the tanks are surrounded by green belt from all sides except one side where the horizontal silos are located.

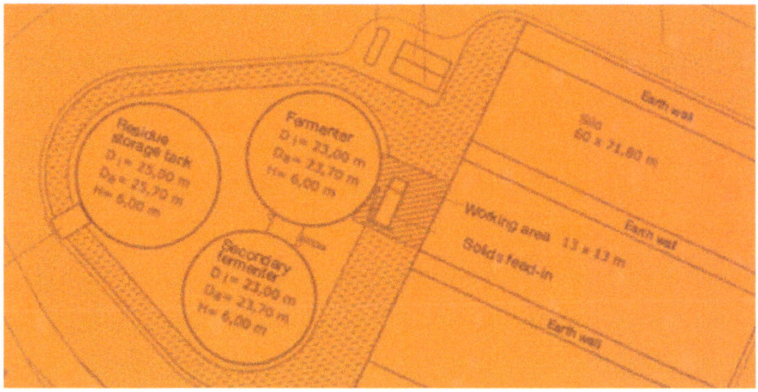

(a) MT-ENERGIE GmbH & Co. KG

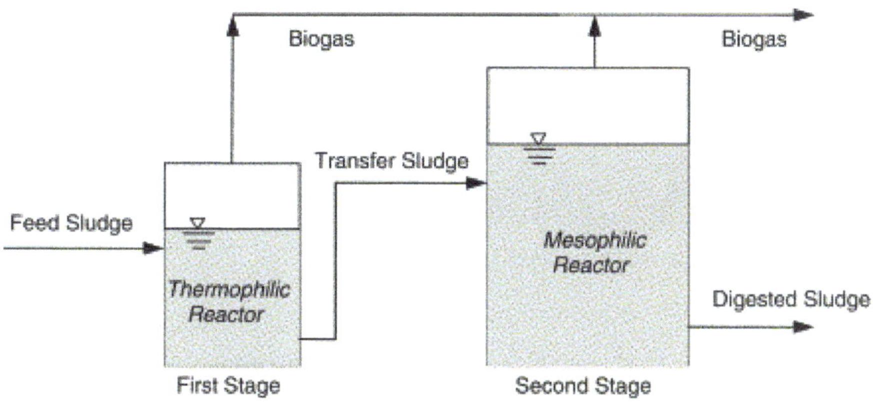

(b) General process scheme of the two-stage anaerobic digestion process (Blumensaat and Keller, 2005)

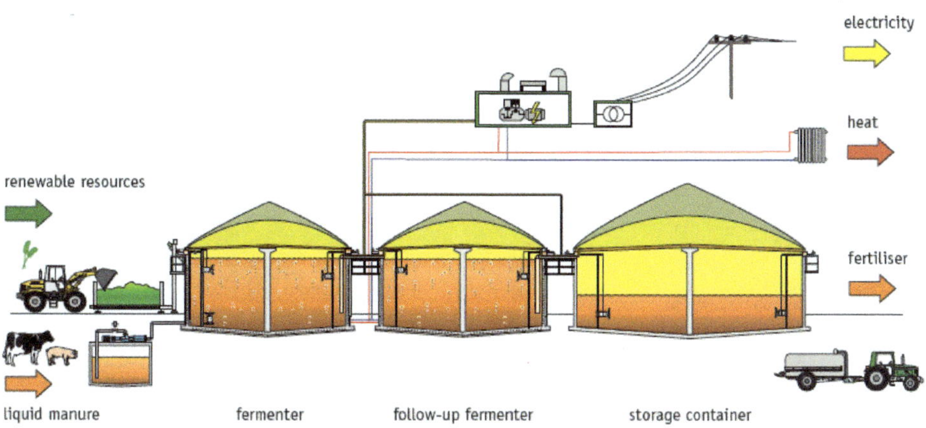

(c) BIOGAS NORD GmbH

Fig. 4. Facility layout for a two-stage biogas plant

5.2 Dimensions marking

The marking of dimensions (Fig. 5) for biodigester unit should be performed prior to start of excavation work. The marking is considered as preparation for excavation and construction works. An operational area of 3 m width around the digester should be considered, where the workers will use this area to achieve the construction works around the tank base in order to prepare the structure of the concrete base, i.e. the bottom of the digester.

Fig. 5. Marking of dimensions for a biodigester

5.3 Excavation works

The depth of digging depends on the specifications of the soil. The inclination of the sides should be 30 cm for each meter depth for the cohesive soil, 60 cm to one meter for the light soil, and 90 cm for the sandy soil.

(a, b) BIOGAS PLANT DESIGN

Fig. 6. Excavation works for household units

The bottom of the pit should be concave, where the center of the digester should be the most concave. Generally, the pit shape depends on the design of the digester, where in the case of round-shaped household units (Figs. 6a and b) a guide wood post is installed in the center of pit bottom. A string is linked to the post and used to set the round-shape of the pit. On the other hand, in the case of commercial biogas plants (Fig. 7), total station, teodolit or laser leveling is used for surveying. For large digesters, i.e. for commercial biogas plants, bulldozers are used to achieve the excavation (Fig. 7).

Fig. 7. Excavation works for a commercial biogas plant

5.4 Preparation of the digester's bottom

The pit's bottom should be cleaned, and the gridiron is built (Fig. 8) using a pre-selected type of iron rods as either 6Ø6 m^{-1} or 6Ø8 m^{-1}. Subsequently, the concrete mixture is poured (Fig. 9 and 10). The water:cement ratio is 0.53 L kg^{-1} and the cement:sand:gravel mass ratio is 1:2.2:3.7. The thickness of the concrete base ranges between 10-25 cm depending on the soil's specifications and the ground water level.

Fig. 8. Structuring the grid irons for a commercial biogas plant (MT-ENERGIE GmbH & Co. KG)

Fig. 9. Pouring the concrete mixture for a commercial biogas plant (BIOGAS NORD GmbH)

Fig. 10. Concrete bottom of a household unit

5.5 Building the digester

In case of commercial biogas plants, the digester is huge as its diameter may reach 25 m; therefore, the concrete structure should be reinforced (Fig. 11). Hence, the iron rods are used to build 2 iron grids to reinforce the digester wall starting from the digester bottom plate. The standard length of iron rods is 12 m. The standard iron rods are cut to shorter iron rods, and they are then used to build up the tank. Subsequently, either wood panels or pre-constructed metal sheets are used to enclose the iron grids and to form a container for the fluid concrete. When the digester wall is built, about one third of the internal wall of the tank is covered by a protection layer in order to protect the internal face of the wall against corrosion.

Fig. 11. Building the digester wall for a commercial biogas plant

In case of household units, burnt-clay bricks are used to build the digester (Fig. 12) and they should be able to tolerate a pressure up to 100 kg cm^{-2} owing to the fact that the walls of the digester are exposed to the pressures of the soil and the moving equipment near to the digester. A mortar of cement and sand mixture by 1:4 is used. The construction works are followed up till the appropriate height, and the entry or exit holes of the pipes are blocked by a filling material.

Fig. 12. Building the digester wall of a household unit using burnt-clay bricks

5.6 Integrating the heating tubes

Building the digester is associated with integrating the heating tubes. Building the wall starts with structuring the iron grids which will be encased by wood panels or pre-constructed metal sheets, and before pouring the concrete, the heating tubes should be integrated (Figs. 11 and 13). The heating tubes are made of polyvinyl chloride (PVC), inside these tubes hot water flows to heat the digester. The water temperature is either 35 ^{o}C or 55 ^{o}C depending on the used bacteria as either mesophilic of thermophilic bacteria, respectively. On the other hand, in other designs the heating tubes are installed on the internal surface area of the digester wall.

Fig. 13. Integrating the heating tubes (BIOGAS NORD GmbH)

5.7 Building the gas holder

A wood or steel structure in form of umbrella is built, and then a mesh network is relayed on the umbrella structure (Figs. 14 and 15). The air supported double membrane cover, which includes the gas holder, is mounted over the structure. The flexible membrane of the gas collector, i.e. holder, moves up and down as a function of the gas pressure. On the other hand, the storage tank covers are numerous: (1) closed cover (concrete, plastic, and tent), (2) straw cover, (3) granulate (perlite), (4) swimming vinyl covering, and (5) open as open lagoons and open storage tanks.

Fig. 14. Wood structure and mesh network that supports the gas holder of digester in a commercial biogas plant (MT-ENERGIE GmbH & Co. KG)

Fig. 15. Mesh network that supports the gas holder of a household unit

5.8 Technology installation

The technology that should be installed includes the filling indicator, tubes, measuring devices and meters, electricity network, fiber cables ...etc. Afterwards, the gas collector should be installed as well as the excess and low pressure safeguard and the air support fan. Figure 16 shows an overview of the technology installation.

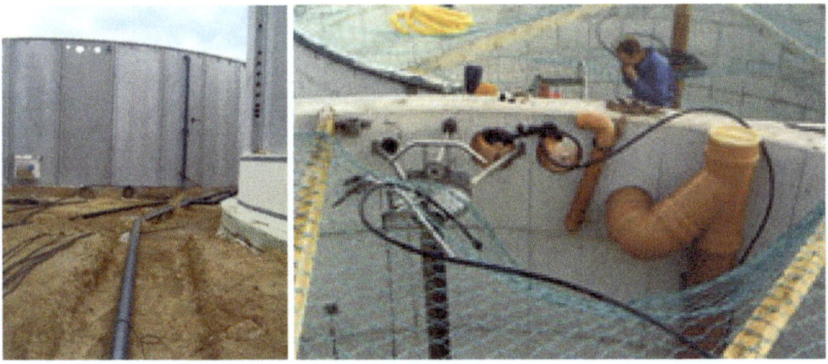

(a) MT-ENERGIE GmbH & Co. KG (b) BIOGAS NORD GmbH

Fig. 16. Technology installation

5.9 Installing the insulation

This is the process of lining the digester by mortar or using sheets of foam as in Figure 17. This is one of the most important construction steps and should be carefully and accurately achieved. In case of lining, the process is performed using mortar containing 1% silica. After the completion of the lining, the digester is painted using the petroleum Albumen. In other designs, the walls are heat insulated with a clad with non-corroding and weather-proof aluminum trapezoidal panels. On the other hand, the rural digesters are coated with layers of dry dirt and asbestos.

Fig. 17. Installing the insulation (BIOGAS NORD GmbH)

6. Information technology and mechanization

6.1 Computer-aided design

A software was developed by Samer (2010) to plan and design biogas plants, specify the dimensions of the different tanks (raw slurry tank, liquid organic matter tank, digester tank, secondary digester tank, and residue storage tank), and compute the amounts of construction materials (iron rods, cement, sand, and gravel) required to build the concrete constructions. Furthermore, the software is able to calculate the capital investment and the fixed costs, the variable costs, and the total costs. Figure 18 shows the user interfaces of the input and output data windows.

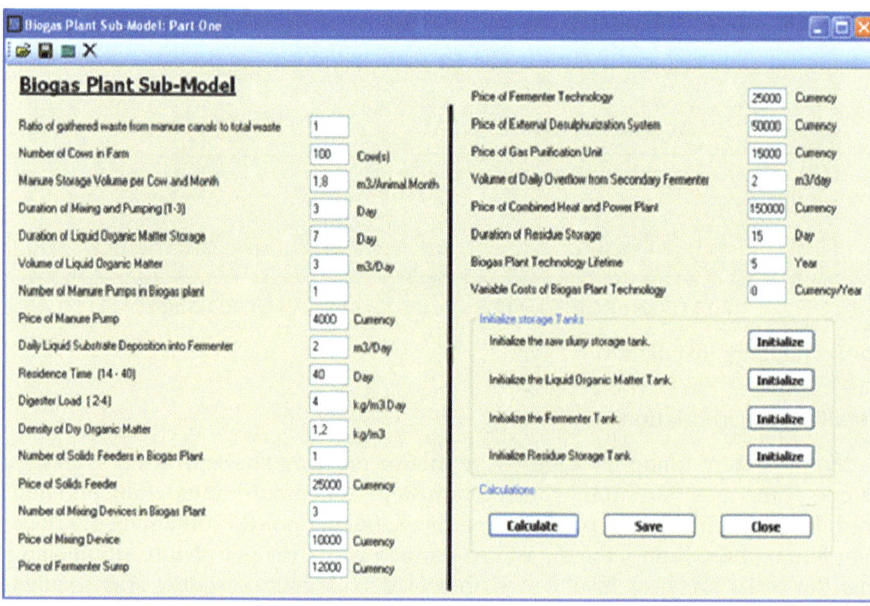

(a) User interface for general input data

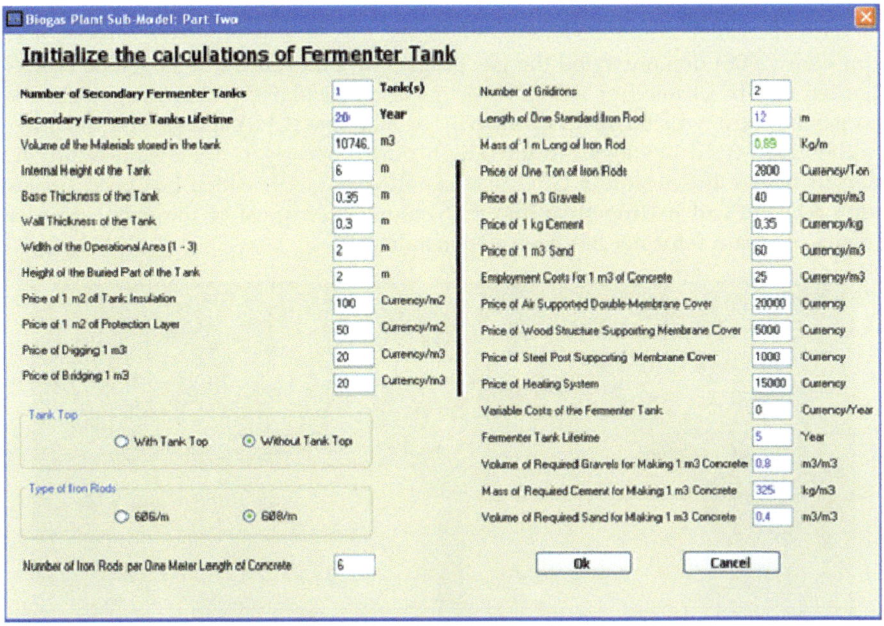

(b) User interface for input data of digester tank

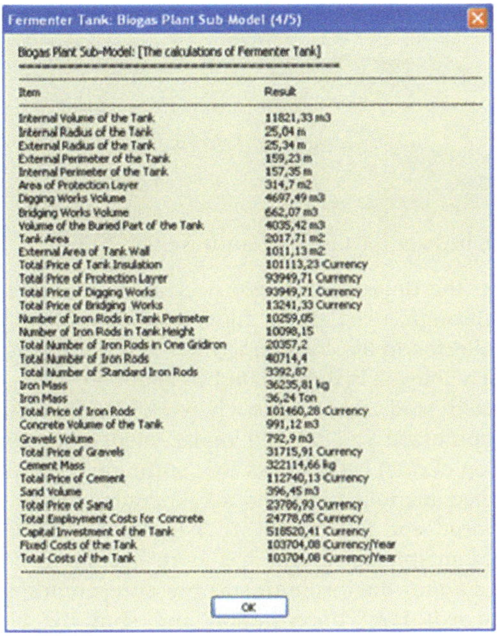

(c) User interface for output data

Fig. 18. Software for planning and designing biogas plants (Samer, 2010)

6.2 Gas processing unit

Chandrasekar (2006) demonstrated the gas processing unit (GPU). In stage one biogas from the digester will be cleaned of moisture droplets, particulates and hydrogen sulfide. The cleaned gas mixture, which consists primarily of methane (CH_4) and carbon dioxide (CO_2), will be then converted in stage 2 to ultra-high purity hydrogen in a steam reformer. As a first step to realize this vision, a GPU was installed (Fig. 19) which has been successfully removing over 99% of hydrogen sulfide (H_2S) along with most of the water droplets and particulates. A steam reformer has been also installed.

Fig. 19. Activated carbon beds of the GPU (Chandrasekar, 2006)

In the GPU biogas from the digester is pressurized to over 3 inches water column by a blower. It then passes through a coalescing filter to remove most of the particulates and water droplets. Water collected in the coalescing filter gets automatically drained out once it reaches a certain level. The biogas is then heated to about 85 ^{o}F in a heater before it passes through two successive activated carbon beds where H_2S is converted into elemental sulfur. The process has been optimized so that bed replacement is needed only once every six months. The configuration of dual beds allows for continuous operation even when one bed is being replaced. The bed manufacturer should be contracted to replace the used beds, thereby obviating the need for the farmer to handle the sulfur. The design requires minimum operation and maintenance and has been set up to be controlled through a computer that will also monitor the incoming gas pressure, control and monitor the blower as well as monitor the exit H_2S concentration and shut the blower/GPU if the exit concentration is greater than the set point. If the GPU shuts down, biogas will automatically feed the engine generator like before to produce electricity. A simple schematic of the GPU is shown in Figure 20.

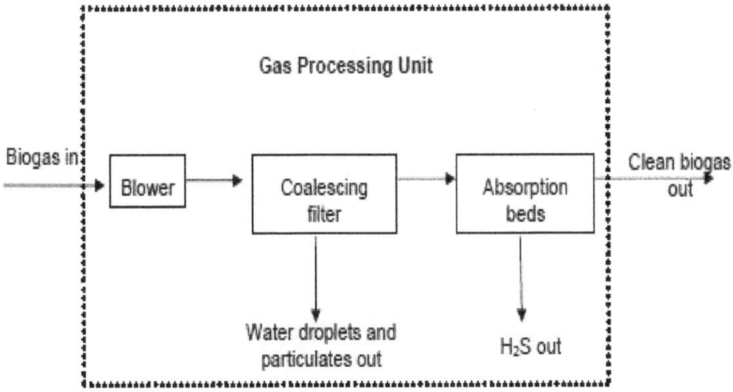

Fig. 20. Schematic of the GPU (Chandrasekar, 2006)

6.3 Mixing technology

The types of mechanical mixing (Fig. 21) are: vertical mixing, horizontal mixing, and side mixing. Submersible motor mixing devices are usually used in commercial biogas plants. Each device is provided by a cable and gear protection system (Fig. 22). Light agitation increases the velocity of digestion, differently from heavy agitation which decreases the velocity of reaction. In digesters with capacities higher than 100 m^3, it is necessary to install equipment to provide agitation of the contents.

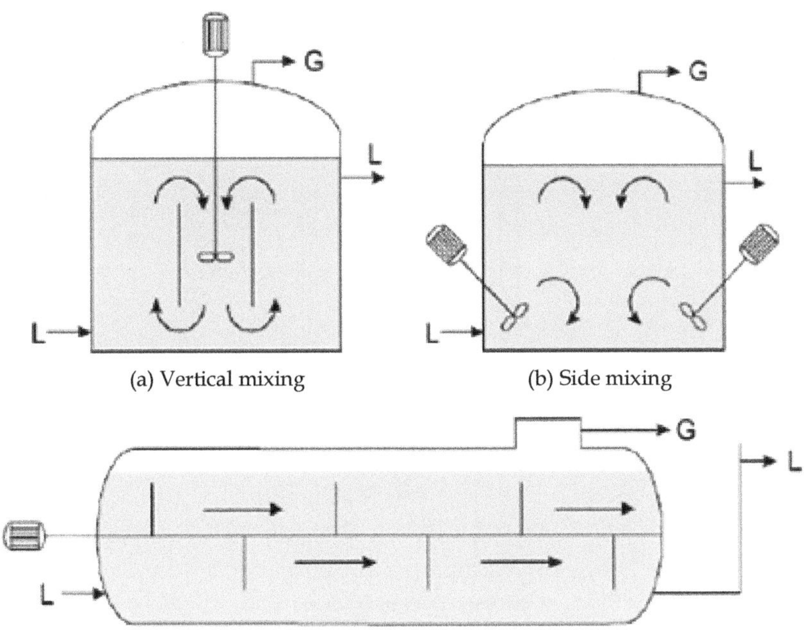

(a) Vertical mixing

(b) Side mixing

(c) Horizontal mixing

Fig. 21. Mechanical mixing

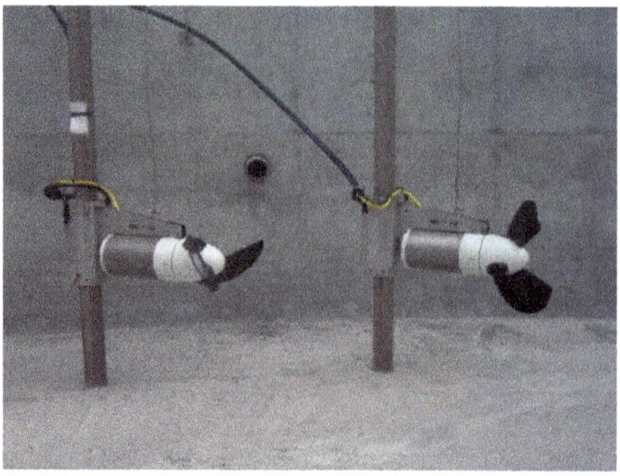

(a) Submersible motor mixing devices

(b) Cable and gear protection system

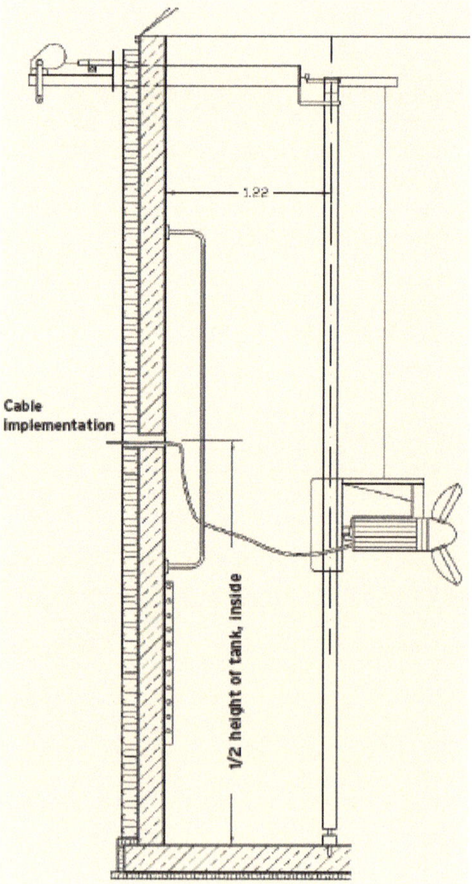

(c) Cable implementation to mixer

Fig. 22. Gear protection system (MT-ENERGIE GmbH & Co. KG)

6.4 Solids feeder

The solids feeder (Fig. 23) is a device that feeds the digester by the solid biowastes, i.e. dry organic matter. The solids feeder consists of a dosing container, weighing cells, identification/weighing system, and a mechanical system. The mechanical system of the solids feeder consists of a hydraulic drive system for the pushing floor, cross feeder auger drive, transport auger drive, transport auger, tamping auger drive, and tamping auger.

(a)

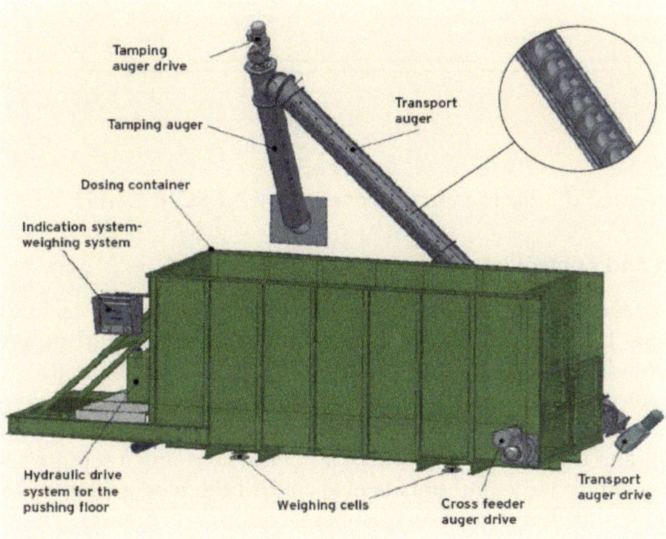

(b)

Fig. 23. Solids feeder (MT-ENERGIE GmbH & Co. KG)

6.5 Cover membrane

The cover membrane consists of a weather protection foil, gasholder membrane or gas collector, clamp hose, level indicator for the gas collector, excess and low-pressure safeguard, and a supporting structure (Fig. 24). A desulphurization unit is integrated into the inner membrane (gasholder) to reduce the hydrogen sulfide (H_2S). The collector is sealed gas-tight with two cone-shaped foils and a clamp rail.

There are three structure types for supporting the cover, which are: (1) a steel post supports directly the cover, (2) a wood structure (incl. a wood post) supports the cover, and (3) a steel post supports a wood structure which in turn supports the cover.

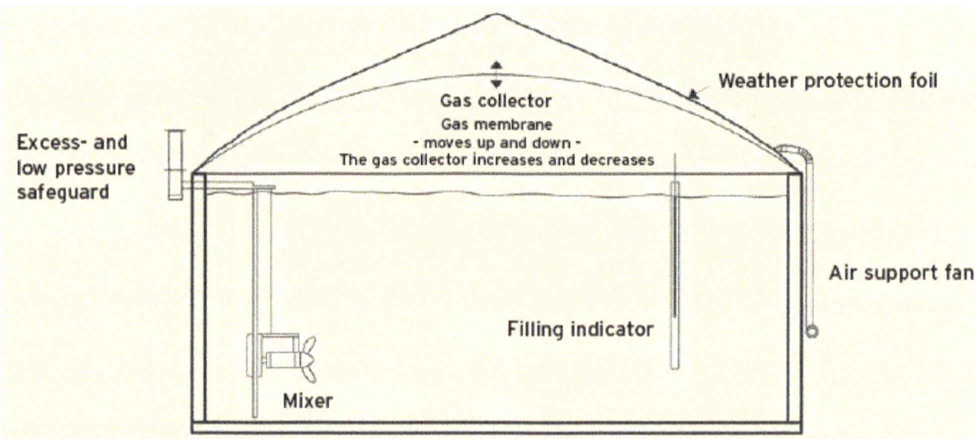

Fig. 24. Air supported double membrane cover (MT-ENERGIE GmbH & Co. KG)

6.6 Monitoring and controlling

The individual facility components are monitored by computer technology even from afar. At the same time technical procedures, as well as input and output quantities are neatly documented. The computer system consists of an on-site control system installed on a computer and sometimes another remote access system on another computer located far away from the site, e.g. in the company's headquarter. The remote computer is connected to the site through routers (ISDN or DSL), allowing a remote control of the site. Any of the aforementioned computers is able control the biogas plant through controlling the central plant control system which is connected to the digester running components (agitator, pump, feeding system …etc.), the cogeneration unit, and the biogas analyzer. A control system allows for real-time local and remote operation and monitoring as well as data gathering (Fig. 25).

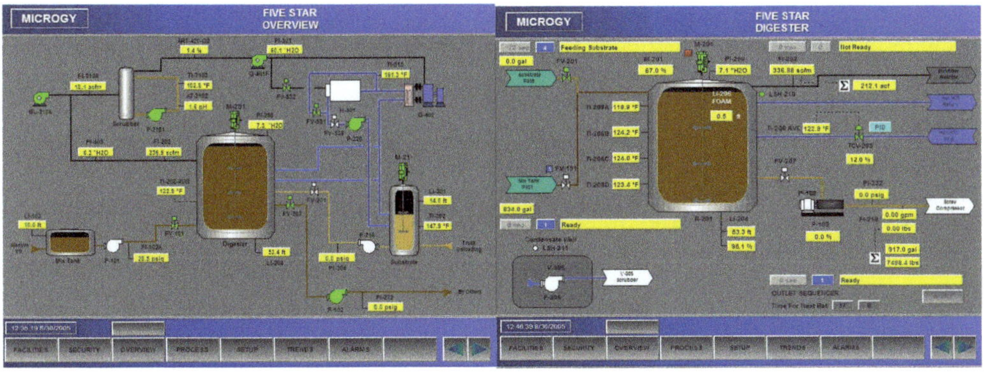

Fig. 25. Monitoring and controlling system (Environmental Power Corp.)

7. Running the biogas plant

Controlled fermentation of biomass in biogas plants produces a gas that can be used to produce electrical and thermal energy on account of its high percentage of methane. The raw materials used in biogas plants or their main substrates, are often liquid manure, agricultural products, and some agro-industrial wastes. The biogas plant may use silage maize as one of its renewable raw materials, with the aid of wheel loader, the maize is fed into either a storage bin or solids feeder, which takes a filling up approximately once a day. Silage maize is rich in energy, and on account of it is high degree of production it is very well suited for use in biogas firms. The storage bin is equipped by hydraulic flow discharger that continuously feed the maize onto a conveyor belt. A scale under the conveyor belt registers the weight of the maize silage. Liquid manure is the most important basic substrate used in biogas plants, after short influence storage in big tank, it is pumped through pipes directly into the blending pumps beside the maize conveyor belt, at the same time the maize fall off from the conveyor belt into separators, which is equipped with two mixing rollers, in this way the maize silage is mixed before fermentation. With this technology, it is possible to supply several fermentation tanks (also known as fermenters and digesters) with fresh substrates even if they are not close together. Liquid wastes from the food industry are the third substrates used in biogas plants, as the availability of such wastes varies considerably, a large storage pit should be installed to integrate this into a whole serving to minimize smells and help to prevent epidemics. The liquid waste is heated with hot water into 70 °C in a tubular heat exchanger using a counter current process. After heating for one hour, the hydrogenation of the substrate is complete so that they can be poured into the fermenters. The fermenter is the place where the biogas is formed, the substrate are continuously stirred in order to prevent layers of materials forming at the top or on the bottom, hot water tubes heat the substrates to between 35 and 55 ^{o}C to accelerate the formation of methane. The substrate is in the fermenter for a period of around 30 days before it is filled into another fermenter for a further 30 days to complete the gas formation process. When fermentation is complete the thin liquid substrate is pumped into two reinforced concrete tanks, where it is stored until it can be brought out onto the fields.

7.1 Gas guidance

If the fermenters are filled regularly with biomass which is air-tight heated and regularly stirred, the biogas will be formed within a matter of days where the formation of biogas is a complex and delicate process. The organic fats which cover high rates contained in the substrates are digested by various kinds of bacteria, this is a starting point for the development of the gas, the contents are continuously stirred, the gas drives slowly to the top of the container and it consists of approximately 50 to 70% methane, carbon dioxide, water vapor, hydrogen, and hydrogen sulfide. As water vapor and hydrogen sulfide are problematic for the utilization of the gas maker, it is necessary to purify the biogas.

The gas is first cleaned from water vapor. The condensed water is collected and a condensation shaft pumps it out. On the other hand, the aggressive trace gas hydrogen sulfide is now extracted from the biogas in a biological desulfurization unit, by introducing air to the container certain bacteria culture which is able to establish colonies on chains. The bacteria decompose the hydrogen sulfide to harmless sulfa and water. The almost unpressurized biogas is then fed into a compressor where it is watered up to 70 mbar pressure later required for burning. In order to completely condense any reaming water vapor freeing biogas of any suspending matters, the biogas is subjected to a washing brine process, this is carried out at almost a freezing point, so that the gas is cooled down to a temperature of 5 degrees. In order to control the purification of the gas, the biogas is constantly tested with an online measuring system which records the amount of methane, hydrogen sulfide, and carbon dioxide. This guarantees a high degree of efficiency and security. In case of any over production of biogas, it is necessary to operate a gas flame for the unburned methane gas that escapes to the atmosphere which is harmful for the environment.

Using 15,000 tons of biomass per year, the plant produces a total of 500 kW of electricity and heat. The optimal gas processing engines can run several years with the minimum of maintenance costs. Up to 30% of the waste heats from the water cooling the engine is used in the heat exchanger and the fermenter so that no additional heat is required, the remaining heat can also be used profitably to heat industrial plants and houses. The electric power generated by the Combined Heat and Power Plant (CHP) is converted to high voltage and then the electricity can be fed into the grid who meets the annual requirement of around 1000 households.

7.2 Instrumentation and control

A biogas plant with an annual operating capacity of 15,000 tons a year requires 3-5 hours' work daily in order to keep the amount of work gone to minimum with a particularly recommended use of an effective measurement of control technology. For safe exchange of data it is also possible for someone who is not on-site to monitor and control the unit, i.e. the unit can be remotely controlled. For example, the agitators can be switched on and off, and all the solid supply equipment can be monitored. Information about malfunctions can be registered on computer service or on the operator's mobile phone, these guarantees a short reaction time when anything unexpected happens.

Biogas plant produces high quality biofertilizers as well as electricity, the nutrients contained in the substrate are retained and more easily available to the plants, as the flow power of the liquid manure and its ammoniacal content have increased, due to fermentation the unpleasant smell of liquid manure and organic wastes have disappeared, as the organic acids have been decomposed. The biogas plant is therefore useful for both the operators and their neighbors.

8. Conclusion

There is flexibility in the types of digester designs available. Digester types and designs are selected for the types of feedstocks to be digested. There are general approaches to tank design, mixing systems, and electrical generation systems. There are different construction quality approaches for household and farm-based in comparison to commercial digesters.

9. References

Bartali, E. (1999). Characteristics and Performances of Construction Materials, In: *CIGR Handbook of Agricultural Engineering: Animal Production and Aquacultural Engineering*, E. Bartali, A. Jongebreur, and D. Moffitt, (eds.), 3-29, ASAE, ISBN 0-929355-98-9, St. Joseph, Michigan, USA

Blumensaat, F. & Keller J. (2005). Modelling of Two-Stage Anaerobic Digestion Using the IWA Anaerobic Digestion Model No. 1 (ADM1). *Water Research*, Vol. 39, No. 1, pp. 171-183, ISSN 0043-1354

BIOGAS PLANT DESIGN. http://bio-gas-plant.blogspot.com/

BIOGAS NORD GmbH, Werningshof 2-4, 33719 Bielefeld, Germany. http://www.biogas-nord.com/

Chandrasekar, A. (2006). Demonstration of Hydrogen Production from Dairy Manure Derived Biogas, *Proceedings of 2006 ASABE Annual International Meeting*, Paper Number: 067065, Portland, Oregon, USA, July 9-12, 2006

Environmental Power Corp., 120 White Plains Rd, Tarrytown, NY 10591, USA. www.environmentalpower.com

Florentino, H. (2003). Mathematical Tool to Size Rural Digesters. Scientia Agricola, Vol. 60, No. 1, pp. 185-190, ISSN 0103-9016

Ghafoori, E. & Flynn P. C. (2007). Optimizing the Size of Anaerobic Digesters. *Transactions of the ASABE*, Vol. 50, No. 3, pp. 1029-1036, ISSN 2151-0032

Godbout, S.; Marquis, A.; Fafard, M. & Picard, A. (2003). Analytical Determination of Internal Forces in a Cylindrical Tank Wall from Soil, Liquid, and Vehicle Loads. *Canadian Biosystems Engineering*, Vol. 45, pp. 5.7-5.14, ISSN: 1492-9058

Gronauer, A. G. & Neser S. (2003). Concepts and Techniques for Sustainable Slurry Management on Farms, Proceedings of Ninth International Symposium on Animal, Agricultural and Food Processing Wastes, Paper Number: 701P1203, North Carolina, USA, October 12-15, 2003

Lindley, J. A. & Whitaker, J. H. (1996). *Agricultural Buildings and Structures*, ASAE, ISBN 0-929355-73-3, St. Joseph, Michigan, USA

MT-ENERGIE GmbH & Co. KG, Vor dem Seemoor 1, 27404 Rockstedt, Germany. http://www.mt-energie.com/

Samer, M. (2011). How to Construct Manure Storages and Handling Systems? *IST Transactions of Biosystems and Agricultural Engineering*, Vol. 1, No. 1, pp. 1-7, ISSN 1913-8741

Samer, M. (2010). A Software Program for Planning and Designing Biogas Plants. *Transactions of the ASABE*, Vol. 53, No. 4, pp. 1277-1285, ISSN 2151-0032

Samer, M. (2008). *An Expert System for Planning and Designing Dairy Farms in Hot Climates*, Dissertation, VDI-MEG, ISSN 0931-6264, Script 472

Samer, M.; Grimm, H.; Hatem, M.; Doluschitz, R. & Jungbluth T. (2008). Mathematical Modeling and Spark Mapping for Construction of Aerobic Treatment Systems and their Manure Handling System, *Proceedings of International Conference on Agricultural Engineering*, Book of Abstracts p. 28, EurAgEng, Hersonissos, Crete, Greece, June 23-25, 2008

17

Enhancing Biogas Production and UASB Start-Up by Chitosan Addition

Chantaraporn Phalakornkule[1,2] and Maneerat Khemkhao[2]
[1]*Department of Chemical Engineering, Faculty of Engineering, King Mongkut's University of Technology North Bangkok,*
[2]*The Joint Graduate School of Energy and Environment, King Mongkut's University of Technology Thonburi, Thailand*

1. Introduction

Anaerobic digesters have been applied for the treatment of wastewater yielding biogas as a value by-product. The biogas from the treatment plant can be utilized for generating heat and electricity. Anaerobic bacteria form granules through cell self-immobilization which then settle out as floc aggregates. These granules are dense microbial consortia packed with different bacterial species and contain millions of organisms per gram of biomass (Liu & Tay, 2002; Liu et al., 2003; Sheng et al., 2010). Granules in anaerobic digestion are important for enhancing process efficiency by increasing biomass hold-up. An anaerobic digester with higher biomass hold-up will be better in terms of COD removal and biogas production.

Granular sludge is a prominent characteristic of upflow anaerobic sludge blanket (UASB) reactors. This type of reactor has a longitudinal structure with a gas/liquid/solid separator at the top, while microbial granules with high settling velocity are formed in a thick biomass blanket zone at the bottom (Lettinga et al., 1983). The performance of UASB systems depends upon the granulation process. Unfortunately, a long start-up period is required for the development of anaerobic granules in UASB reactors since anaerobes are slow-growing bacteria (Liu & Tay, 2002; Show et al., 2006a). When seed sludge is not granulated, the UASB start-up periods are relatively long and washout of finely dispersed sludge particles is a typical problem (Poh & Chong, 2009).

The UASB start-up period can be shortened by enhancing sludge granulation. The development of well-settleable granular sludge is the key factor for successful UASB operation (Show et al., 2006b). Both synthetic and natural polymers are known to promote particle agglomeration and have been used to enhance the formation of anaerobic granules (El-Mamouni et al., 1998; Show et al., 2006a; Show et al., 2006b). Chitosan is a natural flocculant that has been used for the solid–liquid separation treatment of livestock wastewater (Garcia et al., 2009). Recently, chitosan in the form of freely moving polymeric chains has been found to enhance sludge granulation and shorten the start-up period of UASB systems (El-Mamouni et al., 1998; Lertsittichai et al., 2007; Liu et al., 2002; Thaveesri et al., 1995).

2. Chitosan as flocculants

Chitosan has been largely employed in many areas, such as photography, biotechnology, cosmetics, food processing, biomedical products (artificial skin, wound dressing, contact lens, etc.) and in a system for controlled liberation of medicines (capsules and microcapsules). In addition, chitosan has been used as a flocculant for the removal of metallic and colouring ions from industrial effluents by bonding the micro-floc particles together to form larger, denser flakes that are easier to separate (de Alvarenga et al., 2010; Renault et al., 2009).

Chitosan is a natural polysaccharide whose structure is similar to extracellular polymeric substances (ECP). ECP are widely known to assist anaerobic cell aggregation. Polymeric chains of ECP enhance flocculation by bridging microbial cells to form an initial microbial nucleus which is the first step in microbial granulation. There are many hypotheses to explain adhesion and aggregation processes by ECP. For example, in one hypothesis, ECP production is thought to occur prior to adhesion and the appearance of polymer materials at the initial site of contact between microbial cells is believed to be caused by the migration of polymer molecules onto the cell surface. In another hypothesis, ECP production is thought to occur after adhesion. In this case, it is believed that bacterial adhesion provides a favorable physiological condition for ECP excretion (El-Mamouni et al., 1998; Liu et al., 2002; Show et al., 2006a).

Chitosan is obtained by partial deacetylation of chitin (de Alvarenga et al., 2010). Chitin is a β-(1→4)-linked polymer of 2-acetamido-2-deoxy-d-glucose (N-acetyl-d-glucosamine) which exists in the exoskeletons of insects, crustaceans and the cell walls of fungi and algae. Basically, deacetylation involves the replacement of acetyl groups in the molecular chain of chitin by complete amino groups (NH_2). Chitosan is a mixture of straight-chain copolymers of N-acetyl-D-glucosamine and D-glucosamine of varying degrees of deacetylation (DD), i.e., with varying average numbers of D-glucosamine units per 100 monomers (Khan et al., 2002; Sabnis & Block, 1997). Chitosan also has the advantage that it is naturally biodegradable and therefore should have little adverse affect on human health.

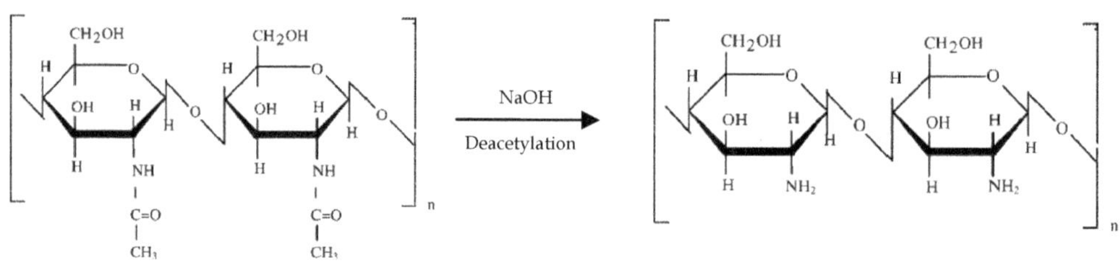

Fig. 1. Deacetylation of chitin to chitosan

Chitosan is insoluble in water, organic solvents and aqueous bases, but it is soluble after stirring in acids such as acetic, nitric, hydrochloric, perchloric and phosphoric acids (de Alvarenga et al., 2010). The glucosamine moieties in chitosan carry free amine groups that are protonated in an acidic environment. The amount and the positions of the glucosamine determine the charge and the charge distribution in the chitosan molecule. Changes in charge density have an effect on the dissolution and binding properties of chitosan (Domard, 1996). The degree of deacetylation also controls the degree of crystallinity and hydrophobicity of chitosan (Vander Lubben et al., 2003). Chitosan enhances the flocculation of sludge, and the flocculation efficiency depends on both DD and molecular weight (MW).

3. The effects of chitosan characteristics and environmental conditions on flocculation of anaerobic sludge

The flocculation efficiency of chitosan is sensitive to its characteristics. The most important characteristics of chitosan for flocculation efficiency are the degree of deacetylation and molecular weight since these are the main factors that affect particle size, particle formation and aggregation in the flocculation process. However, environmental conditions, i.e. pH and ionic strength, are also important in the dissolution and the charge of chitosan for flocculation process.

3.1 Effect of % deacetylation (DD) of chitosan

pH 7 is a typical starting pH in a UASB and most other anaerobic digesters (Lettinga et al., 1980). Kaseamchochoung et al. (2006) investigated the effect of %DD of chitosan on anerobic flocculation by using chitosan with different degrees of deactylation: M85 (DD = 85%) and M70 (DD = 70%) at pH 7. Their experimental procedure was as follows. In the flocculation assay, an initial sludge suspension was transferred into a beaker and a chitosan stock solution was added to achieve a concentration of 0 to 45 mg chitosan/g oven-dried (o.d.) sludge. The suspension was then stirred. The pH of the suspension was adjusted to 5, 6, or 7, with either 1% acetic acid or 3% sodium carbonate, depending on the pH of chitosan added to the suspension. After continuous mixing, the turbidity of supernatant was determined using a turbidimeter. The flocculation was calculated from the decrease in turbidity of supernatant after the treatment with chitosan compared with a reference without chitosan.

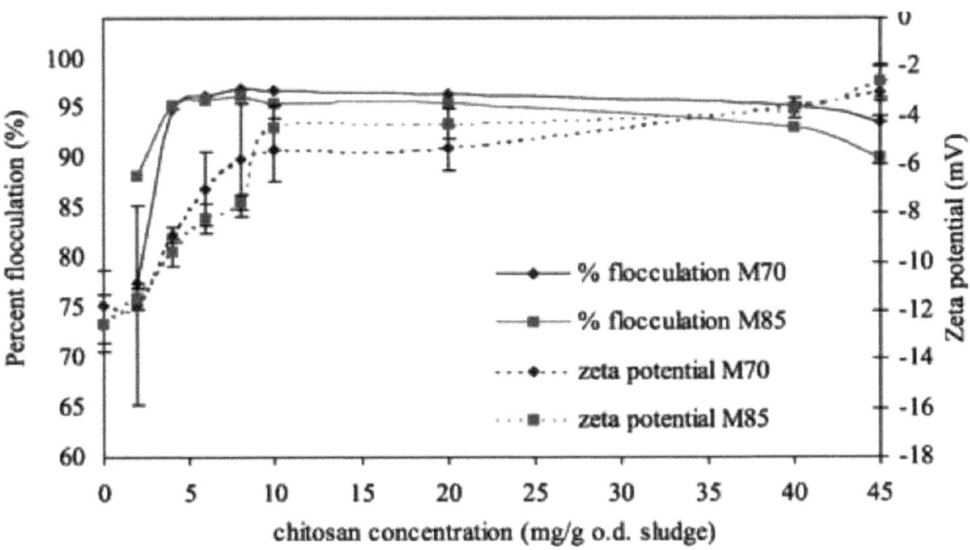

Fig. 2. Flocculation and zeta potential as a function of chitosan concentration in sludge suspension at pH 7 with ionic strength of 0.1 M (from Kasemchochoung et al., 2006. Reprinted with permission from *Water Environment Research*. Volume 78, No. 11, pp. 2211 to 2214, Copyright © 2006 Water Environment Federation, Alexandria, Virginia.)

Kaseamchochoung et al. (2006) found that at a low concentration (2 mg chitosan/g o.d. sludge) chitosan M85 gave approximately 90% flocculation, whereas M70 gave only approximately 80% flocculation (Fig. 2). However, at a concentration of 4 mg chitosan/g o.d. sludge the flocculation efficiencies of M70 and M85 became approximately equal at 95% flocculation and then remained approximately equal up to concentrations of 45 mg chitosan/g o.d. sludge (Fig. 2).

3.2 Effect of chitosan molecular weight

Kaseamchochoung et al. (2006) also studied the effect of molecular weight of chitosan on flocculation. They controlled the deacetylation of chitosan samples at 83 ± 2% and studied two levels of molecular weight ($3.5x10^5$ and $1.4x10^6$ dalton; Da). They found that the low molecular weight chitosan had a higher flocculation efficiency than the high molecular weight chitosan. Following Gregory (1993), they suggested that a possible explanation is that the longer polymers make more surface contacts per molecule and possibly saturate the cell surfaces, leaving no space for other polymers from different cell particles to initiate bridging.

3.3 Effect of environmental pH and ionic strength

Kaseamchochoung et al. (2006) found that the progression of anaerobic digestion in a UASB may cause pH to drop to 6 or even lower. At pH 6 and 7, approximately 90% flocculation was obtained by adding 2 mg chitosan/g o.d. sludge of chitosan M70 and M85. However, at pH 5, approximately 95% flocculation was obtained at the same chitosan concentration (Fig. 3).

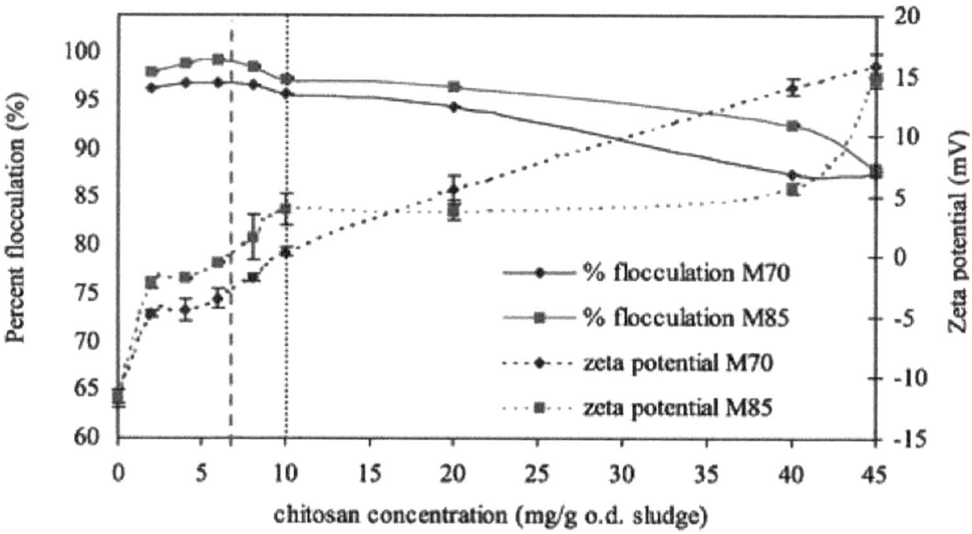

Fig. 3. Flocculation and zeta potential as a function of chitosan concentration in sludge suspension at pH 7 with ionic strength of 0.1 M. Vertical lines indicate the position of the CNP: (............) for M70 and (_ _ _) for M85 (from Kasemchochoung et al., 2006. Reprinted with permission from *Water Environment Research*. Volume 78, No. 11, pp. 2211 to 2214, Copyright © 2006 Water Environment Federation, Alexandria, Virginia.)

Similar results were obtained by Roussy et al. (2004). They studied chitosan efficiency at three different pH values (pH 5, 6.3, and 9). They found that a lower chitosan dosage (87% DD) was required at pH 5, while a significantly higher dosage of chitosan was required at pH 9 to obtain a residual turbidity below a fixed limit of 5 formalin turbidity units. Their explanation was that two possible mechanisms were possible at pH 5—(a) coagulation by charge neutralization and (b) flocculation by entrapment in the polymer network. However, at pH 9 only the latter mechanism is possible, but its effect can only be significant at a high chitosan concentration.

Kaseamchochoung et al. (2006) found that both chitosan M70 and M85 were able to flocculate anaerobic sludge even when the system pH dropped to 5. A small degree of restabilization was observed after the charge neutralization point (CPN). That is, the percentage of flocculation dropped only slightly after the CPN, whereas zeta potential values became positive. A possible explanation given in Kaseamchochoung et al. (2006) is that the charge density of chitosan is greatly influenced by pH (Strand et al., 2001). Because the intrinsic pKa of chitosan is close to 6.5, most amine groups are protonated at pH 5, but become significantly less protonated when the pH increases. The polymer is therefore more highly positively charged at pH 5 than at pH 7. At pH 7, chitosan with 70%DD contains a lower charge density than chitosan with 85%DD, and the performance of chitosan (70%DD) would be noticeably lower at a low chitosan dosage (Fig. 2). Kaseamchochoung et al. (2006) suggested that charge density may play an important role in the flocculation mechanism and that this is not surprising because electrostatic forces are typically the main cause of polyelectrolyte adsorption on an oppositely charged surface. They concluded that chitosan has the potential to be used as an effective cationic bioflocculant, which is able to function either in acidic or neutral conditions, and that only relatively small amounts of chitosan (less than 4 mg/g dried sludge) are required.

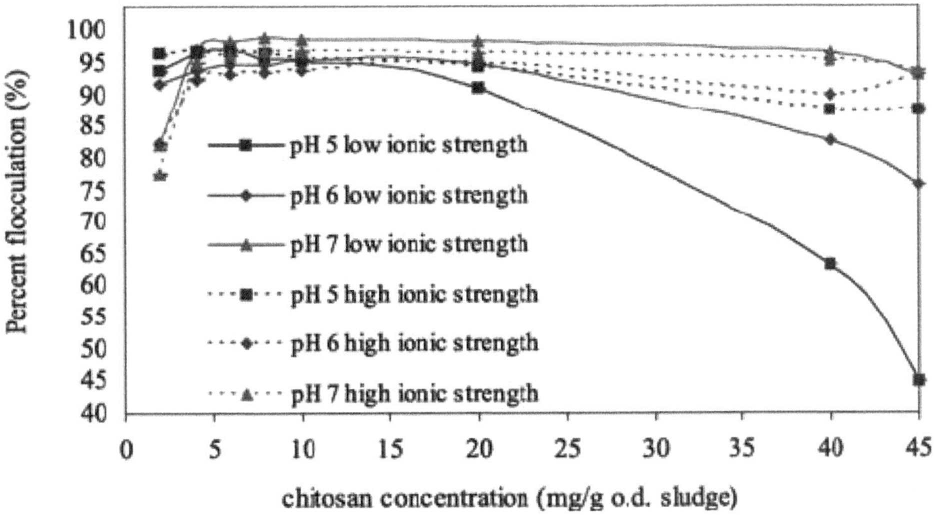

Fig. 4. Percent flocculation as a function of chitosan M70 concentration in sludge suspension at different pH values and ionic strengths (from Kasemchochoung et al., 2006. Reprinted with permission from *Water Environment Research*. Volume 78, No. 11, pp. 2211 to 2214, Copyright © 2006 Water Environment Federation, Alexandria, Virginia.)

In addition to pH, ionic strength of a medium is also a major factor affecting flocculation. Kaseamchochoung et al. (2006) investigated the effect of ionic strength on flocculation by chitosan of high (0.1 M) and low (0.01 M) ionic strength. At pH 7, ionic strength did not signficantly influence the pattern of flocculation by chitosan M70 and the flocculation remained at approximately 95%. In contrast, at pH 5, chitosan M70 performed significantly better in the high-ionic-strength medium. Under the low ionic strength condition, the flocculation dropped from approximately 95% to 45% (Fig. 4). A possible explanation for the effect of salt was obtained from classical theories of colloidal stability (Strand et al., 2001). The extension of the double layer, which causes electrostatic repulsion between charged colloids and the range of repulsion forces, decreases with increasing ionic strength in the surrounding medium. Therefore, bacterial cells should be able to come closer and thus flocculate better in a high ionic strength medium.

4. Effect of chitosan on the performance of UASB treating fruit-processing wastewater

According to Kaseamchochoung et al. (2006), chitosan with 85%DD and MW of $3.5x10^5$ Da yielded the highest flocculation efficiency and versatility to changes in environmental pH and ionic strength.

Lertsittichai et al. (2007) studied the efficiency of chitosan in a UASB reactor treating tropical fruit-processing industry wastewater. The details of their study were as follows. The fruit canning factory wastewater consisted mainly of sugar. The wastewater characteristics were: COD 5,130 to 5,520 mg/L, volatile fatty acid (VFA) 703 to 1,834 mg/L, pH 5 to 6 and ionic strength of 0.028 to 0.036 M.

Two identical UASB reactors with a working volume of 30 L were employed for the comparative study. The startup period was operated at a hydraulic retention time (HRT) of 85 hours, corresponding to an organic loading rate (OLR) of 1.45 g COD/L·d. Chitosan at a concentration of 2 mg/g suspended solids was added to the reactor on the second day and the same amount was added on the 37th operating day. The HRT of both reactors were reduced in a stepwise fashion, at 85, 65, 45, and 35 hours, when the COD removal was higher than 80% for at least 3 times the HRT.

Throughout the operation of the process, the OLR values ranged from approximately 1 to 4 g COD/L·d. Lertsittichai et al. (2007) found that the UASB with chitosan addition gave 9 to 59% lower COD effluent and had a 4 to 10% higher removal efficiency than the control UASB. The low VFA values corresponded to high biogas production because VFA is an intermediate for methane production. The UASB with chitosan addition gave a lower VFA value and a 35% higher biogas production rate than the control (Fig. 5).

Effluent VSS refers to biomass washout. Lertsittichai et al. (2007) found that the biomass washout increased during the initial operation period of both reactors. After 35 days, the biomass washout decreased due to granule formation. The biomass washout from the UASB with chitosan addition was 16 to 68% lower than that from the control. The UASB with chitosan addition was found to consistently have 24 to 37% higher average particle sizes than the control, corresponding to the lower biomass washout.

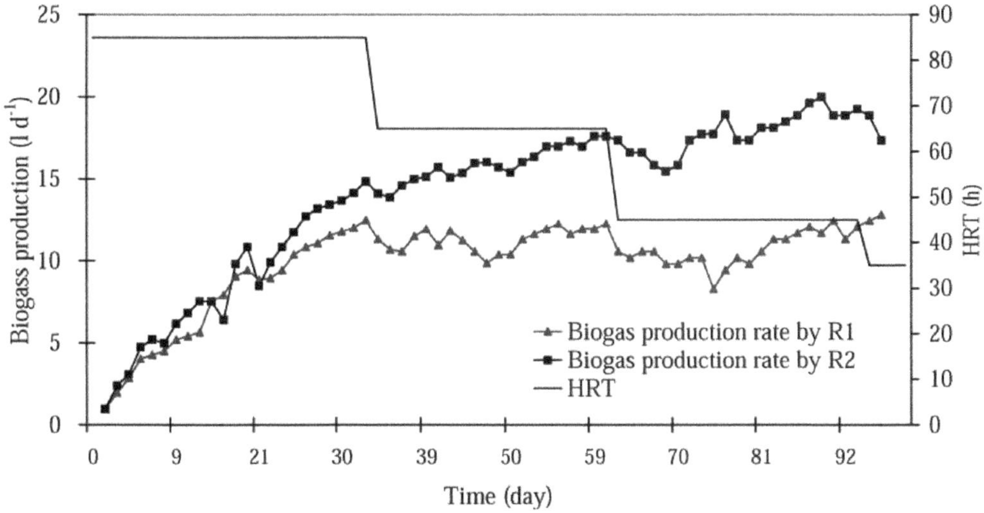

Fig. 5. Biogas production against time (from Lertsittichai et al., 2007). R1 is the control UASB reactor and R2 is the reactor with chitosan addition. Reprinted with permission from *Water Environment Research*. Volume 79, No. 7, pp. 802 to 806, Copyright © 2007 Water Environment Federation, Alexandria, Virginia.

In addition, Lertsittichai et al. (2007) found that the UASB with chitosan addition consistently had a 6 to 41% longer solids retention time (SRT) than the control corresponding to a lower effluent VSS and a higher average particle size. The VSS from the bottom sampling ports of the UASB with chitosan addition was higher than that of control, leading to greater overall sludge density. From their observations, Lertsittichai et al. (2007) concluded that chitosan helped sludge pellet development. They gave the possible explanation that the cell surfaces of bacteria carry negative charges, and the electrostatic interactions between them are repulsive. Therefore, a cationic polymer, such as chitosan, assists the flocculation of the bacteria leading to faster sludge formation and a higher density of sludge retained in the reactor.

Overall, Lertsittichai et al. (2007) used only small amounts of chitosan (two injections with 2 mg chitosan/g suspended solids at each injection). They saw no sign of inhibition to biomass activity. Throughout the course of their experiment at a mesophilic temperature (35°C), the UASB with chitosan addition clearly showed superior performance to the reactor without chitosan, with 9 to 59% lower effluent COD, 4 to 10% higher COD removal, up to 35% higher biogas production rate, and decreased washout of biomass and increased granular size.

5. Investigation of chitosan in different forms

Chitosan is available commercially in three forms: solution, flake and powder. The prices of chitosan in the forms of solution, flake and powder range between 50 to 70 baht/L, 700 to 900 baht/kg and 750 to 2,300 baht/kg, respectively. Chitosan in the form of freely moving polymeric chains has previously been found to enhance sludge granulation and shorten the start-up period of UASB systems (El-Mamouni et al., 1998; Lertsittichai et al., 2007; Liu et al., 2002; Thaveesri et al., 1995).

The effectiveness in enhancing granulation of different forms of chitosan, i.e. solution, bead and powder, has also been studied by Nuntakumjorn et al. (2008). They prepared chitosan solution by dissolving chitosan in acetic acid solution (1% w/v). In preparing chitosan powders, they used a spray dryer to spray-dry chitosan solution (1% w/v). In preparing the chitosan beads, they dropped the chitosan solution (4% w/v) into a solution of KOH and ethanol. The chitosan beads were found to have spherical shape, white color and looked like glutinous pellets. The appearance of the chitosan beads is shown in Fig. 6.

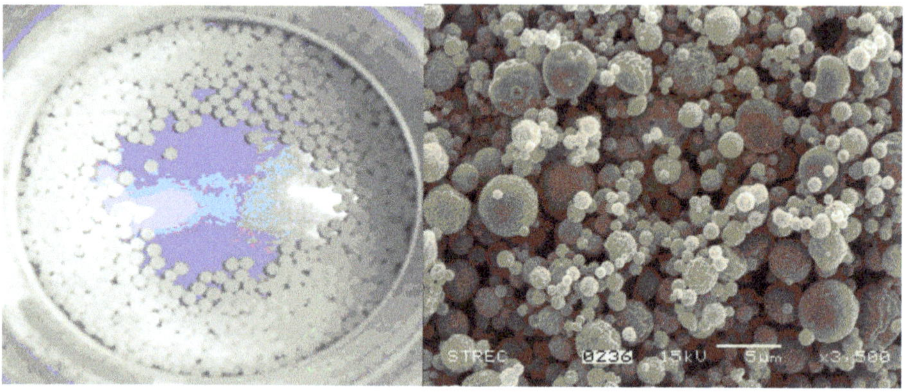

(a) (b)

Fig. 6. (a) Chitosan beads in the KOH/Ethanol solution (b) SEM micrograph with 3500x of chitosan powders (from Nuntakumjorn et al., 2008)

Nuntakumjorn et al. (2008) used two identical reactors, with a working volume of 5.3 L, running in parallel. A sludge suspension with an initial VSS concentration of 12 g VSS/L was inoculated into the reactors. The acclimation of the sludge was carried out until the COD removal was approximately 80%. The reactors were run with a HRT of 1.5 day corresponding to an OLR of 1.45 g COD/L·d. Chitosan in the different forms was introduced into the reactors on the second operating day of the start-up period at dose rates of 2 mg chitosan/g suspended solids.

A summary of the results of Nuntakumjorn et al. (2008) is as follows. When comparing between the UASB with no chitosan addition and the UASB with chitosan addition in the solution form, the UASB with chitosan addition was found to have a 9 to 59% lower effluent COD, 5 to 7% higher COD removal, up to 25% higher biogas production rate, 21 to 39% lower biomass washout, 37% larger particle size and 4 day longer sludge retention time.

When comparing between the UASB with chitosan addition in the solution form and with addition in the bead form, the UASB with chitosan solution was found to have 5 to 17% lower effluent COD, 16 to 45% higher COD removal, 7 to 20% lower biomass washout and 3 to 17% higher biogas production than the UASB with chitosan beads. The reduced effectiveness of chitosan in the bead form might be caused by a lower amount of chitosan in the bead form and by insufficient contact between the chitosan beads and biomass.

When comparing between the UASB with no chitosan addition and the UASB with chitosan addition in the powder form, no differences were found in terms of COD removal, biogas production and biomass washout. The average COD removal of the UASB with chitosan addition was approximately 80% and that without chitosan was approximately 81%. The biogas production rate was 9.85 L/d and 10.23 L/d for the UASB with and without chitosan addition, respectively. Both UASB reactors had biomass washout in the range of 0.6 to 1.5 g VSS/L. Although chitosan powders have net positive charge, the electrostatic interaction between the negatively charged bacteria was not significantly reduced. Nuntakumjorn et al. (2008) concluded that chitosan powders does not enhance the granulation process and UASB performance.

6. Effect of chitosan on microbial diversity in UASB treating POME under thermophilic condition

Palm oil mill effluent (POME) contains high COD and biochemical oxygen demand (BOD). POME consists of a wide range of biological substances from complex biopolymers such as proteins, starches and hemicelluloses to simple sugars and amino acids. POME may also contain dissolved oil and fatty acids, glycerin, crude oil solids and short fibers as well as soluble materials that are harmful to the environment. Since POME is discharged at high temperatures (80-90°C), both mesophilic and thermophilic temperatures have been widely applied for POME treatment by anaerobic digestion.

6.1 Effect of chitosan on UASB treating POME

It has been reported that thermophilic operation of anaerobic reactors provides some advantages over mesophilic operation in areas such as higher rates of substrate degradation and biogas production. However, mesophilic reactors can be preferable because of greater process stability (Mustapha et al., 2003; Poh & Chong, 2009). Operating temperature is a major factor that greatly influences digester performance (Choorit & Wisarnwan, 2007; Poh & Chong, 2009; Yu et al., 2002).

The effects of chitosan as a sludge granulation accelerator during the transition from mesophilic (37°C) to thermophilic condition (57°C) has been investigated by Khemkhao et al. (2011). They used two UASB reactors, with a working volume 5.3 L, both of which they inoculated with mesophilic anaerobic sludge. The sludge was then acclimatized to a thermophilic condition with a stepwise temperature increase of 5°C from 37 to 57°C. The OLR ranged from approximately 2 to 9.5 g COD/L·d. One of the reactors was then injected with a chitosan dosage of 2 mg chitosan g/VSS on the first day of operation and the second reactor was used as a control.

At all times during the operation of the two reactors, the UASB with chitosan addition was found to have 5% higher COD removal efficiency and 16 L/d higher biogas production rate (7.82 L/g VSS removed·d) than that of the control. The methane contents of both reactors were found to be similar, with approximately 78% methane content for UASB with chitosan addition and 76% for the control. The effluent VSS in both reactors was found to increase with increase of OLR. The UASB with chitosan addition was found to have 6 to 23% lower effluent VSS than that of the control. Khemkhao et al. (2011) concluded that the UASB with chitosan addition had consistently better performance than the control.

6.2 Effect of chitosan on microbial diversity in UASB treating POME

The mechanism of anaerobic digestion in methane production consists of a series of complex metabolic interactions between various types of microorganisms in the absence of oxygen. Anaerobic digestion is mediated through processes of hydrolysis, acidogenesis, acetogenesis and methanogenesis. Khemkhao et al. (2011) used 16S rRNA targeted denaturing gradient gel electrophoresis (DGGE) fingerprints to study the microbial communities during anaerobic digestion. They found that bacteria and methanogens could both be detected in the UASB reactors operating both with and without chitosan addition.

6.2.1 Effect of chitosan on bacterial diversity in UASB treating POME

In their experiments, Khemkhao et al. (2011) found that DGGE patterns of bacterial diversity of the three bacterial groups, hydrolytic, acidogenic and acetogenic, persisted at all operating temperatures. However, the distribution of their members among bacteria in each group did show small changes under the different operating conditions. By the end of the operating period, the UASB with chitosan addition was found to contain a lower proportion of hydrolytic bacteria and a higher proportion of acidogenic bacteria than the control. However, the diversity of acetogenic bacteria was found to be similar in the two reactors. Sulfate-reducing bacteria were detected in the control but not in the chitosan reactor.

It is known (Bitton, 1994) that hydrolytic, acidogenic and acetogenic bacteria work together to degrade complex organic matters into acetate, CO_2 and H_2. Hydrolytic bacteria begin the process of degradation by breaking down complex organic molecules such as proteins, cellulose, lignin and lipids into soluble monomer molecules by extracellular enzymes, i.e., proteases, cellulases and lipases. The monomer molecules produced are amino acids, glucose, fatty acids and glycerol. These monomers are then degraded by the acidogenic (acid-forming) group of bacteria which convert them into organic acids, alcohols and ketones, acetate, CO_2, and H_2. The organic acids produced include acetic, propionic, formic, lactic, butyric, and succinic acids. The alcohols and ketones produced are ethanol, methanol, glycerol and acetone. In the final stage, the acetogenic bacteria (acetate and H_2-producing bacteria) convert the fatty acids, alcohols and ketones into acetate, CO_2 and H_2.

6.2.2 Effect of chitosan on archaeal diversity in UASB treating POME

Khemkhao et al. (2011) found that all methanogen DGGE bands observed in the control were also detected in UASB with chitosan addition. The observed acetotrophic methanogens were the family *Methanosarcinaceae* and the species *Methanosaeta soehngenii*. The observed hydrogenotrophic methanogens were the order *Methanomicrobiales*, the genus *Methanolinea* sp. and the species *Methanoculleus marisnigri*. On the other hand, some of the acetotrophic methanogens were observed in the UASB with chitosan addition, but not in the control. These acetotrophic methanogens were the order *Methanosarcinales*, the species *Methanosaeta thermophila* and *Methanosaeta harundinacea*.

Methanogenic archaea are oxygen–sensitive anaerobes (Bitton, 1994). They can grow into individual cells, filamentous chains, cubes and/or sarcina. Methanogens are subdivided into two subcategories: (i) hydrogenotrophic methanogens and (ii) acetotrophic methanogens.

Hydrogenotrophic methanogens convert H_2 and CO_2 into CH_4. Acetotrophic methanogens convert acetate into CH_4 and CO_2. The acetotrophic methanogens grow slower than the acid-forming bacteria. About two-thirds of CH_4 is derived from acetate conversion by acetotrophic methanogens. The other third is the result of H_2 and CO_2 reduction by hydrogenotrophic methanogens.

As stated above (Khemkhao et al., 2011), lower biomass washout was observed from the UASB with chitosan addition than from the control, especially at higher biogas production rates. The DGGE analysis shows that UASB with chitosan addition contains higher populations of *Methanosaeta* species than the control. It can be concluded that the chitosan helped to retain these methanogens, thus resulting in higher populations of acetotrophic methanogens.

Tiwari et al. (2005) and Tiwari et al. (2006) have reported that acetotrophic methanogens significantly accelerate granule development. Higher population of acetotrophic methanogens may in turn lead to higher methane production in the reactors with chitosan addition.

Chitosan has been reported to act like an ECP in enhancing the aggregation of acidogens. As shown in Fig. 7, the aggregated acidogens then form granules with highly elastic outer

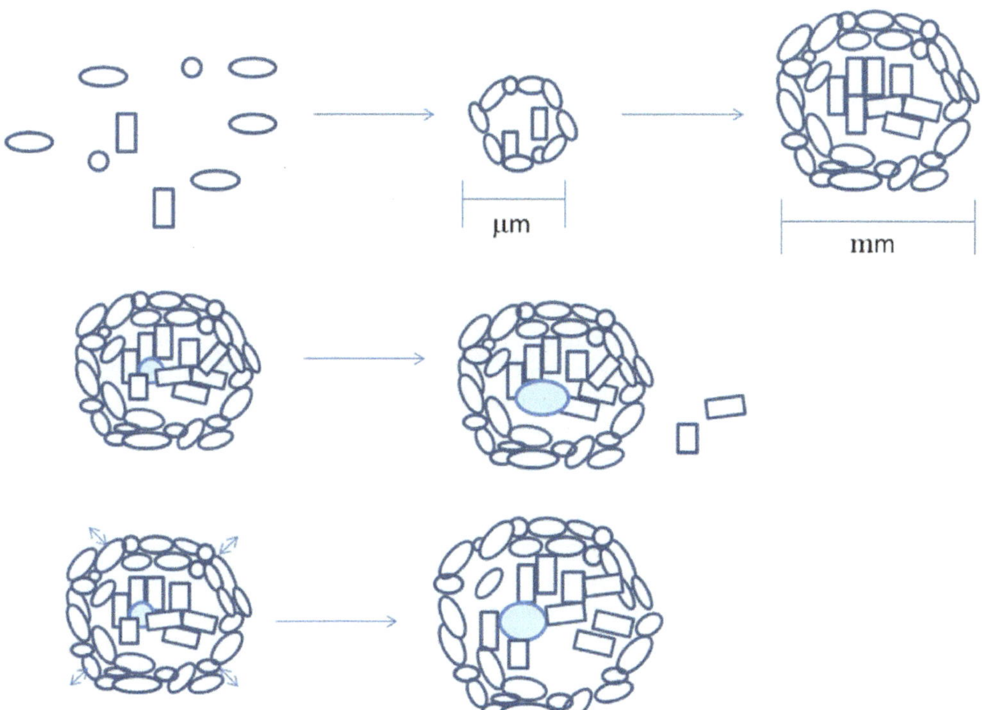

Fig. 7. Scheme of granule formation. Top: Surface tension model according to Thaveesri et al. (1995) and Hulshoff Pol et al. (2004). Middle: Some circumstances in the control reactor. Bottom: Enhanced aggregation by chitosan in UASB with chitosan addition (from Khemkhao et al., 2011)

hydrophilic layers around a core of methanogens. According to Hulshoff Pol et al. (2004) and Thaveesri et al. (1995), the acidogens (round and rod cells) aggregate by forming ECP. Dispersed cells are washed out, while some methanogens (rectangular cells) are enclosed inside, becoming the nucleus of a granule with an outer elastic hydrophilic layer formed by ECP-rich acidogens and an inner core of hydrophobic methanogens. Chitosan has been thought to act like ECP in aggregating anaerobic sludge (El-Mamouni et al., 1998). Therefore it may increase the elasticity of outer hydrophilic layers of the granular samples. In UASB with chitosan addition, the growing methanogens are better protected inside an acidogenic layer and may become less susceptible to adhesion to gas bubbles (filled circles) and consequently may be less washed out from the reactor than those in the control.

The polymer additives appear to play a similar role to naturally secreted ECP in aggregating anaerobic sludge. The addition of polymers to anaerobic systems changes the surface properties of bacteria to promote association of individual cells. Polymer may form a solid and stable three-dimensional matrix within which bacteria multiply and daughter cells are then confined (Liu et al., 2002; Show et al., 2006a; Uyanik et al., 2002).

In addition, Show et al. (2006b) have reported that adding an appropriate dosage of polymer in the seeding stage accelerates the start-up time by approximately 50% and the granule formation by approximately 30%. In addition, granules developed in polymer-assisted reactors exhibited better settleability, strength and methanogenic activity at all OLRs tested. Positively charged polymer forms bridges among the negatively charged bacterial cells through electrostatic charge attraction. The bridging effect would enable greater interaction between biosolids resulting in preferential development and enhancement of biogranulation in UASB reactors (Show et al., 2006a).

In the experiments of Khemkhao et al. (2011), the UASB reactor with chitosan addition was treated with a one-time chitosan dose of 2 mg chitosan/g VSS on the first operating day. The performance of the UASB reactor may be further enhanced by more injections of the chitosan solution. However, the evidence from the one-time chitosan dose of 2 mg chitosan/g VSS on the first operating day was that the initial stage of granulation was very important for forming high quality granules.

7. Conclusion

Chitosan is a biopolymer which can be used to enhance the sludge granulation process and UASB performance. Flocculation efficiency of chitosan was sensitive to its characteristics as well as to the pH and ionic strength of the environment. An increase in the deacetylation of the chitosan from 70 to 85% led to a two-fold reduction in the chitosan concentration necessary to achieve 90% flocculation at pH 7 (Kaseamchochoung et al., 2006).

Chitosan, with a degree of deacetylation of 85% and molecular weight of 3.48×10^5 Da, yielding high flocculation efficiency (85 to 100% flocculation) and broad flocculation region (2 to 45 mg/g suspended solids), was shown to accelerate granulation in a 30-L pilot-scale UASB used to treat wastewater from a tropical fruit-processing industry (Lertsittichai et al., 2007).

For the same amount of chitosan, chitosan in the solution form was shown to be significantly better at enhancing the granulation process and the UASB performance than chitosan in bead or powder forms (Nuntakumjorn et al., 2008).

For POME treatment, the biogas production rate and the COD removal of the UASB with chitosan addition was on an average 16% and 5%, respectively, higher than that of the control. A DGGE analysis indicates that the chitosan helped to retain the methanogens in the genus *Methanosaeta*, thus resulting in higher populations of acetotrophic methanogens. Further investigations are required to determine optimal chitosan dosages and the optimal times to add chitosan under thermophilic conditions (Khemkhao et al., 2011).

8. Acknowledgment

The authors are grateful to the King Prajadhipok and Queen Rambhai Barni Memorial Foundation for financial support to S. Lertsittichai, to Thailand Research Fund (TRF-Master Research Grant, Grant No. MRG-OSMEP505E225) for the financial support to B. Nuntakumjorn and to Thailand Graduate Institute of Science and Technology (TGIST) and the Joint Graduate School of Energy and Environment (JGSEE) for the financial supports to M. Khemkhao. We also would like to acknowledge Faculty of Engineering, King Mongkut's University of Technology North Bangkok for supporting the publication fee. The authors would like to thank Ngaung-Khaem water quality control plant for providing sludge, Suksomboon Palm Oil Co., Ltd. for wastewater samples and Taming Enterprises Co., Ltd. for providing chitosan samples. Special thanks to Dr. Elvin Moore for his critical reading of the manuscript.

9. References

Bitton, G. (1994). Anaerobic Digestion of Wastewater and Sludge. In: *Wastewater Microbiology*, G. Bitton, (Ed), pp. 229-245, Wiley-Liss, Inc., ISBN 229-245, USA

Choorit, W., & Wisarnwan, P. (2007). Effect of Temperature on the Anaerobic Digestion of Palm Oil Mill Effluent. *Electronic Journal of Biotechnology*, Vol. 10, pp. 376-385, ISSN 07173458

de Alvarenga, E. S., Pereira de Oliveira, C., & Roberto Bellato, C. (2010). An Approach to Understanding the Deacetylation Degree of Chitosan. *Carbohydrate Polymers*, Vol. 80, No. 4, pp. 1155-1160, ISSN 0144-8617

Domard, A. (1996). Some Physico-Chemical and Structural Basis for Applicability of Chitin and Chitosan. *Proceeding of the 2nd Asia Pacific Symposium*, pp. 1-12, ISBN 967-942-210-0, Asian Institute of Technology, Bangkok, Thailand

El-Mamouni, R., Leduc, R., & Guiot, S. R. (1998). Influence of Synthetic and Natural Polymers on the Anaerobic Granulation Process. *Water Science and Technology*, Vol. 38, No. 8-9, pp. 341-347, ISSN 0273-1223

Garcia, M. C., Szogi, A. A., Vanotti, M. B., Chastain, J. P., & Millner, P. D. (2009). Enhanced Solid-Liquid Separation of Dairy Manure with Natural Flocculants. *Bioresource Technology*, Vol. 100, No. 22, pp. 5417-5423, ISSN 0960-8524

Gregory, J. (1993). The Role of Colloid Interactions in Solid-Liquid Separation. *Water Science and Technology*, Vol. 27, pp. 1–17, ISSN 0273-1223

Hulshoff Pol, L. W., de Castro Lopes, S. I., Lettinga, G., & Lens, P. N. L. (2004). Anaerobic Sludge Granulation. *Water Research*, Vol. 38, No. 6, pp. 1376–1389, ISSN 0043-1354

Kaseamchochoung, C., Lertsutthiwong, P., & Phalakornkule, C. (2006). Influence of Chitosan Characteristics and Environmental Conditions on Flocculation of Anaerobic Sludge. *Water Environment Research*, Vol. 78, pp. 1061-4303, ISSN 1061-4303

Khan, T. A., Peh, K. K., & Ch'ng, H. S. (2002). Reporting Degree of Deacetylation Values of Chitosan: the Influence of Analytical Methods. *Journal of Pharmacy and Pharmaceutical Science*, Vol. 5, pp. 205–212, ISSN 1482-1826

Khemkhao, M., Nuntakumjorn, B., Techkarnjanaruk, S., & Phalakornkule, C. (2011). Effect of Chitosan on UASB Treating POME during a Transition from Mesophilic to Thermophilic Conditions. *Bioresource Technology*, Vol. 102, No. 7, pp. 4674-4681, ISSN 0960-8524

Lertsittichai, S., Lertsutthiwong, P., & Phalakornkule, C. (2007). Improvement of Upflow Anaerobic Sludge Bed Performance using Chitosan. *Water Environment Research*, Vol. 79, pp. 802-806, ISSN 1061-4303

Lettinga, G., Hobma, S. W., Hulshoff Pol, L., de Zeeuw, W., de Jong, P., Grin, P., & Roersma, R. (1983). Design Operation and Economy of Anaerobic Treatment. *Water Science and Technology*, Vol. 15, pp. 177–195, ISSN 0273-1223

Lettinga, G., van Velsen, A. F. M., Hobma, S. W., de Zeeuw, W., & Klapwijk, A. (1980). Use of the Upflow Sludge Blanket (USB) Reactor Concept for Biological Wastewater Treatment, especially for Anaerobic Treatment. *Biotechnology and Bioengineering*, Vol. 22, pp. 699–734, ISSN 0006-3592

Liu, Y., & Tay, J.-H. (2002). The Essential Role of Hydrodynamic Shear Force in the Formation of Biofilm and Granular Sludge. *Water Research*, Vol. 36, No. 7, pp. 1653-1665, ISSN 0043-1354

Liu, Y., Xu, H.-L., Show, K.-Y., & Tay, J.-H. (2002). Anaerobic Granulation Technology for Wastewater Treatment. *World Journal of Microbiology and Biotechnology*, Vol. 18, No. 2, pp. 99-113, ISSN 1573-0972

Liu, Y., Xu, H.-L., Yang, S.-F., & Tay, J.-H. (2003). Mechanisms and Models for Anaerobic Granulation in Upflow Anaerobic Sludge Blanket Reactor. *Water Research*, Vol. 37, No. 3, pp. 661-673, ISSN 0043-1354

Mustapha, S., Ashhuby, B., Rashid, M., & Azni, I. (2003). Start-up Strategy of a Thermophilic Upflow Anaerobic Filter for Treating Palm Oil Mill Effluent. *Process Safety and Environmental Protection*, Vol. 81, No. 4, pp. 262-266, ISSN 0957-5820

Nuntakumjorn, B., Khumsalud, W., Vetsavas, N., Sujjaviriyasup, T., & Phalakornkule, C. (2008). Comparison of Sludge Granule and UASB Performance by Adding Chitosan in Different Forms. *Chiang Mai Journal of Science*, Vol. 35, No. 1, pp. 95-102, ISSN 0125-2526

Poh, P. E., & Chong, M. F. (2009). Development of Anaerobic Digestion Methods for Palm Oil Mill Effluent (POME) Treatment. *Bioresource Technology*, Vol. 100, pp. 1-9, ISSN 0960-8524

Renault, F., Sancey, B., Badot, P. M., & Crini, G. (2009). Chitosan for Coagulation/Flocculation Processes - An Eco-friendly Approach. *European Polymer Journal*, Vol. 45, No. 5, pp. 1337-1348, ISSN 0014-3057

Roussy, J., Van Vooren, M., & Guibal, E. (2004). Chitosan for the Coagulation and Flocculation of Mineral Colloids. *Journal of Dispersion Science and Technology*, Vol. 25, No. 5, pp. 663–677, ISSN 0193-2691

Sabnis, S., & Block, L. H. (1997). Improved Infrared Spectroscopic Method for the Analysis of Degree of N-deacetylation. *Polymer Bulletin*, Vol. 39, pp. 67–71, ISSN 1436-2449

Sheng, G. P., Yu, H. Q., & Li, X. Y. (2010). Extracellular Polymeric Substances (EPS) of Microbial Aggregates in Biological Wastewater Treatment Systems: A review. *Biotechnology Advances*, Vol. 28, No. 6, pp. 882-894, ISSN 0734-9750

Show, K. Y., Tay, J. H., Stephen, P. E., Tay, T. L., Liu, Y., & Ivanov, V. (2006a). Chapter 2 Factors Affecting Anaerobic Granulation. In: *Waste Management Series*, J. H Tay, S. T. L. Tay, Y. Liu, K. Y. Show, & V. Ivanov (Ed), pp. 35-56, Elsevier, Available from http://www.sciencedirect.com/science/article/pii/S0713274306801042

Show, K. Y., Tay, J. H., Stephen, P. E., Tay, T. L., Liu, Y., & Ivanov, V. (2006b). Chapter 3 Applications of Anaerobic Granulation. In: *Waste Management Series*, J. H Tay, S. T. L. Tay, Y. Liu, K. Y. Show, & V. Ivanov (Ed), pp. 57-84, Elsevier, Available from http://www.sciencedirect.com/science/article/pii/S0713274306801054

Strand, S. P., Vandvik, M. S., Va°rum, K., & Østgaard, K. (2001). Screening of Chitosans and Conditions for Bacterial Flocculation. *Biomacromolecules*, Vol. 2, pp. 126–133, ISSN 1525-7797

Thaveesri, J., Daffonchio, D., Liessens, B., Vandermeren, P., & Verstraete, W. (1995). Granulation and Sludge Bed Stability in Upflow Anaerobic Sludge Bed Reactors in Relation to Surface Thermodynamics. *Applied and Environment Microbiology*, Vol. 61, pp. 3681-3686, ISSN 0099-2240

Tiwari, M. K., Guha, S., Harendranath, C. S., & Tripathi, S. (2005). Enhanced Granulation by Natural Ionic Polymer Additives in UASB Reactor Treating Low-Strength Wastewater. *Water Research*, Vol. 39, No. 16, pp. 3801-3810, ISSN 0043-1354

Tiwari, M. K., Guha, S., Harendranath, C. S., & Tripathi, S. (2006). Influence of Extrinsic on Granulation in UASB Reactor. *Applied Microbiology and Biotechnology*, Vol. 71, pp. 145–154, ISSN 0175-7598

Uyanik, S., Sallis, E. J., & Anderson, G. K. (2002). The Effect of Polymer Addition on Granulation in an Anaerobic Baffled Reactor (ABR): Part I. Process Performance. *Water Research*, Vol. 36, pp. 933-942, ISSN 0043-1354

Vander Lubben, I. M., Kersten, G., Fretz, M. M., Beuvery, C., Verhoef, J. C., & Juginder, H. E. (2003). Chitosan Microparticles for Mucosal Vaccination against Diphtheria, Oral, and Nasal Efficiency Studies in Mice. *Vaccine*, Vol. 21, pp. 1400–1408, ISSN 0264-410X

Yu, H. Q., Fang, H. H. P., & Gu, G. W. (2002). Comparative Performance of Mesophilic and Thermophilic Acidogenic Upflow Reactors. *Process Biochemistry*, Vol. 38, No. 3, pp. 447-454, ISSN 1359-5113

18

Conditioning of Biogas for Injection into the Natural Gas Grid

Frank Burmeister, Janina Senner and Eren Tali
Gaswärme - Institut e.V. Essen, Germany

1. Introduction

In the following sections, recommendations supported by schematics are given for the injection of compliant processed biogas into natural gas grids. Based on the characteristics of the natural gases distributed in Germany and taking into account the applicable

- Laws, technical rules and regulations
- Billing procedures
- and the physical and technical conditions to be taken into account solutions for each individual supply case are given.

In order to feed biogas into a natural gas grid, unwanted components need to be removed from the gas and the burning properties of the gas need to be adjusted to those of the rest of the gas in the grid. In this way, the correct operation of the gas-burning appliances and the accuracy of the billing of retail customers is assured.

The purified biogas is conditioned depending upon the properties of the base gas (Fig.1). In the case of L gas, the calorific value or Wobbe index is realised by adding air, or air and LPG. In the case of H gas characteristics, the addition of LPG is required to adjust the calorific value to that of the usually higher calorific H base gas.

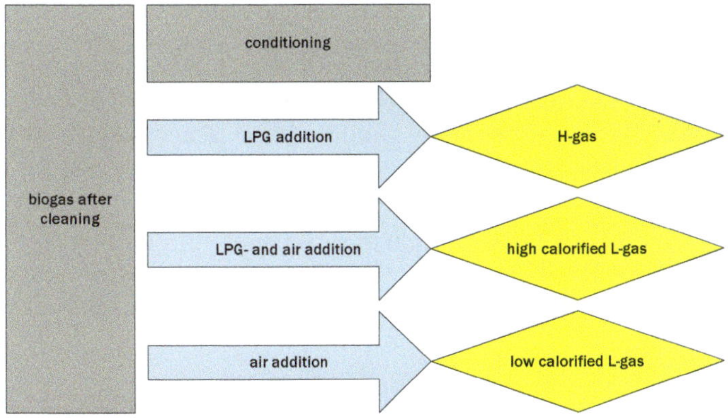

Fig. 1. gas conditioning by air and LPG addition

Schematic recommendations include the answers to the key questions listed below and allow a simple "read out" of the target properties, taking into account the current regulatory requirements. Gas utilities (GU) and operators can already in the planning phase determine the feed options and requirements with the help of the graphs. The key questions are:

1. What qualities and technical characteristics of combustion must processed biogas have at the very least so that it can be fed into grids in which the natural gases typical in Germany are present as base gases, without having to make changes to the grid?
2. What additional aspects need to be considered when feeding processed biogas into the existing natural gas grid, taking into account fairness in the billing process, properties (functionality of end-user equipment) and cost effectiveness?

The following conditions are to be observed in addition to point 1 : Engine applications, natural gas filling stations (methane, K number, condensation of higher hydrocarbons) and industrial customers.

2. Basic concepts and regulations

2.1 Characteristics of the base gas

The term base gas refers to the natural gas provided by the gas utilities (GUs) in the respective coverage areas without the addition of biogenic gas. The classification of the various natural gases distributed in Germany into H and L gases is made according to worksheet DVGW-G 260 (German Association of Gas and Water [DVGW], 2008). The Wobbe Index, which is a measure of the thermal energy released on the burner of a gas appliance or the energy transported through a pipe has a special significance here.

The Wobbe Index is an important variable to assess the interchangeability of fuel gases. When replacing one fuel gas with another, the output of the burner changes in proportion to the ratio of the Wobbe index. Its definition according G 260 is given in equation 1:

$$W_{S,n} = \frac{H_{S,n}}{\sqrt{d}} \quad d = \frac{\rho_{Gas,n}}{\rho_{Luft,n}} \quad \text{With the relative density and the calorific value } H_{S,n} \tag{1}$$

[1]

The upper value of the Wobbe index total range should not be exceeded. A shortfall in the lower value is acceptable under certain conditions and subject to a time limit. Both limits are specified in the German regulations. The nominal value listed in G 260 is used for setting the gas appliances used. Technically, the local variation range could be omitted, since the gas appliances are set to a nominal value. Currently however there are still many appliances set to differing values. The major boundary conditions are dictated by the Wobbe Index total range, the calorific range and the relative density, since conditioning with air and / or liquefied gas influences precisely these variables.

For the calorific value of a gas mixture, equation 2 states:

$$H_{s,n} = \sum_i r_i H_{s,n,i} \quad \text{or} \quad \rho_n = \sum_i r_i \rho_{n,i} \tag{2}$$

[2]

---

[1] Indexing "S" (superior) is the formation with calorific value and "n" standard conditions

[2] ri denotes the volume fraction of component i

When considering equation 2 and table 1, it is clear that even small volumes of higher hydrocarbons affect the parameters of combustion of the gas mixture, due to the greater density and calorific values. The same applies to air and carbon dioxide.

	ρ_n in kg/m³	$H_{S,n}$ in kWh/m³
CH_4 (Methane)	0,7175	11,064
C_2H_6 (Ethane)	1,3551	19,537
C_3H_8 (Propane)	2,010	28,095
C_4H_{10} (Butane)	2,709	37,252
CO_2	1,9767	0
Air	1,293	0

Table 1. Standard density and calorific value of the main components

In addition to the basic requirements for the gas properties, limits for accompanying substances are specified in worksheet G 260, which may not be exceeded.

Gas Accompanying Substances	Indicative maximum	
Hydrocarbons: Condensation point	°C	Soil temperature at the respective Line pressure
Water: dew point	°C	
Fog, dust, liquid		Technically free
Percentage of oxygen - in dry distribution networks - In humid distribution networks	 % %	 3 0,5
Total sulphur		
Annual mean value (excluding odorants)	mg/m³	30
Mercaptan	mg/m³	6
short-term	mg/m³	16
Hydrogen sulphide	mg/m³	5
In exceptional cases, briefly	mg/m³	10

Table 2. Permitted substances in the gas according to DVGW worksheet G 260 (DVGW, 2008)

Important boundary conditions are determined by the dew points, the oxygen content and the sulphur content. Information on the dew points is formulated so that condensation can be excluded. As far as the oxygen content is concerned, the grids in Germany can be regarded as dry and therefore the limit of 3 vol -% is to be applied. It should be noted at this point that at the long-range transport level significantly lower O_2contents, usually in the low ppm range are to be observed (EASEE gas, CPB European Association for the Streamlining of Energy Exchange-gas Common Business Practice, 2005) for cross-border distribution (H gas).

The raw gas must be cleaned, processed (according to G 260) and compressed to the pressure of the grid operator. Under no circumstances should health risks arise from processed gas. For injection into the distribution network of a local GU, the gas must be odorized according to G 280-1(DVGW, 2004). In addition, the presence of certain gas accompanying substances such as H_2S must be monitored regularly. Furthermore, for a time and heat equivalent transfer, the calorific value for billing purposes must also be known.

After processing the raw gases for the public gas supply, these can be used according to G 260 as an exchange gas (G 260 Section 4.4.2) or as additional gas (G 260 4.2, 4.3), (gas for conditioning) and be made available to the grid operator at the transfer interface. (Note: it should be noted that additional gas feeds are only possible in a single pipeline under certain circumstances.)

Put simply, it can be said that the conditions given in worksheets G 260 and G 262 (DVGW, 2007) ensure that the customers' appliances will work correctly. Sensitive industrial processes sometimes require much tighter limits on the gas properties (e.g. glass and ceramics production). The DVGW worksheet G 260 is very often a component of supply contracts and is an expression of the flexibility of the gas sector, which is necessary in the procurement of natural gas, in order to deliver natural gas from various gas fields into the transport and distribution system of the German gas industry and on to the customer. Due to its geographical location, historical and political development, in Germany natural gases from the most diverse of foreign origins as well as natural gases from its own sources are thus forwarded to the customers, with the guarantee of security of supply, functionality of the natural gas applications and fair billing.

In summary, in order to inject biogas into the natural gas grid, the above requirements must be met. In addition to excluding the gas accompanying substances by cleaning and processing the biogas, further conditioning to adjust the Wobbe Index and the calorific value to the target grid is required, depending upon the case in point.

The processed, conditioned biogas is considered to be an exchange gas, if it meets the requirements set out in G 260, G 262 and G 685. Furthermore, during conditioning with liquid gas, limits according to G 486 (DVGW, 1992) need also to be considered.

2.2 "Gas billing" according to worksheet G 685 of the DVGW regulations (DVGW, 2008)

For billing, two parameters need to be determined: The volume flow of the fuel gas under standard conditions ($T = 0\ °C$, $p = 1.01325$ bar) and the calorific value for billing purposes. The determination of the standard volume flow from the operational flow is done using the procedure described in G 685, taking into account the temperature and atmospheric pressure. At pressures greater than 1 bar, real gas behaviour should be taken into account (G 486). The calorific value for billing purposes is determined from the calorific values of the feed for each billing interval (such as 1 month (special contract customers) or 1 year (residential customers)) in a coverage area (the total area, which is supplied by the GU, not necessarily contiguous). If the calorific values of the feed change over time, then these are determined arithmetically or by volume-weighted methods over the month.

If gases with different calorific values are distributed, then the following, according to G 685 (DVGW, 2008), applies:

If gases are fed through geographically separate feed points into a grid or into gas supply areas which cannot be isolated, the calorific value for billing purposes is to be determined according to the regional location of the customer. The gas delivered to customers may not deviate in calorific value by more than 2% from the calorific value for billing purposes. To check this, the mean values and the quantity-weighted average in the downstream network are to be determined.

Since worksheet G 486 is particularly applicable due to the admixing of propane and butane as considered in this report, essential aspects are explained below.

2.3 The worksheet G 486 "Gas quantity measurement, compressibility factors and gas law deviation factors of natural gases, calculation and application" from the DVGW regulations

The determination of gas quantity, or volume is carried out under operating conditions (metering conditions). The result is an operational flow V_B (T_B, P_B) as a function of temperature and pressure. This operational flow needs to be converted to standard conditions (T_N = 0 ° C, p_N = 1.01325 bar) in order to compare volumes and so that it can be used as an input for gas billing. Since the model for an ideal gas is only approximately valid for real gases at low pressures, a compressibility factor Z (T, p, x_i) is introduced into the equation of state for ideal gases. The compressibility factor is mathematically approximated by a series expansion of the molar density (virial approach). The calculation of standard volume is thus given by equation 8. 3:

$$\frac{V(T_N, p_N)}{V(T_B, p_B)} = \frac{T_N p_B}{T_B p_N} \frac{Z_N}{Z_B} \tag{3}$$

The ratio of the compressibility factors is called the gas law deviation factor.

Two methods for calculating compressibility factors are given in G 486 including the supplementary sheets: The standard GERG-88 virial equation and the AGA8-DC92 equation of state. The former requires input parameters of p, T, H_S, N, ρ, x_{CO2} and X_{H2}, the latter the mole fractions. The AGA8 equation of state requires a full analysis by means of a process gas chromatograph.

2.4 Conditioning with liquid gas (propane / butane)

The term liquid gas (Liquefied Petroleum Gas[3] it refers to C3 and C4 hydrocarbons or mixtures thereof. It is generated as a by-product in petroleum refining and as an associated gas from the extraction of oil and natural gas. LPG is gaseous at room temperature under atmospheric conditions, but can be liquefied at low pressures. In liquid form, its specific volume is about 260 times smaller than in the gaseous state. Therefore, large amounts of energy can be transported and stored in relatively small containers.

The transportation of LPG is carried out worldwide by tanker ships, barges, pipelines, by rail tank cars, road tankers or in liquefied gas cylinders. LPG is stored in stationary tank facilities or in gas cylinders. Up to a tank size of 2.9 t capacity, the above-ground installation does not require a permit. From a tank capacity of 2.9 tonnes, the federal emission

[3] The term LPG is not to be understood as "car gas" which has a different propane / butane mixture ratio

regulations need to be considered when granting a permit. The technical conditions for setting up tank installations are defined in TRB 801 No.25 "LPG storage tank facilities".

Commercial LPG consists of at least 95 percent by mass of propane and propene, whereby the propane content must predominate. The remainder may consist of ethane (C2H6), ethene (C_2H_4), butane (C_4H_{10}) and butene (C_4H_8) isomers. The classification for commercial propene, butane and butene is equivalent. Note also the degree of purity according to DIN 51 622 [DIN 1985]: Data on sulphur or sulphur compounds are listed here.

In DIN 51624 "automotive fuels - natural gas requirements and test methods" [8-15]upper limits for the propane/butane mole fractions in natural gas of 6% / 2% in the total mixture and a methane number > 70 are required. EASEE-gas CBP (EASEE, 2005) specifies a hydrocarbon dew point of -2 ° C at 1-70 bar. For the calculations shown below, a typical LPG composition of propane / butane, 95 / 5 is used.

2.5 Aspects of the energy industry act (EnWG) and the gas network access ordinance (GasNZV)

Section 19 of the Energy Industry Act stipulates that the operators of gas distribution networks are obliged, taking into account conditions set out in section 17 for the network connection of LNG facilities, de-centralised generation and storage facilities, other transmission or gas distribution systems and direct pipelines, to determine minimum technical requirements for design and operation, and to publish them on the Internet.

The minimum technical requirements set out according to these sections shall ensure the interoperability of the grids and shall be justified objectively and not be discriminatory. Interoperability includes in particular the technical connection requirements and conditions for grid-compatible gas properties, including gas generated from biomass or other types of gas, as far as they are technically able to be injected into the gas supply grid or transported through this grid without compromising security (Energy Industry Act, 2005). For ensuring technical security section 49 applies: energy facilities shall be constructed and operated so that technical security is guaranteed. Thereby, subject to other legislation, generally accepted technical rules are to be observed. Compliance with the generally recognized technical rules shall be presumed if, in the case of plants for the generation, transmission and distribution of gas, the technical rules of the German Association of the Gas and Water Industry have been complied with.

Conditions for gas grid access are described in the Gas Network Access Ordinance (GasNZV, 2005) in part 11a "special arrangements for injecting biogas into the natural gas grid". The Gas Network Access Ordinance regulates the conditions under which the gas network operators are obliged to grant transportation customers access to the gas networks.

3. Base data

After the basic principles and the definitions of terms, there follows three examples of the calorific value adaptation of biogas, before it is injected into the natural gas grid. In these selected cases, one deals with the conditioning for an H gas grid and the other two with L gas grids, where conditioning is based on the addition of air, and on LPG and air. Other cases are described in the DVGW study "Developing a scientific basis for injecting biogas into natural gas grids."

First, the most important combustion-related characteristic data are listed for the selected base gases, in order to summarise the requirements imposed on the biogas, in particular with respect to the Wobbe index and calorific value.

Then the characteristics of the processed biogas with a methane content of 94 - 99,5 Vol.-% will be compiled in order to determine the conditioning necessary to adapt to the base gas. Processed biogases with these methane levels are generally H gases.

To attain the Wobbe index of an L gas, air must be added, which lowers the calorific value. In the case of L gases with higher calorific values, liquid gas needs to be added. When higher demands are placed on the calorific value (H gases), a liquid gas addition is necessary. A processed biogas containing 99.5 Vol -% methane has a calorific value of 11.0 kWh / m³.

For the calculations of the compositions in the following sections, the values from the following table 3 will be used. The composition of air is taken to be 20.95 Vol.-% oxygen and 79.05 Vol.-% nitrogen. All flow rates are standard flow rates.

	$H_{S,n}$ in kWh/m³	$\overline{v}_{m,n}$ in m³/kmol
CH_4	11,064	22,36
CO_2	0	22,261
N_2	0	22,403
O_2	0	22,392
C_3H_8/C_4H_{10}	28,578	21,904
air	0	22,4

Table 3. Numerical values used for the calculation

3.1 Data on the base gases

For the base gas data, gas properties taken from the DVGW worksheet G 260 appendix 1 and sample values from the GASCALC computer program from e.on Ruhrgas AG have been used. The data are summarized in Table 4. Based on the technical characteristics of combustion of the base gases, the calorific ranges for the processed biogas were determined. Here, the calorific values for the calculations were assumed to be quantity-weighted averages and a + / - 2 percent band was placed around these (Tab. 5). In this way, in the following sections admixtures whose corresponding calorific values lie in this interval will be determined.

Designation	$\varphi_{Methane}$ in Vol.-%	$H_{S,n}$ in kWh/m³	$W_{S,n}$ in kWh/m³	Density ρ in kg/m³	Methane number (+/-2)
North Sea I	88,6	12,2	15,4	0,81	72
Holland II	82,9	10,2	12,8	0,83	86
Weser Ems L Gas	87,81	9,85	12,53	0,80	102

Table 4. Base gases

Designation	Calorific range (+/-2%) in kWh/m³
North Sea I	11,956 - 12,444
Holland II	9,996 - 10,404
Weser Ems L Gas	9,653 - 10,047

Table 5. Calorific value ranges

3.2 Data on processed biogases

As a basis for further considerations, data for the biogas compositions as specified in Table 6 will be used. Further accompanying substances occurring in the biogas have not been taken into account, because they vary too greatly depending on the fermented substrates and type of processing.

Components	Unit	Case 1	Case 2	Case 3	Case 4
CH_4	Vol.-%	94	96	98	99,5
CO_2	Vol.-%	5,6	3,6	1,6	0,1
N_2	Vol.-%	0,3	0,3	0,3	0,3
O_2	Vol.-%	0,1	0,1	0,1	0,1
Calorific value	kWh/m³	10,400	10,622	10,843	11,009
Wobbe - Index $W_{S,n}$	kWh/m³	13,290	13,798	14,326	14,738

Table 6. Examples of the composition of processed biogases

A processing level of 99.5% vol. methane is attainable and is state of the art. The N_2 and O_2 levels are 0.3 and 0.1 vol -% and remain constant for the calculations at these initial methane levels. The CO_2 content conforms to the relevant selected methane content within the specified range.

3.3 Requirements for compliant processed biogas

The composition, the combustion-related characteristic data and the accompanying substances in the "processed biogas" - summarized under the term "gas properties" - are crucial for injection. In particular, these are the data described in the DVGW work sheet G 260 "gas properties", G 262 "use of regeneratively produced gases" and in the DVGW Worksheet G 685 "gas billing" and G 486 and are defined within their respective limits. Table 7 shows the principal conditioning parameters for injecting into a grid.

The methane content in biogas is not subject to any defined restriction. However methane is, according to the latest technical data, the major combustible component and thus largely determines the calorific value and hence the Wobbe index, as long as no liquid gas admixture is included. According to worksheet DVGW-G 262, the maximum permissible CO_2 content is 6 Vol.-%. Since biogas consists mostly of methane and carbon dioxide, processing to at least a minimum methane content of 94 vol -% is assumed. Together with the fractions of oxygen and nitrogen, the biogas has a maximum carbon dioxide content of 5,6 Vol.-%.

Parameter	Abbreviation	Unit	Limit	Comment
Methane volume fraction	φ_{CH4}	Vol.-%	---	Since methane is the main combustible component of biogas, the concentration largely determines the technical combustion characteristics ($H_{S,n}$, $W_{S,n}$), (H_2 not taken into account)
Carbon dioxide volume fraction	φ_{CO2}	Vol.-%	6.	G 262
Oxygen volume fraction	φ_{O2}	Vol.-%	3.	This limit applies to dry networks (G 260)
LPG mole fraction	x_{LPG}	mol.-%	Propane: 3,5 (6) Butane: 1,5	According to G 486-B2 for pressures > 100 bar (<100 bar), for details see here
Nitrogen volume fraction	φ_{N2}	Vol.-%	---	
Hydrogen volume fraction	φ_{H2}	Vol.-%	5.	G 262 (in gases produced by fermentation usually not present)
Relative density	d		0,55-0,75	G 260
Calorific value	$H_{S,n}$	kWh/ m^3	8,4-13,1	Maximum + / - 2% variation to the distributed gas (see G 685, thermal billing)
Wobbe Index L gas H gas	$W_{S,n}$	kWh/ m^3	10,5-13,0 12,8-15,7	Total range

Table 7. Requirements for processed biogas

According G 260, the maximum permissible O2 volume fraction is 3% in dry networks, otherwise 0.5%. German natural gas grids are considered to be dry.

It is important to note that CO_2 and O_2 in combination with moisture can lead to corrosion in pipes, fittings and equipment.

In addition to the above conditions, of course the information on sulphur compounds and other impurities and accompanying substances given in G 260 is also to be noted. The data for the dew point temperatures lie below the requirements defined in G 260 (about 4 - 7 ° C (soil temperature at 1 m depth) at line pressure).

4. Opportunities for the production of compliant gases

In the previous sections, the conditions for injecting processed biogas into a natural gas grid were described in detail. In this chapter, appropriate mixture compositions matching the individual gas types will be determined and the possibilities for conditioning with air and / or propane / butane admixture will be discussed.

Four different methane volume fractions (processing grades) are shown, for a total of three natural gas types as a "target" properties. Depending on the application, LPG admixtures and air admixtures are applied across a large range in order to determine a suitable, practical combination.

The addition of liquid gas or air is to be understood as an additive to the processed biogas (100%). This means that the proportion of the total mixture (100% + X) is lower than the amount added. The depicted LPG addition shows the component added, the value of the Wobbe Index, calorific value and the limits of propane and butane components resulting from the total mixture.

$$x_{LPG\ im\ Gemisch} = \frac{\dot{V}_{LPG\ Zugabe}}{\dot{V}_{LPG\ Zugabe} + \dot{V}_{Biogas}} \tag{4}$$

The type of conditioning selected depends upon economic and technical factors and will ensure that the broadest possible spectrum of combustion values is achieved.

For implementation in practice, it should be noted that the methane volume fractions arising from processing may be subject to fluctuations. Equally, the composition of the LPG may vary and the measuring instruments and the control and regulating equipment will have tolerances, so that error propagation through the system needs to be noted when trying to achieve the desired bandwidth of "target" properties.

4.1 Target properties: North sea I H gas

For the production of compliant H gas with technical combustion characteristics matching North Sea I specifications, conditioning by admixing LPG is examined below.

The figures 2 to 5 show the potential composition of biogas mixtures, based upon methane levels in the processed bio-gas of 94, 96, 98 and 99,5 vol..-%, to which propane/butane (in a ratio of 95 / 5) is added. The necessary LPG admixtures for the desired calorific range for a methane content of 94, 96 and 98 vol -% in the treated biogas, lie above the limits as defined in G 486-B2 and DIN 51624, at 9.4 to 12.6 Vol -%, from 8.1 to 11.3 Vol -% and 6.8 to 9.9 vol -%. For processing to a methane content of 99.5 Vol -%, the limit according to G 486-B2 for pressures <100 bar is numerically satisfied up to a propane / butane admixture of 6.5 vol -%. The applicability criteria as described in section 2 apply. On the basis of this restriction, only an admixture of 5.8 to 6.5Vol.-% of LPG for an initial methane content of 99.5Vol.-% is possible. This would then cover a calorific range of 11.971 to 12.080 kWh/m^3.

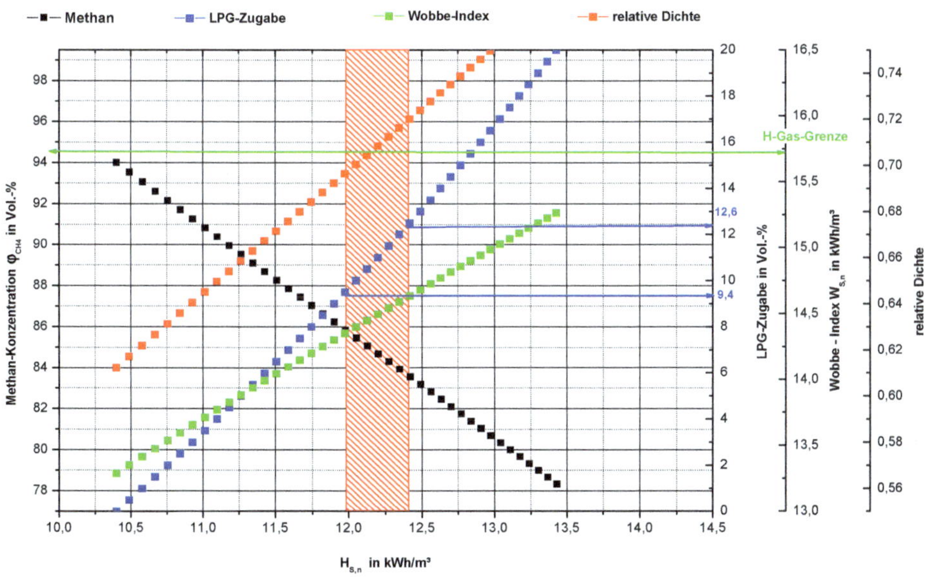

Fig. 2. Possible H gas mixtures by admixing LPG to an initial concentration of 94 Vol. -% methane

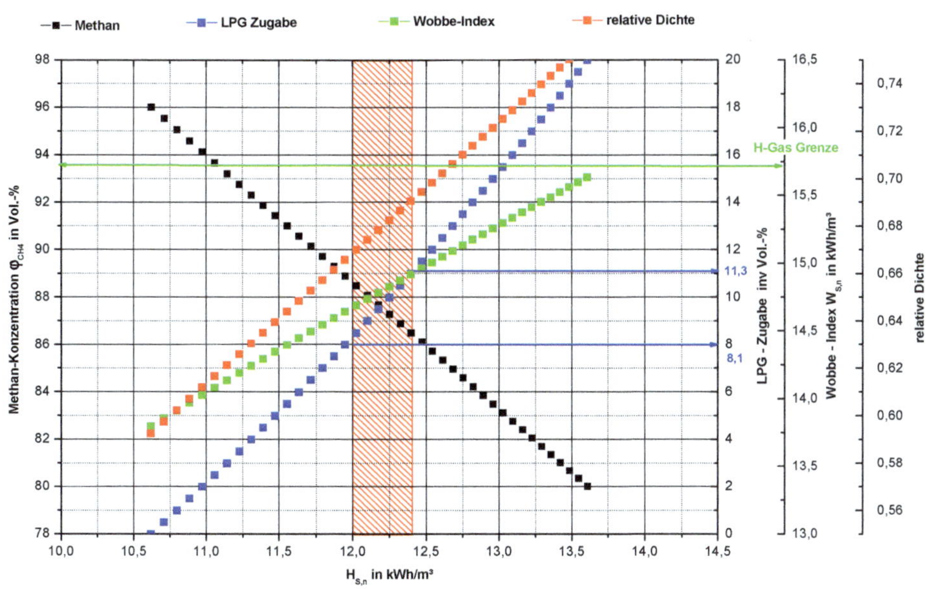

Fig. 3. Possible H gas mixtures by admixing LPG to an initial concentration of 96 Vol. -% methane

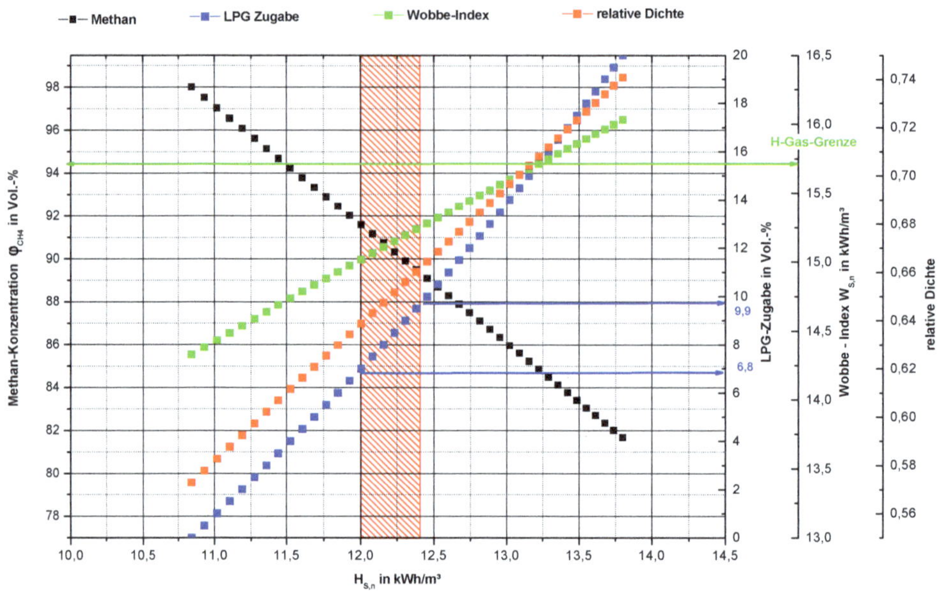

Fig. 4. Possible H gas mixtures by admixing LPG to an initial concentration of 98 Vol. -% methane

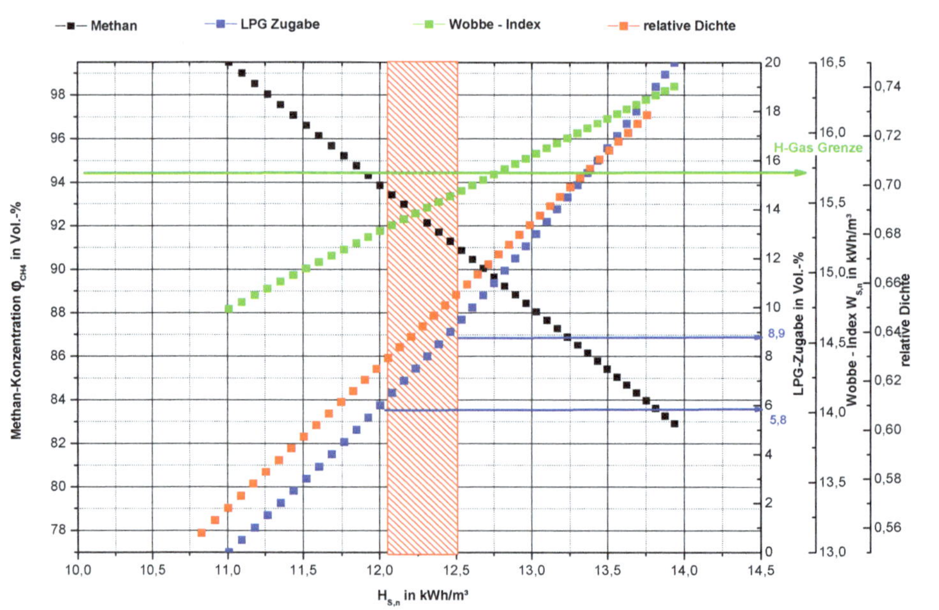

Fig. 5. Possible H gas mixtures by admixing LPG to an initial concentration of 99,5 Vol. -% methane

Table 8 shows a summary of LPG additions necessary to achieve an average target calorific value of approx.12.2 kWh/ m³.

Methane **After processing**	**North Sea I** **$H_{S,n}$ = 12,2 kWh/m³**
Vol.-%	Vol.-%
94	11,1[2]
96	9,7[2]
98	8,3[2]
99,5	7,3[4]

Table 8. LPG quantities necessary to achieve the average target calorific value

4.2 Target properties: Weser ems L gas

For the production of compliant, low calorific L gas, conditioning by the addition of air is described in the following sections.

Figures 6 to 9 inclusive show possible fuel gas mixtures with a calorific value range of 9.653 to 10.047 kWh / m³, which can be achieved by the addition of air.

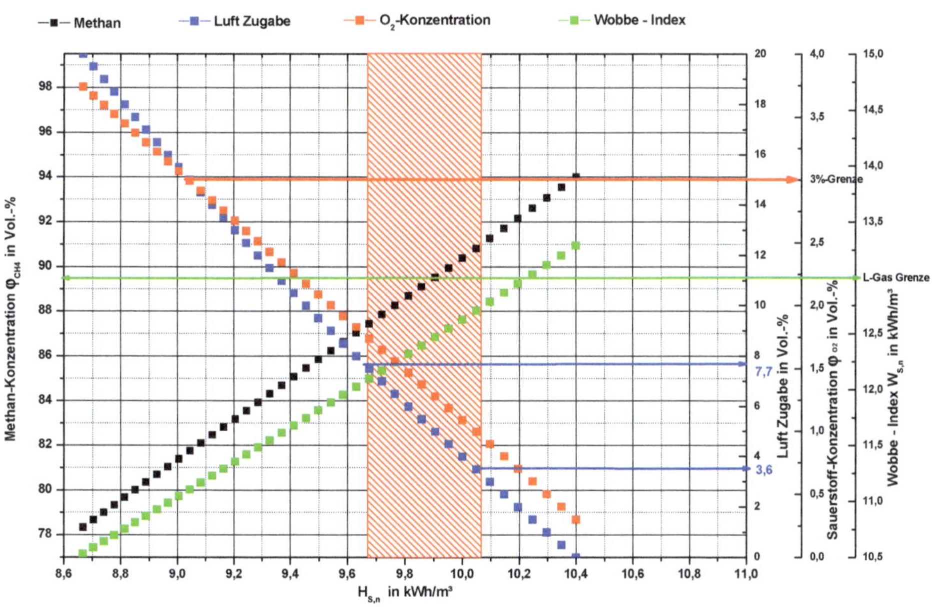

Fig. 6. Possible low calorific L gas mixtures achieved by admixing air to an initial concentration of 94 Vol. -% methane

[4] Exceeds the maximum concentration of propane for p < 100 bar according to G 486-B2

Figure 6 shows the possible mixture compositions, when air is added to an initial methane content of 94 vol -% . When interpreting this, please note that the data points for the Wobbe index, the methane content and the air belong together along the line of constant calorific value. A +/- 2% range has been set for the calorific value limits, based on the calorific value defined in DVGW worksheet G 260. All values with a Wobbe index of less than 13 kWh / m³ and an O_2 content of less than 3 vol -% meet requirements. All other boundary conditions are shown in the table above.

Figure 7 show the possible mixture compositions, when air is added to initial methane content of 99,5 Vol.-%.

The two cases presented with initial methane contents of 94 and 99.5 Vol -% in the biogas, clearly show that increases in methane content also make necessary increased amounts of air, in order to achieve the desired calorific value and Wobbe index.

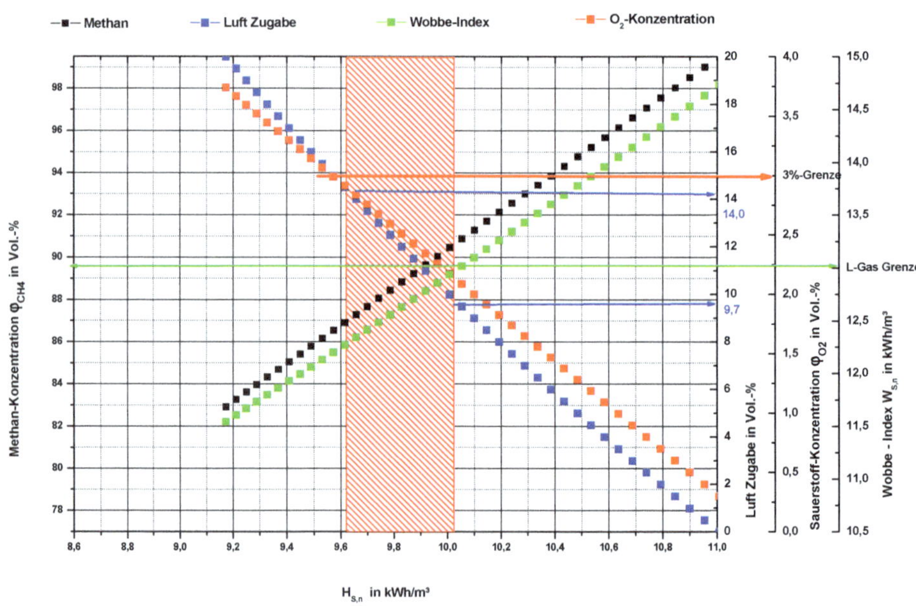

Fig. 7. Possible low calorific L gas mixtures achieved by admixing air to an initial concentration of 99,5 Vol. -% methane

Table 9 shows the respective admixtures.

Methane concentration after processing in vol -%	Weser Ems L Gas $H_{S,n}$ = 9,85 kWh/m³	
	Air added to the biogas in Vol.-%	Wobbe Index in kWh/m³
94,0	5,6	12,379
96,0	7,8	12,492
98,0	10,1	12,589
99,5	11,8	12,664

Table 9. Air admixture to Weser Ems L-Gas and corresponding Wobbe-Index

The higher the initial content of methane in the biogas is, the greater is the approximation to the maximum compliant O_2 content of 3 vol -% from conditioning.

Thus the O_2 levels upon reaching the lower calorific value band are:

- at 94 % vol. methane 1,776% vol. O2
- at 96 % vol. methane 2,177% vol. O2
- at 98 % vol. methane 2,562% vol. O2
- at 99,5 % vol. methane 2,836% vol. O2

Reaching of the required calorific value band (9.653 to 10.047 kWh / m³) is possible from all four initial methane contents.

Figure 8 shows a summary of the air admixture ranges of the four initial methane levels. The red line indicates the maximum permissible volume fraction of 3% of O_2 in the mixture.

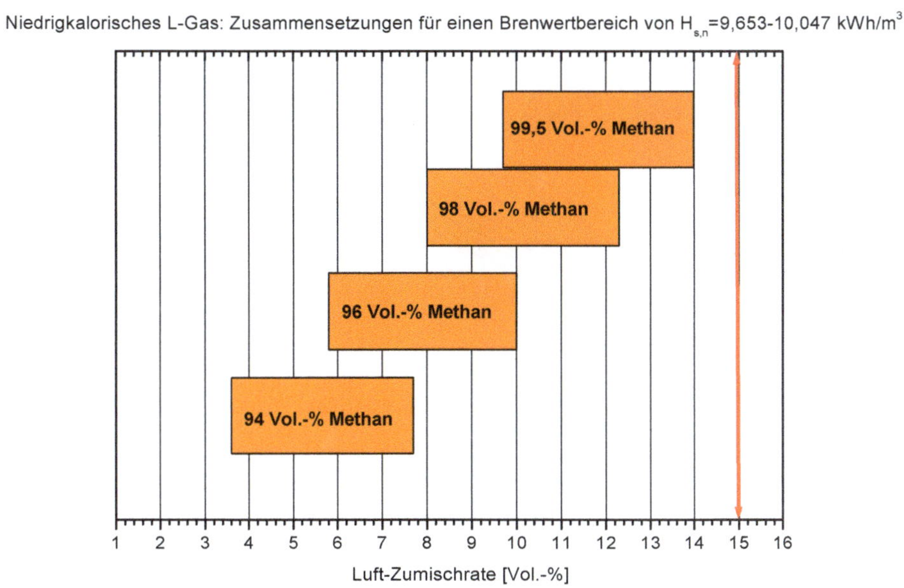

Fig. 8. Rates of air admixture at different initial concentrations of methane

4.3 Target properties: Holland II L gas

For the production of compliant, high calorific L gas, conditioning by the addition of air and LPG is described in the following section.

Please note the following when interpreting the diagrams below: The field of admixtures includes a range of 0 - 20 Vol -% for presentational purposes. In practice, for technical and economic reasons, it is desirable to make the least possible admixtures with a "target" Wobbe Index of 12.4 kWh / m³ for example (setting of the gas appliances). In this context it should be noted that according to G 486 appendix B, the mole fractions of propane are not to exceed 3.5 mol% (6 mol% at p <100 bar) and butane max.1.5 mol% in natural gas, in order make a conversion of standard and operating conditions using the AGA8-DC92 equation of state.

The "field" of the possible mixtures is bounded by the Wobbe Index of 13 kWh / m³, the given calorific value limits, the max. oxygen volume fraction of 3%, and the maximum propane/butane or air admixture. For each value of air addition, there is always a value for the propane / butane addition.

The following figures apply only to the four initial properties of the biogas used.

Figures 9 and 10 show the calorific values and the Wobbe index for an air and LPG admixture of 0 to 20 vol -% to a biogas with an initial methane content of 94 vol -% and 96 vol -%.

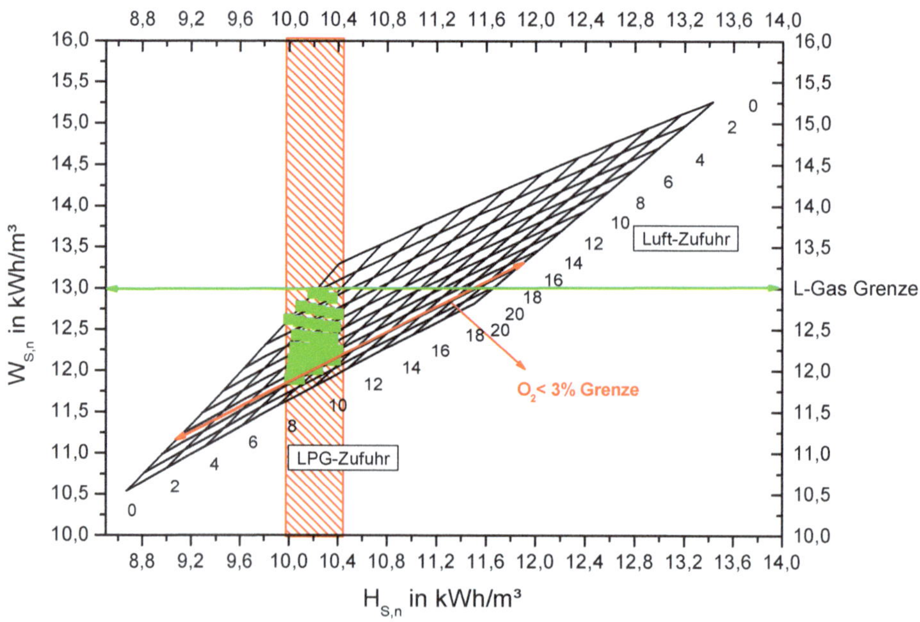

Fig. 9. Possible highly calorific L gas mixtures by admixing air and LPG to an initial concentration of 94 Vol. -% methane.

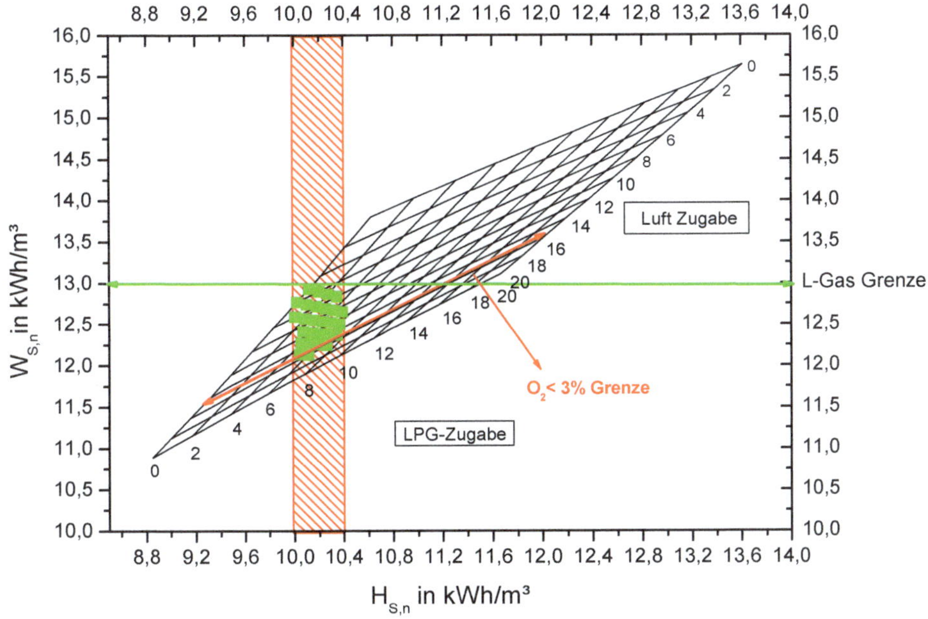

Fig. 10. Possible highly calorific L gas mixtures by admixing air and LPG to an initial concentration of 96 Vol. -% methane.

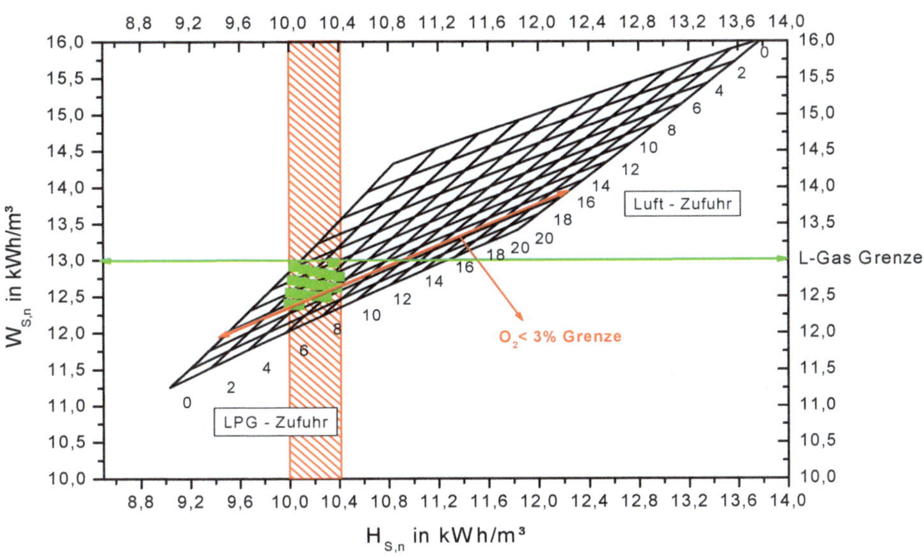

Fig. 11. Possible highly calorific L gas mixtures by admixing air and LPG to an initial concentration of 98 Vol. -% methane

Figures 11 and 12 show the calorific values and the Wobbe index for an air and LPG admixture of 0 to 20 vol -% with an initial methane content of 98 vol -% and 99,5 vol -%.

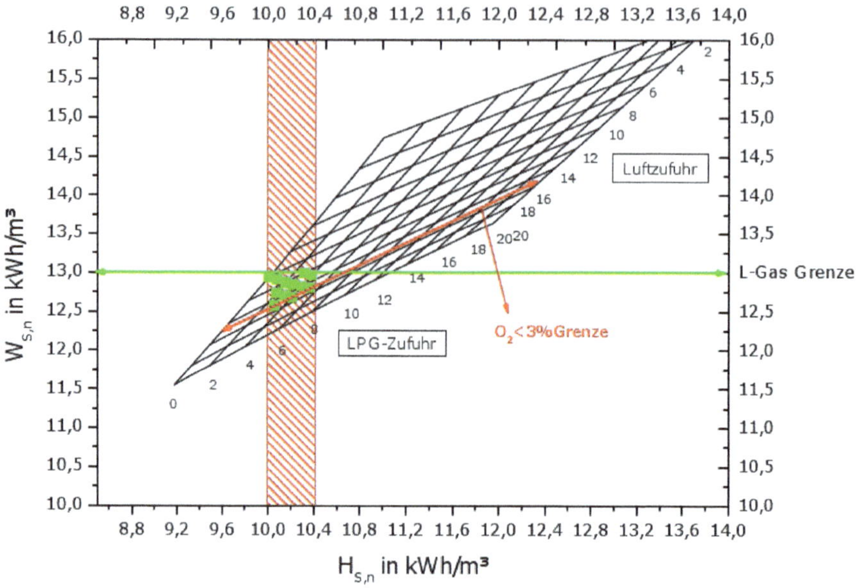

Fig. 12. Possible highly calorific L gas mixtures by admixing air and LPG to an initial concentration of 99,5 Vol. -% methane

The red area represents the required calorific value range from 9.97 to 10.4 kWh / m³. The green dots show the possible, compliant mixtures that lie within all the conditions to be fulfilled.

5. Condensation curves and methane numbers

The conditioning with air and / or liquid gas to adjust the technical combustion characteristics may influence both the methane number and the condensation of higher hydrocarbons in the combustion gas mixture.

The methane number - equivalent to the octane number of petrol - is a statement of the anti-knock properties of fuel combustion in a engine, where the term anti-knock refers to the tendency to uncontrolled and undesirable self-ignition. Methane has by definition a methane number of 100, hydrogen a methane number of 0. A methane number of 80 for example means that the gas mixture associated with this methane number has the same anti-knock properties as a mixture of 80% vol. methane and 20 vol. -% hydrogen. Some inert mixture components such as CO_2 increase the methane number, higher hydrocarbons, reduce it. The calculated methane numbers (Gascalc, E.on Ruhrgas) of L gases are generally greater than those of H gases (nitrogen not factored out). As a lower limit for the smooth operation of modern engines, a methane number of MZ > 70 is considered necessary (DIN 51624, 2008).

For multi-component mixtures such as natural gases, the condensation and boiling curves do not lie together, but span a conditional area, where different gas-liquid compositions are possible. Between the critical pressure and the cricondenbar point with increasing temperature, and between the critical temperature and the criconden therm point with falling pressure, condensate (retrograde condensation) can form when the throttle curve touches the dew line, intersects or the final state lies in the two-phase region (Höner zu Siederdissen & Wundram, 1986).

An admixture of propane / butane to natural gas and processed biogas generally manifests itself in a shift of the dew curve to higher temperatures. According to (Oellrich et al., 1996), in the case of Russian H gas, condensation is only to be expected at temperatures of -35 ° C, while it will occur with Dutch L gas already at -5 ° C. If liquid gas/air is admixed within the limits described in DVGW worksheet G 260, the criconden therm point moves toward +15 ° C or +45 ° C, but at higher pressures. For mixtures of natural gas and processed conditioned biogas, this is to be expected to a lesser degree, since the concentrations of propane / butane are correspondingly smaller.

It should be noted that the calculation of the condensation curves of natural gases requires an analysis that takes into account the higher hydrocarbons, since even small amounts in the ppm range result in a significant shift. Furthermore, the process of condensation is not in itself critical, but the quantity of condensate is the decisive criterion. For large flow rates, a seemingly low volume of condensate can therefore lead to problems (Oellrich et al., 1996).

The following Table 10 and Figure 13 show cases of condensation in conditioning by the addition of LPG, in order to meet the North Sea I specification. The lowest and the highest admixtures were selected for the diagrams. It should be noted that at the highest level of admixing, the restrictions imposed by G 486 were not observed.

Initial concentration of CH_4 in the biogas in Vol.-%	LPG addition to biogas in Vol.-%	Calorific value in kWh/m^3	Wobbe Index in kWh/m^3	rel. Density	Methane number
94,000	9,400	11,960	14,339	0,696	71
94,000	12,600	12,432	14,642	0,721	67
96,000	8,100	11,965	14,654	0,667	72
96,000	11,300	12,442	14,946	0,693	67
98,000	6,800	11,970	14,998	0,637	73
98,000	9,900	12,438	15,268	0,664	67
99,500	5,800	11,971	15,276	0,614	74
99,500	8,900	12,443	15,534	0,642	67

Table 10. Cases of condensation for mixtures of the North See I H gas specification

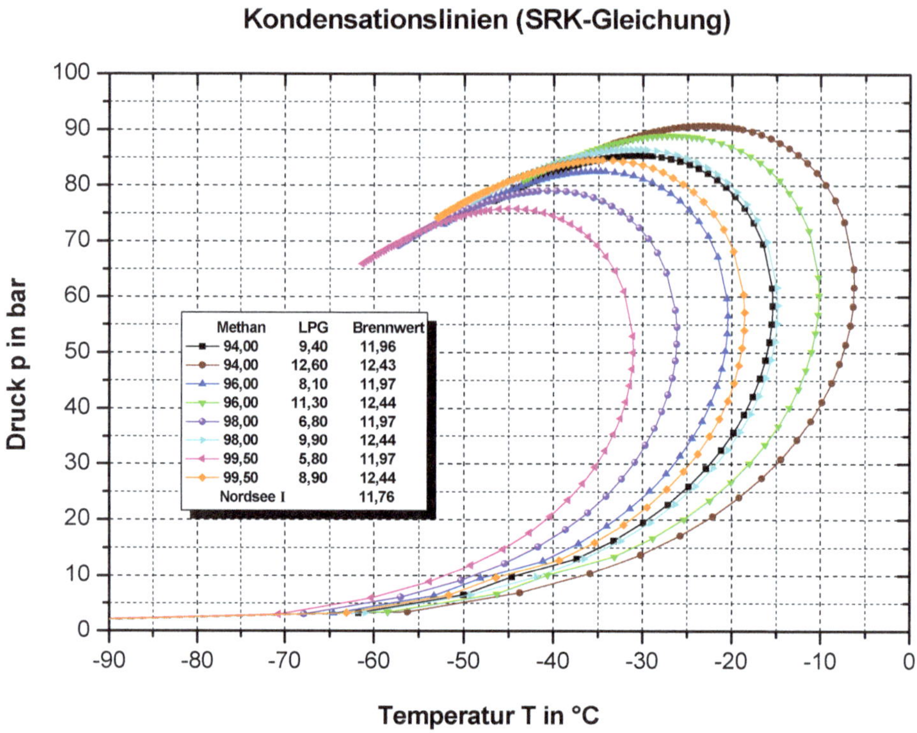

Fig. 13. Condensation curves for mixtures of the North See I H gas specification

In summary, it can be said that the criconden therm points of the H gases in Germany lie below temperatures of -20 ° C. An exception to this are the higher caloric mixtures of the North Sea quality - here, 0 ° C is also possible. However, the mixtures were ignored in the calculation of the limits in G 486 and DIN 51 624, so that it applies mainly to the higher LPG quantities added. Generally, this means that process procedures, in which pressure and temperature lie in the two-phase region, should be avoided.

6. Conclusion

This section shows a summary of all the mixing rates of LPG and / or air to attain the base gas properties under consideration.

6.1 Conditioning: Target H gas

The table 11 shows the determined rates of admixture for LPG to attain the appropriate target calorific value range. With the LPG quantities shown, the respective initial concentrations of methane, the entire calorific value range of the respective base gas quality is covered, with some restrictions.

LPG mixing rates to achieve the target calorific value + / - 2%	
Methane concentration after processing in vol -%	North Sea I H gas $H_{S,n}$ = 11,956 - 12,444 kWh/m³
	LPG in Vol.-%
94,0	9,4 - 12,6
96,0	8,1 - 11,3
98,0	6,8 - 9,9
99,5	5,8 - 8,9
	Restriction of the lower calorific value range by high processing levels

Table 11. Air and LPG additions of the investigated H gas properties

For the practical implementation of the listed quantities of the LPG admixtures, limits as given in Table 12 are to be observed, in accordance with the requirements presented on the need for the use and applicability of SGERG88 and AGA8 procedures and the resulting maximum admixture quantities according to Table 12. Due to these limits, defined in DVGW G 486-B2, it will not be possible in every case to reach the upper calorific value range at higher pressures. In addition, the availability of appropriate measuring technology for higher liquid gas fractions is limited. At the very high degrees of methane processing, there are limitations on attaining the lower calorific value range, since when processing to 99.5 Vol -% methane, an H gas with a calorific value of 11.009 kWh / m³ results.

	Descriptor	**Unit**	**Limit according to G 486 supplementary sheet 2 Appendix B p> 100 bar**	**Limit according to G 486 supplementary sheet 2 Appendix B p> 100 bar**
Propane	x_{C3H8}	mol.-%	3,5	6.
Butane	x_{C4H10}	mol.-%	1,5	1,5

Table 12. Limits according to DVGW G 486 supplementary sheet 2 appendix B

Table 13 shows the maximum possible , compliant LPG admixture, a propane / butane mixture of 95 / 5 Mass .-%. As a result, it is clear that processing to a maximum methane content of 99.5% vol. with the maximum permissible LPG admixtures, a calorific value of maximum 11,361 kWh/m³ (NTP) is possible at pressures above 100 bar, and a maximum of 12,075 kWh/m³ (NTP) at pressures below 100 bar.

Methane content in biogas	LPG as admixture	Propane content in admixture	Butane content in admixture	Calorific value	Wobbe Index	Relative Density
in Vol.-%	in Vol.-%	in mol .-%	in mol .-%	in kWh/m³	in kWh/m³	
94,000	3,7	3,497	0,141	11,044	13,733	0,647
95,000		3,497	0,141	11,151	13,969	0,637
96,000		3,497	0,141	11,257	14,210	0,628
97,000		3,497	0,141	11,364	14,454	0,618
98,000		3,497	0,141	11,471	14,704	0,609
99,000		3,497	0,141	11,577	14,959	0,599
99,500		3,497	0,141	11,631	15,088	0,594
94,000	6,5	5,978	0,241	11,501	14,039	0,671
95,000		5,978	0,241	11,605	14,265	0,662
96,000		5,979	0,241	11,709	14,495	0,653
97,000		5,979	0,241	11,813	14,729	0,643
98,000		5,979	0,241	11,917	14,967	0,634
99,000		5,979	0,241	12,021	15,209	0,625
99,500		5,980	0,241	12,073	15,332	0,620

Table 13. Gas properties with maximum compliant LPG additions

Figure 14 shows the admixture required to achieve the corresponding H gas properties. The limits on the maximum concentration of propane are also shown according to DVGW regulations G 486 supplementary sheet 2 a ppendix B. Admixtures to achieve the properties of North Sea I / North Sea II H gas are, based upon all levels of methane processing, above the limits. A compliant mixture is not possible in this case, or needs to be tested on an individual basis.

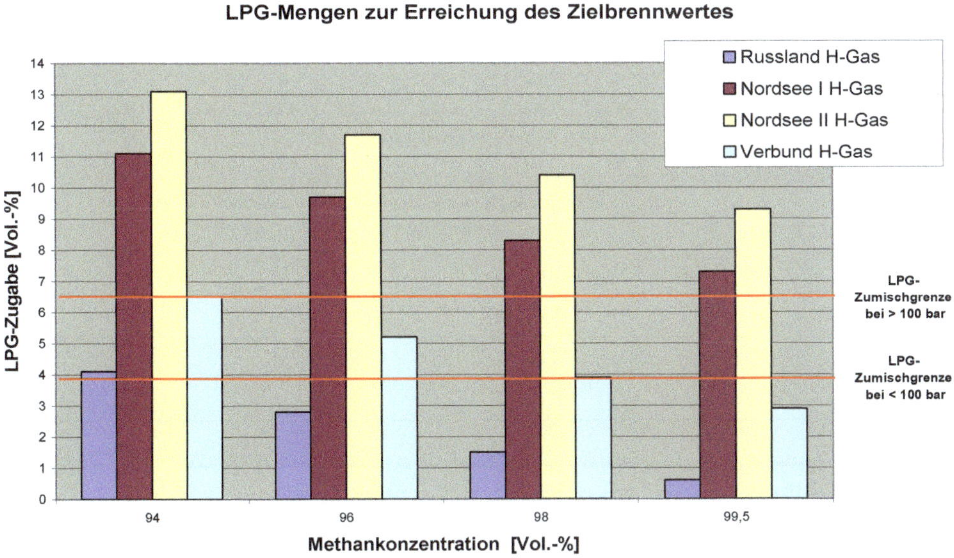

Fig. 14. LPG quantities necessary to achieve the target calorific value

In order to achieve higher calorific values, alternative conditioning measures can be employed. For example, ready-made mixtures with customized propane / butane ratios can be used for conditioning. This can increase the calorific value, but technical and physical effects, such as condensation behaviour, methane number and k-number deviations need to be considered.

6.2 Conditioning: Target L gas

Two different L gas target properties have been described. Because of their basic constitutions, one bio-methane mixture is conditioned with air and the other is conditioned with a combination of air and LPG.

Table 14 shows a summary of the admixtures with which a target calorific value-oriented mixture for the low calorific base gas property can be achieved.

In the case of simple air addition, particular attention should be paid to compliance with the maximum O_2 volume fraction. This should not exceed 3 % vol. in dry networks according to DVGW worksheet G 260. This quantity is reached when adding pure air to the processed biogas, at an admixture of 15 vol -% of air. In the low caloric L gases (e.g. Weser Ems gas), this limit is never reached.

Furthermore, a minimum air addition may also be necessary, in order to achieve the required Wobbe Index according to DVGW worksheet G 260.

Table 15 shows the minimum air addition for the individual processing grades of methane to achieve an L gas compliant Wobbe Index of under 13.0 kWh/m³ (NTP).

Air admixtures to attain the target calorific value + / - 2%	
Methane concentration after processing in vol -%	**Weser Ems L Gas** **$H_{S,n}$ = 9,653 - 10,047 kWh/m³**
	Air admixture **in Vol.-%**
94,0	3,6 - 7,7
96,0	5,8 - 10,0
98,0	8,0 - 12,3
99,5	9,7 - 14,0

Table 14. Air additions to the H gas properties under investigation

Methane in Biogas	**Methane in admixture**	**CO_2 in admixture**	**Air to the Biogas**	**O_2 in admixture**	**Calorific value**	**Wobbe Index**	**rel. Density**
in Vol.-%	**in Vol.-%**	**in Vol.-%**	**in Vol.-%**	**in Vol.-%**	**in kWh/m³**	**in kWh/m³**	
94,000	92,429	5,506	1,700	0,645	10,226	12,999	0,619
96,000	91,778	3,442	4,600	1,208	10,154	12,996	0,611
98,000	91,163	1,488	7,500	1,741	10,086	12,993	0,603
99,500	90,702	0,091	9,700	2,126	10,035	12,988	0,597

Table 15. Minimum quantity of air to attain L gas specification

For the high-caloric L gas mixtures (target properties according to Holland II L gas) the processed biogas is conditioned with air and LPG. Table 16 shows the correlating LPG-air additions, to reach the calorific value range (+ / -2%).

The gray-shaded areas show where a compliant combination of air and LPG additions is impossible. With increasing LPG additions, the necessary addition of air is limited by the maximum O_2 volume fraction of 3 %. If too little LPG is added, only the lower calorific value range can be covered. The broadest coverage of the calorific value range lies in between and is marked by the wider bandwidth of air additions.

Methane concentration in the Vol -%		LPG - addition [Vol -%]				
		0	2.	4	6	8
Holland II $H_{S,n}$ = 9,996 - 10,404 kWh/m³	94	2 - 4	4 - 7	7 - 10	10 - 14	14 - 16
	96	5 - 5	7 - 9	9 - 12	12 - 16	16 - 16
	98	-	10 - 11	11 - 15	14 - 16	-
	99,5	-	12 - 13	13 - 15	16 - 16	-

Table 16. Air addition, depending on the addition of LPG and methane concentration

7. References

DIN 51 624 "Automotive Fuels - Natural Gas Requirements and Test Methods"

DIN 51622 "Liquid Gases: Propane, Propene, Butane, Butene and Mixtures Thereof; Requirements"

DVGW Worksheet G 260 "Gas Properties", January 2000 and May 2008

DVGW Worksheet G 262 "Use of Gases from Renewable Sources in the Public Gas Supply", November 2004

DVGW Worksheet G 486 "Gas Quantity Measurement, Compressibility Factors and Gas Law Deviation Factors of Natural Gases" August 1992

DVGW Worksheet G 685 "Gas Billing incl. First Supplementary Sheet of April 1995 ", April 1993,

DVGW Worksheet G 685-2-B "Second Supplementary Sheet to DVGW Worksheet G 685 - Quantity Splitting within One Billing Cycle", December 2004, DVGW Worksheet G 685-3-B " third supplementary sheet to DVGW Worksheet G 685 - Substitute Values of Billing-Related Gas Data ", December 2004

EASEE-gas, Common Business Practice, No. 2005-001/01 "Harmonisation of Natural Gas Quality"

GASCALC program, E.ON Ruhrgas

Höner zu Siederdissen, Jürgen and Friedrich Wundram "Retrograde Condensation in Natural Gas Transmission Systems", gwf-Gas/Erdgas , 127 (1986)

Jaeschke, M., P. Schley, "Calculation of the Compressibility Factor of Natural Gases with the AGA8-DC92 Equation of State", gwf-Gas/Erdgas 137 (1996) No. 7

Law on Electricity and Gas Supply (Energy Industry Act - EnWG) EnWG Issue Date: 07.07. 2005 Full quote: "Energy Act of 7 July 2005 (Federal Law Gazette I p. 1970 (3621)), as last amended by Article 2 of the Act of 18 December 2007 (Federal Law Gazette I p. 2966)" Version: as last amended by Section 2 G v.18.12. 2007 I 2966

Oellrich, Lothar R., Thorsten Engler, Heribert Kaesler and Jens Nixdorf, "Studies of Retrograde Condensation Behaviour of some European Natural Gases", Reprint from gwf-Gas/Erdgas 137 (1996) issue 1 page 1-6,

Position Paper of the PTB-AG 1.42 "Gas Measuring Equipment" and AG 3.41 "Caloric Values" on the subject of "Injecting Biogas into Networks", of 6.12.2006

Regulations on access to gas supply networks (Gas Network Access Ordinance - GasNZV) of 25 July 2005, last amended by Regulation amending the Gas Network Access Ordinance, the gas network tariff regulations, the incentive regulations and the electricity network tariff regulations of 8 April 2008.

19

Kinetics of Biogas Production from Banana Stem Waste

Norazwina Zainol
Universiti Malaysia Pahang, Malaysia

1. Introduction

Biogas produced in anaerobic digesters consists of methane (50%–80%), carbon dioxide (20%–50%), and trace levels of other gases such as hydrogen, carbon monoxide, nitrogen, oxygen, and hydrogen sulfide. Anaerobic digestion is a biological process in which organic material is decomposed by bacteria in the absence of air. The general technology of anaerobic digestion of complex organic matter is well known and has been applied for over 60 years as part of domestic sewage treatment to stabilize organic wastes. Bal & Dhagat (2001) points out that the anaerobic process is more advantageous than the aerobic process in organic waste treatment because of the high degree of waste stabilization, low production of excess biological sludge, low nutrient requirement and production of methane gas as a useful byproduct. Several studies have been carried out for evaluating kinetic parameters and model equations for anaerobic digestion by Siles et al. (2010), Borja et al. (2005), Jimenez et al. (2004), Raposo et al. (2009), Rincon et al. (2009) and Hu et al. (2002); these are all based on the Monod kinetic model (Monod 1950) and on the revised kinetic model developed by Chen et al. (1980) and Hashimoto et al. (1981).

In the microbiology of methanogenic process four different bacterial groups are identified as being responsible for carrying out the anaerobic digestion of complex organic matter. The first group of bacteria is hydrolytic bacteria which catabolizes carbohydrate, protein, lipid and other minor components of organic matter to fatty acids, H_2 and CO_2. The second group of bacteria is hydrogen producing acetogenic bacteria which catabolizes certain fatty acids and neutral end products to acetate, CO_2 and H_2. The third group of bacteria is homo acetogenic which synthesizes acetate using H_2, CO_2 and formate, and hydrolyzes multicarbon compound to acetic acid. Finally, the fourth group of bacteria i.e. methanogenic bacteria utilizes acetate, carbon dioxide and hydrogen to produce methane. The concerted action of these four bacterial groups ensures process stability during anaerobic digestion of the complex organic matter.

The reactions involved in these steps are given below:

- Phase-I. Solubilization of carbohydrate via hydrolysis

$$[C_6H_{10}O_5]n + nH_2O = nC_6H_{12}O_6 \qquad (1)$$

- Phase-II. Acidogenesis fermentation of glucose to acetate

$$C_6H_{12}O_6 + 2H_2O = 2CH_3COOH + 4H_2 + 2CO_2 \tag{2}$$

- Phase-III. Methanogenic reaction

$$CH_3COOH = CH_4 + CO_2 \tag{3}$$

$$4H_2 + CO_2 = CH_4 + 2H_2O \tag{4}$$

1.1 Chen–Hashimoto kinetic model of anaerobic digestion

Chen-Hashimoto model was used for kinetic analysis of the experimental data. In a completely mixed continuous digester the rates of change of cell mass and substrate concentration are expressed by the following equations:

$$\frac{dX}{dt} = \mu X - \frac{X}{\theta} \tag{5}$$

$$\frac{dS}{dt} = -r + \frac{S_0 - S}{\theta} \tag{6}$$

Where
X is the concentration of cell mass
μ the specific microbial growth rate
θ the hydraulic retention time
S_0 the concentration of substrate in the influent,
S the concentration of substrate in the effluent
r is the volumetric substrate utilisation rate

The relationship between r and μ is defined by the following equation:

$$\mu = \frac{Yr}{X} \tag{7}$$

Where
Y is the yield coefficient (cell mass/substrate mass) and is considered constant (Chen & Hashimoto, 1978). In the steady-state, dX/dt = 0 and dS/dt = 0, hence

$$\mu = \frac{1}{\theta} = D \tag{8}$$

Where
D is the dilution rate

$$r = \frac{S_0 - S}{\theta} \tag{9}$$

and

$$X = Y(S_0 - S) \tag{10}$$

Substituting these expressions in Contois' equation:

$$\mu = \frac{\mu_{max} S}{\beta X + S} \tag{11}$$

Where
μ_{max} is the maximum specific microbial growth rate
β is a dimensionless kinetic parameter

$$\frac{S}{S_0} = \frac{K}{\mu_{max}\theta - 1 + K} \tag{12}$$

Where
K is an dimensionless kinetic parameter.

Eq. 12 shows that effluent substrate concentration depends on the influent substrate concentration.

The minimum retention time indicating when the washout of micro-organisms occurs is numerically equal to the reciprocal of the maximum growth rate:

$$\theta_{min} = \frac{1}{\mu_{max}} \tag{13}$$

There are two different approaches generally used to study the kinetics of biogas production of lignocellulosic waste: one approach is to find the rate-limiting substrate for the kinetic evaluation; another approach is using chemical oxygen demand or volatile solids concentration as an indicator of the substrate concentration, (Chen & Hashimoto, 1978). There are difficulties in using COD or VS as the gross substrate since a portion of the COD or VS is not available to the microbes as substrate. The laboratory test for COD of high strength residues requires at least 100 times dilution which generally yields unreliable data. Also, some of the volatile acids in the effluent are volatilised during the VS determination. Because the volatile acids are precursors of biogas production, their volatilisation during the VS determinations causes errors in the calculated amount of substrate utilised.

Biogas production is directly correlated with COD reduction. Since no oxidising agent is added, the only way COD reduction can occur is through the removal of organic material from the waste, such as through the evolution of methane and carbon dioxide. The other avenues of COD reduction through hydrogen sulpsulphide and hydrogen gas evolution are insignificant (Chen & Hashimoto, 1978). A reduction of 1 g COD is equivalent to the production of 0.35 l of methane at STP. Knowing the COD loading to the reactor and the volume of methane produced, the remaining COD in the digester can be calculated.

The biodegradable COD in the reactor will be directly proportional to $(B_0 - B)$ where B denotes the volume (in litres) of methane produced under normal conditions of pressure and temperature per gram of substrate (COD) added to the digester and B_0 is the volume of methane produced under normal conditions of pressure and temperature per gram of substrate added at infinite retention time or for complete utilization of substrate and B0 will be directly proportional to the biodegradable COD loading (Chen & Hashimoto, 1978). Therefore, from Eq. (12) one obtains:

$$\frac{B_0 - B}{B_0} = \frac{K}{\mu_{max}\theta - 1 + K} \tag{14}$$

From Eq. (14) one obtains:

$$\theta = \frac{1}{\mu_{max}} + \frac{K}{\mu_{max}} \frac{B}{(B_0 - B)} \tag{15}$$

Thus, by first calculating the value of B_0, the graph of θ versus $B/(B_0-B)$ produces a straight line with an intercept of $1/\mu_{max}$ and with a slope of K/μ_{max}. To obtain the parameter B_0 one uses the following equation, which is easily derived from Eq. (14):

$$B = B_0 \left|1 - \frac{K}{\mu_{max}\theta - 1 + K}\right| \quad (16)$$

Since B is the methane production per gram of added COD, the volumetric methane production rate (δ) equals B multiplied by the loading rate:

$$\delta = \frac{BS_0}{\theta} = \frac{B_0 S_0}{\theta}\left|1 - \frac{K}{\mu_{max}\theta - 1 + K}\right| \quad (17)$$

Where
δ has the dimensions of volume methane per volume digester per unit time.

The objective of the present study is to develop kinetic parameters for two-stage biogas production using banana stem waste as substrate.

2. Materials and methods

2.1 Acclimatization

In the acclimatization step, soil sludge containing mixed culture from soil (MCS) will be collected from banana plantation soil using polyvinyl chloride pipe in 10 cm depth from surface to make sure the anaerobic mixed culture available in the sample. The end of the pipe is then closed with rubber stopper and need to be process in 4 hours after collection was made.

Subsequently, 20 ml of the soil sludge is put into each of 20 serum bottles that contained 20 mg of the banana stem waste (BSW). The bottle is then closed with bottle cap or rubber stopper and flushed with nitrogen for 5 minutes. The flushed bottle is then incubated in anaerobic container in ambient temperature (28°C to 30°C) and dark condition for a month. Then, the content in all serum bottles are put into 1 liter anaerobic container which contained of 10 g of BSW. The 1 litre anaerobic container is then closed and flushed with nitrogen for 5 minutes and left for incubation in ambient temperature (28°C to 30°C) for two months.

Afterward, the materials in the 1 litre anaerobic container are put into two anaerobic container (5 litres) which each of container contained 50 g of BSW. It was found by the inventors of the present invention that acclimatization increased the microbe amount in reactor.

Subsequently, the acclimatized MCS will be used as inoculum for biogas production. Biogas production will be done in 10 L anaerobic fed-batch bioreactor with gas outlet (Fig. 1). Bioreactor equipped with temperature controller and agitator to ensure the inoculum and BSW mixed evently. To maintain anaerobic condition in the reactor nitrogen will be purged everytime the inlet and outlet were done. 5000 mg/l inoculum mixed with BSW at HRT of 12 d and OLR 1.5 gTS/l.d and ambient temperature (28°C to 30°C) for biogas production.

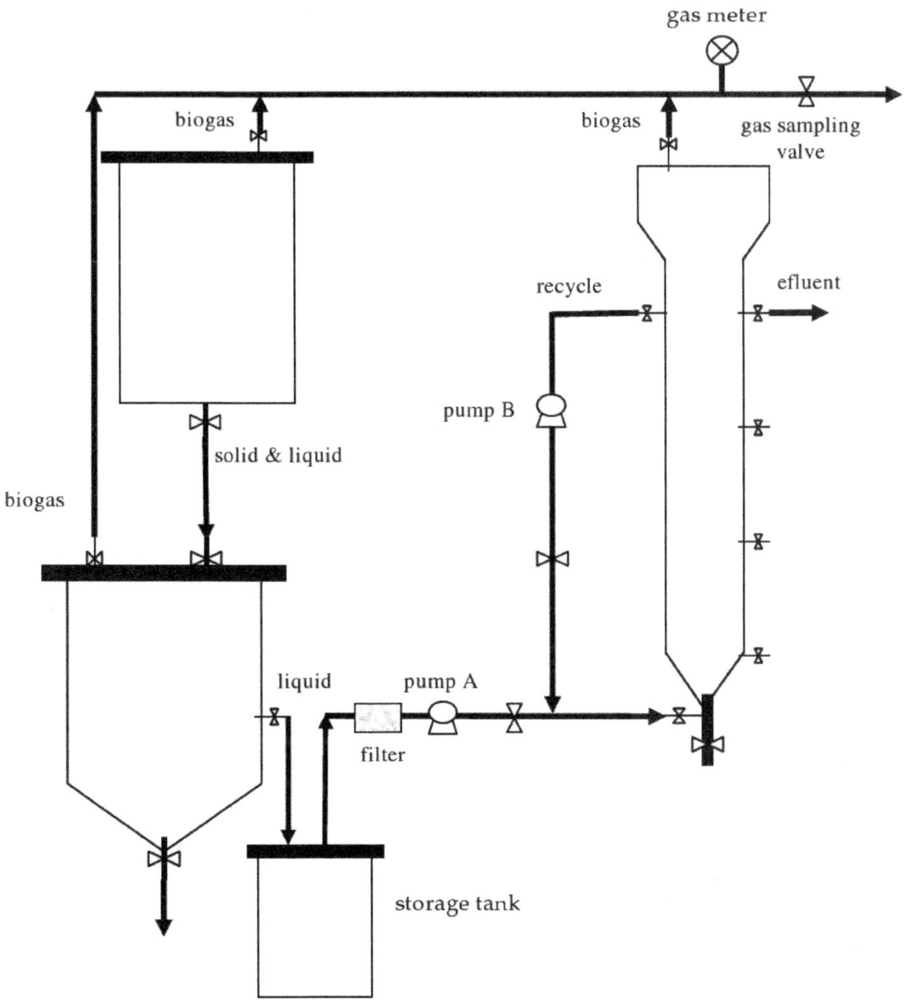

Fig. 1. Two-stages biogas production system

2.2 Experimental set-up

All experiments were done in 20 L anaerobic sequencing batch reactor followed by 10 L fixed bed reactor with gas outlet (Fig. 1). All the reactors were seeded with anaerobic acclimatized banana stem sludge. The anaerobic digestion system was varied at different reaction temperatures using water bath. The HRT and OLR for this system were 9 d and 4 g TS/l.d respectively. The process was conducted at ambient temperature for the first stage and thermophilic temperature for the second stage. Daily withdrawal of an appropriate volume from the reactor corresponding to the determined HRT or OLR was done by a draw-and-fill method. Biogas evolved from the fixed bed reactor was measured and collected in a gas holder by water displacement. Samples were collected and analyzed for performance evaluation.

2.3 Two-stages biogas production system description

2.3.1 Bioreactor description

This system consists of 4 components which are hydrolysis reactor, liquid-solid separator, storage tank and methanogenesis reactor. The dimensions of those four components are listed in Table 1. Detailed of each component are as follows:

	CRR	Solid-liquid separator	Storage tank	BPR
Volume (l)	20	10	10	10
Diameter (in)	12	9	10	Upper-10 Lower-6
Length (in)	15	16	10	35
Maximum pressure(bar)	2	2	-	2
Relieve	0.2	0.2	-	0.2

Table 1. Component dimension in two-stages biogas production

2.3.2 Hydrolysis reactor

Influent for the hydrolysis reactor is banana stem waste slurry. The type of reactor is anaerobic batch reactor. Reactor volume is 20 litre. Initial concentration of the biomass in teh reactor is 5000 mg/l which is mixed culture from acclimatized banana plantation soil. Inlet and outlet from this reactor is drawn manually (Scharer et al. 1981; Gavala et al. 2003).

2.3.3 Solid-liquid separator

The second component is solid-liquid separator. The tank volume is 10 litre. The function of this tank is to separate the solid and liquid from CRR effluent. The effluent from CRR will be sediment at the bottom of conical shaped separator tank. Sludge from separator will be recycled back to CRR and the liquid from separator will be transferred to storage tank (Batstone et al. 2002).

2.3.4 Storage tank

The storage tank with 10 litre function as storage for BPR influent.

2.3.5 Biogas Production Reactor (BPR)

BPR with 10 litre volume function as biogas production reactor. This reactor is anaerobic fixed bed reactor and contained plastic media for biomass support. The initial biomass concentration in the reactor is 5000 mg/l which is acclimatized mixed culture from banana plantation soil.

2.4 Analytical methods

COD concentration was spectrophotometrically analyzed using a HACH spectrophotometer and methods as in Spectrophotometric Instrument Manual. Gas collection was done using daily water displacement. Methane content was analyzed using gas chromatography with thermal conductivity detector (GCTCD) with helium as the carrier gas. Acetic acid concentration (TVA) was determined using HPLC. Substrate concentration was measured as suspended solids according to the Standard Methods for The Examination of Water and Wastewater. 20 ml well-mixed sample was filtered through a weighed standard glass-fiber filter and the residue retained on the filter is dried to a constant weight at 103oC to 105oC. The increase in weight of the filter represents the total suspended solids (APHA 1989).

3. Results and discussion

Chen et al. (1978) developed a kinetic model on substrate utilization based on the Contois model as follows:

$$\frac{\mu_{max}}{\mu} = K\frac{S_0-S}{S} + 1 \tag{18}$$

The kinetic model has been aptly used in many studies, notably in investigations on anaerobic digestion of high strength wastes (Mata-Alvarez et al. 1992; Sales et al. 2000).

Equation (18) can be written as:

$$\frac{1}{\mu} = \frac{1}{\mu_{max}} + \frac{K}{\mu_{max}}\frac{S_0-S}{S} \tag{19}$$

For a completely mixed system $1/\mu = \theta$ and $1/\mu_m = \theta_m$. Therefore

$$\theta = \frac{1}{\mu_{max}} + \frac{K}{\mu_{max}}\frac{S_0-S}{S} \tag{20}$$

here
S is substrate total effluent

The kinetic parameters, μ_{max} and K, were calculated with the aim of studying possible inhibition phenomena. Using the least squares method, the values for the kinetic parameters μ_{max} and K can be obtained from the intercept and the slope of the adjusted lines. Thus, according to Eq. 20, μ_{max} = 1/intercept, and K = slope/intercept. Through linear regression T vs S value of μ_{max} and K could be determined (Fig. 2). Here

$$T=\theta$$

and

$$S= S_0\text{-}S/S.$$

In this study the value of μ_{max} and K calculated were 0.111 d^{-1} and 0.330 g/g respectively.

The kinetics of methane fermentation as proposed by Chen and Hashimoto (1978) is described by

$$\delta = \frac{B_0 S_0}{\theta}\left|1 - \frac{K}{\mu_{max}\theta - 1 + K}\right| \tag{21}$$

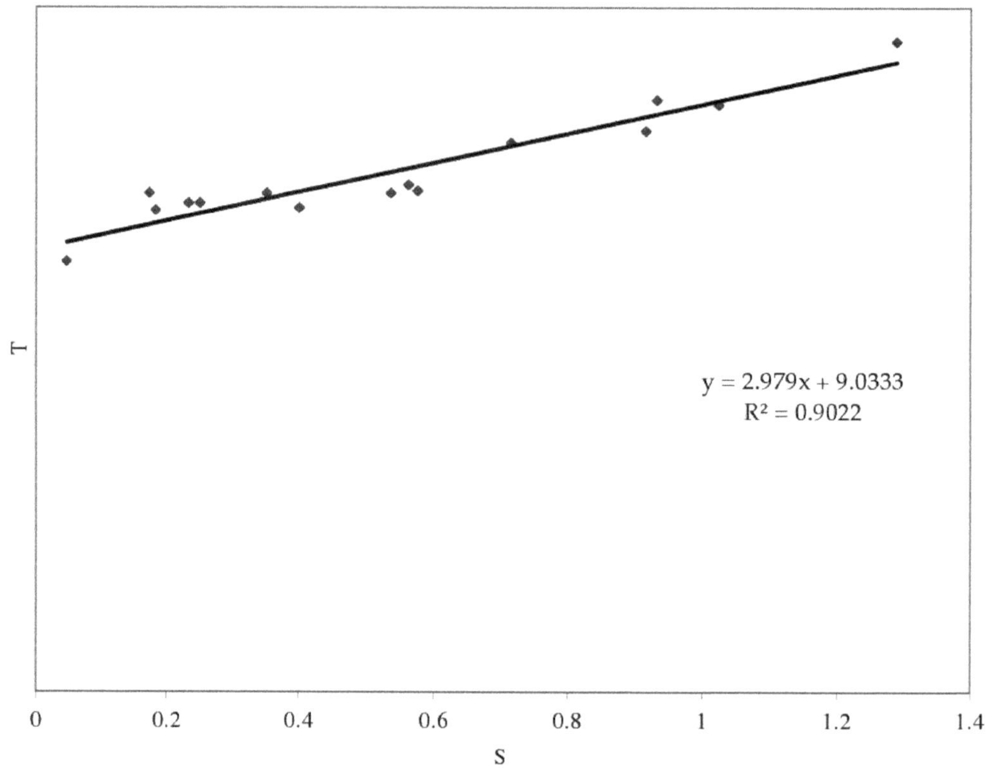

Fig. 2. Determination of μ_{max} and K

Equation (21) states that for a given substrate loading rate (S_0/θ), the daily volumetric methane production depends on the biodegradability of the wastewater (B_0) and the kinetic parameters θ and K (Yeoh, 1997). These values are calculated from the values of methane volume (Table 2), taking into account that the influents used have a COD of 2000 mg/l. Least squares method were used to determine the intercept B_0. Through linear regression Y vs St the value of B_0 could be determined. Here Y=δ and

$$St=\frac{S_0}{\theta}\left|1-\frac{K}{\mu_{max}\theta-1+K}\right|.$$

Fig. 3 shows the regression to determine B_0. The value of B_0 from this study is 0.326 l methane/g COD.

St	Y
0.033	0.085
0.028	0.081
0.047	0.088
0.090	0.101
0.068	0.090
0.081	0.100
0.068	0.100
0.068	0.101
0.071	0.098
0.046	0.090
0.051	0.092
0.105	0.107
0.105	0.109

Table 2. Values of Y and St for linear regression to determine B_0

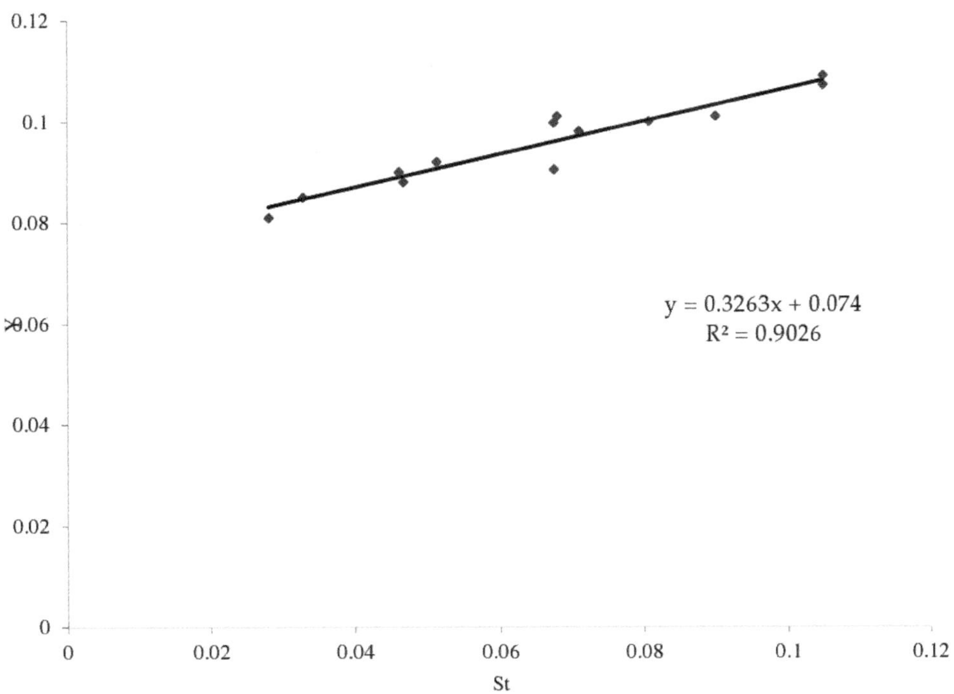

Fig. 3. Determination of B_0

The kinetic values corresponding to the substrate used (banana stem waste) is of the same order of magnitude with those obtained in the mesophilic anaerobic digestion process of some solid wastes such as vegetable waste, banana peel and palm oil (Table 4). From the kinetic constants (K, μ_{max} & B_0), the theoretical daily volumetric methane production values (δ) were calculated by using Eq. (21). It can be seen that the experimental results were reproduced with errors equal to or less than 5% in all cases. As can be seen in Table 3 and Fig. 4, the values of δ increased when influent substrate concentration (COD in) increased. Therefore, the kinetic parameters were found to be influenced by the influent substrate concentration. When the influent substrate concentration increased from 0.835 to 2.463 g COD/l, the δ values also increased from 0.085 to 0.109 l methane/g COD. The mixed culture in biomass also gives effect to methane production. As can be seen in Table 3, when the biomass in the bioreactor concentration (X) increased from 0.23 g/l to 0.93 the δ increased from 0.085 to 0.109 was, therefore, multiplied by a factor of 1.3. A similar behaviour was observed in the anaerobic digestion process of traditional olive mill wastewaters (Borja et al., 1995). In this work, the minimum hydraulic retention time, θ_{min} (days), at which the washout of the micro-organisms occurs was: 9.04 days. These values were calculated by using Eq. (13) and taking into account that this retention time is numerically equal to the reciprocal of the maximum micro-organisms growth rate (μ_{max}).

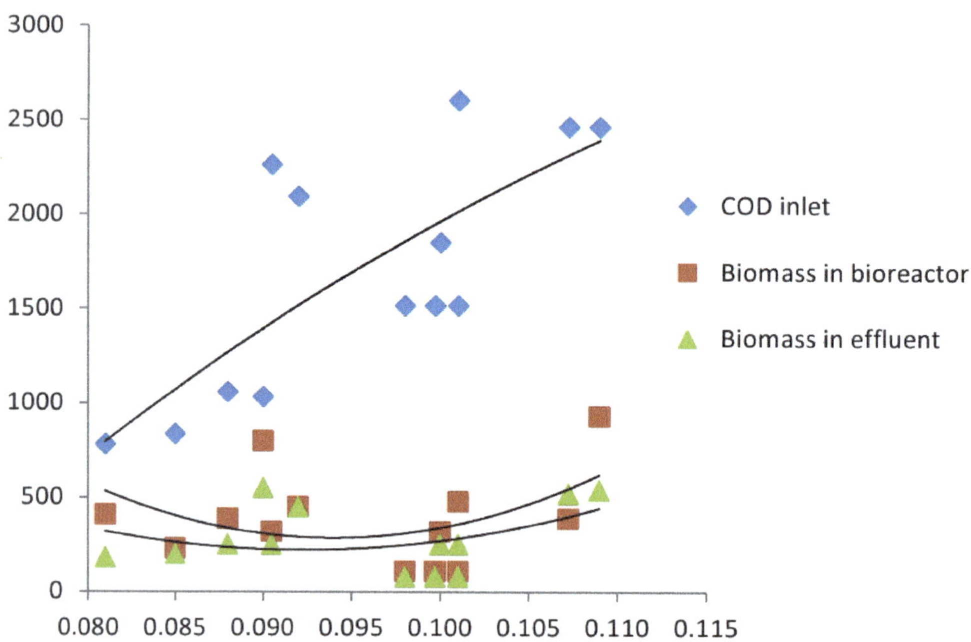

Fig. 4. Graph for influent substrate concentration (COD in), biomass in bioreactor concentration (X) and biomass in effluent concentration (Xe)

COD in	X	Xe	δ
0.835	0.230	0.200	0.085
0.780	0.410	0.180	0.081
1.058	0.390	0.250	0.088
2.604	0.480	0.250	0.101
2.262	0.320	0.250	0.090
1.851	0.320	0.250	0.100
1.515	0.110	0.080	0.100
1.515	0.110	0.080	0.101
1.515	0.110	0.080	0.098
1.032	0.800	0.550	0.090
2.094	0.450	0.450	0.092
2.463	0.390	0.520	0.107
2.463	0.930	0.540	0.109

Table 3. Values of influent substrate concentration (COD in), biomass in bioreactor concentration (X), biomass in effluent concentration (Xe) and daily volumetric methane production(δ)

Study	**Substrate**	**B_0 (l methane/ g COD)**	**K (g/g)**	**μ_{max} (day^{-1})**
This study	Banana stem waste	0.326	0.330	0.111
Gunaseelan (2004)	Banana peel	0.277*	-	0.089
Faisal & Unno (2001)	Palm oil mill wastewater	0.381	-	0.304
Gunaseelan(2007)	Banana peel	0.322*	-	-
Zhao & Viraraghavan (2004)	Treatment plant wastewater	0.366	0.079	1.16
Maya-Altamira et. al (2008)	Vegetable product-peas	0.36	-	-
	Vegetable product-leek & fried onion	0.36	-	-

*l methane/ g VS added

Table 4. Comparison of kinetics parameter to other research

The values of B_0, μ_{max} and K is fixed for different type of wastewater. So the Eq. (21) and values of B_0, μ_{max} and K could be used in scaling up biogas production process. The values of B_0, μ_{max} and K was compared to other research in Table 4. B_0 is a good parameter to determine biodegradability of any particular waste (Torres-Castillo et al. 1995). From Table 1 it was shown that banana stem waste from this research have a good methane production potential and comparable with other agricultural waste such as POME and sugar cane waste. μ_{max} is microorganism maximum growth rate and the value of μ_{max} in this study is considered low compared to other research. This is because the value of kinetic parameter μ_{max} and K is dependent on each other. High value of K could lower down the value of μ_{max} and vice versa. However the value of μ_{max} reported in this research is still within acceptable range.

4. Conclusion

A kinetic model for studying the anaerobic digestion process of banana stem waste was proposed on the basis of the two-stages process data obtained. The two-stages system comprising two bioreactors for acidogenic and methanogenic phases respectively. The experiments were conducted with hydraulic retention time (HRT) of 9 d corresponding to organic loading rate (OLR) of 4 gTS/l.d. The parameters obtained represent and predict the activity of the mixed culture in the biogas production of this waste. Kinetic evaluation of the experimental data provided μ_{max}(maximum microorganism growth rate) and K(kinetic constant) values as 0.111 d^{-1} and 0.330 g/g respectively based on COD. The waste biodegradability (B_0) was graphically evaluated to be 0.326 l methane/g COD.

5. References

APHA. (1989). *Standard Methods for the Examination of Water and Wastewater, 17th Edition,* American Public Health Association, Washington, DC.

Bal, A.S. & Dhagat, N.N. (2001). Upflow Anaerobic Sludge Blanket Reactor—A Review. *Indian J Environ Health, Vol.* 43, No. 2, pp. 1-82.

Batstone, D.J.; Keller, J.; Angelidaki, I.; Kalyuzhnyi, S.; Pavlostathis, S.; Rozzi, A.; Sanders, W.; Siegrist, H. & Vavilin, V. (2002). The IWA Anaerobic Digestion Model No1. *Wat. Sci. Tech.* Vol. 45, No. 10, pp. 63-73.

Borja, R.; Mart´ın, A.; Alonso, V.; Garc´ıa, I. & Banks, C.J. (1995). Influence Of Different Aerobic Pretreatments On The Kinetics of Anaerobic Digestion of Olive Mill Wastewater. *Water Res.,* Vol. 29, pp. 489–495.

Borja, R.; Raposo, F.; Sánchez, E. & Martín, A. (2005). A Kinetic Evaluation of The Anaerobic Digestion of Two-Phase Olive Mill Effluent in Batch Reactors. *Journal of Chemical Technology and Biotechnology,* Vol. 80, No. 3, pp. 241–250.

Chen, Y. & Hashimoto, A.G. (1978). Kinetic of Methane Fermentation. *Biotechnol. Bioeng.* Vol. 8, pp. 269-282.

Chen, Y. R.; Varel, V. H. & Hashimoto, A.G. (1980). Methane Production from Agricultural Residues. A Short Review, *J. Ind. Eng. Chem.* Vol. 19, No. 12, pp. 471.

Faisal, M. & Unno, H. (2001). Kinetic Analysis of Palm Oil Mill Wastewater Treatment by A Modified Anaerobic Baffled Reactor. *Biochemical Engineering Journal,* Vol. 9, No. 1, pp. 25-31.

Gavala, H. N.; Skiadas, I. V.; Bozinis, N. A. & Lyberatos, G. (1996). Anaerobic Codigestion Of Agricultural Industries Wastewaters. *Water Science and Technology,*Vol. 34, No. 11, pp. 67-75.

Gunaseelan, V. N. (2004). Biochemical Methane Potential of Fruits and Vegetable Solid Waste Feedstocks. *Biomass and Bioenergy,* Vol. 26, pp. 389 - 399.

Gunaseelan, V. N. (2007). Regression Models of Ultimate Methane Yields of Fruits and Vegetable Solid Wastes, Sorghum and Napiergrass on Chemical Composition. *Bioresource Technology*, Vol. 98, pp. 1270–1277.

Hashimoto, A. G.; Chen, Y. R. & Varel, V. H. (1981). *Theoretical Aspect of Anaerobic Fermentation: State-Of-The-Art, in Livestock Wastes: A Renewable Resource*, American Society of Agricultural Engineers, St. Joseph, MI.

Hu, W. C.; Thayanithy, K. & Forster, C.F. (2002). A Kinetic Study of The Anaerobic Digestion of Ice-Cream Wastewater. *Process Biochem*. Vol. 37, pp. 965–971.

Jiménez, E.; Gierczak, T.; Stark, H.; Burkholder, J. B. & Ravishankara, A. R. (2004). Reaction of OH with HO_2NO_2 (Peroxynitric Acid): Rate Coefficients between 218 and 335 K and Product Yields at 298 K. *J. Phys. Chem. A*, Vol. 108. pp. 1139-1149

Mata-Alvarez, J.; Llabrés, P.; Cecchi, F. & Pavan, P.(1992). Anaerobic Digestion of The Barcelona Central Food Market Organic Wastes: Experimental Study. *Bioresource Technology*, Vol. 39, No. 1, pp. 39-48.

Maya-Altamira, L.; Baun, A.; Angelidaki, I. & Schmidt, J.E. (2008). Influence of Wastewater Characteristics on Methane Potential in Food-Processing Industry Wastewaters. *Water Research*, Vol. 42, pp. 2195 - 2203.

Monod, J. (1950). The Technique of Continuous Culture Theory and Applications. *Ann. Inst. Pasteur*, Vol. 79, pp. 390-410.

Raposo, F.; Borja, R.; Martín, M. A.; Martín, A.; de la Rubia, M.A. & Rincón, B.(2009). Influence of Inoculum-Substrate Ratio On The Anaerobic Digestion Of Sunflower Oil Cake In Batch Mode: Process Stability And Kinetic Evaluation. *Chemical Engineering Journal*, Vol. 149, No. 1–3, pp. 70–77.

Rincón, B.; De La Rubia, M. A.; Raposo, F. & Borja, R. (2009). Evaluation of The Hydrolytic-Acidogenic Step of a Two-Stage Mesophilic Anaerobic Digestion Process of Sunflower Oil Cake. *Bioresource Technology*, Vol. 100, No. 18, pp. 4133–4138.

Sales, F.G.; Pereira, J.A.F.R. & Abreu, C.A.M. (2000). *Estudo Hidrodinâmico de um Reator de Leito de Lama Trifásico*. XXVIII Congresso Brasileiro de Sistemas Particulados, Teresópolis, Brazil.

Scharer, J. M.; Moo-Young, M. & Fujita, M. (1981). Methane Production by Anaerobic Digestion of Animal Manures. *Biotechnology Letters*, Vol. 3, No. 5, pp. 231-237.

Siles, J. A.; Brekelmans, J.; Martín, M. A.; Chica, A. F. & Martín, A. (2010). Impact of Ammonia and Sulphate Concentration on Thermophilic Anaerobic Digestion. *Bioresource Technology*, Vol.10, No. 123, pp. 9040-9048.

Torres-Castillo, R., Llabrés-Luengo, P. & Mata-Alvarez, J. (1995). Temperature Effect on Anaerobic Digestion of Bedding Straw in A One Phase System at Different Inoculum Concentration. *Agriculture, Ecosystems & Environment*, Vol. 54, No 1-2,pp. 55-66.

Yeoh, B. G. (1997). Two-Phase Anaerobic Treatment of Cane-Molasses Alcohol Stillage. *Water Science and Technology*, Vol. 36, No. 6-7, pp. 441-448.

Zhao, H. W. & Viraraghavan, T. (2004). Analysis of The Performance of An Anaerobic Digestion System at The Regina Wastewater Treatment Plant. *Bioresource Technology*, Vol. 95, No. 3, pp. 301-307.

Permissions

The contributors of this book come from diverse backgrounds, making this book a truly international effort. This book will bring forth new frontiers with its revolutionizing research information and detailed analysis of the nascent developments around the world.

We would like to thank Dr. Sunil Kumar, for lending his expertise to make the book truly unique. He has played a crucial role in the development of this book. Without his invaluable contribution this book wouldn't have been possible. He has made vital efforts to compile up to date information on the varied aspects of this subject to make this book a valuable addition to the collection of many professionals and students.

This book was conceptualized with the vision of imparting up-to-date information and advanced data in this field. To ensure the same, a matchless editorial board was set up. Every individual on the board went through rigorous rounds of assessment to prove their worth. After which they invested a large part of their time researching and compiling the most relevant data for our readers. Conferences and sessions were held from time to time between the editorial board and the contributing authors to present the data in the most comprehensible form. The editorial team has worked tirelessly to provide valuable and valid information to help people across the globe.

Every chapter published in this book has been scrutinized by our experts. Their significance has been extensively debated. The topics covered herein carry significant findings which will fuel the growth of the discipline. They may even be implemented as practical applications or may be referred to as a beginning point for another development. Chapters in this book were first published by InTech; hereby published with permission under the Creative Commons Attribution License or equivalent.

The editorial board has been involved in producing this book since its inception. They have spent rigorous hours researching and exploring the diverse topics which have resulted in the successful publishing of this book. They have passed on their knowledge of decades through this book. To expedite this challenging task, the publisher supported the team at every step. A small team of assistant editors was also appointed to further simplify the editing procedure and attain best results for the readers.

Our editorial team has been hand-picked from every corner of the world. Their multi-ethnicity adds dynamic inputs to the discussions which result in innovative outcomes. These outcomes are then further discussed with the researchers and contributors who give their valuable feedback and opinion regarding the same. The feedback is then collaborated with the researches and they are edited in a comprehensive manner to aid the understanding of the subject.

Apart from the editorial board, the designing team has also invested a significant amount of their time in understanding the subject and creating the most relevant covers. They scrutinized every image to scout for the most suitable representation of the subject and create an appropriate cover for the book.

The publishing team has been involved in this book since its early stages. They were actively engaged in every process, be it collecting the data, connecting with the contributors or procuring relevant information. The team has been an ardent support to the editorial, designing and production team. Their endless efforts to recruit the best for this project, has resulted in the accomplishment of this book. They are a veteran in the field of academics and their pool of knowledge is as vast as their experience in printing. Their expertise and guidance has proved useful at every step. Their uncompromising quality standards have made this book an exceptional effort. Their encouragement from time to time has been an inspiration for everyone.

The publisher and the editorial board hope that this book will prove to be a valuable piece of knowledge for researchers, students, practitioners and scholars across the globe.

List of Contributors

B. Amigun and A. S. Badmos
Renewable Energy Group, National Biotechnology Development Agency (NABDA)- an Agency Under the Federal Ministry of Science and Technology, Abuja, Nigeria

W. Parawira
Department of Applied Biology, Kigali Institute of Science and Technology (KIST), Kigali, Rwanda

J. K. Musango
Gauteng City-Region Observatory, a Partnership Between University of Johannesburg, University of Witwatersrand and Gauteng Provincial Government, Wits, South Africa

A. O. Aboyade
Process Engineering Department, Stellenbosch University, Stellenbosch, South Africa

Babajide A. Adelekan
Federal College of Agriculture Ibadan, Institute of Agricultural Research & Training, Ibadan, Nigeria

Annika Björn, Paula Segura de La Monja, Anna Karlsson, Jörgen Ejlertsson and Bo H. Svensson
Department of Thematic Studies, Water and Environmental Studies, Linköping University, Sweden

Jorge Del Real Olvera and Alberto Lopez-Lopez
Environmental Technology Unit, Centre of Research and Assistance in Technology and Design of the State of Jalisco, (CIATEJ), Mexico

Bárbara Rincón and Rafael Borja
Instituto de la Grasa (CSIC), Avda. Padre García Tejero, Sevilla, Spain

Elvia Ines Garcia-Peña, Alberto Nakauma-Gonzalez and Paola Zarate-Segura
Bioprocesses Department, Unidad Profesional Interdisciplinaria de Biotecnología, IPN, Mexico City, Mexico

Peter Hass
IPSOS Industrial Consulting GmbH, Berlin, Germany

Nora Niemetz, Karl-Heinz Kettl and Michael Narodoslawsky
Graz University of Technology, Austria

Manfred Szerencsits
Ökocluster, Austria

Derbal Kerroum, Bencheikh-LeHocine Mossaab and Meniai Abdesslam Hassen
University Mentouri Constantine, Algeria

Marta Kisielewska
University of Warmia and Mazury in Olsztyn, Poland

Chen Haitao, Li Lixia, Wang Hanyang and Liu Lixue
Northeast Agricultural University, China

Marianna Makádi, Attila Tomócsik and Viktória Orosz
Research Institute of Nyíregyháza, RISF, CAAES, University of Debrecen, Hungary

Krystyna Seifert, Michał Thiel, Ewelina Wicher, Marcin Włodarczak and Marek Łaniecki
Faculty of Chemistry, A. Mickiewicz University, Poznań, Poland

Roman Zagrodnik and Mikołaj Stodolny
Faculty of Chemistry, A. Mickiewicz University, Poznań, Poland

Agnes Godfrey Mwakaje
Institute of Resource Assessment, University of Dar es Salaam, Tanzania

M. Samer
Cairo University, Faculty of Agriculture, Department of Agricultural Engineering, Egypt

Chantaraporn Phalakornkule
Department of Chemical Engineering, Faculty of Engineering, King Mongkut's University of Technology North Bangkok, Thailand
The Joint Graduate School of Energy and Environment, King Mongkut's University of Technology Thonburi, Thailand

Maneerat Khemkhao
The Joint Graduate School of Energy and Environment, King Mongkut's University of Technology Thonburi, Thailand

Frank Burmeister, Janina Senner and Eren Tali
Gaswärme - Institut e.V. Essen, Germany

Norazwina Zainol
Universiti Malaysia Pahang, Malaysia